Lecture Notes of the Institute for Computer Sciences, Social Informatics and Telecommunications Engineering 433

More information about this series at https://link.springer.com/bookseries/8197

Honghao Gao · Jun Wun · Jianwei Yin ·
Feifei Shen · Yulong Shen · Jun Yu (Eds.)

Communications and Networking

16th EAI International Conference, ChinaCom 2021
Virtual Event, November 21–22, 2021
Proceedings

Editors
Honghao Gao
Shanghai University
Shanghai, China

Jianwei Yin
Zhejiang University
Hangzhou, China

Yulong Shen
Xidian University
X'ian, China

Jun Wun
Fudan University
Shanghai, China

Feifei Shen
Tsinghua University
Beijing, China

Jun Yu
Hangzhou Dianzi University
Hangzhou, China

ISSN 1867-8211   ISSN 1867-822X (electronic)
Lecture Notes of the Institute for Computer Sciences, Social Informatics
and Telecommunications Engineering
ISBN 978-3-030-99199-9   ISBN 978-3-030-99200-2 (eBook)
https://doi.org/10.1007/978-3-030-99200-2

© ICST Institute for Computer Sciences, Social Informatics and Telecommunications Engineering 2022
This work is subject to copyright. All rights are reserved by the Publisher, whether the whole or part of the material is concerned, specifically the rights of translation, reprinting, reuse of illustrations, recitation, broadcasting, reproduction on microfilms or in any other physical way, and transmission or information storage and retrieval, electronic adaptation, computer software, or by similar or dissimilar methodology now known or hereafter developed.
The use of general descriptive names, registered names, trademarks, service marks, etc. in this publication does not imply, even in the absence of a specific statement, that such names are exempt from the relevant protective laws and regulations and therefore free for general use.
The publisher, the authors and the editors are safe to assume that the advice and information in this book are believed to be true and accurate at the date of publication. Neither the publisher nor the authors or the editors give a warranty, expressed or implied, with respect to the material contained herein or for any errors or omissions that may have been made. The publisher remains neutral with regard to jurisdictional claims in published maps and institutional affiliations.

This Springer imprint is published by the registered company Springer Nature Switzerland AG
The registered company address is: Gewerbestrasse 11, 6330 Cham, Switzerland

Preface

We are delighted to introduce the proceedings of the 16th European Alliance for Innovation (EAI) International Conference on Communications and Networking in China (ChinaCom 2021). This conference brought together researchers, developers, and practitioners around the world who are interested in communications and networking from the viewpoint of big data, cloud computing, sensor networks, software-defined networks, and so on.

The technical program of ChinaCom 2021 consisted of 52 papers, including 47 full papers and 5 workshop papers in oral presentation sessions at the main conference tracks. The conference sessions were as follows: Session 1 - Scheduling and Transmission Optimization in Edge Computing; Session 2 - Complex System Optimization in Edge Computing; Session 3 - Network Communication Enhancement; Session 4 - Signal Processing and Communication Optimization; Session 5 - Deep Learning and Vehicular Communication; Session 6 - Edge Computing and Deep Learning; Session 7 - Finite Blocklength and Distributed Machine Learning; Session 8 - Deep Learning and Network Performance Optimization; and Session 9 - Edge Computing and Reinforcement Learning. Apart from high-quality technical paper presentations, the technical program also featured a keynote speech and a technical workshop. The keynote speech delivered by Antonio Iera from the University of Calabria, Italy, introduced the concept of 'social digital twins', while the Workshop on Data Intensive Services based Application (DISA) aimed to encourage academic researchers and industry practitioners to present and discuss all methods and technologies related to research and experiences in a broad spectrum of data-intensive services-based applications.

Coordination with the steering chair, Imrich Chlamtac, was essential for the success of the conference. We sincerely appreciate the constant support and guidance. It was also a great pleasure to work with such an excellent organizing committee team for their hard work in organizing and supporting the conference. In particular, we are grateful to the Technical Program Committee who completed the peer-review process for the technical papers and helped to put together a high-quality technical program. We are also grateful to Conference Manager Lucia Sladeckova for her support and all the authors who submitted their papers to the ChinaCom 2021 conference including the DISA workshop.

We strongly believe that the ChinaCom conference provides a good forum for all researchers, developers, and practitioners to discuss all science and technology aspects that are relevant to communications and networking. We also expect that the future

ChinaCom conferences will be as successful and stimulating as this year's, as indicated by the contributions presented in this volume.

November 2021

Honghao Gao
Jun Wu
Jianwei Yin
Feifei Gao
Yulong Shen
Jun Yu
Chunguo Li
Yueshen Xu

Organization

Steering Committee

Chair

Imrich Chlamtac	University of Trento, Italy

Members

Changjun Jiang	Tongji University, China
Qianbin Chen	Chongqing University of Posts and Telecommunications, China
Jian Wan	Zhejiang University of Science and Technology, China
Honghao Gao	Shanghai University, China
Jun Wu	Tongji University, China

Organizing Committee

General Co-chairs

Honghao Gao	Shanghai University, China
Jun Wu	Fudan University, China
Jianwei Yin	Zhejiang University, China
Feifei Gao	Tsinghua University, China
Yulong Shen	Xidian University, China
Jun Yu	Hangzhou Dianzi University, China

International Advisory Committee

Honggang Wang	University of Massachusetts Dartmouth, USA
Yang Yang	ShanghaiTech University, China
Han-Chieh Chao	National Dong Hwa University, Taiwan

Technical Program Committee Chairs

Chunguo Li	Southeast University, Bangladesh
Yueshen Xu	Xidian University, China

Web Chair

Xiaoxian Yang   Shanghai Polytechnic University, China

Publicity and Social Media Chairs

Rui Li   Xidian University, China
Yucong Duan   Hainan University, China

Workshops Chair

Yuyu Yin   Hangzhou Dianzi University, China

Sponsorship and Exhibits Chair

Honghao Gao   Shanghai University, China

Publications Chair

Youhuizi Li   Hangzhou Dianzi University, China

Local Chair

Weng Yu   Minzu University of China, China

Technical Program Committee

Hongcheng Yan   China Academy of Space Technology, China
Xiang Chen   Sun Yat-sen University, China
Shijun Liu   Shandong University, China
Tien-Wen Sung   Fujian University of Technology, China
Wei Du   Wuhan University of Technology, China
Lei Shi   Hefei University of Technology, China
Fu Chen   Central University of Finance and Economics,
                                       China
Kai Peng   Huaqiao University, China
Deli Qiao   East China Normal University, China
Yi Xie   Sun Yat-sen University, China
Jinglun Shi   South China University of Technology, China
Taiping Cui   Chongqing University of Posts and
                                       Telecommunications, China
Fei Dai   Yunnan University, China
Xumin Pu   Chongqing University of Posts and
                                       Telecommunications, China
Chengchao Liang   Carleton University, Canada
Krishna Kambhampaty   North Dakota State University, USA
Liang Xiao   Xiamen University, China

Contents

Data Intensive Services Based Application (DISA) Workshop

Research on Airborne Wireless Sensor Network Based on Wi-Fi
Technology 3
  Zou Liang

Millimeter Wave Hybrid Precoding Based on Deep Learning 13
  Qing Liu and Ken Long

Open-Set Recognition Algorithm of Signal Modulation Based on Siamese
Neural Network 28
  *Pengcheng Liu, Yan Zhang, Mingjun Ma, Zunwen He,
  and Wancheng Zhang*

Accurate Frequency Estimator of Real Sinusoid Based on Maximum
Sidelobe Decay Windows 40
  Zhanhong Liu, Lei Fan, Jiyu Jin, Renqing Li, Jinyu Liu, and Nian Liu

Transfer Learning Based Algorithm for Service Deployment Under
Microservice Architecture 52
  Wenlin Li, Bei Liu, Hui Gao, and Xin Su

Scheduling and Transmission Optimization in Edge Computing

Joint Opportunistic Satellite Scheduling and Beamforming for Secure
Transmission in Cognitive LEO Satellite Terrestrial Networks 65
  Xiaofen Jiao, Yichen Wang, Zhangnan Wang, and Tao Wang

GAN-SNR-Shrinkage-Based Network for Modulation Recognition
with Small Training Sample Size 80
  Shuai Zhang, Yan Zhang, Mingjun Ma, Zunwen He, and Wancheng Zhang

Joint Resource Allocation Based on F-OFDM for Integrated
Communication and Positioning System 91
  Ruoxu Chen, Xiaofeng Lu, and Kun Yang

UAV Formation Using a Dynamic Task Assignment Algorithm
with Cooperative Combat 102
  Ying Wang, Yonggang Li, Zhichao Zheng, Longjiang Li, and Xing Zhang

Complex System Optimization in Edge Computing

Research and Implementation of Multi-Agent UAV System Simulation
Platform Based on JADE 115
    Zhichao Zheng, Yonggang Li, and Ying Wang

A Complex Neural Network Adaptive Beamforming for Multi-channel
Speech Enhancement in Time Domain 129
    Tao Jiang, Hongqing Liu, Yi Zhou, and Lu Gan

Robust Transmission Design for IRS-Aided MISO Network
with Reflection Coefficient Mismatch 140
    Ran Yang, Ning Wei, Zheng Dong, Hongji Xu, and Ju Liu

Network Communication Enhancement

Distributed Deep Reinforcement Learning Based Mode Selection
and Resource Allocation for VR Transmission in Edge Networks 153
    Jie Luo, Bei Liu, Hui Gao, and Xin Su

Joint Optimization of D2D-Enabled Heterogeneous Network Based
on Delay and Reliability Constraints 168
    Dengsong Yang, Baili Ni, Haidong Wang, and Baoxiang Wei

Channel Allocation for Medical Extra-WBAN Communications in Hybrid
LiFi-WiFi Networks 182
    Novignon C. Acakpo-Addra, Dapeng Wu, and Andrews A. Okine

Cache Allocation Scheme in Information-Centric Satellite-Terrestrial
Integrated Networks 192
    Jie Duan, Xianjing Hu, Hao Liu, and Zhihong Zhang

Signal Processing and Communication Optimization

Resource Allocation in Massive Non-Orthogonal Multiple Access System 209
    Wen Zhang, Jie Zeng, and Zhong Li

Distortionless MVDR Beamformer for Conformal Array GNSS Receiver 220
    Han Li, Di He, Xin Chen, Jiaqing Qu, and Lieen Guo

HEVC Rate Control Optimization Algorithm Based on Video
Characteristics 235
    Qiang Li and Jun Nie

Performance Analysis of Radar Communication Shared Signal Based
on OFDM 250
    Zeyu Liu, Ying Zhang, and Xinmin Luo

Joint Power Allocation and Passive Beamforming Design for IRS-Assisted
Cell-free Networks 264
    Chen He, Xie Xie, Yangrui Dong, and Shun Zhang

Deep Learning and Vehicular Communication

Text Error Correction Method in the Construction Industry Based
on Transfer Learning 277
    Zhenguo Hou, Weitao Yang, Haiying He, Peicong Zhang, Ziyu Wang,
    and Xiaosheng Ji

A Beam Tracking Scheme Based on Deep Reinforcement Learning
for Multiple Vehicles 291
    Binyao Cheng, Long Zhao, Zibo He, and Ping Zhang

A Dynamic Transmission Design via Deep Multi-task Learning
for Supporting Multiple Applications in Vehicular Networks 306
    Zhixing He, Mengyu Ma, Chao Wang, and Fuqiang Liu

Sub-carrier Spacing Detection Algorithm in 5G New Radio Systems 322
    Tong Li, Hang Long, and Li Huang

Pre-handover Mechanism in the Internet of Vehicles Based on Named
Data Networking 335
    Gaixin Wang, Zhanjun Liu, and Qianbin Chen

Edge Computing and Deep Learning

Selective Modulation and Cooperative Jamming for Secure
Communication in Untrusted Relay Systems 351
    Li Huang, Xiaoxu Wu, and Hang Long

Deep CSI Feedback for FDD MIMO Systems 366
    Zibo He, Long Zhao, Xiangchen Luo, and Binyao Cheng

Joint Computation Offloading and Wireless Resource Allocation
in Vehicular Edge Computing Networks 377
    Jiao Zhang, Zhanjun Liu, Bowen Gu, Chengchao Liang,
    and Qianbin Chen

Non-coherent Receiver Enhancement Based on Sequence Combination 392
  Xiaoxu Wu, Tong Li, and Hang Long

MCAD-Net: Multi-scale Coordinate Attention Dense Network for Single
Image Deraining 405
  Pengpeng Li, Jiyu Jin, Guiyue Jin, Jiaqi Shi, and Lei Fan

Finite Blocklength and Distributed Machine Learning

6G mURLLC over Cell-Free Massive MIMO Systems in the Finite
Blocklength Regime 425
  Zhong Li, Jie Zeng, Wen Zhang, Shidong Zhou, and Ren Ping Liu

Cloud-Edge Collaboration Based Data Mining for Power Distribution
Networks 438
  Li An and Xin Su

Robust Sound Event Detection by a Two-Stage Network in the Presence
of Background Noise 452
  Jie Ou, Hongqing Liu, Yi Zhou, and Lu Gan

Deep Learning and Network Performance Optimization

Routing and Resource Allocation for Service Function Chain
in Service-Oriented Network 465
  Ziyu Liu, Zeming Li, Chengchao Liang, and Zhanjun Liu

Service-Aware Virtual Network Function Migration Based on Deep
Reinforcement Learning 481
  Zeming Li, Ziyu Liu, Chengchao Liang, and Zhanjun Liu

Dual-Channel Speech Enhancement Using Neural Network Adaptive
Beamforming 497
  Tao Jiang, Hongqing Liu, Chenhao Shuai, Mingtian Wang, Yi Zhou,
  and Lu Gan

Edge Computing and Reinforcement Learning

Fog-Based Data Offloading in UWSNs with Discounted Rewards:
A Contextual Bandit 509
  Yuchen Shan, Hui Wang, Zihao Cao, Yujie Sun, and Ting Li

Deep Deterministic Policy Gradient Algorithm for Space/Aerial-Assisted
Computation Offloading 523
  Jielin Fu, Lei Liang, Yanlong Li, and Junyi Wang

Performance Analysis for UAV-Assisted mmWave Cellular Networks 538
   Jiajin Zhao, Fang Cheng, Linlin Feng, and Zhizhong Zhang

Aerial Intelligent Reflecting Surface for Secure MISO Communication
Systems 553
   Zhen Liu, Zhengyu Zhu, Wanming Hao, and Jiankang Zhang

Author Index 567

Data Intensive Services Based Application (DISA) Workshop

Research on Airborne Wireless Sensor Network Based on Wi-Fi Technology

Zou Liang[⊠]

School of Communication and Information Engineering, Chongqing University of Posts and
Telecommunications, Chongqing, China
719240742@qq.com

Abstract. Due to the limitation of wired airborne communication network, it
has become a trend to explore airborne wireless communication network. With
the rapid development of wireless sensing technology, the International Telecommunication Union (ITU) issued the Wireless avionics Internal Communication
(WAIC) standard. This standard is mainly used to detect the health condition and
emergency of the internal and external structures of the aircraft. In this paper,
based on the analysis of the research status of wireless airborne internal communication system at home and abroad, aiming at the internal structure of C919
narrow-body single-channel aircraft, a wireless communication scheme based on
WI-FI technology and the corresponding wireless network architecture for the
real-time monitoring of aircraft status in accordance with WAIC standards are
proposed. The simulation results show that the wireless sensor network based on
Wi-Fi technology is suitable for WAIC system.

Keywords: Wireless sensor network · The Wi-Fi technology · The network
architecture · WAIC

1 Introduction

Safety and economy have always been the goal of civil aircraft. The development of electronic communication technology has greatly improved the degree of automation and
safety of civil aircraft. Currently, various onboard electronic devices communicate via
cable connections to the data bus. Cables are an important part of existing aircraft. The
weight of cables and connectors between cables is about 2%–5% of the total weight
of the aircraft, and the cable-related cost per aircraft is $2200 per kg [1]. The adoption
of wireless communication can reduce the time and cost of cable wiring and installa-
tion. Studies show that about 30% of cables on airplanes can be replaced by wireless
[1]. In addition, the WAIC system can support some new applications without the need
for rewiring, and provides more opportunities to monitor more systems without adding
weight to the aircraft. Wireless technology can also be used to monitor areas where
cables cannot be executed, such as engine rotor bearing monitoring. Second, WAIC systems can also improve security. It is estimated that on average, more than 1077 missions

H. Gao et al. (Eds.): ChinaCom 2021, LNICST 433, pp. 3–12, 2022.
https://doi.org/10.1007/978-3-030-99200-2_1

were aborted due to wiring problems causing two aircraft fires per month [2]. In order to improve these aspects, using wireless communication system to replace part of wired communication is one of the development directions of the future aviation. Wireless avionics internal communication network refers to the wireless connection between two or more points on an aircraft, which does not provide air-to-ground, air-to-air communication [1]. For example, sensors mounted on the wings or engines can communicate with systems in the aircraft, etc. [3].

The key difference between WAIC network and ground wireless network lies in its different application requirements. Aircraft application environment is more complex and harsh, with higher requirements on reliability and safety [4]. In this paper, according to the application requirements of various electronic equipment in the aircraft, the appropriate wireless technology scheme and network architecture are selected to support airborne wireless communication, to provide theoretical and experimental verification for WAIC applications with high data rate, and to achieve the purpose of real-time monitoring of aircraft status.

2 Previous Work

In 2010, the world communication Congress (WRC-10) adopted the M.2197 standard *Technical Characteristic and Operational Objectives for Wireless Avionics Intra-communications (WAIC) report* [1], which mainly provides the technical characteristics and operational objectives of the WAIC system and identifies four different transmission modes for the WAIC network. They are high rate internal transmission mode (HI, High inside), high rate external transmission mode (HO, high outside), low rate internal transmission mode (LI, Low Inside) and low rate external transmission mode (LO, Low outside) [1]. The 2013 World Radiocommunication Conference (WRC-13) adopted report M.2283, which describes the application characteristics of WAIC networks and network spectrum requirements [8]. Since the release of WAIC system, industry and academia have carried out active exploration and research. WAIC is defined as a safety system for a new generation of aircraft. Both Boeing 737MAX and China C919 use wireless tire pressure and brake temperature measurement systems. In 2013, the Wireless Cabin Project began in Europe to develop in-flight wireless communication technologies, such as health monitoring systems for engines and landing gear. Then the NASA-organized FBW Consortium used wireless sensor technology on the LANDING gear of the B757 to complete a demonstration of an aircraft health monitoring system.

In academic circles, the research in recent two years mainly focuses on the combination of WAIC network and industrial wireless sensor network [3, 4, 5, 6, 7, 8] and the establishment of WAIC network communication model [9, 10]. In the literature [3], assessed the LTE MBB for high data rate WAIC application in the transmission of data rate, delay and reliability of the network performance, the results showed that MBB can meet the demand of communication, but its robustness in airborne environment technology is not enough, must improve robustness to meet under the environment of the airborne network resilience to random access interference. In reference [5], the network performance of the NB-IoT for low data rate WAIC applications was evaluated in terms of latency and reliability, and the results showed that the NB-IoT technology met the requirements of the WAIC system in a non-interference environment. However, in the case of aircraft interference, wireless technology must be discussed in the presence of interference to met WAIC system requirements. In literature [6], UWB technology is used to solve the wireless scheme of WAIC low data rate application, and UWB filter module is used in this scheme to conduct experiments. After evaluation and measurement, it is clear that the UWB filter bandwidth is well within the distance limit and can be applied to the WAIC system even with passenger influence. In reference [7], the performance and feasibility of LLDN, Wirelesshart, WISA and ISA100.11A in avionics applications were analyzed from the aspects of maximum allowable packet loss rate on the premise of flight. Literature [8] takes IEEE802.15.4 and IEEE802.11a/G standards as examples to analyze the media access mechanism of protocols, mainly including communication overhead and modulation efficiency. In reference [9], the experimental results of measuring the wireless channel of the WAIC system in a real aircraft interior environment are given, which show that the LOS environment is preferable to the Rayleigh fading channel and suggest the use of relay nodes to achieve the reliability of the wireless communication system. In literature [10], a model was established based on multiple factors including power factor, modulation efficiency and spectrum, and the technical characteristics of GMSK, QPSK, 16-PSK, 8-FSK and other modulation methods were analyzed in detail.

At present, there are few published papers on wireless aviation internal sensor network technology in China, and most of them are theoretical. Literature [11] analyzes the current research on airborne wireless communication in China, summarizes the research status of WAIC system, and analyzes the combination of WAIC system and industrial wireless sensor network on the basis of studying relevant foreign literature. Literature [12] proposed the communication protocol framework of WAIC network based on the general airborne network communication 5-layer model, and gave the future research direction.

3 Selection and Experimetion Evaluation of Wireless Technology

3.1 Choice of Wireless Technology

In order to meet the requirements of airborne wireless communication sensor network based on WAIC system, it is necessary to choose the wireless technology solution according to the structural characteristics of aircraft when constructing wireless network.

Airborne wireless communication network is mainly used to monitor the health status and emergencies of various parts inside and outside the aircraft. Therefore, different data transmission parts and data rates need to be distinguished during the construction of airborne wireless communication network. Only high data rate applications are considered in this paper, so only HI and HO applications are analyzed and evaluated.

At present, there are many kinds of wireless transmission protocols, such as NB-iot, Wi-Fi (IEEE802.11a/b/g), Zigbee (IEEE802.15.4), ISA100.11A and Bluetooth, etc. Different technologies have different characteristics and are suitable for different requirements and application scenarios. In this paper, bluetooth, Wi-Fi (IEEE802.11ax/g) and Zigbee (IEEE802.15.4) wireless technologies are selected for comparison. The reason is that bluetooth networking is simple and its power consumption is between Wi-Fi and

Zigbee. For Wi-Fi, it is suitable for big data and long-distance transmission. Zigbee has low power consumption, low cost and large network capacity. Table 1 gives a comparison of data rates between WAIC HI and WAIC HO applications and these three wireless technologies, for example: the data rate comparison between WAIC HI application and Wi-Fi shows that the data rate of Wi-Fi is 9.6 Gbps and the maximum data rate of WAIC HI application is 4.8 Mbps. Therefore, the data rate of Wi-Fi meets the requirements of WAIC HI application. Table 1 shows that Wi-Fi and Bluetooth meet the rate requirements for WAIC high data rate applications in terms of maximum data transfer rates.

Table 1. Comparison of data rates between wireless technology and WAIC requirements [13, 14, 15]

Wireless technology	WAIC HI	WAIC HO
Bluetooth	24 Mbps/4.8 Mbps	24 Mbps/1 Mbps
ZigBee	250 Kbps/4.8 Mbps	250 Kbps/1 Mbps
Wi-Fi	9.6 Gbps/4.8 Mbps	9.6 Gbps/1 Mbps

In the technical reports that have been submitted by the ITU, there is no direct requirement for delays in the WAIC system. Therefore, it is necessary to make reasonable assumptions based on wired transmission. In literature [16], the worst end-to-end delay calculation of external environment monitoring system in avionics wired network is 23.86 ms. From this calculated value, a reasonable assumption can be made. We can assume that the external application delay of HO is 10 ms to approximate the delay from the terminal node to the gateway node, so that 10 ms can be regarded as the reasonable delay requirement of HO application. For HI class, 0.5 s is considered to be a reasonable delay time, because HI class is mostly periodic data and monitoring data [2].

Table 2. Delay comparison between wireless technology and WAIC requirements [17, 18]

Wireless technology	WAIC HI	WAIC HO
Bluetooth	0.014 s/0.5 s	0.014 s/0.01 s
ZigBee	0.015 s/0.5 s	0.015 s/0.01 s
Wi-Fi	13.6 μs/0.5 s	13.6 μs/0.01 s

As you can see from Table 2, Wi-Fi meets WAIC system requirements in terms of latency requirements. After comparing the data rate and delay, it can be concluded that Wi-Fi technology meets WAIC system requirements. In addition, Wi-Fi is suitable for long-distance and big data transmission, which can avoid the disadvantages of channel congestion, communication delay and data loss caused by too many nodes to a certain extent. Therefore, Wi-Fi technology is more suitable for operation on airplanes.

3.2 Theoretical Evaluation

In terms of reliability, it can be reflected by the bit error rate. WAIC applications are known to have a bit error rate of 10^{-6} [1] according to the M.2197 standard proposed by the World Radiocommunication Congress, but this limit can be relaxed for some non-mission-critical applications. The calculation formula of ber is [2]:

$$BER = \frac{2}{log_2M}\left(1 - \frac{1}{\sqrt{M}}\right)erfc\left(\sqrt{\frac{3log_2ME_b}{2(M-1)N_0}}\right) \tag{1}$$

Where M is the number of modulation symbols and *erfc* represents the complementary error function. The E_b/N_0 of different modulation schemes can be calculated by formula (1). As shown in Table 3 below [3]:

Table 3. E_b/N_0 values of different modulation schemes

Modulation method	QPSK	16-QAM	64-QAM
E_b/N_0(dB)	10.53	18.74	23.53

The results obtained in Table 3 are converted into signal-to-noise ratio by the formula to study and analyze the performance of aircraft internal communication. According to the calculated E_b/N_0 value, the SNR value can be calculated as [3]:

$$SNR = \frac{E_b}{N_0} + 10log\frac{f_b}{B}dB \tag{2}$$

Where f_b is the total net average data rate in unit of *bps*. According to ITU-R M.2283-0 report [8], f_b of HI class WAIC application is 18.4 MHz, and that of HO class WAIC application is 12.3 MHz. B is the bandwidth in *Hz*. According to the above report, it can be concluded that the bandwidth of HI class WAIC application B is 40 MHz, and that of HO class WAIC application B is 65 MHz.

SNR values of different modulation schemes can be calculated according to Formula (2), and the calculated results are shown in Table 4:

Table 4. SNR values for high data rate applications under different modulation schemes

Modulation method	QPSK	16-QAM	64-QAM
SNR (HI/dB)	8.59	16.34	21.13
SNR (HO/dB)	2.96	11.72	16.51

It can be seen from Table 4 that the required signal-to-noise ratio is proportional to the modulation scheme. The results of these calculations theoretically indicate the reliability requirements for WAIC applications. Based on theoretical analysis, it can be seen that Wi-Fi technology can support the transmission requirements of all high data rate applications of WAIC.

8 Z. Liang

3.3 Analysis of Simulation Results

In order to verify the feasibility of Wi-Fi technology in WAIC system, MATLAB software is used to simulate it. Packets are transmitted from a terminal device to a gateway node, which processes the packets and sends them to the backbone network. Vary the distance from the terminal to the gateway node to measure how far Wi-Fi technology can travel while ensuring reliability in the cabin environment. WAIC applications located inside the aircraft are somewhat affected by channel fading due to a non-line-of-sight path (NLoS) environment. For gateways and sensors placed outside the fuselage, the LoS condition is considered as there are no obstacles blocking the way. The SNR and SINR obtained under different transmission conditions are different from those of sensors placed internally, and should therefore be evaluated separately.

According to WAIC standard, the maximum transmission power of high data rate application is 30 dBm, the maximum path loss of HI application is 2.5 dB, and the maximum path loss of HO application is 2 dB [8].

The Wi-Fi core network consists of two computers and a wireless network card, EDUP AC1610. One computer is installed with a network adapter, which acts as a terminal node and sends high-data-rate packets, and one computer acts as a gateway node and receives packets. Network cable EDUP AC160 supports 2.4 GHz and 5.8 GHz. Because the 5.8 GHz band has less interference, it can effectively improve the wireless transmission rate and meet the requirements of high data rate avionics applications. The wi-fi channel bandwidth is set to 40 MHz. Interference signals are realized by software GNURadio [19]. During the experiment, the effect of different modulation order can be observed by changing the distance between two computers.

The first measurement is the transmission performance of WAIC HI application in Wi-Fi technology. There is no obstacle between the terminal node and the network management node, and it is in the straight channel of LoS position. The figure below shows the changes in SNR values measured using Wi-Fi technology to transmit data in the cabin interior environment. As shown in Fig. 1, Wi-Fi technology meets the operating requirements of HI applications over the entire transmission range without interference.

Next, when sending high data rate packets, interference signals are also sent, and obstacles are added between terminal nodes and gateway nodes for simulation. The interference signal is mainly RA signal. Radio Altimeters (RA) are high-transmitting power Radio devices that provide altitude measurements for aircraft and operate on the same spectrum as the WAIC system and are therefore a major source of operational interference for the WAIC system. Figure 2 shows the performance of Wi-Fi technology in the cabin in the presence of RA signal interference. As can be seen from Fig. 2, with the increase of transmission distance, SINR value decreases and transmission performance gradually decreases. High order modulation cannot meet the operating conditions of HI class applications. It can be seen from the figure that in the case of interference, the transmission distance of Wi-Fi technology for HI application is approximately limited to 9 m.

The difference between the results obtained in the experiment is due to the existence of interference. The cabin is a special transmission environment, which can also be regarded as a closed transmission environment. The production materials of the cabin, the seats in the cabin, and the movement of personnel are all sources of interference to

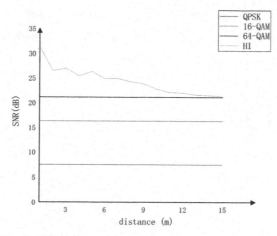

Fig. 1. SNR values measured in engine room environment

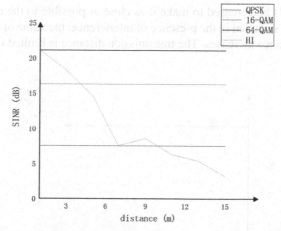

Fig. 2. SINR values measured in the presence of RA interference

wi-fi technology transmission. It can be seen from Fig. 1 and Fig. 2 that in the case of no interference, the SNR requirement can be met in the whole transmission range. When interference exists, high-order modulation cannot meet the SNR requirement, and when others meet the SNR requirement, the transmission distance is reduced to a few meters.

For HO applications, terminal nodes and gateway nodes are located in LoS environment to change the data transmission rate. The measured SNR values are shown in Fig. 3. It can be seen from the figure that Wi-Fi technology can meet the transmission performance of HO class and HI class in the case of no interference.

Figure 4 is the result obtained after adding interference signals, showing the transmission performance of Wi-Fi technology outside the cabin in the presence of interference, as shown in Fig. 4. In the flight of real aircraft, there are many interference signals outside the cabin and the interference ability is strong. Therefore, in the test, the transmission of

10   Z. Liang

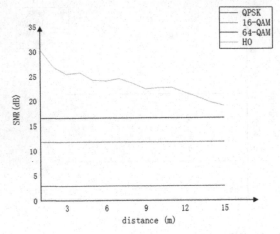

Fig. 3. SNR values measured in extravehicular environment

interference signals is increased to make it as close as possible to the real interference situation. It can be seen that in the presence of interference, the value of SINR decreases sharply as the distance increases. The transmission distance is limited to about 7 m.

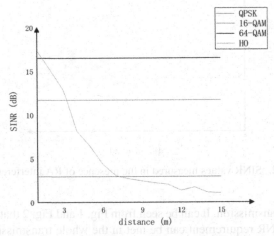

Fig. 4. Measured SINR value in the presence of RA interference

In conclusion, according to the experiments, when there is no interference, Wi-Fi technology can meet the SNR and data rate requirements of high data rate applications. However, when interference is introduced, the transmission performance will be severely degraded, so the transmission distance needs to be limited to meet the high data rate transmission requirements. The reliability of Wi-Fi technology can be achieved by changing the distance between the terminal and the gateway node. Through the above tests, the distance between the terminal node and the gateway node can be set to 7 m, which can meet the transmission performance requirements of HI and HO applications.

4 Hybrid Network Architecture

This section describes a network architecture based on Wi-Fi technology. The purpose of this network architecture is to replace the cables between terminals and AFDX switches, or gateway nodes, with wireless connections. Figure 5 shows a hybrid network architecture deployed in an aircraft. The sensor in the figure is responsible for collecting the monitoring data of the relevant position, and then transmits it to the gateway node (AFDX switch) through Wi-Fi wireless technology. After processing the data by the AFDX switch, the data is transmitted to the cab through the AFDX backbone network for the driver to view the data and analyze the health status of the monitored position.

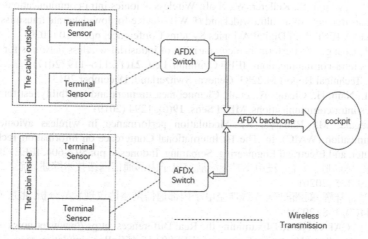

Fig. 5. Hybrid network architecture based on Wi-Fi technology

5 Conclusion

In this paper, airborne wireless sensor networks are studied. WAIC system provides a new method for airborne internal communication. This paper first evaluates the performance of three wireless technologies for high data rate wireless avionics internal communication (WAIC) applications. Through theoretical analysis and simulation experiments, wi-fi technology can meet the requirements of WAIC high data rate applications. In addition, the proposed hybrid network architecture based on Wi-Fi technology demonstrates the application scenarios of WAIC systems, which reduces cabling complexity in the cabin. Finally, the reliability of airborne wireless sensor network technology is analyzed. It can be concluded that wi-fi technology can also meet WAIC system requirements in the presence of interference. In the future, some methods can be studied to reduce transmission delay, and robust technology can also be considered to improve the recovery ability of the network in the presence of interference.

References

1. ITU-R. Technical Characteristic and Operational Objectives for Wireless Avionics Intra-Communications. Technical Report M.2197, Switzerland (2010)
2. Field, S., Arnason, P., Furse, C.: Smart wire technology for aircraft application. In: The 5th Joint NASA/FAA/DoD Conference Aging Aircraft, pp. 1156–1161 (2001)
3. Baltaci, A., Zoppi, S., Kellerer, W., et al.: Evaluation of cellular technologies for high data rate WAIC applications. In: IEEE International Conference on Communications, pp. 1–6 (2019)
4. Park, P., Di Marco, P., Nah, J., et al.: Wireless avionics intra-communications: a survey of benefits, challenges, and solutions. IEEE Internet Things J. **7**(99), 1–23 (2020)
5. Baltaci, A., Zoppi, S., Kellerer, W., et al.: Evaluation of cellular IoT for energy-constrained WAIC applications. In: IEEE 2nd 5G World Forum, pp. 359–364 (2019)
6. Baltaci, A., Zoppi, S., Kellerer, W., et al.: Wireless avionics intra-communications (WAIC) QoS measurements of an ultra wideband (UWB) device for low-data rate transmissions. In: 2020 AIAA/IEEE 39th Digital Avionics Systems Conference, pp. 1–10 (2020)
7. Park, P., Chang, W.: Performance comparision of industrial wireless networks for wireless avionics intra-communications. IEEE Commun. Lett. **21**(1), 116–119 (2017)
8. ITU-R. Technical Report M.2283, Geneva, Switzerland, December 2013
9. Bang, I., Nam, H., Chang, W., et al.: Channel measurement and feasibility test for wireless avionics intra-communications. MDPI Sens. **19**(6), 1294 (2019)
10. Suryanegara, M., Raharya, N.: Modulation performance in wireless avionics intra-communication (WAIC). In: The 1st International Conference on Information Technology, Computer, and Electrical Engineering. Semarang, Indonesia, pp. 597–604. IEEE (2014)
11. 范祥辉, 陈长胜, 史岩. 民用飞机无线航空电子内部通信网络技术综述. 航空工程进展 (09),129–135 (2020)
12. 李士宁,范祥辉,刘洲洲等. 无线航空电子内部通信网络协议现状与分析. 北京邮电大学学报 **44**(3), 1–8 (2021)
13. Lusky, I.: CAT vs NB-IoT-Examining the Real Differences. https://www.iot-now.com
14. Ouyang, G., Zhao, Y.: An application of IEEE802.15.4/ZigBee simulation model. In: 2016 3rd International Conference on Systems and Informatics (ICSAI), pp. 674–679 (2016)
15. IEEE. IEEE Technology Report on Wake-up Radio (2017)
16. Kang, K., Nam, M., Sha, L.: Worst case analysis of packet delay in avionics system for environment monitoring. IEEE Syst. J. **9**(4), 1354–1362 (2015)
17. IEEE. IEEE Standard for Low-Rate Wireless Network (2020)
18. IEEE. Wireless LAN Medium Access Control and Physical Layer Specifications (2020)
19. GNURadio. The Free & Open Software Radio System. https://www.gniradio.org/

Millimeter Wave Hybrid Precoding Based on Deep Learning

Qing Liu[✉] and Ken Long

School of Communication and Information Engineering, Chongqing University of Posts and
Telecommunications, Chongqing, China
s180101008@stu.cqupt.edu.cn

Abstract. To overcome the high energy consumption, and insufficient use of
spatial information in traditional hybrid precoding algorithms, a hybrid precoding algorithm based on deep learning was proposed for millimeter wave massive
MIMO system. Firstly, the hybrid precoding design problem was transformed into
an exhaustive search problem for the analog precoder/combiner using the equivalent channel matrix. Then, a deep learning model was constructed to learn how to
optimize the cascaded hybrid precoder by an improved convolutional neural network model. Finally, the optimized cascaded hybrid precoder was used to predict
the analog precoding/combiner matrix directly, and the digital precoding/combiner
matrix was obtained by applying singular value decomposition (SVD) to the equivalent channel matrix. Simulation results show that the performance of the proposed
cascaded hybrid precoder is close to that of the pure digital precoder and can maximize the achievable rate to enhance the spectral efficiency of the millimeter wave
massive MIMO system.

Keywords: Millimeter wave · Deep learning · Hybrid precoding · MIMO

1 Introduction

The need to transmit large amounts of data in a short period of time has exploded in recent
years, and it is expected that wireless data traffic may grow by more than 10,000 times by
2030 [1]. Millimeter wave communication is an alternative solution to improve spectral
efficiency and frequency resource shortage [2]. In the past decade, hybrid beamforming
technology has been widely used to support high data rates and energy efficiency. Channel estimation in millimeter-wave is challenging due to the training overhead and strict
restrictions on the use of Radio Frequency (RF) chains by a large number of transmitting
and receiving antennas [3].

Inspired by massive multiple input multiple output (MIMO), millimeter wave massive MIMO system is considered as a potential technology to improve system throughput.
Hybrid precoding [1] emerged as a result of multi-multiplexing massive data streams
and achieving more accurate beamforming in millimeter wave massive MIMO. In [4], a
hybrid beamforming based on continuous interference elimination was proposed, which

H. Gao et al. (Eds.): ChinaCom 2021, LNICST 433, pp. 13–27, 2022.
https://doi.org/10.1007/978-3-030-99200-2_2

can provide high performance at low complexity. Its main idea was to decompose the total realizable rate optimization problem with non-convex constraints into a series of simple sub-rate optimization problems. The author in [5] designed a hybrid precoding with low complexity for multi-user millimeter-wave system by configuring a hybrid precoder. However, the previously proposed hybrid precoding scheme is based on singular value decomposition (SVD), resulting in high communication complexity and requiring complex bit allocation strategy. In addition, the proposed scheme based on Geometric Mean Decomposition (GMD) [6] can avoid the bit allocation problem, but it still faces great challenges in solving the non-convex constraints of the analog precoder and utilizing the structural characteristics of millimeter wave system.

In the context of millimeter wave massive MIMO systems, although a lot of research has been done to enhance the performance of hybrid precoding, there are still many problems, the two main challenges are high cost and poor system performance. In the past few years, a number of hybrid precoding methods have been proposed to reduce costs or improve precoding performance [7–9]. At the same time, in order to achieve high spectral efficiency and reduce cost, the author in [10] proposed an alternating minimization scheme for effective design of hybrid precoder. In [11], a hybrid precoding method was proposed based on beam space SVD by using low-dimensional beam space channel state information (CSI) processed by a compressed sensing (CS) detector. These traditional methods cannot make full use of the structural characteristics of millimeter-wave systems, while the traditional low-cost schemes are realized at the cost of reducing the system's mixed precoding. Therefore, it is urgent to propose a new method to enhance the hybrid precoding performance of millimeter wave massive MIMO systems.

This paper intends to design a millimeter-wave hybrid precoding method based on deep learning, which jointly optimizes the channel measurement vector and designs a hybrid beamforming vector to achieve a near-optimal data rate with negligible training overhead.

2   System Model and Problem Description

2.1   System Model

The fully connected hybrid architecture depicted in Fig. 1, where a transmitter employs N_t antennas and N_t^{RF} RF (Radio Frequency) chains is communicating via N_s streams with a receiver which has N_r antennas and N_r^{RF} RF chains. The transmitter pre-codes the transmitted signal using a $N_t^{RF} \times N_s$ baseband precoder $\mathbf{F}_{BB}$ and a $N_t \times N_t^{RF}$ RF beam-former $\mathbf{F}_{RF}$, while the receiver combines the received signal using the $N_r \times N_r^{RF}$ RF combiner $\mathbf{W}_{RF}$ and the $N_r^{RF} \times N_s$ baseband combiner $\mathbf{W}_{BB}$. $\mathbf{F}_{RF}$ is implemented in the analog domain using RF circuits, every entry of the RF precoders is assumed to have a constant-modulus, i.e., $\left| [\mathbf{F}_{RF}]_{m,n} \right|^2 = N_t^{-1}$. Further, the total power constraint is enforced by normalizing the baseband precoder $\mathbf{F}_{BB}$ to satisfy $\|\mathbf{F}_{RF}\mathbf{F}_{BB}\|_F^2 = N_s$ [12].

Due to the limited scattering of the millimeter wave channel, the single-user model is taken as the research object, if only L paths reach the receiving antenna, the extended

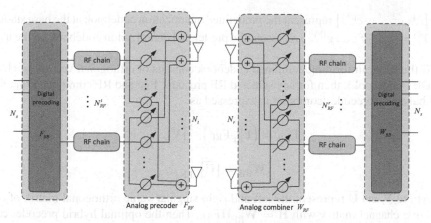

Fig. 1. Fully connected hybrid precoding model

Saleh-Valenzudel geometric channel model [12] is simplified as

$$\mathbf{H} = \sum_{\ell=1}^{L} \alpha_\ell a_r(\phi_{r,\ell}, \theta_{r,\ell}) a_t^H(\phi_{t,\ell}, \theta_{t,\ell}) \tag{1}$$

where α_ℓ denote the complex path gain of the ℓ th path. The angles $\phi_{r,\ell}, \theta_{r,\ell}$ represent the ℓ th path azimuth and elevation angles of arrival (AOAs) at the receive antennas while $\phi_{t,\ell}, \theta_{t,\ell}$ represent the ℓ th path angles of departure from the transmit array. $a_t(\cdot)$ and $a_r(\cdot)$ represent the sending and receiving array response vector. According to the arrangement of antenna elements, common antenna arrays can be divided into uniform linear array (ULA) and uniform planar array (UPA). The array response vector of ULA and UPA arrays are defined in [13]. The array antenna arrangement adopted in this paper is ULA.

2.2 Problem Description

The overall goal of this paper is to directly design hybrid precoders/combinators to maximize the rate of system implementation. If the channel is known and the RF beam-forming/combinator vector is selected from the predefined quantization codebook, the hybrid precoding/combinator design problem can be expressed as

$$\{\mathbf{F}_{BB}^*, \mathbf{F}_{RF}^*, \mathbf{W}_{BB}^*, \mathbf{W}_{RF}^*\} = \arg\max \log_2 \left| \mathbf{I} + \mathbf{R}_n^{-1} \mathbf{W}^H \mathbf{H} \mathbf{F} \mathbf{F}^H \mathbf{H}^H \mathbf{W} \right|$$

$$s.t.\ \mathbf{F} = \mathbf{F}_{BB}\mathbf{F}_{RF}$$

$$\mathbf{W} = \mathbf{W}_{BB}\mathbf{W}_{RF}$$

$$[\mathbf{F}_{RF}]_{:,n_t} \in \mathcal{F}, \forall n_t \tag{2}$$

$$[\mathbf{W}_{RF}]_{:,n_r} \in \mathcal{W}, \forall n_r$$

$$\|\mathbf{F}_{RF}\mathbf{F}_{BB}\|_F^2 = \mathbf{N}_s$$

where $\mathbf{R}_n$ represent the noise covariance matrix with $\mathbf{R}_n = (N_s \sigma_n^2 / P_T) \mathbf{W}^H \mathbf{W}$, where P_T denote the total transmit power and σ_n^2 denote noise power. Further,

$\mathcal{F} = \{c^1, c^2, \ldots, c^{|\mathcal{F}|}\}$ represent the predefined quantization codebook at the base station and $\mathcal{W} = \{g^1, g^2, \ldots g^{|\mathcal{W}|}\}$ represent the predefined quantization codebook on the user side.

If the RF beamforming/combined codebook consists of orthogonal vectors (such as the DFT codebook), then for any selected RF precoder $\mathbf{F}_{RF}$ and RF combiner $\mathbf{W}_{RF}$, the best baseband precoder/combiner is expressed as

$$\mathbf{F}_{BB}^* = \left(\mathbf{F}_{RF}^H \mathbf{F}_{RF}\right)^{-\frac{1}{2}} \left[\overline{\mathbf{V}}\right]_{:,1:N_s} \tag{3}$$

$$\mathbf{W}_{BB}^* = \left[\overline{\mathbf{U}}\right]_{:,1:N_S} \tag{4}$$

where $\overline{\mathbf{V}}$ and $\overline{\mathbf{U}}$ represent the left and right singular value orthogonal matrix of the effective channel matrix with $\overline{\mathbf{H}} = \mathbf{W}_{RF}^H \mathbf{H} \mathbf{F}_{RF}$. Then the optimal hybrid precoder can be found through exhaustive search of candidate RF beamforming/combination vectors, which means the hybrid precoding design problem can be transformed into an exhaustive search problem of RF precoding/combination matrix

$$\{\mathbf{F}_{RF}^*, \mathbf{W}_{RF}^*\} = \arg\max \log_2 \left| \mathbf{I} + \text{SNR} \mathbf{W}_{RF}^H \mathbf{H} \mathbf{F}_{RF} \times \left(\mathbf{F}_{RF}^H \mathbf{F}_{RF}\right)^{-1} \mathbf{F}_{RF}^H \mathbf{H}^H \mathbf{W}_{RF} \right|$$

$$s.t. \, [F_{RF}] :, n_t \in \mathcal{F}, \forall n_t \tag{5}$$

$$[W_{RF}] :, n_r \in \mathcal{W}, \forall n_r$$

3 Problem Solving

3.1 Millimeter Wave Compression Channel Sensing

This paper proposes a novel neural network architecture that can directly find the RF precoding/combination vector in the hybrid precoding architecture, and the vector can maximize the best achievable rate. According to the system model in Sect. 2.1, the achievable rate of the encoder/combiner is expressed as

$$R = \log_2 \left| \mathbf{I} + \mathbf{R}_n^{-1} \mathbf{W}^H \mathbf{H} \mathbf{F} \mathbf{F}^H \mathbf{H}^H \mathbf{W} \right| \tag{6}$$

where $\mathbf{F} = \mathbf{F}_{RF} \mathbf{F}_{BB}$, $\mathbf{W} = \mathbf{W}_{RF} \mathbf{W}_{BB}$.

Let $\mathbf{P}$ and $\mathbf{Q}$ denote the $N_t \times M_t$ and $N_r \times M_r$ channel measurement matrix adopted by both the transmitter and receiver to sense the channel $\mathbf{H}$, with M_t and M_r representing the number of transmit/receiver measurements. If the pilot symbol is equal to 1, then the received measurement matrix, $\mathbf{Y}$, can be written as

$$\mathbf{Y} = \sqrt{P_T} \mathbf{Q}^H \mathbf{H} \mathbf{P} + \mathbf{Q}^H \mathbf{V} \tag{7}$$

where $[\mathbf{V}]_{m,n} \sim \mathcal{N}(0, \sigma_n^2)$ denote receive measurement noise, and when vectorizing this matrix, we get

$$y = \sqrt{P_T}\left(\mathbf{P}^T \otimes \mathbf{Q}^H\right) h + v_q \tag{8}$$

where $y = \text{vec}(\mathbf{Y})$, $v = \text{vec}(\mathbf{Q}^H \mathbf{V})$, $h = \text{vec}(\mathbf{H})$.

3.2 Deep Learning Model

In Sect. 2.2, the hybrid precoding design problem has been transformed into the solution problem of analog precoding/combination matrix or the binary classification problem of analog precoding/combination matrix. Since convolutional neural network has significant effect on classification problems, this paper constructs a deep learning model based on the improved convolutional neural network.

The performance of deep learning models can be improved by increasing the depth and width of the network (the depth refers to the number of layers of the network, and the width refers to the number of channels at each layer), but it will lead to a very large number of network parameters. However, the massive parameters not only easily produce overfitting, but also greatly increase the amount of calculation. The Inception module can reduce the dimension of large-sized matrices and aggregate visual information in different sizes to facilitate feature extraction from different scales and achieve better performance. The cascade hybrid precoder structure proposed in this paper is mainly composed of two parts, as shown in Fig. 2.

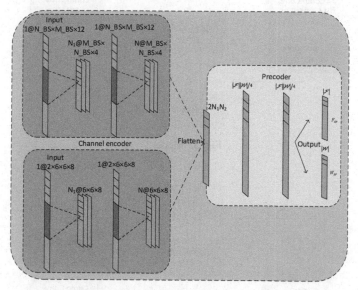

Fig. 2. Cascade hybrid precoding model

The first part is the channel encoder, which consists of two parallel convolution modules. This module is inspired by inception module in GoogLeNet architecture, and the convolution in the inception module is modified to a two-dimensional complex convolution. The first module is shown in Fig. 3, which contains three complex convolution blocks with kernel sizes of M_BS, 12 and 10, respectively. Where N_BS represents the number of antennas at the base station, and M_BS represents the channel measurement vector. Complex convolution blocks create a complex convolution kernel, which is convolved with a complex input layer to produce a complex output tensor. The complex convolution block has complex weights during initialization, which means that the input of the neural network is essentially complex channel coefficients. Therefore, the signal contains real and imaginary feature maps after the convolution operation. Two blocks with kernel sizes of 12 and 10 are connected behind the complex con-volution block to reduce the dimensionality. The output of the block is in series at a depth dimension that preserves the inclination characteristic. The second module is shown in Fig. 4, which contains three complex convolution blocks with kernel sizes of 6, 5 and 4, respectively. In the two modules, different kernel sizes are selected from low to high, with low kernel extracting local features and high kernel extracting global features.

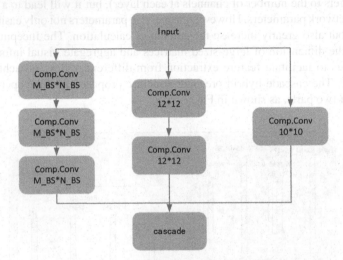

Fig. 3. Model 1

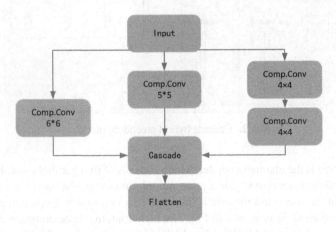

Fig. 4. Model 2

The second part is the precoder that consists of two fully connected layers and two output layers, which takes the output of the channel encoder as input. Through two fully connected layers and fully connected layer maps the learned distributed feature representation to the sample label space for classification. The number of neurons in the two fully connected layers are equal to $|\mathcal{F}||\mathcal{W}|/4$, sigmoid activation and Batch Norm standardization are performed after each fully connected layer. The size of the two output layers is equal to the number of transmit and receive codebooks $|\mathcal{F}|$ and $|\mathcal{W}|$, which means the number of candidate beamforming/combination vectors, and the activation function is Sigmoid.

3.3 Neural Network Training

In the training stage, the cascade hybrid precoder performs end-to-end training in a supervised manner. The data set of millimeter-wave channels and the corresponding RF beamforming/combination matrix are first constructed, and then the cascade hybrid precoder is trained to be able to predict the index of RF precoding/combination vectors given the input channel vector.

In this paper, the neural network is trained based on the channel vector of hybrid architecture and the corresponding data set of RF beamforming/combination vector, and the target RF precoding/combination matrix is calculated based on the approximately optimal Gram-Schmidt hybrid precoding algorithm in [11]. The training and learning process of cascaded hybrid pre-coding neural network is shown in Table 1.

Table 1. The training and learning process of cascaded hybrid pre-coding

Algorithm 1 Neural Network Training
1: Set data parameters, start the data generator, and import training data and test data
2: Design a cascaded precoding neural network structure, set the convolution kernel, learning rate, loss function, number of training rounds epoch and batch size
3: Loop
4: for i=1:1: epoch
5: Update neural network weights according to Adam optimization algorithm
6: end for
7: Output the trained cascade precoding neural network

3.4 Neural Network Prediction

The hybrid precoding neural network architecture is divided into two parts in the prediction stage. The channel encoder is implemented in the RF circuit, and the weights of the two parallel convolution modules of the channel encoder network are used as the weights of the analog/RF measurement matrix at the transmitter and receiver. The precoder uses the output of the channel encoder as input to directly predict the index of the rf beamforming/combination vector of the hybrid architecture.

After the neural network is trained, the trained neural network is used to design the hybrid precoding matrix. The hybrid precoding/combination matrix is predicted by the neural network first, and then the baseband precoding/combination matrix is calculated by the formula (3) and (4). The prediction process is shown in Table 2.

Table 2. The prediction process of cascaded hybrid pre-coding

Algorithm 2 Neural Network Prediction
1: Start the trained cascade precoding neural network
2: Predict the mixed precoding matrix $\mathbf{F_{RF}}$ and the combined matrix $\mathbf{W_{RF}}$
3: Estimate the channel matrix $\tilde{\mathbf{H}} = \mathbf{QYP}^H$
4: Perform singular value decomposition on the channel matrix $\tilde{\mathbf{H}} = \mathbf{U\Sigma V}^H$
5: Calculate baseband precoding $\mathbf{F_{BB}}$ and combination matrix $\mathbf{W_{BB}}$ according to equations (3) and (4)
6: Output the matrix $\mathbf{F_{RF}}$, $\mathbf{W_{RF}}$, $\mathbf{F_{BB}}$ and $\mathbf{W_{BB}}$

The loss function can evaluate the difference between the predicted value and the real value of the model, and the better the loss function, the better the performance of the model. In this paper, the binary cross entropy function is adopted as the loss function of two multi-label classification tasks. The total loss function is the arithmetic mean of binary cross entropy of pre-coding and combination tasks, which is expressed as

$$Loss = -\frac{1}{N} \sum_{i=1}^{N} \left[y_{true} \log y_{pred} + (1 - y_{true}) \log(1 - y_{pred}) \right] \tag{9}$$

4 Simulation Analysis

4.1 Simulation Settings

In this paper, ray-tracing mode is adopted for simulation, and the data set is the universal Deep MIMO [14] data set that can be obtained publicly to generate simulation parameters, as shown in Table 3. The size of the training dataset [14] is 31200, and the size of the prediction dataset is 5000. During the experiment, Keras library and TensorFlow were used, and Adam optimizer was used to optimize the model.

Table 3. The adopted Deep MIMO dataset parameters

Parameter	Value
Number of base stations	4
Number of base station/user terminal antennas	64
System bandwidth/GHz	0.5
Receive noise figure/dB	5
Antenna spacing/wavelength	0.5
Total transmit power/dBm	5,10,15,20,25,30
Number of subcarriers	1024
OFDM sampling coefficient	1
OFDM limitations	1
Number of channel paths	3
Transmitter/receiver RF link number	3
Base station/user side codebook size	64

4.2 Achievable Rate

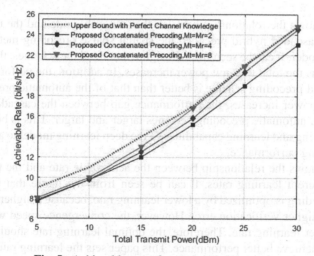

Fig. 5. Achievable rate of cascade hybrid precoder

Figure 5 shows the relationship between the achievable rate of the proposed concatenated precoding and the total transmit power when the channel measurement value is equal to 2, 4, and 8. It can be seen from the figure that the performance of the proposed cascaded precoder is close to the upper limit under a reasonable transmit power value. Additionally, the figure illustrates that the training overhead of the cascaded precoder compared with the traditional hybrid precoding method is significantly reduced. For example, with only 4 channel measurements at both transmitter and receiver, i.e., 16 pilots, the proposed cascaded precoder based on deep learning almost reaches the upper limit of spectral efficiency, which replaces the exhaustive search method required 4096 pilots.

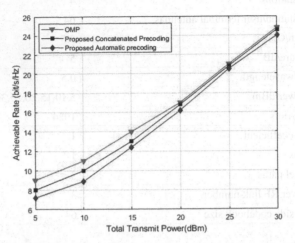

Fig. 6. The achievable rates of several hybrid precoding algorithms

Figure 6 shows the relationship between the achievable rate and the total transmit power of the cascaded hybrid precoder, the spatial sparse precoding method, and the automatic precoder. It can be seen from the figure that in all the schemes, the achievable rate increases as the total transmit power increases. In addition, the performance of the cascaded hybrid precoding scheme is better than that of the automatic precoder. When total transmit power increases, the performance gap between the cascaded precoding scheme and the automatic precoding becomes larger and larger. This is because of the excellent mapping and learning capabilities of the deep learning model to achieve better hybrid precoding performance.

Figure 7 shows the relationship between the achievable rate and the total transmit power for different learning rates. It can be seen from the figure that the proposed cascaded precoding is optimized by a lower learning rate, because a higher learning rate will lead to a higher verification error. However, the convergence speed will be slower due to the lower learning rate. Therefore, the optimal learning rate should be selected reasonably to achieve better performance. This paper sets the learning rate to 0.01.

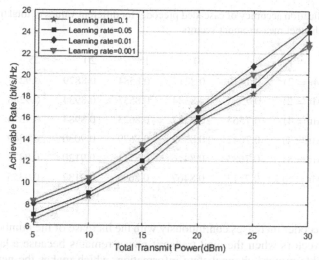

Fig. 7. The achievable rates of the cascaded hybrid precoding method based on deep learning for different total transmit power values and different learning rates

4.3 Predict Accuracy

Since the performance of the deep learning model in various tasks requires quantitative indicators to be evaluated, the sample accuracy of the prediction indicators is calculated in this paper to evaluate the prediction performance of the proposed algorithm. Let $y_{true}^{(i)}$ denote the set of label index, $y_{pred}^{(i)}$ denote the set of prediction index, and the accuracy of sample i is defined as

$$a_i = \frac{\left| y_{true}^{(i)} \cap y_{pred}^{(i)} \right|}{\left| y_{true}^{(i)} \right|} \tag{10}$$

where $\cap$ denote the intersection of two sets and $|\cdot|$ denote cardinality of the set. The sample accuracy can be expressed as

$$a = \frac{\sum_{i=1}^{N} a_i}{N_{samples}} \tag{11}$$

where $N_{samples}$ represent the number of samples.

Under different total transmitting powers, the classification accuracy of the simulated precoding vector at the base station end and the simulated combination vector at the client end is shown in Table 4.

It can be seen from Table 2 that the accuracy for the transmit and receive predicted beams are almost the same since we treat them as equally important tasks when training the auto-precoder neural network model. Besides, the classification accuracy of the cascaded precoder increases continuously with the increase of the total transmit power when the number of channel measurement vectors remains. The classification accuracy

Table 4. Classification accuracy of cascaded precoder for different values of total transmit power and number of channel measurement vectors

P_T(dBm)	5	10	15	20	25	30
Tx acc (Mt = Mr = 2)	0.7449	0.8217	0.8584	0.8879	0.9217	0.9373
Rx acc (Mt = Mr = 2)	0.7553	0.8344	0.8537	0.8934	0.9288	0.9412
Tx acc (Mt = Mr = 4)	0.7565	0.8387	0.8623	0.8984	0.9316	0.9354
Rx acc (Mt = Mr = 4)	0.7642	0.8392	0.8679	0.9030	0.9369	0.9437
Tx acc (Mt = Mr = 8)	0.7683	0.8416	0.8711	0.9120	0.9374	0.9448
Rx acc (Mt = Mr = 8)	0.7713	0.8467	0.8693	0.9137	0.9355	0.9451

of cascaded precoder increases continuously with the increase of the number of channel measurement vectors when the total transmit power remains because a larger number of pilots can obtain more channel state information, which makes the neural network use more features for classification during training. However, when the number of pilots continues to increase, the classification accuracy tends to stabilize, whose value stabilizes at 0.94. This is because no more channel state information can be extracted when the number of pilots exceeds a certain threshold, which means the channel state information has reached the maximum. Therefore, the classification accuracy will no longer continue to increase.

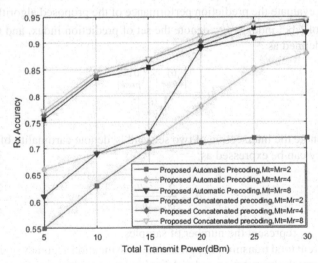

Fig. 8. Comparison of classification accuracy between cascaded hybrid precoding and automatic precoding

Figure 8 shows the relationship between the classification accuracy of the proposed cascaded precoder and auto-precoder [15] methods with different total transmit power. It can be seen from the figure that the classification accuracy of these two schemes increases with the increase of the total transmission power, and the classification accuracy of the cascaded precoder is better than the classification accuracy of the auto-precoder. In addition, the stability of cascaded hybrid precoder is better than that of automatic precoder. Classification accuracy is significantly improved especially with less pilots. For example, for a transmit power of 20 dBm, the accuracy of the auto-precoder is 0.77, and the accuracy of the cascaded precoding is increased to 0.9030, which improves by around 17.27%. This shows that the performance of the cascaded precoder is superior than the auto-precoder.

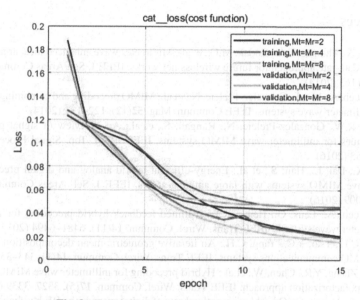

Fig. 9. Training vs Validation Graph for 15 epochs

Figure 9 shows the relationship between the size of the cascaded precoding loss function and the number of iterations. It can be seen from the figure that the loss function curve changes smoothly on the training set and the test set, without excessive fluctuations or jaggedness, which indicate that the learning rate set is reasonable, and it is close to a constant value under a limited number of iterations. There is no obvious gap between the loss function of the training set and the loss function of the test set, indicating that the network does not appear to be under-fitting and overfitting.

5 Conclusions

This paper proposes a cascade hybrid precoding algorithm based on deep learning for millimeter wave hybrid precoding. Cascaded hybrid precoder learns how to optimize channel sensing vectors to concentrate sensing power in the direction most desired, and how to predict RF beamforming and combination vectors of hybrid architectures directly from the received sensing vectors. Compared with the traditional method, the proposed deep learning method can directly predict the simulated precoding/combination vector under the condition of unknown channel state information, which overcomes the huge training cost required to select the optimal beamforming vector in the large antenna array millimeter wave system. At the same time, the data rate is close to the best data bit rate. Compared with some existing deep learning schemes, the performance of cascade hybrid precoding is closer to that of all-digital precoding algorithm, and it can accurately predict the analog precoding matrix/combination matrix with fewer channel measurement vectors.

References

1. Ghosh, A., Thomas, T.A., Cudak, M.C., et al.: Millimeter-wave enhanced local area systems: a high-data-rate approach for future wireless net-works. IEEE J. Sel. Areas Commun. **32**(6), 1152–1163 (2014)
2. Alkhateeb, A., Mo, J., Gonzalez-Prelcic, N., et al.: MIMO precoding and combining solutions for millimeter-wave systems. IEEE Commun. Mag. **52**(12), 122–131 (2014)
3. Heath, R.W., Gonzalez-Prelcic, N., Rangan, S., et al.: An overview of signal processing techniques for millimeter wave MIMO systems. IEEE J. Sel. Top. Signal Process. **10**(3), 436–453 (2016)
4. Gao, X., Dai, L., Han, S., et al.: Energy-efficient hybrid analog and digital precoding for mmWave MIMO systems with large antenna arrays. IEEE J. Sel. Areas Commun. **34**(4), 998–1009 (2016)
5. Alkhateeb, A., Leus, G., Heath, R.W.: Limited feedback hybrid precoding for multi-user millimeter wave systems. IEEE Trans. Wirel. Commun. **14**(11), 6481–6494 (2015)
6. Chen, C.E., Tsai, Y.C., Yang, C.H.: An iterative geometric mean decomposition algorithm for MIMO communications systems. IEEE Trans. Wirel. Commun. **14**(1), 343–352 (2014)
7. Jin, J., Zheng, Y.R., Chen, W., et al.: Hybrid precoding for millimeter wave MIMO systems: a matrix factorization approach. IEEE Trans. Wirel. Commun. **17**(5), 3327–3339 (2018)
8. Zhang, E., Huang, C.: On achieving optimal rate of digital precoder by RF-baseband codesign for MIMO systems. In: 2014 IEEE 80th Vehicular Technology Conference (VTC2014-Fall), pp. 1–5. IEEE (2014)
9. Wang, G., Ascheid, G.: Joint pre/post-processing design for large milli-meter wave hybrid spatial processing systems. In: European Wireless 2014; 20th European Wireless Conference, VDE, pp. 1–6 (2014)
10. Yu, X., Shen, J.C., Zhang, J., et al.: Alternating minimization algorithms for hybrid precoding in millimeter wave MIMO systems. IEEE J. Sel. Top. Signal Process. **10**(3), 485–500 (2016)
11. Chen, C.H., Tsai, C.R., Liu, Y.H., et al.: Compressive sensing (CS) assisted low-complexity beamspace hybrid precoding for millimeter-wave MIMO systems. IEEE Trans. Signal Process. **65**(6), 1412–1424 (2016)
12. Alkhateeb, A., El Ayach, O., Leus, G., et al.: Channel estimation and hybrid precoding for millimeter wave cellular systems. IEEE J. Sel. Top. Signal Process. **8**(5), 831–846 (2014)
13. El Ayach, O., Rajagopal, S., Abu-Surra, S., et al.: Spatially sparse pre-coding in millimeter wave MIMO systems. IEEE Trans. Wirel. Commun. **13**(3), 1499–1513 (2014)
14. Alkhateeb, A.: DeepMIMO: a generic deep learning dataset for millimeter wave and massive MIMO applications. arXiv preprint arXiv:1902.06435 (2019)
15. Li, X., Alkhateeb, A.: Deep learning for direct hybrid precoding in millimeter wave massive MIMO systems. In: 2019 53rd Asilomar Conference on Signals, Systems, and Computers, pp. 800–805. IEEE (2019)

16. Huang, H., Song, Y., Yang, J., et al.: Deep-learning-based millimeter-wave massive MIMO for hybrid precoding. IEEE Trans. Veh. Technol. **68**(3), 3027–3032 (2019)
17. Bao, X., Feng, W., Zheng, J., et al.: Deep CNN and equivalent channel based hybrid precoding for mmWave massive MIMO systems. IEEE Access **8**, 19327–19335 (2020)

Open-Set Recognition Algorithm of Signal Modulation Based on Siamese Neural Network

Pengcheng Liu, Yan Zhang, Mingjun Ma, Zunwen He, and Wancheng Zhang[✉]

Beijing Institute of Technology, BIT, Beijing, People's Republic of China
zhangwancheng@bit.edu.cn

Abstract. To realize the open-set recognition of various signal modulations, we propose a three-stage recognition algorithm based on the siamese neural network. Firstly, the convolutional neural network (CNN) is employed to extract the features of the original signal to obtain the multi-dimensional feature vector, thereby constructing the feature data set. Then, a double-threshold method basedon average valuesis designed to distinguish signals with known modulation types and unknown ones. Finally, unknown modulation types are recognized by the siamese neural network. We proposed an iterative algorithm based on the siamese neural network to obtain modulation types of the unknown signals. Experimental results show that the proposed algorithm model can be used for open-set recognition and is suitable for applications in complex wireless communication environments.

Keywords: CNN · Modulation recognition · Open-set recognition · Siamese network

1 Introduction

With the popularization of 5th Generation Mobile Communication Technology (5G) and the development of 6th Generation Mobile Communication Technology (6G), the electromagnetic spectrum has become more crowded. Complex signals have a wider frequency range and more diverse modulation types. The realization of the unknown signal detection and recognition system becomes a more challenging task. To solve this problem, one fundamental work is to apply automatic modulation classification (AMC) to identify the modulation of the signals. The classic AMC methods use likelihood methods [1–3] and traditional feature analysis [4, 5]. However, the calculation based on the likelihood method is complicated, and once it is simplified the classification performance will decrease.

Deep learning has achieved rapid development and great success in AMC. O'Shea et al. used GNU Radio to simulate 11 types of modulated signals in [6] and used CNN to explore the effect of hyperparameters on recognition. In [7], the authors combined the spectrum with CNN to improve the accuracy of AMC.

However, the above methods are all closed-set identification problems. In the actual non-cooperative communication environment, a variety of samples with unknown modulation types may be collected. If the closed-set recognition system is still used, the

© ICST Institute for Computer Sciences, Social Informatics and Telecommunications Engineering 2022
Published by Springer Nature Switzerland AG 2022. All Rights Reserved
H. Gao et al. (Eds.): ChinaCom 2021, LNICST 433, pp. 28–39, 2022.
https://doi.org/10.1007/978-3-030-99200-2_3

system will mistakenly identify the test sample of the unknown class as belonging to one of the known closed set classes, and the accuracy rate will decrease. To solve this problem, open-set recognition is introduced. In [9], the authors proposed a method based on generative adversarial network and anomaly detection, which realized the open-set recognition by reconstructing and discriminating the network and achieved good results within additive white Gaussian noise (AWGN) channels. However, only digital modulation types are considered in [9] while the analog modulation is not taken into account. Moreover, only a few modulation types, including three closed set signals and three open-set signals, can be recognized. In practice, the multipath fading channel should also be considered because it has a great influence on the recognition results.

In this paper, we propose a three-stage recognition method for the open-set recognition of signal modulation. The output of the two convolutional layers and the two fully connected layers are used as the extracted features. Feature comparison analysis is performed to obtain the best features needed. We designed a double-threshold method for screening unknown signals. This method combined with appropriate feature processing methods can further improve the robustness and recognition effect of the proposed recognition algorithm. Then, the siamese neural network is applied to the open-set scene. The signal in the closed set can be recognized and then the open-set classification of the signal can be realized.

2 Signal Model and Open-Set Recognition System Model

2.1 Signal Model

The signal received on the receiver can be expressed as

$$x(t) = h(t) * s(t) + n(t) \tag{1}$$

Among them, $*$ is the convolution operation, $n(t)$ represents the AWGN, $h(t)$ represents the channel impulse response, $s(t)$ is the modulated signal, and $x(t)$ is the received signal which is usually represented by IQ formatted data. The purpose of modulation recognition is to give the class of $s(t)$. Therefore, this problem can be reduced to a classification problem. Since the interference introduced by $h(t)$ and $n(t)$, the change of $x(t)$ will be unpredictable. This also poses a huge challenge to the open-set recognition of modulation types.

2.2 Open-Set Recognition System Model

The closed-set recognition and open-set recognition tasks of signal modulation are illustrated in Fig. 1. As shown in Fig. 1(a), in the open scene, the traditional closed-set recognition may cause errors: different categories are classified into one category, such as category 1 and category 2, and the same category is classified into two categories, such as category 4.

The open-set recognition tasks of signal modulation can be defined as follows. Give a training signal set as

$$D = \{(x_i, \hat{y}_i)\}_{i=1}^{q}. \tag{2}$$

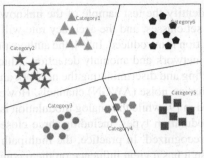

(a) Closed-set recognition of signal modulation

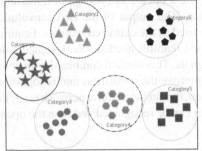

(b) Open-set recognition of signal modulation

Fig. 1. Closed-set recognition and open-set recognition tasks of signal modulation.

In (2), x_i is a training signal sample. q represents the number of samples in the training set. $\hat{y}_i \in \mathbf{Y}_{closed}$ is the known modulation label corresponding to the training signal sample x_i. $\mathbf{Y}_{closed} = \{1, \ldots, K\}$ is the modulation type category set.

In the open-set recognition problem, a signal sample is taken from the open space of an unknown modulation signal

$$D_0 = \{(x_i, \hat{y}_i)\}_{i=1}^{\infty} \tag{3}$$

where $\hat{y}_i \in \mathbf{Y}_{open}$ and $\mathbf{Y}_{open} = \{1, 2, \ldots, K, K + 1, \ldots, M\}$ is the label of the possible modulation types. $M > K$ means there exist some unknown modulation types in the considered signal samples. Our task is to realize the open-set recognition of all modulation types in $\mathbf{Y}_{open}$.

3 Three-Stage Open-Set Recognition Algorithm

In this section, the proposed open-set recognition algorithm of modulated signals based on siamese network will be introduced.

The first stage is feature extraction in part A. Then, the double-threshold method is designed to screen the signal in part B. At the third stage, the closed-set and open-set recognitions of the modulation types are realized in part C. The algorithm flow chart is shown in Fig. 2.

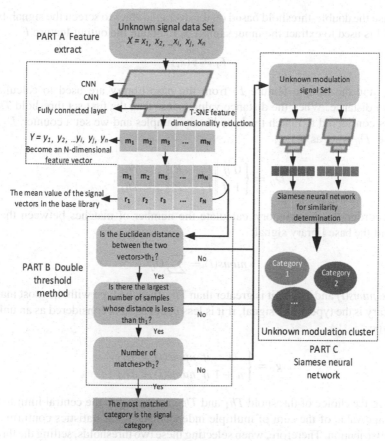

Fig. 2. The three-stage open-set recognition algorithm flow chart.

In part A, the signal is extracted with different layers of CNN, and then clustered after t-distributed stochastic neighbor embedding (T-SNE) dimensionality reduction processing. Take the neural network layer with the best clustering effect as the output layer of the feature extraction module. The feature extraction module converts the input IQ signal into a computable N-dimensional signal feature. Here, we use $[m_1, m_2 ... m_N]$ and $[r_1, r_2 ... r_N]$ to represent the feature vector. N can be set according to the size of the data set and the type of signal.

In part B, we use the double-threshold method based on the average values to screen the signal. The signal to be recognized is compared with a base library signal, and two different thresholds are used to classify this signal into a known modulation type and an unknown one. The source of the base library signal is composed of M samples selected under different signal-to-noise ratios (SNRs) for each type of signal in the data set. By adding samples with different SNRs to the base library, we solve the problem of sample bias caused by a single base library signal. Although this will bring a certain amount of calculation to the calculation of Euclidean distance, the effect is improved compared to a single base library.

We use the double-threshold based on the average values to screen the signal. Firstly, the model is used to extract the input sample x_i to obtain the output feature F_i.

$$F_i = v(x_i) \tag{4}$$

Then, F_i and the sample feature F_j from the base library are used to calculate the Euclidean distance. When the distance value is less than the first set threshold Th_1, the sample is considered to match the base library samples and we set a counter, D_{ij}, as 1. otherwise, D_{ij} is set as 0, i.e.,

$$D_{ij} = \begin{cases} 0 \ if \ \|F_i - F_j\|_2 > Th_1 \\ 1 \ if \ \|F_i - F_j\|_2 < Th_1 \end{cases}. \tag{5}$$

Then, according to the category, calculate the number of matches between the input signal and the base library signal.

$$nums(i) = \sum D_{ij} \tag{6}$$

Compare $nums(i)$ and Th_2, if it is greater than Th_2, then the type with the most matching base library is the type of the signal, if it is less than Th_2, it is considered as an unknown signal type.

$$g_i = \begin{cases} k \quad if \ nums(i) > Th_2 \\ n+1 \ if \ nums(i) < Th_2 \end{cases} \tag{7}$$

Regarding the choice of threshold Th_1 and Th_2, according to the central limit theorem, the average value of the sum of multiple independent signal statistics conforms to the normal distribution. Therefore, when selecting these two thresholds, setting the threshold range based on the average value will make the screening of unknown modulation signals more accurate. It will be proved in the experiment of the double-threshold recognition method based on the average value.

At the beginning of the recognition task, only the existing training data is used as the base library. As the recognition task progresses, a certain number of unknown signals can gradually accumulate. However, it should be noted that at this time the type of unknown signals is still uncertain.

In part C, in order to solve the problem of being unable to distinguish different unknown signals, the siamese neural network is introduced to complete the recognition task. The distance between two different inputs is calculated by the siamese neural network. The output distance is finally obtained by the sigmoid function to obtain a value between (0, 1). The appropriate threshold is used to judge the two inputs are in the same category or not.

For each unknown signal confirmed by the double-threshold method, the siamese neural network will classify these signals based on similarity.

In summary, the following cyclic iterative algorithm process is used to complete the category classification of the input signals, as shown in Algorithm 1.

Algorithm 1

Algorithm : Signal modulation type recognition algorithm

Step 1:

Input: Unknown signal set $X = x_1, x_2 ... x_i, x_j, x_n$

Output: Unknown modulation signal set output by double threshold method $K = \{x_1, x_2, x_3 ... x_Q\}$, Q is the number of elements in the set of K.

1. Unknown signal set $X = x_1, x_2 ... x_i, x_j, x_n$ goes through CNN and T-SNE feature dimensionality reduction, get the set $Y = y_1, y_2 ... y_i, y_j, y_n$.

2. Comparing the set Y vector with the mean value of the signal feature vector in the base library, the Euclidean distance between the two vectors is greater than the threshold th_1, then put it into K, otherwise, proceed to 3.

3. Whether the number of samples smaller than th_1 is the largest, if it is not the largest, put the signal into K, if it is the largest, proceed to step one 4.

4. Determine whether the number of signal matches is greater than th_2. If yes, consider the modulation type of the most matched category, that is, the modulation type of the signal. If not, put the signal into K.

Step 2:

Input: The set of unknown modulation signals output by the double-threshold method based on the average value method $K = \{x_1, x_2, x_3 ... x_Q\}$, Q is the number of elements in the set in K, and the threshold is L.

Output: Final recognition and classification of different types of unknown modulation signal sets $W_1, W_2, W_3 ... W_p$, p is the total number of position signal sets.

1.   Create a new signal set W_1:
2. $p = 1$;
3.   For $i = 1, 2 ... Q$

4.   Take a signal X_i from the signal set K;
5.   For $j = 1, 2 ... p$;
6.   Compare X_i with the set W_j
          similarity array $L_{W(j)}$
7.   End
8.   Calculate the maximum similarity L_{Wmax} in the array $L_{W(j)}$, and the corresponding array number is marked as m;
9.   If $L_{Wmax} >$ Threshold L
10.   Put the signal sample X_i into the signal set W_m;
11.   Else
12.   Create a new signal set W_{p+1}, and put X_i into W_{p+1}
13.   End
14. End

From the actual modulation signal analysis, it is found that there will be confusion between different QAM modulations. Therefore, in the open-set identification process, a separate identification module for QAM is introduced to classify different QAM signals. The specific analysis will be further detailed in Sect. 4.

4   Experimental Test and Result Analysis

In the experiment, we use a CNN with two convolutional layers and two fully connected layers to extract features from a part of the training set. Select the training set under the same SNR. This training set contains 11 types of samples, each type of sample has 100 signals. The output of different layers reduced by T-SNE is shown in Fig. 3, where the horizontal and vertical axis is the normalized distance.

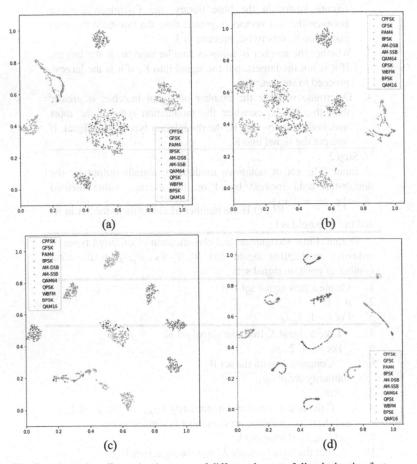

Fig. 3. Dimensionality reduction map of different layers of discriminative features.

Figure 3(a) is the output of the first convolutional layer. It can be seen from the dimensionality reduction graph that the features extracted by the model at this stage can be used for preliminary classification, but there are still a variety of signal features, and the signals are clustered together. It is difficult to distinguish them; Fig. 3(b) is the output of the last convolutional layer. Compared with Fig. 3(a), the distance between some modulation types has become clearer and obvious. The features of PAM4 and BPSK are closely connected in Fig. 3(a). In Fig. 3(b), there is an obvious classification boundary, which shows that as the number of convolutional layers continues to deepen, the representativeness of the features becomes stronger and stronger. Figure 3(c) is the output of the penultimate fully connected layer, which is the best feature in the general sense. Compared with Fig. 3(b), its classification effect has changed significantly. QPSK and 8PSK signals have been completely separated. The distance between different modulation signals has been clearly distinguished; Fig. 3(d) is the output of the last fully connected layer, which is the vector before entering the softmax. Because it is directly supervised by the label during training, the classification boundary of this layer is the most obvious, but the data space is compressed too tightly, and using the output of this layer as a feature is more likely to cause overfitting. Therefore, this paper finally chooses the output of the penultimate fully connected layer as the extracted signal feature.

The experimental results are evaluated on the dataset RadioML.2016.10a [6], The number of samples in the data set is 22000, and each sample is a 2×128 vector. For channel simulation, the dynamic channel model hierarchical block uses frequency offset, sampling rate offset, AWGN, multipath, and fading. The data set is divided into 80% of the training set and 20% of the test set. Epochs = 1000 and batch size = 500. In the training phase, the validation set is used for parameter adjustment, such as batch size, optimizer, learning rate, etc.

Use the double-threshold method based on the average value proposed in this paper. When training the model, only 9 types of dataset at different SNRs are used, and the BPSK and BFSK data are removed from the training set to test the open-set recognition effect of the algorithm. The experiment not only needs to identify the existing 9 types of signal modulation types in the training set but also needs to verify the modulation of the two open-set types of signals. In the current stage of the experiment, both BPSK and BFSK signals are marked as unknown signals in the experiment. The experimental results are shown in Fig. 4.

The vertical axis is the accuracy rate, and the horizontal axis is the signal-to-noise ratio. In Fig. 4, the green curve named "max1thresh" is the experimental result of the threshold method based on the maximum value method in [10]. The pink curve named "max2thresh" is the experimental result obtained by improving the method in [10] using the double-threshold method. It is obvious that the use of the double-threshold method after the SNR reaches a certain level can significantly improve the accuracy of classification. The definition of accuracy here: the ratio of the number of unknown modulation type identified to the number of the actual unknown modulation types, the black curve named "average1thresh" uses the average threshold method but does not use the double threshold, which is a certain improvement compared to the maximum threshold method. However, the classification accuracy performance is worse than that of the double-threshold method based on the maximum threshold method. The blue curve named "average2thresh" is the double-threshold method based on the average method

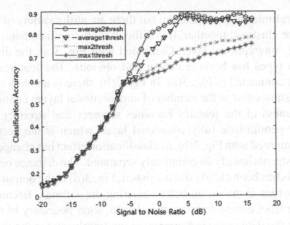

Fig. 4. Recognition results of different methods under open-set requirements (Color figure online)

proposed in this article. Its classification accuracy is higher than other methods, and the highest accuracy rate is 89.18% when the SNR is 14 dB.

Specific analysis of the recognition effect of each category plots all test data with a signal-to-noise ratio above 0 dB, and the confusion matrix is shown in Fig. 5.

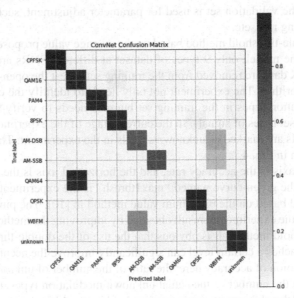

Fig. 5. Confusion matrix of the average value method and double threshold method classification results

In Fig. 5, the darker the color bar on the right represents the higher recognition accuracy, the ordinate represents the true category of the sample, and the abscissa represents the predicted category of the sample. The unknown categories are BPSK and BFSK. For the 9 types of signals included in the training data set, the average accuracy rate can reach 86%. For a total of 11 types of signal input, the recognition accuracy rate is 77.14%. It can also be seen that the classification of QAM64 and QAM16 is the main factor that affects the classification results, and according to what we have learned, QAM64 and QAM16 signals can be distinguished by the zero center normalized instantaneous amplitude compactness (ZCNC).

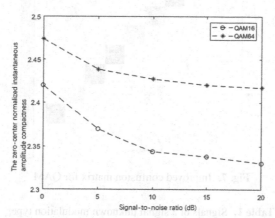

Fig. 6. The zero center normalized instantaneous amplitude compactness (ZCNC) of QAM64 and QAM16.

It can be seen from Fig. 6 that the ZCNC of QAM16 is smaller than that of 64QAM. Therefore, the two QAM signals can be distinguished by calculating this parameter threshold. Setting the threshold value to 2.42 can distinguish 85% of the two signals at different SNRs, and the effect is shown in Fig. 7.

Finally, the siamese neural network is used to reclassify the experimental results and analysis of the open-set category, and each signal is regarded as an unknown signal. The open-set recognition experiment of a single unknown signal with SNR higher than 0 dB is performed well. The experimental results are shown in Table 1.

It can be seen from Table 1 that for a single unknown modulation signal, the probability of identifying it as an unknown signal exceeds 85%. When the input is recognized as an unknown signal, it is clustered through the siamese network, and the main proportion of the set size is used to roughly determine the type of unknown signal. Then, take out BFSK and BPSK as unknown signals for testing. The experimental results of the average similarity calculated by the siamese neural network are shown in Table 2.

The number of final output sets is 4, of which the BPSK largest cluster set accounts for 72.5%, and the number of secondary cluster accounts for 27.5%. The recognition effect for BFSK is even better, with the largest cluster accounting for 93%, and the secondary cluster only accounting for 7%. The final output can also be filtered by adding a limit to the number of samples in each cluster to make it clear that the final unknown signal includes two types. Compared with [9], the types of digital modulation types and analog modulation types for recognition are increased, and multipath and fading factors are taken into account to realize open-set recognition in more complex environments.

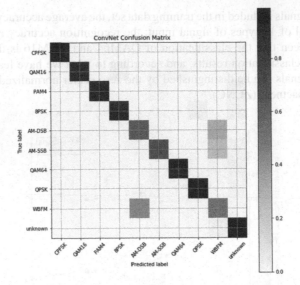

Fig. 7. Improved confusion matrix for QAM

Table 1. Signals of a signal unknown modulation type.

Actual modulation of signal	Probability of being recognized as an unknown signal	Percentage of main clusters	Percentage of secondary clusters
BPSK	87.31%	88.50%	11.50%
QPSK	85.14%	92.45%	7.55%
8PSK	86.67%	94.90%	5.10%
16QAM	87.84%	94.73%	5.27%
64QAM	86.49%	92.13%	7.87%
BFSK	85.50%	95.22%	4.78%
CPFSK	89.33%	96.47%	3.53%
PAM4	91.79%	92.79%	7.21%
WBFM	84.75%	90.90%	9.10%
AM-SSB	90.51%	97.15%	2.85%
AM-DSB	89.62%	93.21%	6.79%

Table 2. The similarity between unknown modulation signals.

Similarity	BPSK	BFSK
BPSK	0.5645	0.0010
BFSK	0.0010	0.7494

5   Conclusions

In this paper, we proposed a three-stage algorithm for open-set modulation recognition. First, we examined the selection of the appropriate network output as the discriminative feature. Then, a double-threshold method based on the average values was proposed. It can divide the input signals into known types and unknown ones, and complete the preliminary identification of unknown signals. Subsequently, the siamese neural network was employed to measure the similarity between two different signals, and the similarity threshold was used to finally complete the classification of the unknown modulation type. Experimental results showed that the proposed three-stage modulation recognition algorithm can maintain an 87% recognition rate for traditional closed-set recognition signals when the SNR was greater than 0 dB. It can also effectively identify 84% of signals with unknown modulation types.

Acknowledgement. This work was supported by the National Key R&D Program of China under Grant 2020YFB1804901, the National Natural Science Foundation of China under Grant 61871035, and the National Defense Science and Technology Innovation Zone.

References

1. Wang, F., Dobre, O.A., Chan, C., Zhang, J.: Fold-based Kolmogorov–Smirnov modulation classifier. IEEE Signal Process. Lett. **23**(7), 1003–1007 (2016)
2. Zhu, D., Mathews, V.J., Detienne, D.H.: A likelihood-based algorithm for blind identification of QAM and PSK signals. IEEE Trans. Wirel. Commun. **17**(5), 3417–3430 (2018)
3. Chen, W., Xie, Z., Ma, L., Liu, J., Liang, X.: A faster maximum likelihood modulation classification in flat fading non-Gaussian channels. IEEE Commun. Lett. **23**(3), 454–457 (2019)
4. Kharbech, S., Dayoub, I., Zwingelstein-Colin, M., Simon, E.P.: On classifiers for blind feature-based automatic modulation classification over multiple-input-multiple-output channels. IET Commun. **10**(7), 790–795 (2016)
5. Wang, Y., Yang, J., Liu, M., Gui, G.: LightAMC: lightweight automatic modulation classification via deep learning and compressive sensing. IEEE Trans. Veh. Technol. **69**(3), 3491–3495 (2020)
6. O'Shea, T.J., Corgan, J., Clancy, T.C.: Convolutional radio modulation recognition networks (2016). https://arxiv.org/abs/1602.04105
7. Zeng, Y., Zhang, M., Han, F., Gong, Y., Zhang, J.: Spectrum analysis and convolutional neural network for automatic modulation recognition. IEEE Wirel. Commun. Lett. **8**(3), 929–932 (2019)
8. Bu, K., He, Y., Jing, X., Han, J.: Adversarial transfer learning for deep learning-based automatic modulation classification. IEEE Signal Process. Lett. **27**, 880–884 (2020)
9. Hao, Y., Liu, Z., Guo, F., Zhang, M.: Open-set recognition of signal modulation based on generative adversarial networks. Syst. Eng. Electron. **41**(11), 2619–2624 (2019)
10. Guo, Y.: Automation modulation recognition of the communication signals based on deep learning. Master's thesis, China Academy of Engineering Physics. https://kns.cnki.net/KCMS/detail/detail.aspx?dbname=CMFD201901&filename=1019025043.nh

Accurate Frequency Estimator of Real Sinusoid Based on Maximum Sidelobe Decay Windows

Zhanhong Liu⬭, Lei Fan(✉)⬭, Jiyu Jin, Renqing Li, Jinyu Liu, and Nian Liu

School of Information Science and Engineering, Dalian Polytechnic University, Dalian, China
fanlei@dlpu.edu.cn

Abstract. The estimation of real sinusoid frequency is a significant problem in many scientific fields. The positive- and negative-frequency components of a real sinusoid interact with each other in the frequency spectrum. This leads to estimation bias. In this paper, we proposed an algorithm which is based on maximum sidelobe decay (MSD) windows. Firstly, the coarse frequency estimate is obtained by using Discrete Fourier Transform (DFT) and MSD windows. Then the negative-frequency component is removed by frequency shift. At last, the fine frequency estimation is performed by a high-precision frequency estimation algorithm. Simulation results show that the proposed algorithm has higher accuracy and better frequency estimation performance than AM algorithm, Candan algorithm, and Djukanovic algorithm.

Keywords: Frequency estimation · Real sinusoid · Windows · DFT

1 Introduction

The problem of sinusoid signal frequency estimation is quite significant in many scientific fields, such as communication, audio system, radar, power system, sonar, measurement and instrument. For example, in communication field, there may be a deviation between the oscillation frequency generated by the crystal oscillator and the nominal frequency, and there is generally a relative motion between the transmitting end and the receiving end of communication, resulting in the Doppler frequency shift. Carrier frequency offset is quite common in communication systems. Therefore, the carrier frequency offset must be estimated correctly. In music signal processing, standard music signals including human songs and musical instrument sounds are usually modeled as the sum of multiple time-varying sinusoidal signals, and the number of these sinusoidal signals is usually unknown. Next, multiple sinusoidal signals in music signals need to be detected. The common method is to continuously search and estimate the frequency of a single sinusoidal signal in the short-time Fourier transform domain. When using linear frequency modulated continuous wave radar for ranging, it is important to exactly estimate the

This work are supported by the 2021 scientific research projects of Liaoning Provincial Department of Education under Grant LJKZ0515, LJKZ0519, LJKZ0518.

H. Gao et al. (Eds.): ChinaCom 2021, LNICST 433, pp. 40–51, 2022.
https://doi.org/10.1007/978-3-030-99200-2_4

frequency of linear frequency modulated continuous wave signal. Because the accuracy of frequency estimation directly determines the ranging accuracy. In power system, a significant parameter of power quality is frequency. And the variation of frequency is the result of dynamic imbalance between the generation and load, which needs accurate estimation. Accurate frequency estimation is the premise of power grid stability and normal operation of electrical equipment.

As we all know, frequency estimation algorithms are mainly divided into two categories in the case of additive white Gaussian noise, time domain algorithms [1–9] and frequency domain algorithms [10–20]. Time domain algorithms include the least mean square algorithm, the autocorrelation algorithm and the maximum likelihood algorithm. The accuracy of the maximum likelihood algorithms is very high, reaching the Cramer Rao lower bound (CRLB). Nevertheless, due to the large amount of computation, they are difficult to be used in real-time applications [1]. The frequency domain algorithms are primarily based on DFT. These algorithms have many advantages, such as fairly little computation and important signal-to-noise ratio (SNR) gain [18–22]. Therefore, they are suitable for real-time applications. For the DFT based algorithms, in additive broadband Gaussian noise, the peak position of the Discrete Time Fourier Transform (DTFT) of the whole signal represents the maximum likelihood frequency estimation of a sinusoid [10]. The location of the peak can be located through a two-step search process. The first step can be named as coarse-search, is to determine the maximum amplitude of DFT samples by a simple maximum search procedure. The second step is the fine-search which obtains the relative frequency deviation between the true frequency and the rough estimate by means of certain interpolation methods. The difference between different interpolation algorithms lies only in the second step. Aboutanios and Mulgrew (A&M) algorithm uses two spectral lines located exactly halfway between the maximum spectral line and its two neighbors for accurate estimation [18]. Candan algorithm achieves accurate estimation by using the maximum spectral line and two spectral lines on the left and right of the maximum spectral line [19].

All the above algorithms are used to estimate the frequency of complex sinusoid. There are also many practical situations and applications related to the real sinusoid model. For the real sinusoid, estimation bias is caused by the spectral superposition of its positive-frequency and negative-frequency components. As we can see in [23], multiplying the received signal by a window function is a simple and effective method to reduce the spectrum leakage. In [24], by adding Kaiser window to suppress the sidelobe of complex sinusoid, the estimation bias is reduced. However, there are still significant deviations in the windowed data. The time domain frequency estimators based on autocorrelation are the most accurate [9]. Although the algorithm in [9] is more precise than other time-domain estimation algorithms, this algorithm has root-mean-square error (RMSE) saturation under high SNR. Candan obtains the fine frequency estimation with arbitrary window functions [25]. In [26], Djukanovic derived a method based on the Candan method [25] and AM method [18], and the negative-frequency component is shifted via modulation before the fine estimation.

The MSD window is a kind of commonly used window function, which belongs to cosine windows. The sidelobe decay rate of MSD window is as high as $6(2H - 1)$

dB/octave, when its number of terms H is constant. Thus, a frequency estimation algorithm based on MSD windows in this paper is proposed to accurately estimate the frequency of real sinusoid. The coarse frequency estimate is obtained with MSD windows. Then the estimation bias is reduced by removal of the negative frequency component which is done by frequency shift. Finally, the fine frequency estimation is performed by a high-precision frequency estimation algorithm which uses two DTFT spectrum lines and the maximum DFT spectrum line. Computer simulations are conducted, and the results of simulation experiments show that the algorithm we proposed in this paper has higher accuracy and better frequency estimation performance than AM algorithm [18], Candan algorithm [25], and Djukanovic algorithm [26].

2 Proposed Algorithm

An accurate real sinusoid frequency estimation algorithm based on MSE windows is mainly described in this part. In the background of additive white Gaussian noise, the following signal model is considered

$$x(n) = A\cos(2\pi f_0 n + \phi) + z(n) \quad n = 0, 1, 2, \dots, N - 1 \quad (1)$$

In the above formula A represents the amplitude of the sinusoid, and f_0 is the frequency, ϕ is the initial phase. $z(n)$ represents zero-mean additive white Gaussian noise with variance σ_z^2. In addition, $A > 0$ and $0 < f_0 < 1/2$. The SNR can be defined as $SNR = \frac{A^2}{2\sigma_z^2}$. The frequency estimation of the CRLB is [27]

$$\text{var}(\hat{f_0}) \geq \frac{12}{(2\pi)^2 SNR \cdot N(N^2 - 1)} \quad (2)$$

We know that the spectrum of a real sinusoid contains both positive and negative frequency components, and the estimation bias is caused by spectral superposition of the positive- and negative-frequency complex components. A known strategy to cut down frequency deviation is to multiply the real sinusoid by a window function. This is a simple method to reduce spectrum leakage [23].

We propose an algorithm by removing the negative-frequency component to reduce the estimation bias. The windowing method shown in Table 1 is used to acquire a rough frequency estimation for the received signal. Then the negative-frequency component can be moved to the low-pass band, and the negative-frequency component accounts for only the DC (direct-current) component of the modulated signal. After the removal of DC component, in this paper, the high-precision algorithm proposed is used to accurately estimate the positive-frequency component of complex sinusoid.

Step1: Coarse frequency estimation

Cosine window has been widely used in many documents [9, 23–25, 28]. Cosine windows include maximum sidelobe decay (MSD) window. The method proposed in this paper multiply the signal by MSD windows. The expressions of H-term MSD window are as follows [29]

$$w(n) = \sum_{h=0}^{H-1} (-1)^h a_h \cos\left(2\pi \frac{h}{N} n\right) \quad n = 0, 1, 2, \dots, N - 1 \quad (3)$$

The number of window coefficients a_h is H, $H \geq 1$. The coefficients expression of the H-term MSD window can be shown as [29]

$$a_h = \frac{C_{2H-2}^{H-h-1}}{2^{2H-3}} \quad h = 0, 1, \ldots, N-1 \tag{4}$$

Multiplying $x(n)$ by a cosine window, we have

$$x_w(n) = x(n) \cdot w(n) \quad n = 0, 1, \ldots, N-1 \tag{5}$$

Then we perform DFT on $x_w(n)$, and search the index number k_m of the maximum DFT spectral line.

We use Δf to represent the DFT frequency resolution. When $p = \pm 0.1$, the DFT samples $X_w(0.1)$ and $X_w(-0.1)$ which are $\pm 0.1 \Delta f$ away from the maximum DFT spectral line can be calculated as

$$X_w(p) = \sum_{n=0}^{N-1} x_w(n) e^{-j2\pi \frac{k_m+p}{N}} \tag{6}$$

In [30], the estimation formula of the normalized frequency shift δ is shown as follows:

$$\hat{\delta} = r \cdot Y(\delta) \tag{7}$$

where $Y(\delta)$ is expressed as

$$Y(\delta) = \frac{N}{\pi} \tan^{-1} \left[\frac{(|X_w(i) - X_w(-i)|) \cdot \sin(\pi i/N)}{|X_w(i) - X_w(-i)|) \cdot \cos(\frac{\pi i}{N}) - 2\cos(\pi i) \cdot |X_w(0)|} \right] \tag{8}$$

and r is calculated as

$$r = \frac{\cos(\pi i) \cdot a_0 - \cos(\frac{\pi i}{N}) \cdot \sin(\pi i) \cdot \frac{1}{\pi} \cdot \sum_{h=0}^{H-1} (-1)^h \frac{a_h \cdot i}{i^2 - h^2}}{\left\{ \left[\frac{1}{\pi} \sin(\pi i) + i\cos(\pi i) \right] \sum_{h=0}^{H-1} \frac{(-1)^h a_h}{i^2 - h^2} - \frac{2i^2}{\pi} \sin(\pi i) \sum_{h=0}^{H-1} \frac{(-1)^h a_h}{(i^2 - h^2)^2} \right\} \cdot \sin(\frac{\pi i}{N})} \tag{9}$$

For two term MSD window, we have $r_{2MSD} = 134.97$. We show the iterative process of obtaining the coarse frequency estimate $\hat{f}_0^f$ based on MSD windows in Table 1.

Step 2: Removal of the negative frequency component

Move the frequency of the received signal $x(n)$ as follows

$$x_m(n) = x(n) e^{j2\pi f_0^c n} \tag{10}$$

The results of (11) in a circular shift of the spectrum of $x(n)$ by f_0^c Hz. Because f_0^c is pretty near to the signal's real frequency f_0, the negative frequency component can be shifted to the low-pass band. However, the most important part of energy located at the

Table 1. Coarse frequency estimation algorithm.

Algorithm: Coarse frequency estimation with MSD windows
1   Get $x_w(n) = x(n) \cdot w(n)$ $n = 0,1,...,N-1$
2   Perform N-point DFT of $x_w(n)$
3   Search the index number k_m of the maximum spectral line
4   Calculate $X_w(0.1)$ and $X_w(-0.1)$, via $X_w(p) = \sum_{n=0}^{N-1} x_w(n)e^{-j2\pi\frac{k_m+p}{N}}$, $p = \pm 0.1$
5   Calculate $\hat{\delta}_1$ with $X_w(0)$, $X_w(0.1)$, $X_w(-0.1)$, via (7)
6   Calculate $X_w(\hat{\delta}_1)$, $X_w(\hat{\delta}_1+0.1)$, and $X_w(\hat{\delta}_1-0.1)$, via $X_w(p) = \sum_{n=0}^{N-1} x_w(n)e^{-j2\pi\frac{k_m+p}{N}}$, $p = \hat{\delta}_1$, $\hat{\delta}_1\pm 0.1$
7   Calculate $\hat{\delta}_2$ with, $X_w(\hat{\delta}_1)$, $X_w(\hat{\delta}_1+0.1)$ and $X_w(\hat{\delta}_1-0.1)$, via (7)
8   The coarse frequency estimate is $\hat{f}_0^c = (k_m + \hat{\delta}_1 + \hat{\delta}_2)\Delta f$

DC component of $x_m(n)$ [25]. Then the negative-frequency component can be removed from the received signal as

$$x_r(n) = x_m(n)e^{-j2\pi f_0^c n} - A_n e^{-j2\pi f_0^c n} \tag{11}$$

where A_n represents the amplitude of the negative frequency component of the real sinusoid, and is expressed as follows

$$A_n = \overline{x_m(n)} \tag{12}$$

where the $\overline{x_m(n)}$ represents the mean value of $x_m(n)$. And from the derivation of (10), the final expression of $x_r(n)$ is as follows

$$x_r(n) = x(n) - A_n e^{-j2\pi f_0^c n} \tag{13}$$

Step 3: Fine frequency estimation
We perform DFT on the reconstructed received signal $x_r(n)$, and then search the index number k_r of the maximum spectral line. When $l = \pm 0.1$, we have

$$X_r(l) = \sum_{n=0}^{N-1} x_r(n)e^{-j2\pi\frac{k_r+l}{N}} \quad l = \pm 0.1 \tag{14}$$

In [20], the estimation formula of the normalized frequency offset δ can be shown as follows

$$\hat{\delta} = \frac{N}{\pi}\tan^{-1}\left[\frac{(|X_r(i) - X_r(-i)|) \cdot \sin(\pi i/N)}{(|X_r(i) - X_r(-i)|) \cdot \cos(\frac{\pi i}{N}) - 2\cos(\pi i) \cdot |X_r(0)|}\right] \tag{15}$$

We show the iterative process of obtaining fine frequency estimate $\hat{f}_0^f$ in Table 2.

Table 2. Fine frequency estimation algorithm.

Algorithm: Fine frequency estimation algorithm	
1	Perform N-point DFT of $x_r(n)$
2	Search the maximum spectral line k_r
3	Calculate $X_r(0.1)$ and $X_r(-0.1)$, via $X_r(l) = \sum\limits_{n=0}^{N-1} x_r(n)e^{-j2\pi\frac{k_r+l}{N}}$ $l = \pm 0.1$, $l = \pm 0.1$
4	Calculate $\hat{\delta}_3$ with $X_r(0)$, $X_r(0.1)$, $X_r(-0.1)$, via (15)
5	Calculate $X_r(\hat{\delta}_3)$, $X_r(\hat{\delta}_3+0.1)$, and $X_r(\hat{\delta}_3-0.1)$, via $X_r(l) = \sum\limits_{n=0}^{N-1} x_r(n)e^{-j2\pi\frac{k_r+l}{N}}$, $l = \hat{\delta}_3$, $\hat{\delta}_3 \pm 0.1$
6	Calculate $\hat{\delta}_4$ with $X_r(\hat{\delta}_3)$, $X_r(\hat{\delta}_3+0.1)$ and $X_r(\hat{\delta}_3-0.1)$, via (15)
7	The fine frequency estimate is $\hat{f}_0^f = (k_r + \hat{\delta}_3 + \hat{\delta}_4)\Delta f$

3 Simulation Results

For this section, we test and verity the performance of the proposed algorithm, and carry out simulation analysis in the presence of additive white Gaussian noise. At the same time, in order to find out the performance difference between the proposed algorithm and the competitive algorithm, this part of the experiments we compare the proposed algorithm with the AM algorithm [18], Candan algorithm [25] with Kaiser window, and the Djukanovic algorithm [26]. The simulation experiments are mainly divided into three categories: RMSE of frequency estimation versus SNR, RMSE versus the signal frequency f_0 and RMSE versus the initial phase ϕ.

The first category: When $f_0 = 0.1917$, $\phi = \pi/7$, and $N = 128$, we evaluate the RMSE of the proposed algorithm versus SNR, varied from -10 to 50 dB, in steps of 2 dB. Figure 1 shows the RMSE of the algorithm we proposed, AM algorithm, Candan algorithm with Kaiser window ($\beta = 5$), and the Djukanovic algorithm under variable SNR. The RMSE of the proposed algorithm based on 2-term MSD window is quite near to that of Djukanovic algorithm. Meanwhile, the proposed algorithm and Djukanovic algorithm surpass the other algorithms in terms of accuracy and RMSE saturation. The proposed algorithm is closer to CRLB than Djukanovic algorithm from the local enlarged view. AM algorithm shows RMSE saturation when SNR is higher than 10 dB. Although Candan algorithm with Kaiser window ($\beta = 5$) conforms to the CRLB trend in the whole range of SNR, the RMSE of this algorithm is relatively large, and its curve is not close to CRLB. Simulation experiments of the proposed algorithm with 3-term MSD window are conducted and the results are very similar to those when the signal is multiplied by 2-term MSD window. Therefore, only the results of the proposed method with 2-term MSD window are shown in the figure.

From the above simulation image, it can be seen that when SNR is variable the performance of the proposed algorithm is better than that of AM algorithm, Candan algorithm with Kaiser window ($\beta = 5$), and Djukanovic algorithm.

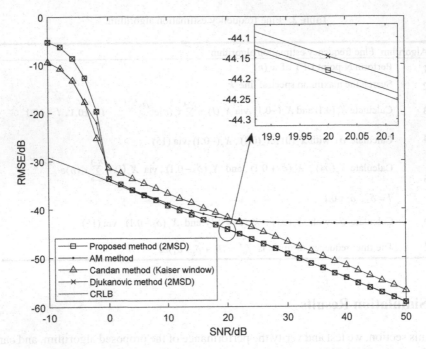

Fig. 1. RMSE versus SNR at $f_0 = 0.1917$, $\phi = \pi/7$ and $N = 128$

The second category: When $\phi = \pi/7$ and $N = 128$, we evaluate the RMSE of different algorithms versus the sinusoid frequency f_0. And f_0 is variable, taken in interval (0, 1/2) in 50 points.

Figure 2 shows the RMSE of the algorithm we proposed, AM algorithm, Candan algorithm with Kaiser window ($\beta = 5$), and the Djukanovic algorithm under variable sinusoid frequency f_0 for $SNR = 30$ dB. We can see that the RMSE of the algorithm we proposed in this paper based on 2-term MSD window is pretty near to that of Djukanovic algorithm. Meanwhile, the algorithm proposed in this paper and the Djukanovic algorithm are still closer to CRLB than the other estimators and their curves are relatively flat at high SNR. But the proposed algorithm is closer to CRLB than Djukanovic algorithm from the local enlarged view. The error of AM algorithm is large at high SNR. That is because AM algorithm is a complex sinusoid frequency estimation algorithm, and it can't effectively deal with the deviation caused by the superposition of positive- and negative-frequency components of real sinusoid. Although Candan algorithm reduces the deviation caused by the superposition of positive- and negative-frequency components of a real sinusoidal signal, the RMSE of this algorithm is about 2.5 dB higher than CRLB.

From the above simulation image, it can be seen that when f_0 is variable the performance of the proposed algorithm is better than that of AM algorithm, Candan algorithm with Kaiser window ($\beta = 5$), and Djukanovic algorithm. And the proposed algorithm is not sensitive to sinusoid frequency.

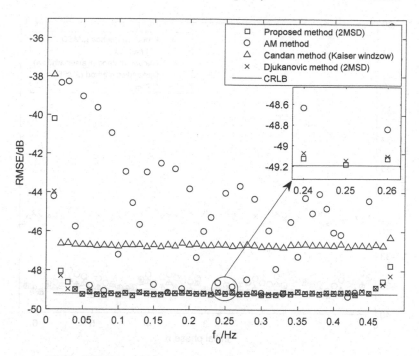

Fig. 2. RMSE versus f_0 with $N = 128$, $\phi = \pi/7$ and $SNR = 30$ dB

The third category: when $f_0 = 0.1917$, $N = 128$, and $SNR = 10$ dB and 50 dB are considered, we evaluate the RMSE of different algorithm versus initial phase ϕ. And ϕ is varied from 0 to 2π. The RMSE curves are calculated for 50 ϕ values.

Figure 3 shows the RMSE of the proposed algorithm, AM algorithm, Candan algorithm with Kaiser window ($\beta = 5$), and the Djukanovic algorithm under variable initial phase ϕ for $SNR = 10$ dB. We can see that Candan algorithm with Kaiser window ($\beta = 5$) has the largest RMSE in the four algorithms, although the algorithm reduces the error caused by the superposition of positive and negative frequency components of real sinusoidal signal, and the curve is not close to CRLB. The RMSE curve of the AM algorithm is also higher than CRLB. Because it is a complex sinusoid frequency estimation algorithm, it cannot effectively deal with the deviation caused by the superposition of positive frequency and negative frequency components of real sinusoid and has a large error in estimating the real signal frequency. The proposed algorithm and Djukanovic algorithm are closer to CRLB than the other two estimators.

When $SNR = 50$ dB and other conditions remain unchanged, Fig. 4 shows the RMSE of the algorithm we proposed and other three algorithms versus initial phase ϕ. We can see that the RMSE curve of AM method changes periodically, but it still has a large RMSE. The RMSE curve of Candan algorithm with Kaiser window ($\beta = 5$) is relatively flat, but it is about 2 dB higher than the CRLB, and the estimation bias is also large. However, even under the condition of higher SNR, the RMSE curves of the proposed algorithm and Djukanovic algorithm are very flat and close to CRLB.

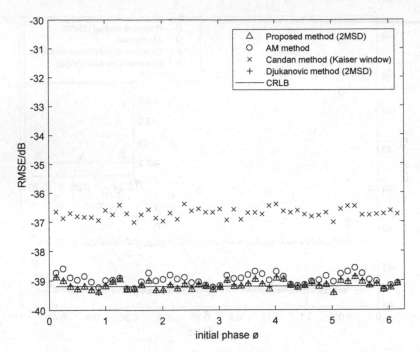

Fig. 3. RMSE versus ϕ with $N = 128, f_0 = 0.1917$ and $SNR = 10\,\text{dB}$

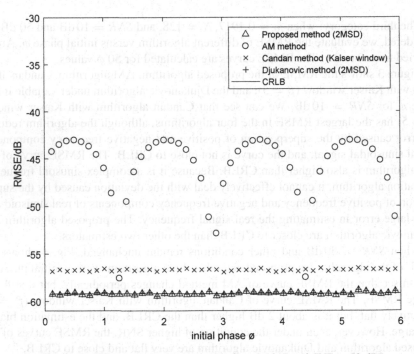

Fig. 4. RMSE versus ϕ with $N = 128, f_0 = 0.1917$ and $SNR = 50\,\text{dB}$

From the above two simulation images, it can be seen that when initial phase ϕ is variable, the performance of the algorithm we proposed is better than AM algorithm, Candan algorithm with Kaiser window ($\beta = 5$), and its performance is similar to that of the Djukanovic algorithm. And the proposed algorithm is not susceptible to initial phase.

Table 3 shows the computational complexity of different algorithms. It can be seen that the proposed algorithm has the same complexity as the other three algorithms. However, according to the simulation results of Fig. 1 and Fig. 2 show that the algorithm proposed in this paper has better frequency estimation performance and higher accuracy.

Table 3. Computational complexity of different methods

Method	Computational complexity
AM [18]	$O(N \log_2 N)$
Candan [25]	$O(N \log_2 N)$
Djukanović [26]	$O(N \log_2 N)$
Proposed	$O(N \log_2 N)$

4 Conclusion

Owing to the spectrum superposition of positive- and negative-frequency components, in the process of real sinusoid frequency estimation, the estimation result will produce estimation error. In order to deal with this problem, a high-precision estimation algorithm based on MSD windows is proposed. Through coarse estimation with MSD window, removal of the negative frequency component with frequency shift and fine estimation, an accurate frequency estimate can be obtained. Computer simulations are conducted, and from the results we can see that the RMSE of the proposed algorithm is closer to CRLB than AM algorithm, Candan algorithm and Djukanovic algorithm. The proposed algorithm and the other three algorithms have the same computational complexity. And the proposed algorithm is not susceptible to the sinusoid frequency and the initial phase. The proposed algorithm can reduce the estimation bias due to the frequency spectrum superposition of real sinusoid and can be used in practical applications.

References

1. Rife, D.C., Boorstyn, R.R.: Single-tone parameter estimation from discrete-time observations. IEEE Trans. Inform. Theory 55(9), 591–598 (1974)
2. Fu, H., Kam, P.Y.: MAP/ML estimation of the frequency and phase of a single sinusoid in noise. IEEE Trans. Signal Process. 55(3), 834–845 (2007)
3. Dutra, A.J.S., de Oliveira, J.F.L.: High-precision frequency estimation of real sinusoids with reduced computational complexity using a model-based matched-spectrum approach. Digit. Signal Process. 34(1), 67–73 (2014)

4. Lui, K.W., So, H.C.: Modified Pisarenko harmonic decomposition for single-tone frequency estimation. IEEE Trans. Signal Process. **56**(7), 3351–3356 (2008)
5. Tu, Y.Q., Shen, Y.L.: Phase correction autocorrelation-based frequency estimation method for sinusoidal signal. Signal Process. **130**, 183–189 (2017)
6. Cao, Y., Wei, G.: An exact analysis of modified covariance frequency estimation algorithm based on correlation of single-tone. Signal Process. **92**(11), 2785–2790 (2012)
7. Elasmi-Ksibi, R., Besbes, H.: Frequency estimation of real-valued single-tone in colored noise using multiple autocorrelation lags. Signal Process. **90**(7), 2303–2307 (2010)
8. Lui, K.W.K., So, H.C.: Two-stage autocorrelation approach for accurate single sinusoidal frequency estimation. Signal Process. **88**(7), 1852–1857 (2008)
9. Cao, Y., Wei, G.: A closed-form expanded autocorrelation method for frequency estimation of a sinusoid. Signal Process **92**(4), 885–892 (2012)
10. Rife, D.C., Vincent, G.A.: Use of the discrete Fourier transform in the measurement of frequencies and levels of tones. Bell Syst. Tech. J. **49**(2), 197–228 (1970)
11. Liang, X., Liu, A.: A new and accurate estimator with analytical expression for frequency estimation. IEEE Commun. Lett. **20**(1), 105–108 (2016)
12. Djukanović, S., Popović, T.: Precise sinusoid frequency estimation based on parabolic interpolation. In: 2016 24th Telecommunications Forum (TELFOR), Belgrade, Serbia, pp. 1–4 (2016)
13. Quinn, B.G.: Estimation of frequency, amplitude, and phase from the DFT of a time series. IEEE Trans. Signal Process. **45**, 814–817 (1997)
14. Yang, C., Wei, G.: A noniterative frequency estimator with rational combination of three spectrum lines. IEEE Trans. Signal Process. **59**(10), 5065–5070 (2011)
15. Jacobsen, E., Kootsookos, P.: Fast, accurate frequency estimators. IEEE Signal Process. Mag. **24**, 123–125 (2007)
16. Candan, C.: A method for fine resolution frequency estimation from three DFT samples. IEEE Signal Process. Lett. **18**(6), 351–354 (2011)
17. Liao, J.-R., Chen, C.-M.: Phase correction of discrete Fourier transform coefficients to reduce frequency estimation bias of single tone complex sinusoid. Signal Process. **94**, 108–117 (2014)
18. Aboutanios, E., Mulgrew, B.: Iterative frequency estimation by interpolation on Fourier coefficients. IEEE Trans. Signal Process. **53**(4), 1237–1242 (2005)
19. Candan, C.: Analysis and further improvement of fine resolution frequency estimation method from three DFT samples. IEEE Signal Process. Lett. **20**(9), 913–916 (2013)
20. Fan, L., Qi, G.Q.: Frequency estimator of sinusoid based on interpolation of three DFT spectral lines. Signal Process. **144**, 52–60 (2018)
21. Serbes, A.: Fast and efficient sinusoidal frequency estimation by using the DFT coefficients. IEEE Trans. Commun. **67**(3), 2333–2342 (2019)
22. Andria, G., Savino, M.: Windows and interpolation algorithms to improve electrical measurement accuracy. IEEE Trans. Instrum. Meas. **38**(4), 856–863 (1989)
23. Rife, D.C., Boorstyn, R.R.: Multiple tone parameter estimation from discrete-time observations. Bell Syst. Tech. J. **55**(9), 1389–1410 (1976)
24. Chen, S., Li, D.: Accurate frequency estimation of real sinusoid signal. In: 2010 2nd International Conference on Signal Processing Systems, vol. 3, pp. v3370–v3372 (2010)
25. Candan, C.: Fine resolution frequency estimation from three DFT samples: case of windowed data. Signal Process. **114**, 245–250 (2015)
26. Djukanović, S.: An accurate method for frequency estimation of a real sinusoid. IEEE Signal Process. Lett. **23**(7), 915–918 (2016)
27. Qi, G.Q.: Detection and Estimation: Principles and Applications. Publishing House of Electronics Industry (2011)

28. Liu, J., Fan, L., Li, R., He, W., Liu, N., Liu, Z.: An accurate frequency estimation algorithm by using DFT and cosine windows. In: Gao, H., Fan, P., Wun, J., Xiaoping, X., Yu, J., Wang, Y. (eds.) ChinaCom 2020. LNICSSITE, vol. 352, pp. 688–697. Springer, Cham (2021). https://doi.org/10.1007/978-3-030-67720-6_47

29. Belega, D., Petri, D.: Frequency estimation by two- or three-point interpolated Fourier algorithms based on cosine windows. Signal Process. **117**, 115–125 (2015)

30. Fan, L., Qi, G.Q.: Frequency estimator of sinusoid by interpolated DFT method based on maximum sidelobe decay windows. Signal Process. **186**, 108–125 (2021)

Transfer Learning Based Algorithm for Service Deployment Under Microservice Architecture

Wenlin Li[1(✉)], Bei Liu[1,2], Hui Gao[3], and Xin Su[4]

[1] School of Communication and Information Engineering,
Chongqing University of Posts and Telecommunications, Chongqing, China
S190101011@stu.cqupt.edu.cn
[2] Beijing National Research Center for Information Science and Technology,
Tsinghua University, Beijing, China
liubei@mail.tsinghua.edu.cn
[3] Beijing University of Posts and Telecommunications, Beijing, China
huigao@bupt.edu.cn
[4] Tsinghua University, Beijing, China
suxin@tsinghua.edu.cn

Abstract. In recent years, with the large-scale deployment of 5G network, research on 6G networks has gradually begun. In the 6G era, new service scenarios, such as Broad Coverage and High Latency Communication (BCHLC) will be introduced into the network, further increasing the complexity of network management. Furthermore, the development of edge computing and microservice architectures enables services to be deployed in a container on the edge clouds closer to the user side, significantly solving the problems. However, how to deploy services on edge clouds with limited resources is still an unresolved problem. In this paper, we model the problem as a Markov Decision Process (MDP), then propose a Deep Q Learning (DQN) based service deployment algorithm to optimize the delay and deployment cost of the services. Furthermore, a Multi-Category Joint Optimization Transfer Learning (MCJOTL) algorithm is proposed in this paper to address the problem of slow convergence of the DQN algorithm, which can adapt to different service scenarios in future networks faster. The simulation results show that the proposed algorithm can effectively improve training efficiency and service deployment effects.

Keywords: Service deployment · Edge computing · DQN · Transfer learning

1 Introduction

With the rapid development of wireless communication, the number of mobile applications and services continues to grow. The current network is facing considerable difficulties in dealing with the exponentially increasing service demands

Supported by the National Key R&D Program of China under Grant 2020YFB1806702.

H. Gao et al. (Eds.): ChinaCom 2021, LNICST 433, pp. 52–62, 2022.
https://doi.org/10.1007/978-3-030-99200-2_5

of mobile users. In the future, the 6G network will have more stringent requirements on delay, throughput, and other aspects. Taking cloud virtual reality (VR) as an example, it is expected to achieve an end-to-end delay within 10 ms [1]. At the same time, in order to adapt to the richer scenarios in the future 6G era, literature [2] further proposes the fourth service scenario, Broad Coverage and High Latency Communication (BCHLC) based on the three service scenarios of enhanced Mobile Broadband (eMBB), massive Machine Type Communication (mMTC) and Ultra Reliable Low Latency Communication (URLLC) in 5G network.

Recently, with continuous research of mobile edge computing (MEC) technology, edge cloud has shown great potential in deploying services on the edge of the network that requires large amounts of resource consumption and delay-sensitive services. Compared with mobile devices and remote clouds, the edge clouds allow users to use the strong computing power of the cloud platform without causing a high delay in communication with remote clouds [3], thereby significantly reducing the data traffic to and from the core network and meeting the requirements of delay-sensitive services. At the same time, with the rise of microservice architecture, services can be dynamically deployed on the edge clouds in a container, which makes future networks more flexible and more scalable [4].

Compared with the traditional remote cloud, the edge clouds are composed of resource-constrained machine clusters. Therefore, it is sential to deploy the services on them appropriately. However, ensuring the effective use of resources on the edge clouds, and the delay of the services, are challenging research problems [5]. Literature [6] proposes a distributed MEC-based service deployment platform, and it introduces a protocol to help service providers deploy services on MEC nodes to reduce transmission delay. Literature [7] proposes a Deep Reinforcement Learning (DRL) based dynamic service orchestration algorithm, taking into account the dynamic migration of the service, which minimizes the orchestration cost and improves the quality of service deployment.

However, these studies do not fully consider the transferability of the models. Therefore, they are difficult to realize the rapid changing of the dynamic network environment. Compared with the above researches, we study the problem of deploying services on edge clouds under microserice architecture. We consider delay and deployment cost to achieve better service deployment effect. First, we model the problem as a Markov Decision Process (MDP), and propose a Deep Q Learning (DQN) based service deployment algorithm. At the same time, to adapt to the ever-changing communication scenarios in the 6G network, we introduce a Multi-Category Joint Optimization Transfer Learning (MCJOTL) algorithm to speed up the model and can adapt to different service scenarios faster. The simulation results show that the algorithm can faster adapt to different network environments and obtain better service deployment effects.

The rest of the paper is organized as follows. First, we describe the system model in Sect. 2. Section 3 introduces the proposed MCJOTL algorithm of service deployment based on DQN and Transfer Learning. Simulation results are presented in Sect. 4. Finally, we conclude this paper in Sect. 5.

2 System Model

We consider a service development scenario in Fig. 1, which contains a central cloud, N edge clouds and K users, i.e., $\mathcal{N} = \{1, 2, \ldots, N\}$, $\mathcal{K} = \{1, 2, \ldots, K\}$. At the same time, each edge cloud has limited amounts of multi-dimensional resources and belongs to different operators. Suppose there are O operators in total, i.e., $\mathcal{O} = \{1, 2, \ldots, O\}$, and the resource prices provided by different operators are different. Then, the set of service requests generated by K users is denoted as $\mathcal{J} = \{1, 2, \ldots, J\}$, and the services reach the network in a uniformly distributed form. Finally, the global controller located on the central cloud determines the deployment location of the services.

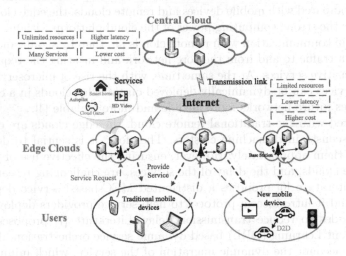

Fig. 1. 6G service deployment scenarios

2.1 Cloud Model

We assume that there are unlimited traffic, storage, and computing resources on the central cloud. While on the edge clouds, multi-dimensional resources are limited, and when new services reach or the services deployed on edge clouds end their lives, they will report their remaining resources to the global controller. Therefore, we use a four-tuple $E(f_{e,i}, s_{e,i}, h_{e,i}, o_i)$, $i \in \mathcal{N}$, $o_i \in \mathcal{O}$ to describe the edge clouds, which are the remaining traffic, storage, and computing resources on the edge cloud i, and the operator to which it belongs.

At the same time, there are differences in resource prices provided by operators, which can be expressed as $\{l_{f,i}, l_{s,i}, l_{h,i}\}$, $i \in \mathcal{O}$, representing the prices of traffic, storage, and computing resources provided by an operator i. Especially, it represents the price of resources provided on the central cloud at the time $i = 0$.

2.2   Service Model

In this scenario, we consider the following four categories of services: mMTC, eMBB, URLLC, and BCHLC [2]. Each service category has large different requirements in terms of traffic, capacity, and computing resource, and there are W types of service in the network that belong to the above four categories. Therefore, we use a five-tuple $B(t_{s,i}, t_{e,i}, f_{s,i}, s_{s,i}, h_{s,i})$, $i \in \mathcal{J}$ to describe the services, which are the time when the service i starts and ends, and the required traffic, storage, and computing resources of service i under normal circumstances. Considering the dynamic nature of the service process, we further assume the maximum instantaneous computing resource demand of the service i as $h_{max,i}$. At the same time, each service needs to correspond to a user, and also needs to be deployed on one cloud, so the total amount of service deployed on the clouds at a time t is:

$$J_t = \sum_{i=1}^{J} u_i, \text{where } u_i = \begin{cases} 1 & \text{if } t_{s,i} \le t \le t_{e,i} \\ 0 & \text{otherwise} \end{cases} \tag{1}$$

In the process of deploying services on the clouds, the remaining traffic, capacity, and computing resources in the clouds will be considered. At the same time, in order to save the occupation of storage resources in the edge clouds, all services are provided to users in a multi-threaded way. So, when the same type of service is deployed on the edge cloud, the storage space occupied by them will be the maximum value of this type of service, i.e.,

$$C_1 : \sum_{i=1}^{M_{j,t}} f_{s,i} \le f_{e,j} \quad C_2 : \sum_{k=1}^{W} \max_k(s_{s,j,k}) \le s_{e,j}$$
$$C_3 : \sum_{i=1}^{M_{j,t}} h_{s,i} \le h_{e,j} \quad C_4 : M_{0,t} + \sum_{j=0}^{N} M_{j,t} = J_t \tag{2}$$

where $j \in \mathcal{N}$, $\max_k(s_{s,j,k})$ represents the maximum storage resource occupied by services of the type k deployed on the edge cloud j, $M_{j,t}$ is the number of services deployed on the edge cloud j at the time t, and $M_{0,t}$ represents the number of services deployed on the central cloud at the time t.

2.3   Delay Model

After deployment, the service process may include two steps: the service i is calculated on the cloud, and the cloud transmits the calculation result to the user. Therefore, two kinds of delays are generated in the above process, i.e., the calculation delay $r_{c,i}$ and the transmission delay $r_{t,i}$.

If the service is deployed on the edge cloud, it will directly provide services to the users. For the wireless channel, assuming that the bandwidth allocated to the service i after development is W_i, and the signal-to-noise ratio between the user j and the edge cloud k of the service is $\frac{S}{N}_{j,k}$, then the maximum transmission rate of service i is

$$C_{i,k} = W_i log_2(1 + \frac{S}{N}_{j,k}) \tag{3}$$

Then the transmission delay of the service i can be expressed as

$$r_{t,i} = \frac{f_{s,i}}{C_{i,k}} \tag{4}$$

Otherwise, if the service is deployed on the central cloud, the required data will be transferred by the edge clouds when provides services. And there is no need to occupy the remaining traffic resources in the edge cloud during the transfer process. Therefore, the total delay includes the delay between the central and edge clouds and between edge clouds and users. Then the transmission delay of service i can be written as:

$$r_{t,i} = \min_k \left\{ \frac{f_{s,i}}{C_{i,k}} + r_{t,c,k} \right\} \tag{5}$$

where $r_{t,c,k}$ represents the transmission delay between the central cloud and the edge cloud k.

At the same time, we express the calculation delay as the ratio between the maximum computing resource demand of the service i and the allocated computing resources, i.e.,

$$r_{c,i} = \frac{h_{max,i}}{h_{s,i}} \tag{6}$$

Then after the service i is deployed, the total delay in providing services is

$$r_i = r_{t,i} + r_{c,i} \tag{7}$$

2.4   Cost Model

When a service deployed on the cloud provides services to users, it will incur a specific cost. Considering that due to different operators, there are differences in the rental costs of resources such as traffic on different edge clouds.

In this scenario, the service cost i consists of three parts: traffic cost $c_{f,i}$, storage rental cost $c_{s,i}$, and computing resource rental cost $c_{h,i}$. First, the cost of traffic is calculated according to the actual usage. That is

$$c_{f,i} = f_{s,i} l_{f,j}, i \in \mathcal{J}, j \in \mathcal{O} \tag{8}$$

where j is the operator of the edge cloud k. However, the rental cost of storage and computing is calculated according to the usage time and the amount of resource allocation, i.e.,

$$\begin{cases} c_{s,i} = s_{s,i} l_{s,j}(t_{e,i} - t_{s,i}) \\ c_{h,i} = h_{h,i} l_{h,j}(t_{e,i} - t_{s,i}) \end{cases}, \; i \in \mathcal{J}, i \in \mathcal{O} \tag{9}$$

Consider that if there is no service deployed on the current edge cloud, the edge cloud should be in a dormant state. Then, after the service is deployed for the first time, the edge cloud needs to enter the working state, and the startup cost c_0 of the edge cloud needs to be added.

So, the total cost of the service deployment is

$$c_i = c_{f,i} + c_{s,i} + c_{h,i} + c_0 \tag{10}$$

At the same time, because the central cloud has unlimited traffic, storage, and computing resources, edge clouds have little traffic, storage, and computing resources. Therefore, the resource prices on edge clouds provided by operators are greater than the resource prices on the central cloud, i.e.,

$$l_{f,i} > l_{f,0}, l_{s,i} > l_{s,0}, l_{h,i} > l_{h,0}, i \neq 0 \tag{11}$$

Therefore, if the service provider cannot pay the deployment cost on edge clouds, the service must be deployed on the central cloud, i.e.,

$$C_5 : c_i \leq c_{\text{tolerate},i} \tag{12}$$

where $c_{\text{tolerate},i}$ represents the maximum deployment cost that the provider of service can tolerate.

2.5 Problem Formation

Finally, in the actual service deployment process, we will comprehensively consider the delay and the deployment cost to achieve better development effects. So we constructed the following optimization problem

$$\begin{aligned} min\,\{\alpha r_i + \beta c_i\} \\ \text{s.t. } C_1, C_2, C_3, C_4, C_5 \end{aligned} \tag{13}$$

where α and β are constants.

The above optimization problem is NP-hard. In order to obtain the optimal solution to this problem, we need to search the entire combination space. However, the combination space will increase exponentially as users and edge clouds increase, bringing great difficulties.

3 Service Deployment Algorithm

For each network, it is difficult not easy to find the optimal service deployment strategy. The reasons are as follows: First, in future network environment, the cloud environment is dynamic due to the continuous change of service status; Second, the interaction between different services is relatively complex. Therefore, for this dynamic environment, we use MDP to model the problem.

3.1 Markov Decision Process

An MDP model can be expressed as $[S, A, R, S']$, where S, A and R are the state space, action space, and reward function, respectively. In the state $s \in S$, the agent will get different rewards when taking actions. Then, according to the state transition function, the state moves from s to $s' \in S$ [8]. The goal of each step is to maximize the reward value. Our model is designed as follows:

State Space. We use nine states included in the service that are waiting to be deployed and the remaining traffic, storage, and computing resources of all edge clouds as the model's state. Specifically, the state $s_n \in S$ of the model at episode n is designed as:

$$s_n = [ser_n; clo_{n,1}, \ldots, clo_{n,N}]^T \tag{14}$$

Which ser_n represents the service ID, continuous service time, traffic, storage, computing, and maximum computing resource demands, type, user of the service, and maximum deployment cost that the service provider can tolerate; $clo_{n,i}$ represents the remaining traffic, capacity, and computing resources in the edge cloud i.

Action Space. According to the state information, the global controller needs to determine the location of service, including edge clouds and central cloud, that is $a \in \{0, 1, 2, \ldots, N\}$, where 0 represents deploying the service on the central cloud.

Reward. When a specific action transforms the state s into another state s', the network will be rewarded immediately. We design the reward as follows

$$R = -(\alpha r_i + \beta c_i) \tag{15}$$

among them, α and β are constants, which are used to adjust the ratio between delay and deployment cost [9].

3.2 DQN Based Service Deployment Algorithm

In order to support the dynamic changes of services and edge clouds, we introduced DQN to optimize the deployment strategy of the services. First, we define the agent's action-value function (Q function) as the maximum expected return achievable after taking some action a in state s, i.e.,

$$Q^*(s, a) = max_\pi(r_n | \pi, s_n = s, a_n = a) \tag{16}$$

where π is the joint policy. Then, we use the following method to update the Q function based on the Bellman equation, i.e.,

$$Q^*(s, a) = r_n + \gamma[\max_{a'} Q^*(s', a') | s, a] \tag{17}$$

It can be observed from (14) that the size of the state space increases exponentially with the arrival of services and the continuous changes of the edge cloud states. In order to solve the problem of dimensionality, we use a deep neural network (DNN) to approximate the Q-function. We define the training loss function of training the DNN as

$$L(\theta_n) = (y_n - Q(s, a; \theta_n))^2 \tag{18}$$

where n represents the number of episodes, y_n represents the target Q-value, θ_n represents the network weight of the Q-network in the nth episode. Then, the gradient of $L(\theta_n)$ is

$$\nabla_{\theta_n} L(\theta_n) = (y_n - Q(s, a; \theta_n)) \nabla_{\theta_n} Q(s, a; \theta_n) \tag{19}$$

Based on this, we can use a gradient-based optimizer to train the Q-network. Algorithm 1 gives the details of the algorithm.

Algorithm 1. DQN based service deployment algorithm.

Input: The next service's status that needs to be deployed, edge clouds' status and channel status according to (14)

Output: Deployment location of service on the clouds

1: Initialize state s and experience replay buffer D.
2: **for** episode = 0, 1, 2, ... **do**
3:    Select an action $a^n = i, i \in \mathcal{N}$ based on the greedy policy, where i represents the number of edge clouds.
4:    **if** restrictions $C_1 \sim C_5$ can be satisfied when the service deployed on the edge cloud i **then**
5:        Deploy the service on the edge cloud i.
6:    **else**
7:        Deploy the service on the central cloud.
8:    **end if**
9:    Calculate the delay, deployment cost and reward according to (7), (10) and (15)
10:    Update s' and store the experience (s, a, r, s') in the experience replay buffer D.
11:    Compute $\nabla_{\theta_n} L(\theta_n)$ by minimizing $L(\theta_n)$ and update weights of the Q-network according to (18) and (19).
12: **end for**

3.3   MCJOTL Based Service Deployment Algorithm

Although the DQN algorithm can achieve great service deployment effects, the network is constantly changing due to the continuous changes of the remaining resources in the edge clouds. In addition, the DQN algorithm usually takes a long time to converge, so it is not easy to adapt to the future network.

Therefore, we propose a transfer learning-based MCJOTL algorithm to avoid large-scale training networks from scratch during service deployment. In our MCJOTL algorithm, the DQN results we trained in the previous subsection are used as the source model. We freeze all layers in the DQN model except the last few layers, then add two linear layers at the end, and only the last few layers and the two linear layers can be trained. Thus, each model consists of a frozen pre-trained body and a linear layer body that must be retrained. Finally, the two bodies are merged by linear weighting to improve the overall performance.

At the same time, considering that there are great differences in various types of services in different scenarios. For example, in a cloud gaming scenario, eMBB services may have a large number of requirements, so its heavy traffic needs need to be prioritized during deployment. However, other types of services, such as URLLC, may have a small amount of requirements. Therefore, the actual requirements of some types of services may be ignored when one model is used for training all types of services in different scenarios. So, our proposed MCJOTL algorithm trains different transfer learning models for different categoties of service in every scenarios. At the same time, each model improves the training effect by sharing replay buffer including state space, reward functions, etc.

Therefore, when the global controller receives service requests, it can select an appropriate transfer learning model according to the service category to which it belongs to determine the optimal service deployment location in MCJOTL. Such as when eMBB service arrives, the model correspond to eMBB will be used. At the same time, the parameters of the model to which the eMBB belongs will also be updated. Thus, the MCJOTL algorithm can better adapt to the most categories of services in different scenarios. By using this algorithm, we can well solve the demand between different categories of services in different scenarios. The simulation results of this algorithm are shown in the following section.

4   Simulation Result

In this section, we demonstrate the effectiveness of the proposed MCJOTL algorithm in terms of deployment cost and delay. We performed the simulation on a 64-bit computer with 32 GB of RAM and Intel i7 2.6 GHz. First, we considered 16 edge clouds, 32 users, and 50 services. In order to evaluate the performance of the algorithm, we used a method similar to the literature [10] to generate datas. For example, we set the delay between the central cloud and edge clouds between 10 and 20 ms. Then, the spectrum bandwidth, service's traffic, storage, computing requirements, etc. are set according to the specific characteristics of the services.

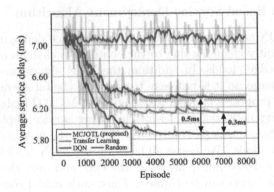

Fig. 2. Variation curve of service delay with episode

We selected DQN, traditional transfer learning, and random deployment algorithms for comparison. The DQN is composed of five linear layers. In transfer learning and our MCJOTL algorithm, the transfer source comes from the DQN algorithm trained for 50,000 episodes. At the same time, to achieve the training effect faster, all models in the simulation result have only been trained for 8,000 episodes. Figure 2 shows the performance of the MCJOTL algorithm in terms of delay. It can be seen that under the same number of episodes, the MCJOTL algorithm can achieve better service deployment effects by using more prior knowledge and adapting to different service scenarios. Figure 3 shows the algorithm's performance in terms of deployment costs, which shows that the MCJOTL algorithm can also achieve better performance effects in terms of deployment costs.

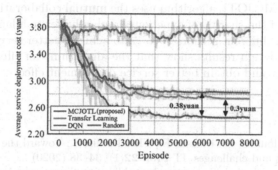

Fig. 3. Variation curve of deployment cost with episode

Finally, Fig. 4 shows the curve of the average delay and deployment cost to the number of edge clouds in the case of 32 users and 50 services. With the increase in the number of edge clouds, delay and deployment costs have gradually improved. However, as the number of edge clouds increases, the improvement gradually slows down in the end.

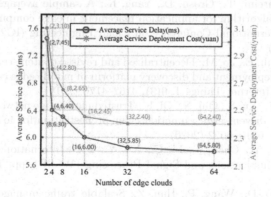

Fig. 4. The change curve of average service deployment cost and service delay with the number of edge clouds

5 Conclusion

In this paper, we studied a deployment algorithm for services. Based on the dynamically changing edge clouds, we model the service deployment problem as an MDP to find an optimal strategy that minimizes delay and development costs. First, we use the DQN algorithm to realize the intelligent deployment of the services. Then, for the slow convergence of DQN, we introduce the MCJOTL algorithm, which trains different models according to the categories of the service. So, when new service arrives, the algorithm uses the corresponding model according to the category of the service to achieve the optimal deployment effects. Beyong this, the MCJOTL algorithm uses the mutual collaboration between different models and the sharing of experience replay buffer to adapt to the service characteristics in different scenarios, so as to achieve better service deployment effects. The simulation results show that the algorithm can effectively improve training efficiency and obtain better service deployment effects.

References

1. Tomkos, I., Klonidis, D., Pikasis, E., Theodoridis, S.: Toward the 6G network era. Opportunities and challenges. IT Prof. **22**(1), 34–38 (2020)
2. Wang, H.: 6G vision: unified network enabling intelligent megalopolis. ZTE Technol. J. **25**(06), 55–58 (2019)
3. Guo, T., Zhang, H., Huang, H., Guo, J., He, C.: Multi-resource fair allocation for composited services in edge micro-clouds. In: 2019 IEEE International Conference on Parallel & Distributed Processing with Applications, Big Data & Cloud Computing, Sustainable Computing & Communications, Social Computing & Networking (ISPA/BDCloud/SocialCom/SustainCom), Xiamen, pp. 405–412 (2019)
4. Wang, S., Guo, Y., Zhang, N., Yang, P., Zhou, A., Shen, X.: Delay-aware microservice coordination in mobile edge computing: a reinforcement learning approach. IEEE Trans. Mob. Comput. **20**(3), 939–951 (2021)
5. Badri, H., Bahreini, T., Grosu, D., Yang, K.: A sample average approximation-based parallel algorithm for application placement in edge computing systems. In: 2018 IEEE International Conference on Cloud Engineering (IC2E), Orlando, pp. 198–203 (2018)
6. Nguyen, T., Huh, E., Jo, M.: Decentralized and revised content-centric networking-based service deployment and discovery platform in mobile edge computing for IoT devices. IEEE Internet Things J. **6**(3), 4162–4175 (2019)
7. Guo, S., Dai, Y., Xu, S., Qiu, X., Qi, F.: Trusted cloud-edge network resource management: DRL-driven service function chain orchestration for IoT. IEEE Internet Things J. **7**(2), 6010–6022 (2020)
8. Wang, X., Xu, Y., Chen, J., Li, C., Xu, Y.: 2020 International Conference on Wireless Communications and Signal Processing (WCSP), pp. 195–200, Nanjing (2020)
9. Liu, L., Niyato, D., Wang, P., Han, Z.: Scalable traffic management for mobile cloud services in 5G networks. IEEE Trans. Netw. Serv. Manag. **15**(4), 1560–1570 (2018)
10. Samanta, A., Tang, J.: Dyme: dynamic microservice scheduling in edge computing enabled IoT. IEEE Internet Things J. **7**(7), 6164–6174 (2020)

Scheduling and Transmission Optimization in Edge Computing

Joint Opportunistic Satellite Scheduling and Beamforming for Secure Transmission in Cognitive LEO Satellite Terrestrial Networks

Xiaofen Jiao, Yichen Wang(✉), Zhangnan Wang, and Tao Wang

School of Information and Communications Engineering, Xi'an Jiaotong University,
Xi'an 710049, Shaanxi, China
{jiaoxiaofen,wang19960729,wangtaot}@stu.xjtu.edu.cn,
wangyichen0819@mail.xjtu.edu.cn

Abstract. Security is a critical issue for cognitive LEO satellite terrestrial networks (CLSTNs) in the future 6G era. To guarantee the secure transmission of both the satellite and terrestrial users under limited power budgets, we propose a joint opportunistic satellite scheduling and beamforming (BF) scheme for the CLSTN. Specifically, we formulate an optimization problem to maximize the secrecy energy efficiency (SEE) of the satellite user under the secrecy constraint of the base station (BS) user, the signal-to-interference plus noise ratio (SINR) requirements of the intended users and the limited power budgets of the transmitters. Since the formulated problem is complex and nonconvex, we first consider the achievable SEE of a given scheduled satellite and transform the original SEE maximization problem into a convex problem based on the Dinkelbach's method and the difference of two-convex functions (D.C.) approximation method. Consequently, we propose a three-tier algorithm to jointly determine the satellite scheduling vector and the BF vectors of the satellite and BS. Simulation results verify the superiority of our proposed scheme over the existing schemes.

Keywords: Physical layer security · Cognitive LEO satellite terrestrial network · Secrecy energy efficiency · Satellite scheduling scheme

1 Introduction

In the upcoming 6G era, large-scale low earth orbit (LEO) satellite constellations will be integrated with terrestrial systems to facilitate global coverage. However,

This work was supported in part by the National Natural Science Foundation of China under Grant 61871314 and in part by the Key Research and Development Program of Shaanxi Province under Grant 2019ZDLGY07-04.

H. Gao et al. (Eds.): ChinaCom 2021, LNICST 433, pp. 65–79, 2022.
https://doi.org/10.1007/978-3-030-99200-2_6

due to the dramatic growth of data traffic and the scarcity of spectrum resource, the transmission performance of satellite terrestrial integrated networks (STINs) cannot be well guaranteed. To solve this issue, cognitive radio technology can be incorporated into STINs to construct the cognitive LEO satellite terrestrial networks (CLSTNs) and improve the spectrum utilization efficiency [1]. In the downlink/uplink of CLSTNs, the spectrum can be shared between the LEO satellite systems and the terrestrial systems. Since 6G wireless networks are expected to be an elegant solution for secure and ubiquitous future communication, security should be considered as a critical performance requirement of CLSTNs [2].

To guarantee secure transmission, the confidential messages of CLSTNs should be protected to prevent the eavesdroppers from intercepting it. Traditionally, cryptographic technique is utilized at upper layers to achieve security. However, this method may become unreliable as the processing speed of eavesdroppers is increasing. Recently, physical layer security (PLS) without any need for key distribution and service management has aroused a research upsurge. In physical layer, security is guaranteed when intended users have better signal-to-interference plus noise ratios (SINRs) than the eavesdroppers by exploiting technologies such as jamming or transmit beamforming (BF).

A great deal of effort has been devoted to investigating the PLS in multibeam satellite systems [3–7]. Under the SINR constraints of the satellite users (SUs), robust BF schemes were proposed to maximize the minimum secrecy rate (SR) among all the intended users [3,4]. The authors in [5] investigated the optimal BF design to maximize the minimum secrecy energy efficiency (SEE) of SUs. To further degrade the transmission quality of the wiretap channel, artificial noise (AN) was inserted into the transmit signal to satisfy SR requirements of SUs while minimizing the transmit power of the satellite [6]. The authors in [7] explored a novel dual-beam dual-frequency scheme to enhance the PLS of the desired channel and impair the wiretap channel at the same time. In the above works, [5,6] assumed that perfect channel state informations (CSIs) of eavesdroppers are available while [3,4,7] considered a more realistic scenario where CSIs are imperfect.

In CSTN, the terrestrial base station (BS) can serve as a green jamming source to interfere with eavesdroppers, which further improves the security of transmissions [8–13]. The authors in [9] exploited the jamming from BS and proposed a transmit power minimization scheme while satisfying the SINR requirements of SU, eavesdroppers and terrestrial BS user (BU). To satisfy the requirements of service qualities and SR constraints of SUs, a joint BF and jamming scheme was proposed, which aimed to minimize total transmit power of the satellite and the BS [10]. Individual BF scheme of the satellite and cooperative BF scheme of the satellite and the BS were investigated in [11] to achieve maximal achievable SR of the SU. In [12], the full duplex receivers receive and jam simultaneously to confuse eavesdroppers in an AN-aided BF scheme. The authors in [13] utilized an intelligent reflecting surface to reflect the jamming signals and therefore profitably enhanced its interference effect. However, the above works were devoted to maximizing or guaranteeing SU's secure transmission demands without taking BU's security requirements into account. Also, the above works merely focused on the geostationary orbit (GEO) satellite terrestrial system where there is only one satellite, the existing schemes cannot be adopted directly in CLSTNs where the number of satellites is large. Hence, a satellite scheduling scheme should be properly designed to ensure the secure transmission in CLSTNs.

The conventional satellite scheduling strategies include highest elevation angle/maximal service time priority satellite scheduling schemes [14]. It is proved that multi-satellite scheduling and round-robin satellite scheduling scheme have different impact on PLS in LEO systems [15]. Some researchers have investigated the secrecy performance of opportunistic user scheduling scheme and optimal relay scheduling scheme in the integrated satellite terrestrial relay systems (ISTRSs) [16–18]. The authors in [16] analysed the secrecy outage probability (SOP) of opportunistic user scheduling scheme in ISTRS. [17] and [18] explored the SOP of optimal relay scheduling scheme while [18] incorporated the hardware impairments of transceiver nodes into study additionally. In general, opportunistic satellite scheduling scheme has not been investigated in the existing schemes. Secure transmission in CLSTNs is still an open yet challenging problem, which motivates the work in this paper.

In this paper, we propose a joint opportunistic satellite scheduling and BF scheme to achieve the energy-efficient secure transmission of the SU in the CLSTN. We formulate the maximization problem to maximize the SEE of the SU under transmit power budgets of the satellite and the BS while satisfying the SR constraint of the BU and the SINR requirements of the SU and BU in the CLSTN. To tackle the formulated nonconvex problem, we first consider the case when the scheduled satellite is given. Then, the Dinkelbach's method and the difference of two-convex functions (D.C.) approximation method are exploited to transform the SEE maximization problem into a convex one and solve the problem iteratively. Consequently, a three-tier algorithm is proposed to jointly determine the satellite scheduling vector and BF vectors of the satellite and BS. Finally, the effectiveness of the proposed algorithm is verified by simulation.

The rest of this paper is organized as follows. Section 2 introduces the system model and the optimization problem formulation. Section 3 presents the three-tier algorithm to solve the original problem. Simulation results and relevant analyses are provided in Sect. 4. Section 5 concludes this paper.

2   System Model and Problem Formulation

As shown in Fig. 1, we consider the downlink transmission of a CLSTN which consists of a primary network and a secondary network. The two networks occupy the same wireless spectrum resource in the CLSTN. In the primary network, the satellite transmits confidential messages to the single-antenna SU via N beams while a nearby single-antenna eavesdropper, which is denoted as SE, attempts to decode the information intended to SU. In the secondary network, the BS with M antennas serves a single-antenna BU. BU is wiretapped by another eavesdropper which is denoted as BE. The signals sent by the satellite and BS can also be

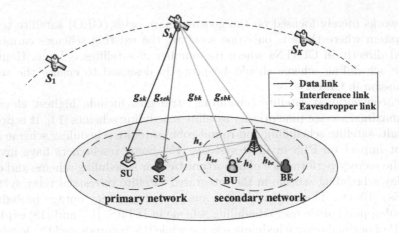

Fig. 1. System model.

utilized as green jamming to the eavesdroppers. There are K satellites in the SU's observable range, SU can only be served by the scheduled satellite within one time slot [14]. The secure transmission is guaranteed by joint opportunistic satellite scheduling and BF of the satellite and BS. In this paper, the CSI of each link is assumed to be perfectly known [6,7].

2.1 Channel Model

In this paper, quasi-static slow-fading is adopted to model the links between each satellite and the users. The links in different beams undergo independent fading. Considering the effects of free space path loss, satellite antenna gain and small-scale fading, the downlink channel coefficient of LEO can be expressed as:

$$g = C_L \sqrt{b(\varphi)} \widetilde{g} \tag{1}$$

where $C_L = \frac{\lambda}{4\pi d}$ describes the free space path loss with the carrier wavelength λ and the distance from the satellite to the user d. $b(\varphi)$ denotes the beam gain coefficient and can be approximated as [13]:

$$b(\varphi) = b_{\max} \left(\frac{J_1(u)}{2u} + 36 \frac{J_3(u)}{u^3} \right)^2 \tag{2}$$

where $u = 2.07123 \frac{\sin \varphi}{\sin(\varphi_{3dB})}$, $b_{\max}$ denotes the maximal gain of the satellite antenna and $J_1(.)$ and $J_3(.)$ describes the first and third order of the first-kind Bessel functions, respectively. φ denotes the angle between beam center and the SU's position with respect to the satellite and φ_{3dB} indicates the 3-dB angle.

The small-scale fading of the satellite channel $\widetilde{g}$ is modeled as Lutz distribution [19], which consists of a combination of Rician fading and lognormally shadowed Rayleigh fading. The phase in Lutz model is uniformly distributed over $[0, 2\pi)$ and the overall probability density function (PDF) of the received power in Lutz model is expressed as:

$$p(S) = (1 - A)p_{Rice}(S) + A \int_0^\infty p_{Rayl}(S|S_0)P_{ln}(S_0)dS_0. \qquad (3)$$

The PDF of the received power S is expressed as $p_{Rice}(S) = ce^{-c(S+1)}I_0(2c\sqrt{S})$ when S obeys Rician distribution. When S obeys log-normally shadowed Rayleigh distribution, its PDF is expressed as $p_{Rayl}(S|S_0) = \frac{1}{S_0}e^{-\frac{S}{S_0}}$, where S_0 describes the short-term mean received power, which follows the distribution of $p_{ln}(S_0) = \frac{10}{\sqrt{2\pi}\sigma \ln 10}\frac{1}{S_0}\exp\left(-\frac{(10 \log S_0 - \mu)^2}{2\sigma^2}\right)$. The percentage of shadowed state A, the mean μ and deviation σ^2 of the lognormally shadowed Rayleigh distribution and the Rice-factor c all depend on the elevation angle of the satellite and the terrestrial environment. The specific values of the four parameters A, μ, σ^2 and c can be obtained by referring to the measured value in [19] or the fitting formula and the figure in [20].

2.2   Signal Model

Denote the signals intend for the SU and BU by x and s, respectively. Assume that $E(x^2) = 1$ and $E(s^2) = 1$, where $E(\cdot)$ denotes the expectations operator on random variables. Let $\mathbf{g}_{sk}, \mathbf{g}_{sek}, \mathbf{g}_{bk}, \mathbf{g}_{bek} \in \mathbb{C}^{N \times 1}$ and $\mathbf{h}_s, \mathbf{h}_{se}, \mathbf{h}_b, \mathbf{h}_{be} \in \mathbb{C}^{M \times 1}$ denote the links of S_k-SU, S_k-SE, S_k-BU, S_k-BE, BS-SU, BS-SE, BS-BU, BS-BE, respectively, where S_k denotes the k-th satellite, $k = 1, 2...K$. When the k-th satellite is scheduled, the signals received at SU, SE, BU and BE are respectively given as:

$$y_{sk} = \mathbf{g}_{sk}^H \mathbf{w}x + \mathbf{h}_s^H \mathbf{v}s + n_s \qquad (4)$$

$$y_{sek} = \mathbf{g}_{sek}^H \mathbf{w}x + \mathbf{h}_{se}^H \mathbf{v}s + n_{se} \qquad (5)$$

$$y_{bk} = \mathbf{h}_b^H \mathbf{v}s + \mathbf{g}_{bk}^H \mathbf{w}x + n_b \qquad (6)$$

$$y_{bek} = \mathbf{h}_{be}^H \mathbf{v}s + \mathbf{g}_{bek}^H \mathbf{w}x + n_{be} \qquad (7)$$

where $\mathbf{w} \in \mathbb{C}^{N \times 1}$ and $\mathbf{v} \in \mathbb{C}^{M \times 1}$ are the BF weight vectors of S_k and BS, respectively. $n_i \in \mathcal{CN}(0, \sigma_i^2), i \in \{s, se, b, be\}$ denotes the Gaussian white noise at SU, SE, BU and BE. $\sigma_i^2 = \kappa BT$ with the noise bandwidth B, noise temperature T and Boltzmann constant $\kappa \approx 1.38 \times 10^{-23} J/K$. Then the received SINRs at SU, SE, BU and BE are respectively given by:

$$\gamma_{sk} = \frac{|\mathbf{g}_{sk}^H \mathbf{w}|^2}{|\mathbf{h}_s^H \mathbf{v}|^2 + \sigma_s^2} \qquad (8)$$

$$\gamma_{sek} = \frac{|\mathbf{g}_{sek}^H \mathbf{w}|^2}{|\mathbf{h}_{se}^H \mathbf{v}|^2 + \sigma_{se}^2} \qquad (9)$$

$$\gamma_{bk} = \frac{|\mathbf{h}_b^H \mathbf{v}|^2}{|\mathbf{g}_{bk}^H \mathbf{w}|^2 + \sigma_b^2} \qquad (10)$$

$$\gamma_{bek} = \frac{|\mathbf{h}_{be}^H \mathbf{v}|^2}{|\mathbf{g}_{bek}^H \mathbf{w}|^2 + \sigma_{be}^2}. \qquad (11)$$

Then, the achievable SR of SU and BU can be written as:

$$C_{sk} = \max\{\log_2(1 + \gamma_{sk}) - \log_2(1 + \gamma_{sek}), 0\} \qquad (12)$$
$$C_{bk} = \max\{\log_2(1 + \gamma_{bk}) - \log_2(1 + \gamma_{bek}), 0\}. \qquad (13)$$

As a consequence, the SEE of the SU which is defined as the amount of achievable secret bits per unit energy and bandwidth can be expressed as:

$$\eta_{sk} = \frac{C_{sk}}{||\mathbf{w}||^2 + P_{ck} + ||\mathbf{v}||^2 + P_b} \qquad (14)$$

where P_{ck} and P_b denotes the constant transmit circuit power consumed by S_k and BS, respectively.

2.3   Problem Formulation

To realize the energy-efficient secure transmission of the SU in the CLSTN, we formulate a SEE maximization problem under the limited power budgets of the satellite and BS while satisfying the SR constraint of the BU, the SINR requirements of the SU and BU by jointly scheduling the opportunistic satellite and optimizing the BF weight vectors of the scheduled satellite and the BS. The optimization problem can be expressed as:

$$\max_{\mathbf{C},\mathbf{v},\mathbf{w}} \mathbf{C}^T \boldsymbol{\eta} \qquad (15)$$
$$\text{s.t. } \mathbf{C}^T \mathbf{B} \geq \Gamma_b \qquad (16)$$
$$\mathbf{C}^T \boldsymbol{\gamma}_s \geq \Lambda_s \qquad (17)$$
$$\mathbf{C}^T \boldsymbol{\gamma}_b \geq \Lambda_b \qquad (18)$$
$$||\mathbf{w}||^2 \leq P_S \qquad (19)$$
$$||\mathbf{v}||^2 \leq P_B \qquad (20)$$
$$\sum_{i=1}^{K} c_i = 1, c_i \in \{0,1\} \qquad (21)$$

where $\mathbf{C} = [c_1, c_2...c_K]^T$ is the scheduling vector of the LEO satellites, $\boldsymbol{\eta} = [\eta_{s1}, \eta_{s2}...\eta_{sK}]^T$ denotes the SEE vector of the SU, $\mathbf{B} = [C_{b1}, C_{b2}...C_{bK}]^T$ represents the SR vector of the BU, $\boldsymbol{\gamma}_s = [\gamma_{s1}, \gamma_{s2}...\gamma_{sK}]^T$ denotes the SINR vector of the SU and $\boldsymbol{\gamma}_b = [\gamma_{b1}, \gamma_{b2}...\gamma_{bK}]^T$ denotes the SINR vector of the BU. Γ_b is the SR threshold of BU to realize the secure transmission and Λ_s and Λ_b denote the SINR requirements of the SU and BU, respectively. P_S and P_B are the power budgets of the scheduled satellite and BS, respectively. The constraint (21) is to ensure that only one satellite can be scheduled at one time slot, where $c_k = 1$ denotes the k-th satellite is scheduled to transmit the confidential signal.

3   Solution of the Optimization Problem

Since there are only K possible values of $\mathbf{C}$, the exhaustive method can be used to search for the optimal value of $\mathbf{C}$. For a fixed $\mathbf{C}$, assume the p-th satellite is scheduled, then, $c_p = 1$ and $c_i = 0, i \in \{1, 2...p-1, p+1, ...K\}$. In this case, the original problem can be written as:

$$\max_{\mathbf{v}, \mathbf{w}} \frac{\log_2(1+\gamma_{sp}) - \log_2(1+\gamma_{sep})}{||\mathbf{w}||^2 + P_{cp} + ||\mathbf{v}||^2 + P_b} \tag{22}$$

$$\text{s.t. } \log_2\left(\frac{1+\gamma_{bp}}{1+\gamma_{bep}}\right) \geq \Gamma_b \tag{23}$$

$$\gamma_{sp} \geq \Lambda_s \tag{24}$$

$$\gamma_{bp} \geq \Lambda_b \tag{25}$$

$$||\mathbf{w}||^2 \leq P_S \tag{26}$$

$$||\mathbf{v}||^2 \leq P_B \tag{27}$$

The optimization problem in (22)–(27) is nonconvex due to the fractional form of SEE and the presence of the difference of two logarithmic functions [22]. The Dinkelbach's method can be adopted to reformulate the fractional form problem to the equivalent subtractive form by introducing an auxiliary variable η_{sp} [23]. Besides, we denote $\mathbf{H}_s = \mathbf{h}_s\mathbf{h}_s^H$, $\mathbf{H}_{se} = \mathbf{h}_{se}\mathbf{h}_{se}^H$, $\mathbf{H}_b = \mathbf{h}_b\mathbf{h}_b^H$, $\mathbf{H}_{be} = \mathbf{h}_{be}\mathbf{h}_{be}^H$, $\mathbf{G}_{sp} = \mathbf{g}_{sp}\mathbf{g}_{sp}^H$, $\mathbf{G}_{sep} = \mathbf{g}_{sep}\mathbf{g}_{sep}^H$, $\mathbf{G}_{bp} = \mathbf{g}_{bp}\mathbf{g}_{bp}^H$, $\mathbf{G}_{bep} = \mathbf{g}_{bep}\mathbf{g}_{bep}^H$ and introduce relaxations $\mathbf{W} = \mathbf{w}\mathbf{w}^H$ and $\mathbf{V} = \mathbf{v}\mathbf{v}^H$, then, the problem can be transformed as:

$$f(\eta_{sp}) = \max_{\mathbf{W}, \mathbf{V}} f_1(\mathbf{W}, \mathbf{V}) - f_2(\mathbf{W}, \mathbf{V}) - \eta_{sp}(Tr(\mathbf{W}) + P_{cp} + Tr(\mathbf{V}) + P_b) \tag{28}$$

$$\text{s.t. } r_1(\mathbf{W}, \mathbf{V}) - r_2(\mathbf{W}, \mathbf{V}) \geq \Gamma_b \tag{29}$$

$$Tr(\mathbf{G}_{sp}\mathbf{W}) - \Lambda_s Tr(\mathbf{H}_s\mathbf{V}) \geq \Lambda_s \sigma_s^2 \tag{30}$$

$$Tr(\mathbf{H}_b\mathbf{V}) - \Lambda_b Tr(\mathbf{G}_{bp}\mathbf{W}) \geq \Lambda_b \sigma_b^2 \tag{31}$$

$$Tr(\mathbf{W}) \leq P_S, rank(\mathbf{W}) = 1, \mathbf{W} \geq 0 \tag{32}$$

$$Tr(\mathbf{V}) \leq P_B, rank(\mathbf{V}) = 1, \mathbf{V} \geq 0 \tag{33}$$

where

$$f_1(\mathbf{W}, \mathbf{V}) = \log_2(Tr(\mathbf{G}_{sp}\mathbf{W}) + Tr(\mathbf{H}_s\mathbf{V}) + \sigma_s^2) + \log_2(Tr(\mathbf{H}_{se}\mathbf{V}) + \sigma_{se}^2) \tag{34}$$

$$f_2(\mathbf{W}, \mathbf{V}) = \log_2(Tr(\mathbf{H}_s\mathbf{V}) + \sigma_s^2) + \log_2(Tr(\mathbf{G}_{sep}\mathbf{W}) + Tr(\mathbf{H}_{se}\mathbf{V}) + \sigma_{se}^2) \tag{35}$$

$$r_1(\mathbf{W}, \mathbf{V}) = \log_2(Tr(\mathbf{H}_b\mathbf{V}) + Tr(\mathbf{G}_{bp}\mathbf{W}) + \sigma_b^2) + \log_2(Tr(\mathbf{G}_{bep}\mathbf{W}) + \sigma_{be}^2) \tag{36}$$

$$r_2(\mathbf{W}, \mathbf{V}) = \log_2(Tr(\mathbf{G}_{bp}\mathbf{W}) + \sigma_b^2) + \log_2(Tr(\mathbf{H}_{be}\mathbf{V}) + Tr(\mathbf{G}_{bep}\mathbf{W}) + \sigma_{be}^2). \tag{37}$$

The optimal solution of problem (22)–(27) is denoted as $(\mathbf{W}^*, \mathbf{V}^*)$, which can be acquired from the problem given in (28)–(33) if and only if $f(\eta_{sp}) = 0$ [21]. The transformed problem (28)–(33) is a difference of two-convex functions (D.C.) problem. On the basis of [22], the Frank-and-Wold algorithm can be applied to transform the nonconvex optimization problem into a convex problem and obtain the optimal solution through iterative procedure. We apply the Taylor formula to transform $f_2(\mathbf{W}, \mathbf{V})$ and $r_2(\mathbf{W}, \mathbf{V})$ into approximate linear functions, which is the so-called D.C. approximation method. The gradients of $f_2(\mathbf{W}, \mathbf{V})$ and $r_2(\mathbf{W}, \mathbf{V})$ are given by:

$$df_2(\mathbf{W}, \mathbf{V}) = \frac{1}{\ln 2}\left(\frac{Tr(\mathbf{H}_s d\mathbf{V})}{Tr(\mathbf{H}_s\mathbf{V}) + \sigma_s^2} + \frac{Tr(\mathbf{H}_{se}d\mathbf{V}) + Tr(\mathbf{G}_{sep}d\mathbf{W})}{Tr(\mathbf{G}_{sep}\mathbf{W}) + Tr(\mathbf{H}_{se}\mathbf{V}) + \sigma_{se}^2} \right) \quad (38)$$

$$dr_2(\mathbf{W}, \mathbf{V}) = \frac{1}{\ln 2}\left(\frac{Tr(\mathbf{G}_{bp}d\mathbf{W})}{Tr(\mathbf{G}_{bp}\mathbf{W}) + \sigma_b^2} + \frac{Tr(\mathbf{G}_{bep}d\mathbf{W}) + Tr(\mathbf{H}_{be}d\mathbf{V})}{Tr(\mathbf{H}_{be}\mathbf{V}) + Tr(\mathbf{G}_{bep}\mathbf{W}) + \sigma_{be}^2} \right). \quad (39)$$

Then, according to the first-order Taylor series expansion, we have:

$$f_2(\mathbf{W}, \mathbf{V}) \leq f_2(\widetilde{\mathbf{W}}, \widetilde{\mathbf{V}}) + \frac{1}{\ln 2}\left(\frac{Tr\left(\mathbf{H}_{se}\left(\mathbf{V} - \widetilde{\mathbf{V}}\right)\right) + Tr\left(\mathbf{G}_{sep}(\mathbf{W} - \widetilde{\mathbf{W}})\right)}{Tr(\mathbf{G}_{sep}\widetilde{\mathbf{W}}) + Tr(\mathbf{H}_{se}\widetilde{\mathbf{V}}) + \sigma_{se}^2} \right)$$
$$+ \frac{1}{\ln 2}\frac{Tr\left(\mathbf{H}_s(\mathbf{V} - \widetilde{\mathbf{V}})\right)}{Tr(\mathbf{H}_s\widetilde{\mathbf{V}}) + \sigma_s^2} \quad (40)$$

$$r_2(\mathbf{W}, \mathbf{V}) \leq r_2(\widetilde{\mathbf{W}}, \widetilde{\mathbf{V}}) + \frac{1}{\ln 2}\left(\frac{Tr\left(\mathbf{G}_{bep}(\mathbf{W} - \widetilde{\mathbf{W}})\right) + Tr\left(\mathbf{H}_{be}(\mathbf{V} - \widetilde{\mathbf{V}})\right)}{Tr(\mathbf{H}_{be}\widetilde{\mathbf{V}}) + Tr(\mathbf{G}_{bep}\widetilde{\mathbf{W}}) + \sigma_{be}^2} \right)$$
$$+ \frac{1}{\ln 2}\frac{Tr\left(\mathbf{G}_{bp}(\mathbf{W} - \widetilde{\mathbf{W}})\right)}{Tr(\mathbf{G}_{bp}\widetilde{\mathbf{W}}) + \sigma_b^2} \quad (41)$$

where $(\widetilde{\mathbf{W}}, \widetilde{\mathbf{V}})$ is in the domin of the functions $f_2(\mathbf{W}, \mathbf{V})$ and $r_2(\mathbf{W}, \mathbf{V})$. By dropping the rank-1 constraints and substituting (40), (41) into (28)–(33), the problem (28)–(33) can be transformed as:

$$\max_{\mathbf{W}, \mathbf{V}} \quad f_1(\mathbf{W}, \mathbf{V}) - f_2(\widetilde{\mathbf{W}}, \widetilde{\mathbf{V}}) - \frac{Tr\left(\mathbf{G}_{sep}(\mathbf{W} - \widetilde{\mathbf{W}})\right)}{\ln 2\left(Tr(\mathbf{G}_{sep}\widetilde{\mathbf{W}}) + Tr(\mathbf{H}_{se}\widetilde{\mathbf{V}}) + \sigma_{se}^2\right)}$$

$$- \frac{Tr\left(\mathbf{H}_s(\mathbf{V} - \widetilde{\mathbf{V}})\right)}{\ln 2\left(Tr(\mathbf{H}_s\widetilde{\mathbf{V}}) + \sigma_s^2\right)} - \frac{Tr(\mathbf{H}_{se}(\mathbf{V} - \widetilde{\mathbf{V}}))}{\ln 2(Tr(\mathbf{G}_{sep}\widetilde{\mathbf{W}}) + Tr(\mathbf{H}_{se}\widetilde{\mathbf{V}}) + \sigma_{se}^2)}$$
$$- \eta_{sp}(Tr(\mathbf{W}) + P_{cp} + Tr(\mathbf{V}) + P_b) \quad (42)$$

$$\text{s.t.} \quad r_1(\mathbf{W}, \mathbf{V}) - r_2(\widetilde{\mathbf{W}}, \widetilde{\mathbf{V}}) - \frac{Tr(\mathbf{G}_{bep}(\mathbf{W} - \widetilde{\mathbf{W}}))}{\ln 2(Tr(\mathbf{H}_{be}\widetilde{\mathbf{V}}) + Tr(\mathbf{G}_{bep}\widetilde{\mathbf{W}}) + \sigma_{be}^2)}$$

$$- \frac{Tr(\mathbf{G}_{bp}(\mathbf{W} - \widetilde{\mathbf{W}}))}{\ln 2(Tr(\mathbf{G}_{bp}\widetilde{\mathbf{W}}) + \sigma_b^2)} - \frac{Tr(\mathbf{H}_{be}(\mathbf{V} - \widetilde{\mathbf{V}}))}{\ln 2(Tr(\mathbf{H}_{be}\widetilde{\mathbf{V}}) + Tr(\mathbf{G}_{bep}\widetilde{\mathbf{W}}) + \sigma_{be}^2)} \geq \Gamma_b (43)$$

$$Tr(\mathbf{G}_{sp}\mathbf{W}) - \Lambda_s Tr(\mathbf{H}_s\mathbf{V}) \geq \Lambda_s \sigma_s^2 \quad (44)$$

$$Tr(\mathbf{H}_b\mathbf{V}) - \Lambda_b Tr(\mathbf{G}_{bp}\mathbf{W}) \geq \Lambda_b \sigma_b^2 \quad (45)$$

$$Tr(\mathbf{W}) \leq P_S, \mathbf{W} \geq 0 \quad (46)$$

$$Tr(\mathbf{V}) \leq P_B, \mathbf{V} \geq 0. \quad (47)$$

It can be proved the problem (42)–(47) is convex, then the mathematical tool CVX could be used to solve the problem iteratively. Specifically, we denote $(\mathbf{W}^n, \mathbf{V}^n)$ as the solution at n-th iteration, then $(\mathbf{W}^{n+1}, \mathbf{V}^{n+1})$ can be obtained by solving the following convex problem:

$$\max_{\mathbf{W},\mathbf{V}} f_1(\mathbf{W},\mathbf{V}) - f_2(\mathbf{W}^n,\mathbf{V}^n) - \eta_{sp}(Tr(\mathbf{W}) + P_{cp} + Tr(\mathbf{V}) + P_b)$$

$$- \frac{Tr(\mathbf{H}_{se}(\mathbf{V}-\mathbf{V}^n))}{\ln 2(Tr(\mathbf{G}_{sep}\mathbf{W}^n)+Tr(\mathbf{H}_{se}\mathbf{V}^n)+\sigma_{se}^2)} - \frac{Tr(\mathbf{H}_s(\mathbf{V}-\mathbf{V}^n))}{\ln 2(Tr(\mathbf{H}_s\mathbf{V}^n)+\sigma_s^2)}$$

$$- \frac{Tr(\mathbf{G}_{sep}(\mathbf{W}-\mathbf{W}^n))}{\ln 2(Tr(\mathbf{G}_{sep}\mathbf{W}^n)+Tr(\mathbf{H}_{se}\mathbf{V}^n)+\sigma_{se}^2)} \tag{48}$$

$$\text{s.t.} \quad r_1(\mathbf{W},\mathbf{V}) - r_2(\mathbf{W}^n,\mathbf{V}^n) - \frac{Tr(\mathbf{G}_{bep}(\mathbf{W}-\mathbf{W}^n))}{\ln 2(Tr(\mathbf{H}_{be}\mathbf{V}^n)+Tr(\mathbf{G}_{bep}\mathbf{W}^n)+\sigma_{be}^2)} \tag{49}$$

$$- \frac{Tr(\mathbf{G}_{bp}(\mathbf{W}-\mathbf{W}^n))}{\ln 2(Tr(\mathbf{G}_{bp}\mathbf{W}^n)+\sigma_b^2)} - \frac{Tr(\mathbf{H}_{be}(\mathbf{V}-\mathbf{V}^n))}{\ln 2(Tr(\mathbf{H}_{be}\mathbf{V}^n)+Tr(\mathbf{G}_{bep}\mathbf{W}^n)+\sigma_{be}^2)} \geq \Gamma_b$$

$$(44)\text{--}(47). \tag{50}$$

Based on the above analysis, we propose a three-tier iterative algorithm to obtain the optimal solution of our problem as summarized in Algorithm 1, which consists of three functions, namely main function, outer_iteration function and inner_iteration function. In main function, we obtain the optimal value of the satellite scheduling vector $\mathbf{C}$ by traversing the value of $\mathbf{C}$ through setting $c_p = 1$ and $c_i = 0, i \in \{1, 2...p - 1, p + 1, ...K\}$ and activating the outer_iteration function to calculate the achievable maximal SEE corresponding to the given $\mathbf{C}$. The outer_iteration function will be called K times to search for the optimal $\mathbf{C}$ and obtain the corresponding BF vectors $(\mathbf{w}_p^*, \mathbf{v}_p^*)$. When the outer_iteration function is activated, the maximal number of iterations i_{max}, the minimum tolerance error ε, the iteration index i and the achievable SEE η_{sp}^i at i-th iteration are initialized. Based on the given η_{sp}^i, the problem (28)–(33) is solved by D.C. approximation method to obtain the matrices $(\mathbf{W}_{sp}^{i+1}, \mathbf{V}_{sp}^{i+1})$ according to the inner_iteration function. The updated matrices $(\mathbf{W}_{sp}^{i+1}, \mathbf{V}_{sp}^{i+1})$ will be used to update the value of η_{sp}^{i+1} according to the Dinkelbach's method until $i > i_{max}$ or the criterion $|\eta_{sp}^{i+1} - \eta_{sp}^i| < \varepsilon$ is satisfied. Then the BF vectors $(\mathbf{w}_p^*, \mathbf{v}_p^*)$ can be obtained through corresponding method. When inner_iteration function is activated, the matrices $(\mathbf{W}_p^j, \mathbf{V}_p^j)$ are updated by formula (48)–(50) until the maximum number of inner_iteration j_{max} is reached or the value of $f^{j+1} = f_1(\mathbf{W}_p^{j+1}, \mathbf{V}_p^{j+1}) - f_2(\mathbf{W}_p^{j+1}, \mathbf{V}_p^{j+1}) - \eta_{sp}^i(Tr(\mathbf{W}_p^{j+1}) + P_{cp} + Tr(\mathbf{V}_p^{j+1}) + P_b)$ satisfies the criterion $|f^{j+1} - f^j| < \varepsilon$. The convergence of outer_iteration and inner_iteration functions have already been approved in [21,23].

Denote by $(\mathbf{W}^*, \mathbf{V}^*)$ the optimal solution of the problem (28)–(33). Note that $(\mathbf{W}^*, \mathbf{V}^*)$ might not be rank-1. If $(\mathbf{W}^*, \mathbf{V}^*)$ are of rank-1, the singular value decomposition method is employed to acquire the optimal BF vectors $(\mathbf{w}^*, \mathbf{v}^*)$. Otherwise, the Gaussian randomization method is utilized to find an approximate solution [6].

Algorithm 1 :The Proposed Three-tier Algorithm.

Input: K, $g_i, h_i, i \in sj, sej, bj, bej, j \in 1, 2...K$.

Output: Optimal scheduling vector $\mathbf{C}$ with the BF vectors $(\mathbf{w}^*, \mathbf{v}^*)$.

  Function: *Main*

1: Initialize $p = 1$.

2: Set $\mathbf{C} = \mathbf{C}_p$ with $c_p = 1$ and $c_i = 0, i \in \{1, 2...p-1, p+1, ...K\}$, call **Function** *Outer_Iteration* to calculate the achievable SEE η_{sp}^* and corresponding $(\mathbf{w}_p^*, \mathbf{v}_p^*)$.

3: Set $\eta(\mathbf{C}_p) = \eta_{sp}^*, p = p + 1$.

4: **if** $p \leq K$ **then**

5:    goto step 2.

6: **else**

7:    Calculate $\mathbf{C}_t = \arg\max_{\mathbf{C}} \eta(\mathbf{C})$.

8: **end if**

9: Obtain the maximal SEE η^* with the satellite scheduling vector $\mathbf{C} = \mathbf{C}_t$ and corresponding BF vectors $(\mathbf{w}^*, \mathbf{v}^*) = (\mathbf{w}_t^*, \mathbf{v}_t^*)$.

  end

  Function: *Outer_Iteration*

10: Initialize the maximal number of iterations i_{max} and minimum tolerance error ε.

11: Set $\eta_{sp}^i = 0, i = 0$.

12: Call **Function** *Inner_Iteration* with η_{sp}^i and ε to calculate $(\mathbf{W}_{sp}^{i+1}, \mathbf{V}_{sp}^{i+1})$.

13: Update $\eta_{sp}^{i+1} = \frac{\log_2(1+\gamma_{sp}) - \log_2(1+\gamma_{sep})}{Tr(\mathbf{W}_{sp}^{i+1}) + P_{cp} + Tr(\mathbf{V}_{sp}^{i+1}) + P_b}$.

14: **if** $|\eta_{sp}^{i+1} - \eta_{sp}^i| \geq \varepsilon$ and $i \leq i_{max}$ **then**

15:    Update $i = i + 1$, goto step 12.

16: **else**

17:    $\eta_{sp}^* = \eta_{sp}^{i+1}, (\mathbf{W}_{sp}^*, \mathbf{V}_{sp}^*) = (\mathbf{W}_{sp}^{i+1}, \mathbf{V}_{sp}^{i+1})$.

18: **end if**

19: **if** rank$(\mathbf{O}) = 1, \mathbf{O} \in \{\mathbf{W}_{sp}^*, \mathbf{V}_{sp}^*\}$ **then**

20:    Use Singular Value Decomposition Method to get $\mathbf{o}, \mathbf{o} \in \{\mathbf{w}_p^*, \mathbf{v}_p^*\}$.

21: **else**

22:    Use Gaussian Randomization Method to get $\mathbf{o}, \mathbf{o} \in \{\mathbf{w}_p^*, \mathbf{v}_p^*\}$.

23: **end if**

24: Return η_{sp}^* and $(\mathbf{w}_p^*, \mathbf{v}_p^*)$.

  end

  Function: *Inner_Iteration*

25: Initialize the maximal number of iterations j_{max}, $(\mathbf{W}_p^0, \mathbf{V}_p^0) = (\mathbf{0}, \mathbf{0})$ and $f^0 = 0$.

26: Set $j = 0$.

27: Compute $(\mathbf{W}_p^{j+1}, \mathbf{V}_p^{j+1})$ of problem (48)-(50) with $(\mathbf{W}_p^j, \mathbf{V}_p^j)$ and η_{sp}^i.

28: Compute $f^{j+1} =$
   $f_1(\mathbf{W}_p^{j+1}, \mathbf{V}_p^{j+1}) - f_2(\mathbf{W}_p^{j+1}, \mathbf{V}_p^{j+1}) - \eta_{sp}^i(Tr(\mathbf{W}_p^{j+1}) + P_{cp} + Tr(\mathbf{V}_p^{j+1}) + P_b)$.

29: **if** $|f^{j+1} - f^j| \geq \varepsilon$ and $j \leq j_{max}$ **then**

30:    Update $j = j + 1$, goto step 27.

31: **else**

32:    Return $(\mathbf{W}_p^{j+1}, \mathbf{V}_p^{j+1})$.

33: **end if**

  end

---

4   Simulation Results

In this section, we provide simulation results to evaluate the performance of our proposed scheme. To conduct comparison, the joint highest elevation angle priority satellite scheduling and BF scheme (JHEASSB) and the joint maximal service time priority satellite scheduling and BF scheme (JMSTSSB) are also presented as counterparts.

The simulated scenario and related parameters are set as follows. We assume that the terrestrial links experience Rayleigh distribution. The distance between the SU and the SE, the BU and the BE are set as 0.1D, 0.1D and 0.11D, respectively, where D denotes the beam diameter of the satellite. The simulated satellite constellation is an iridium satellite-like constellation. The power budgets of the satellite and the BS are set as 35 dBm and 42 dBm, respectively. The elevation angles of the satellites in the visual range of the SU are determined at a fixed time. We set the latitude and the longitude coordinates of the SU as $(75°, 95°)$. According to the scenario we established, there are four satellites in the SU's visual range. The elevation angles of the satellites are $20.0568°$, $20.1066°$, $27.3138°$ and $31.1373°$, respectively. The scheduled satellite will transmit the confidential signal to the SU via N beams which have the smallest angles between the beam center and SU's position with respect to the satellite among all beams. Unless otherwise specified, the simulation is carried out under the above conditions. Other simulation parameters are summarized in Table 1 [24]. In addition, all of the simulation curves are averaged over 500 random channel realizations.

Table 1. Main simulation parameters.

Parameter	Value
No. of satellites	66
No. of orbital planes	6
Inclination of the orbital plane	86.4°
No. of beams per satellite	48
Beam diameter	400 km
Minimum elevation angle	19°
Altitude of the orbit	780 km
Carrier frequency	20 GHz
Noise bandwidth	500 MHz
Noise temperature	300 K
3 dB angle	14°

Figure 2 depicts the SEE versus the number of beams N utilized for the secure transmission. We can observe that the proposed scheme achieves the maximal

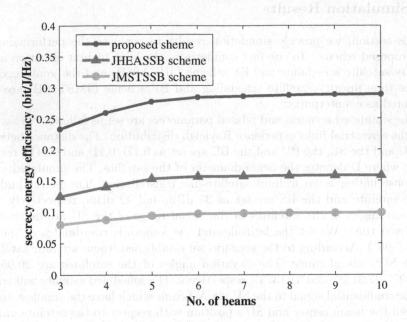

Fig. 2. Achievable SEE versus number of beams.

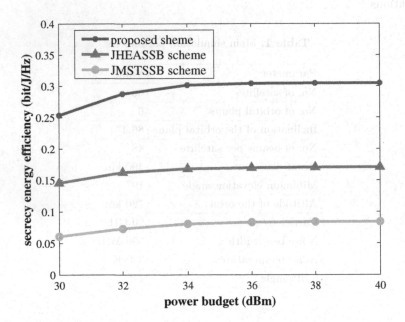

Fig. 3. Achievable SEE versus power budget of satellite.

SEE as compared with the JHEASSB scheme and the JMSTSSB scheme. In our proposed scheme, the achievable SEE is improved about 100% as compared with the JHEASSB and more than 150% as compared with the JMSTSSB, which demonstrates the efficiency of our proposed scheme. The reason for this phenomenon is that the link between the SU and the satellite which has the maximum elevation angle does not always have the best transmission quality, the opportunistic satellite scheduling scheme can ensure the link between the scheduled satellite and SU has the best transmission quality and improve the SEE by BF effectively. As the number of beams for BF increases, all the schemes reach their own SEE floors and remain unchanged. The reason is that when more beam resources are available, the achievable SEE tends to increase. However, the further away from the SU, the smaller beam gain of the beam to SU, thus the beams which far away form the SU are of little use for the secure transmission and the SEE tends to keep constant eventually.

Figure 3 depicts the SEE performance versus the power budget P_{cp} of the proposed scheme, the JHEASSB scheme and the JMSTSSB scheme when the number of beams utilized for secure transmission is set as 6. As the power budget increases, the SEE increases until converges. The SEE of the three schemes are all improved with an increasing transmit power budget P_{cp} in the 30–36 dBm region, which indicates that when more energy resource available, the larger SEE can be obtained within certain power budget bounds. Morever, it can be observed that the proposed scheme obtains the maximal SEE as compared with the other two schemes, which proves the superiority of our proposed scheme.

5   Conclusions

This paper is devoted to enhancing the secrecy performance in the CLSTN. Taking the security performance and energy efficiency into account, we propose a joint opportunistic satellite scheduling and BF scheme to realize the secure transmission of both the SU and BU. An optimization problem is formulated to maximize the SEE of the SU under the SR constraint of the BU, the SINR requirements of the SU and BU and the limited power budgets of the satellite and BS. Considering the case when the scheduled satellite is given, we can transform the SEE maximization problem into a convex one by exploiting the Dinkelbach's method and the D.C. approximation method. Then, a three-tier algorithm is proposed to obtain the optimal satellite scheduling vector and the BF vectors of the satellite and BS jointly. Simulation results confirm the validity of our proposed scheme.

References

1. Hu, J., Li, G., Bian, D., Gou, L., Wang, C.: Optimal power control for cognitive LEO constellation with terrestrial networks. IEEE Commun. Lett. 24(3), 622–625 (2020)

2. Yang, P., Xiao, Y., Xiao, M., Li, S.: 6G wireless communications: vision and potential techniques. IEEE Netw. **33**(4), 70–75 (2019)
3. Lin, Z., Lin, M., Ouyang, J., Zhu, W., Panagopoulos, A.D., Alouini, M.: Robust secure beamforming for multibeam satellite communication systems. IEEE Trans. Veh. Technol. **68**(6), 6202–6206 (2019)
4. Schraml, M.G., Schwarz, R.T., Knopp, A.: Multiuser MIMO concept for physical layer security in multibeam satellite systems. IEEE Trans. Inf. Forensics Secur. **16**, 1670–1680 (2021)
5. Lin, Z., Yin, C., Ouyang, J., Wu, X., Panagopoulos, A.D.: Robust secrecy energy efficient beamforming in satellite communication systems. In: 2019 IEEE International Conference on Communications (ICC), May 2019, pp. 1–5. IEEE, Shanghai (2019)
6. Zheng, G., Arapoglou, P., Ottersten, B.: Physical layer security in multibeam satellite systems. IEEE Trans. Wirel. Commun. **11**(2), 852–863 (2012)
7. Wang, P., Ni, Z., Jiang, C., Kuang, L., Feng, W.: Dual-beam dual-frequency secure transmission for downlink satellite communication systems. In: 2019 IEEE Globecom Workshops (GC Wkshps), December 2019, pp. 1–6. IEEE, Waikoloa (2019)
8. An, K., Lin, M., Ouyang, J., Zhu, W.: Secure transmission in cognitive satellite terrestrial networks. IEEE J. Sel. Areas Commun. **34**(11), 3025–3037 (2016)
9. Li, B., Fei, Z., Chu, Z., Zhou, F., Wong, K., Xiao, P.: Robust chance-constrained secure transmission for cognitive satellite terrestrial networks. IEEE Trans. Veh. Technol. **67**(5), 4208–4219 (2018)
10. Lin, M., Lin, Z., Zhu, W., Wang, J.: Joint beamforming for secure communication in cognitive satellite terrestrial networks. IEEE J. Sel. Areas Commun. **36**(5), 1017–1029 (2018)
11. Du, J., Jiang, C., Zhang, H., Wang, X., Ren, Y., Debbah, M.: Secure satellite-terrestrial transmission over incumbent terrestrial networks via cooperative beamforming. IEEE J. Sel. Areas Commun. **36**(7), 1367–1382 (2018)
12. Cui, G., Zhu, Q., Xu, L., Wang, W.: Secure beamforming and jamming for multibeam satellite systems with correlated wiretap channels. IEEE Trans. Veh. Technol. **69**(10), 12348–12353 (2020)
13. Xu, S., Liu, J., Cao, Y., Li, J., Zhang, Y.: Intelligent reflecting surface enabled secure cooperative transmission for satellite-terrestrial integrated networks. IEEE Trans. Veh. Technol. **70**(2), 2007–2011 (2021)
14. Li, J., Xue, K., Liu, J., Zhang, Y.: A user-centric handover scheme for ultra-dense LEO satellite networks. IEEE Wirel. Commun. Lett. **9**(11), 1904–1908 (2020)
15. Ding, X., Zhang, G., Qu, D., Song, T.: Security-reliability tradeoff analysis of spectrum-sharing aided satellite-terrestrial networks. In: 2019 IEEE Globecom Workshops (GC Wkshps), December 2019, pp. 1–6. IEEE, Waikoloa (2019)
16. Guo, K., An, K., Tang, X.: Secrecy performance for integrated satellite terrestrial relay systems with opportunistic scheduling. In: 2019 IEEE International Conference on Communications Workshops (ICC Workshops), May 2019, pp. 1–6. IEEE, Shanghai (2019)
17. Cao, W., Zou, Y., Yang, Z., Zhu, J.: Secrecy outage probability of hybrid satellite-terrestrial relay networks. In: 2017 IEEE Global Communications Conference (GLOBECOM 2017), December 2017, pp. 1–5. IEEE, Singapore (2017)
18. Wu, H., Zou, Y., Zhu, J., Xue, X., Tsiftsis, T.: Secrecy performance of hybrid satellite-terrestrial relay systems with hardware impairments. In: 2019 IEEE International Conference on Communications (ICC), May 2019, pp. 1–6. IEEE, Shanghai (2019)

19. Lutz, E., Cygan, D., Dippold, M., Dolainsky, F., Papke, W.: The land mobile satellite communication channel-recording, statistics, and channel model. IEEE Trans. Veh. Technol. **40**(2), 375–386 (1991)
20. Glisic, S.G., Talvitie, J.J., Kumpumaki, T., Latva-aho, M., Iinatti, J.H., Poutanen, T.J.: Design study for a CDMA-based LEO satellite network: downlink system level parameters. IEEE J. Sel. Areas Commun. **14**(9), 1796–1808 (1996)
21. Jiang, Y., Zou, Y., Ouyang, J., Zhu, J.: Secrecy energy efficiency optimization for artificial noise aided physical-layer security in OFDM-based cognitive radio networks. IEEE Trans. Veh. Technol. **67**(12), 11858–11872 (2018)
22. Kha, H.H., Tuan, H.D., Nguyen, H.H.: Fast local D.C. programming for optimal power allocation in wireless networks. In: 2011 IEEE Global Communications Conference (GLOBECOM 2011), December 2011, pp. 1–5. IEEE, Houston (2011)
23. Chen, P., Ouyang, J., Zhu, W.-P., Lin, M., Shafie, A.E., Al-Dhahir, N.: Artificial-noise-aided energy-efficient secure beamforming for multi-eavesdroppers in cognitive radio networks. IEEE Syst. J. **14**(3), 3801–3812 (2020)
24. Gao, Z., Liu, A., Liang, X.: The performance analysis of downlink NOMA in LEO satellite communication system. IEEE Access **8**, 93723–93732 (2020)

GAN-SNR-Shrinkage-Based Network for Modulation Recognition with Small Training Sample Size

Shuai Zhang, Yan Zhang, Mingjun Ma, Zunwen He$^{(\boxtimes)}$, and Wancheng Zhang

Beijing Institute of Technology, BIT, Beijing, People's Republic of China
hezunwen@bit.edu.cn

Abstract. Modulation recognition plays an important role in non-cooperative communications. In practice, only a small number of samples can be collected for training purposes. The limited training data degrade the accuracy of the modulation recognition networks. In this paper, we propose a novel network to realize the modulation recognition on basis of the few-shot learning. Generative adversarial networks (GANs) and a signal-to-noise ratio (SNR) augment module are introduced to expand the training dataset. In addition, a preprocessing module and residual shrinkage networks are used to improve the capability of characterizing signal features and the anti-noise performance. The proposed network is evaluated using the RML2016.10a dataset. It is illustrated that the proposed network outperforms the baseline method and the method without data augment with a small number of training samples.

Keywords: Modulation recognition · GAN · SNR · Few-shot learning

1 Introduction

Non-cooperative communication plays an important role in radio resource management, communication spectrum monitoring, and other civilian and military fields. Modulation recognition, which is between signal detection and demodulation, has important research value in the field of non-cooperative communication. It is also of great significance to signal demodulation, information extraction, and signal detection in communication systems.

Traditional modulation recognition techniques can be divided into two categories: likelihood-based methods and signal-characteristic-based methods. Likelihood-based methods [1–3] have good performance in recognition accuracy. The main ideas of these methods are to realize the recognition of signals according to the modulation mode of signals with the help of probability theory and Bayesian theory. However, they suffer from the problem of computational complexity, i.e., the technical calculation complexity is very high in the recognition process. The signal-characteristic-based methods which are mainly based on manually extracted characteristics of signal processing are generally more widely employed. In these methods, signal features, such as instantaneous feature

H. Gao et al. (Eds.): ChinaCom 2021, LNICST 433, pp. 80–90, 2022.
https://doi.org/10.1007/978-3-030-99200-2_7

quantity [4], spectral correlation [5], wavelet transform [6], etc., are firstly designed and extracted, and then classification rules are designed for modulation recognition of signal features. However, the recognition accuracy of manually designed signal features and classification rules is usually limited in complex channel environments.

In recent years, modulation recognition techniques based on neural networks and deep learning have been proposed to automatically learn classification rules from feature data to improve recognition accuracy. In the field of modulation recognition, feature transformation based on Convolutional Neural Networks, Recurrent Neural Networks, or Convolutional, Long-Short-Term-Memory, Deep-Neural-Network is proposed in [7–9], and they have achieved good recognition rates. Although the deep-learning-based methods outperform the traditional manual extracted-based methods in recognition performance, they require a large number of labeled samples for training. If the number of labeled samples is insufficient, the recognition performance of the deep-learning-based methods will degrade sharply.

Therefore, few-shot learning is introduced to solve the modulation recognition problem with few training samples. Capsule networks [10], modulated filters, a manual-feature-based method [11], and a data preprocessing method [12] are proposed in order to improve recognition rates. Various networks based on the GAN have been applied in [13] to generate the samples for training purposes. However, these methods assume that training data contains all SNR cases which is difficult to satisfy in practice.

In this paper, we propose a GSS (GAN-SNR-Shrinkage) network structure for modulation recognition based on a small sample set. An SNR augment module and a deep convolutional GAN (DCGAN) module are introduced to this network. According to the characteristics of the modulated signal, we extend the data according to the SNR by the designed SNR augment module. This module does not need complicated operations and ensures the feasibility of extended data.

Then, we preprocess the signal by shifting and splicing in the time-domain, which increases the dimensions of data to be beneficial for the recognition. Finally, we train an effective classifier that is specially designed including residual shrinkage block to realize modulation recognition of the input data.

We verify our network performance on dataset RML2016.10a [14]. Multiple cross-experiments and comparison of results show that the proposed network has higher accuracy than the baseline method and the method without data augment under the condition of small sample size.

The rest of this paper is organized as follows: Sect. 2 and Sect. 3 introduce the system model and the structure of the network. In Sect. 4, we present the test results on the RML2016.10a and analyze the advantages of our network. Section 5 presents the conclusion.

2 System Model

In the general communication channel model with the additive white Gaussian noise (AWGN), signals of different modulation modes are expressed as

$$s(t) = h(t) * f(t) + n(t) \tag{1}$$

where is the modulated signal, which may be analog signal or digital signal. is the channel impulse response and represents the convolution operation. is the additive white Gaussian noise with a zero mean and a variance of . is the received signal.

To train an effective recognition network, received signals under different SNR conditions should be collected. The collected dataset can be expressed as

$$\{s_{original}\} = \bigcup_{i=1,2,\dots,n} \{s_{ri}\} \qquad (2)$$

where is the dataset with SNR and are the different SNR values.

3   GSS Network Structure

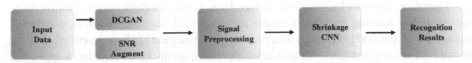

Fig. 1. Proposed GSS network structure.

The proposed network structure of the system is shown in Fig. 1 is the input of the network including different SNRs. DCGAN is set up as the first data enhancement module, which is used to generate some pseudo samples with an input. These pseudo samples are directly generated on the I/Q dimension to expand the number of training samples.

Then, the original signal is also processed through the SNR augment module to generate new samples, which is able to transform the data of high SNRs into the data of low SNRs by adding artificial Gaussian noise, and to increase the samples quantity. Finally, the entire training samples include.

$$\{s_{train}\} = \{f_{DCGAN}(z)\} \bigcup \{f_{SNR}(s_{original})\} \bigcup \{s_{original}\}. \qquad (3)$$

Moreover, a signal preprocessing module [12] is applied in the third part of the GSS network structure. On the one hand, the I/Q expression is transferred into the amplitude-phase (AP) expression. On the other hand, signal samples are regularly shifted and spliced to improve the feature expression ability of the samples.

The last module is the specially designed CNN module with residual shrinkage block. Residual shrinkage block has shown a good effect in mechanical fault signal diagnosis [14]. Therefore, we introduce it into our network structure to replace the original CNN for the modulation recognition.

3.1   DCGAN

DCGAN model mainly consists of a generating network G and a discriminant network D. G is responsible for generating samples. G accepts a random noise z, and the generated samples are denoted as G(z). D is responsible for judging whether the sample is real or generated, represented by 1 and 0, respectively. The process of model training constitutes a dynamic game between G and D, and the cross-entropy loss function [15] in the training process is constructed as

$$\min_{G} \max_{D} V(D, G) = E_{x \sim pdata(x)}[\ln D(x)] + E_{x \sim P_z(z)}[\ln(1 - D(G(z)))] \qquad (4)$$

where is the real data used for training, and the distribution of data is. represents the distribution of noise, and G(z) represents the samples generated by G.

To generator G, the optimization problem is

$$\min_{G} V(D, G) = E_{x \sim P_z(z)}[\ln(1 - D(G(z)))]. \qquad (5)$$

On the other hand, D wants to have a strong discriminating ability, the mixed loss function of D can be depicted as

$$\max_{D} V(D, G) = E_{x \sim pdata(x)}[\ln D(x)] + E_{x \sim P_z(z)}[\ln(1 - D(G(z)))]. \qquad (6)$$

Then, the discriminator D will update its parameters by descending its gradient.

$$\nabla_{\theta_d} \frac{1}{m} \sum_{l=1}^{m} [\ln D(x^{(l)}) + \ln(1 - D(G(z^{(l)})))] \qquad (7)$$

where is training data batch size, is the noise vector in the minibatch.

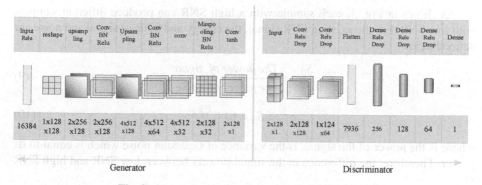

Fig. 2. Network structure parameters of DCGAN.

The network structure and parameter settings of G and D are shown in Fig. 2. In DCGAN, the input of G is random noise, and its output is a vector with dimensions 2 × 128 × 1. Meanwhile, the input of D is a vector with dimensions 2 × 128 × 1, and its output is the discriminant result (1 represents sample is real, in contrast, 0 means sample is generated). The training of DCGAN is realized by alternate optimization of D and G. DCGAN can generate any amounts of samples at different SNRs.

3.2 SNR Augment

The input of the SNR augment module is the data under the existing SNRs. These samples are enhanced to produce more samples with a lower SNR, and the amounts of generated samples present a stepwise type. The generated samples are represented as

$$\{s_{SNR}\} = \bigcup_{n=r_1,r_2,\dots,r_n} \bigcup_{i \in \{r_1,\dots,n\}} \{s_i\}. \tag{8}$$

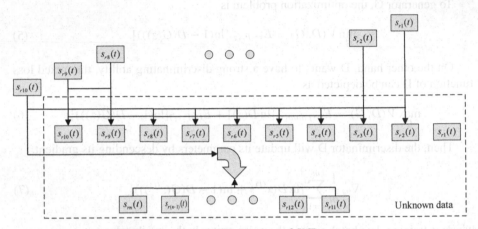

Fig. 3. Description of the process of SNR augment.

As shown in Fig. 3, each sample with a high SNR can produce different samples with low SNR. For the existing SNR signal, the energy and SNR of signal and noise are defined below

$$\frac{S_0}{\sigma^2} = \frac{The\ power\ of\ signa}{The\ power\ of\ noise} \tag{9}$$

$$\sigma^2 = S_0 * 10^{(-\frac{SNR}{10})} \tag{10}$$

where is the power of the signal, is the variance of Gaussian noise which is equal to its power. Therefore, the difference in the signal powers between low SNR and high SNR is

$$\left| \sigma_{high}^2 - \sigma_{low}^2 \right| = \left| S_0 * (10^{(-\frac{SNR_{high}}{10})} - 10^{(-\frac{SNR_{low}}{10})}) \right|. \tag{11}$$

Through the stepwise data enhancement method of the SNR augment, low -SNR samples can be obtained from high -SNR samples, and the reliability and validity of the samples can be guaranteed. When the SNR is below 0 dB, the signal itself is already greatly polluted by the noise. Thus, the signal samples whose SNRs are below 0 dB will not be used to carry out such corresponding data enhancement.

3.3 Signal Preprocessing

According to the characteristics of the signal, the signal preprocessing module is used before entering CNN to make contributions for recognition, and the specific processing process is shown in Fig. 4.

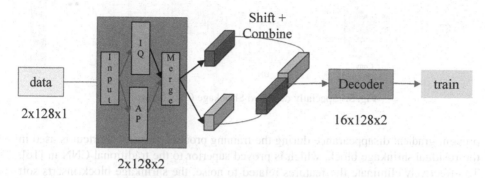

Fig. 4. Structure of signal preprocessing.

In Fig. 4, the amplitude-phase format of the data is.

$$\begin{cases} A = \sqrt{I^2 + Q^2} \\ P = \arctan(\frac{Q}{I}) \end{cases} \quad (12)$$

The dimension of the input data is expanded to.

$$s_l = [I_l, Q_l]^T \bigcup [A_l, P_l]^T = \left[\begin{bmatrix} I_l \\ Q_l \end{bmatrix}, \begin{bmatrix} A_l \\ P_l \end{bmatrix} \right] \quad (13)$$

In this part, the input I/Q is originally one channel, and the data are converted to amplitude-phase data and then regrouped into two channels. In addition, the sample size has been expanded through cross-shift combinations, and the combined samples are used as training data.

3.4 Shrinkage CNN

Samples expanded by the previous module will be input into the CNN for classification. The convolution kernels in CNN have good feature extraction capability and fewer parameters. BN (Batch Normalization) is set to reduce the internal covariant shift and prevent gradient disappearance during the training process. Residual shrinkage blocks are inserted into the designed CNN. The structure of Specially designed CNN is shown in Fig. 5.

The function of the first two different convolution layers is to extract horizontal and vertical features, and the purpose of BN is to reduce the internal covariant shift and

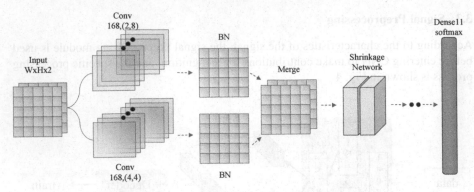

Fig. 5. Specially designed Shrinkage CNN Network.

prevent gradient disappearance during the training process. Identity shortcut is used in the residual shrinkage block, which is proved superior to the traditional CNN in [16]. To effectively eliminate the features related to noise, the shrinkage block inserts soft threshold [17] into the structure as a nonlinear transformation layer as.

$$y = \begin{cases} x - \tau, x > \tau \\ 0, -\tau \leq x \leq \tau \\ x + \tau, x < -\tau \end{cases} \qquad (14)$$

where is the input vector of the soft threshold module, is the output vector, is the threshold.

In addition, specially designed sub-networks are used in the network to determine the threshold adaptively [18], so that each signal can have its own set of thresholds as.

$$\tau = \alpha * \underset{u,v,w}{average} |x_{u,v,w}| \qquad (15)$$

where ,, and are indexes of width, height, and channel of the feature map, respectively.

The discriminator outputs a dimensional vector of, which is turned into by the softmax function. is the number of the signal modulation types. The expression of is

$$p_j = \frac{e^{n_j}}{\sum_{q=1}^{k+1} e^{n_q}}, j \in \{1, 2, ...k\} \qquad (16)$$

4   Experimental Results

4.1   Dataset

The RML2016.10a dataset [14] is widely used for model performance verification in modulation recognition. It contains 11 different modulation modes, i.e., BPSK, QPSK, 8PSK, QAM16, QAM64, CPFSK, GFSK, 4PAM, WBFM, AM-DSB, and AM-SSB. Both analog modulations and digital modulations are included. Each signal contains 2200 baseband I/Q data with 128 sampling points.

These samples are collected from simulated signals through wireless channels and are affected by multipath fading, sampling rate offset and center frequency offset. The SNR of the data ranges from −20 dB to 18 dB, with 2 dB intervals. We randomly select 20% samples as the test set, and the evaluation of the model is carried out under the same test set.

4.2 Results

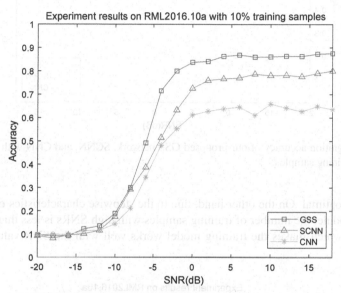

Fig. 6. Recognition accuracy of our proposed GSS network, SCNN, and CNN with 10% of the number of training samples.

Considering the few-shot learning requirement, only 10% of the number of samples are selected as the training sample. We verify the performance of the GSS network and compare it with a baseline CNN classifier and the specially designed CNN (SCNN) [12] which achieves high- accuracy in modulation recognition. Figure 6 shows the accuracy of different networks at varying SNR. Based on our experiment results, our proposed method can achieve the highest recognition accuracy of 88.1% at 6dB, which is 6.4% and 23.7% higher than the baseline CNN and the SCNN. In addition, the GSS network shows great advantages over the other two methods starting from −8 dB. The accuracy rate of our method has reached a high level when the SNR is above −4 dB.

Figure 7 shows the recognition accuracy of our proposed GSS network, SCNN, and CNN with 20% of the number of training samples. As a result of Fig. 6 and Fig. 7, interestingly, the three recognition rate curves do not rise all the time. It is noted that there are some slight fluctuations in the GSS curve when the SNR value is between 0 dB to 18 dB. The reason may come from two aspects. On the one hand, the number of training samples is small, so the fitting effect of the network on the test set at each SNR

88   S. Zhang et al.

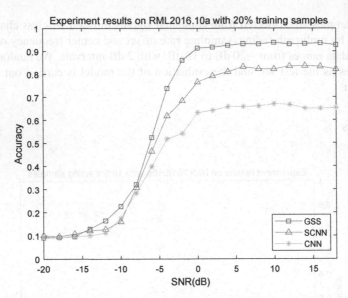

Fig. 7. Recognition accuracy of our proposed GSS network, SCNN, and CNN with 20% of the number of training samples.

may not be optimal. On the other hand, due to the stepwise characteristics of the SNR augment module, the number of training samples with high SNRs is less than that with low SNRs, which makes the training model works well with the SNR values around 6 dB.

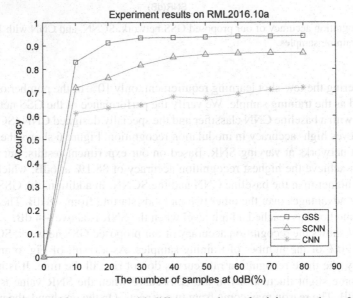

Fig. 8. Recognition accuracy with different numbers of training samples at 0 dB.

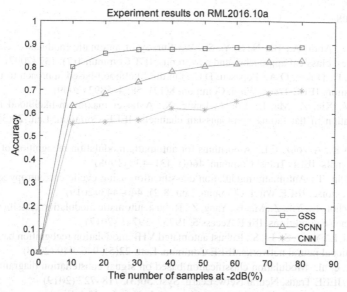

Fig. 9. Recognition accuracy with different numbers of training samples at −2 dB.

Then, we also do comparison experiments when SNR is fixed and the number of samples changes. Figure 8 and Fig. 9 illustrate the recognition accuracy of the above models at 0 dB and −2 dB. The GSS network achieves 91.3% accuracy with only 20% of the samples when the SNR is 0 dB, whereas the maximum accuracy rate of SCNN and CNN is 87.3% and 70.1%. Although the recognition rate of GSS at −2 dB is lower than that of 0 dB, it is still much higher than the other two methods. In addition, with the increase of the training sample size, the recognition accuracy of the GSS network increases faster than the other two methods.

5 Conclusion

In this paper, we have proposed the GSS network for automatic modulation recognition with a small training sample set. An SNR augment module, as well as DCGAN, have been introduced to realize the expansion of the training dataset. Specially designed CNN with the signal preprocessing module has been used to achieve recognition of signals. It has been proved that the proposed method has a high recognition rate when the training data are in a small size. Experiments on the RML2016.10a dataset have shown that the proposed GSS network can achieve a recognition rate of over 90% with only 20% training samples.

Acknowledgements. This work was supported by the National Key R&D Program of China under Grant (2019YFE0196400), the National Natural Science Foundation of China under Grant (61871035), and the National Defense Science and Technology Innovation Zone.

References

1. Dobre, O., Abdi, A., Bar-Ness, Y., Su, W.: Survey of automatic modulation classification techniques: classical approaches and new trends. IET Commun 1(2), 137 (2007)
2. Hameed, F., Dobre, O.A., Popescu, D.C.: On the likelihood-based approach to modulation classification. IEEE Trans. Wirel. Commun. 8(12), 5884–5892 (2009)
3. Chen, W., Xie, Z., Ma, L., Liu, J., Liang, X.: A faster maximum-likelihood modulation classification in flat fading non-Gaussian channels. IEEE Commun. Lett. 23(3), 454–457 (2019)
4. Nandi, A.K., Azzouz, E.E.: Algorithms for automatic modulation recognition of communication signals. IEEE Trans. Commun. 46(4), 431–436 (1998)
5. Ma, J., Qiu, T.: Automatic modulation classification using cyclic correntropy spectrum in impulsive noise. IEEE Wirel. Commun. Lett. 8(2), 440–443 (2019)
6. Wu, Z., Zhou, S., Yin, Z., Ma, B., Yang, Z.: Robust automatic modulation classification under varying noise conditions. IEEE Access 5, 19733–19741 (2017)
7. Li, R., Li, L., Yang, S., Li, S.: Robust automated VHF modulation recognition based on deep convolutional neural networks. IEEE Commun. Lett. 22(5), 946–949 (2018)
8. Peng, S., et al.: Modulation classification based on signal constellation diagrams and deep learning. IEEE Trans. Neural Netw. Learn. Syst. 30(3), 718–727 (2019)
9. Hong, D., Zhang, Z., Xu, X.: Automatic modulation classification using recurrent neural networks. In: 2017 3rd IEEE International Conference on Computer and Communications (ICCC), pp. 695–700 (2017)
10. Li, L., Huang, J., Cheng, Q., Meng, H., Han, Z.: Automatic modulation recognition: a few-shot learning method based on the capsule network. IEEE Wirel. Commun. Lett. 10(3), 474–477 (2021)
11. Zhang, D., Ding, W., Liu, C., Wang, H., Zhang, B.: Modulated autocorrelation convolution networks for automatic modulation classification based on small sample set. IEEE Access 8, 27097–27105 (2020)
12. Zhang, H., Huang, M., Yang, J., Sun, W.: A data preprocessing method for automatic modulation classification based on CNN. IEEE Commun. Lett. 25(4), 1206–1210 (2021)
13. Gong, J., Xu, X., Lei, Y.: Unsupervised specific emitter identification method using radio-frequency fingerprint embedded InfoGAN. IEEE Trans. Inf. Forensics Secur. 15, 2898–2913 (2020)
14. O'Shea, T.J., Corgan, J., Clancy, T.C.: Convolutional radio modulation recognition networks. In: Jayne, C., Iliadis, L. (eds.) EANN 2016. Communications in Computer and Information Science, vol. 629, pp. 213–226. Springer, Cham (2016). https://doi.org/10.1007/978-3-319-44188-7_16
15. De Boer, P.T., Kroese, D.P., Mannor, S., et al.: A tutorial on the cross-entropy method. Ann. Oper. Res. 134(1), 19–67 (2005)
16. Zhang, Z., Li, H., Chen, L.: Deep residual shrinkage networks with self-adaptive slope thresholding for fault diagnosis. In: 2021 7th International Conference on Condition Monitoring of Machinery in Non-Stationary Operations (CMMNO) (2021). In press
17. He, K., Zhang, X., Ren, S., Sun, J.: Deep residual learning for image recognition. In: Proceedings of IEEE Conference on Computer Vision and Pattern Recognition, Seattle, WA, USA, 27–30 June 2016, pp. 770–778 (2016)
18. He, K., Zhang, X., Ren, S., Sun, J.: Identity mappings in deep residual networks. In: Leibe, B., Matas, J., Sebe, N., Welling, M. (eds.) ECCV 2016. LNCS, vol. 9908, pp. 630–645. Springer, Cham (2016). https://doi.org/10.1007/978-3-319-46493-0_38

Joint Resource Allocation Based on F-OFDM for Integrated Communication and Positioning System

Ruoxu Chen[1] , Xiaofeng Lu[1]([✉]) , and Kun Yang[2]

[1] School of Communication Engineering, Xidian University, Xi'an 710071, China
luxf@xidian.edu.cn
[2] School of Information and Communication Engineering,
University of Electronic Science and Technology of China, Chengdu, China

Abstract. OFDM signal can provide communication and ranging services at the same time, but it is limited by sub-carriers and power resources, and cannot meet service requirements adaptively. In order to better improve the communication and positioning performance, a joint communication and ranging resource allocation strategy based on F-OFDM for integrated communication and positioning system is proposed. Firstly, F-OFDM is used to construct a communication and positioning integrated network under protocol interference. Then, link grouping is performed, and a joint sub-carrier and power optimization model is established based on independent sub-bands. The simulation results show that the resource allocation strategy proposed in this paper can significantly improve the ranging performance of the system with satisfying the communication performance.

Keywords: OFDM ranging · F-OFDM · Resource allocation · Data transmission rate · Equivalent Fisher information (EFI)

1 Introduction

Accurate localization of mobile devices is becoming increasingly important for many emerging scenarios, such as indoor navigation and IoT applications [1]. In recent years, with the development of communication technology, the idea of using communication facilities as positioning infrastructure to meet the dual needs of communication and positioning has attracted much attention [2,3]. In the current wireless communication system, the waveform technology is dominated by OFDM, so there is a lot of research on OFDM ranging technology [4,5]. In [6], a TOA (Time of Arrival) estimation method by using OFDM sub-carrier phase difference is proposed. Literature [7–9] analyzes the accuracy and influence factors when OFDM signal is used as a ranging signal, and shows that

This work was supported in part by the National Natural Science Foundation of China under Grant U1705263.

ranging accuracy is one of the fundamental reasons affecting positioning performance. Improving ranging accuracy is an effective means to enhance positioning performance. Therefore, on the existing communication infrastructure, we can not only use OFDM signals for high-speed and high-capacity communication, but also achieve high-precision ranging. Since communication and ranging using OFDM technology are realized by processing the received and sent signals, the resource allocation in the system will affect the propagation quality of the signal, which in turn will affect the communication and ranging performance. However at the beginning, the resource allocation in the existing communication system [10,11] only serves to improve the communication performance, and the literature [12,13] only considers to optimize the positioning accuracy of the system. There are very few studies combining the two at present.

In view of the above problems, this paper studies the impact of joint resource allocation about communication and positioning. At the same time, in order to overcome the shortcoming of OFDM that can only be configured with fixed frequency band parameters, the F-OFDM technology that can adapt to communication and positioning services is adopted [14]. Therefore, we build an integrated communication and positioning network based on F-OFDM, realize the joint design of communication and positioning depend on independent sub-bands, and propose a JCP-OSOP (optimal sub-carrier optimal power under joint communication and positioning) allocation strategy.

2   System Model

In this section, we introduce the wireless multi-hop communication and positioning integrated network system model. At the same time, we also make the analysis of the communication and ranging performance of the network.

2.1   Network Model

As shown in Fig. 1(a), we abstract it as the connected graph $G = (V, L)$ in Fig. 1(b), where $V = \{v_1, v_2, \cdots, v_n\}$ is the set of all vertices and $L = \{l_{12}, l_{13}, \cdots, l_{mn}\}$ is the set of all edges in the connected graph, representing the wireless access nodes and transmission links in the network of Fig. 1(a) respectively. In a multi-hop network, interference may occur between any two links due to the broadcast characteristics of the wireless channel. Interference determines whether multiple nodes in the network can work at the same time, and the interference is crucial to the study of data transmission in the network. Our system model adopts the protocol interference mentioned in [15], assuming that there is interference within two hops among the network node. Therefore, based on the connected graph, we use the edges in Fig. 1(b) as vertices, and connect the vertices corresponding to the two interference edges to obtain the conflict graph shown in Fig. 1(c).

When grouping the vertices of the graph, it can be realized by the vertex coloring algorithm of the graph. We can group the links of the conflict graph to get the link group set $LG = \{(l_{12}), (l_{23}, l_{45}), (l_{14}), (l_{15}), (l_{16})\}$, which means that only the link group (l_{23}, l_{45}) does not have interference and can reuse resources, and other links cannot share resources with each other.

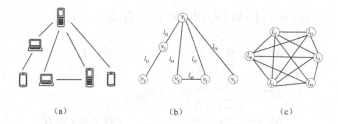

(a)   (b)   (c)

Fig. 1. Wireless multi-hop network model and interference conflict graph.

2.2 Communication and Ranging Performance Analysis

In communication and positioning integrated network, we use the data transmission rate on the link as the communication performance, at the same time, the number of sub-carriers is set to N, the channel bandwidth it set to B, and the data transmission rate is expressed as follows:

$$r_{ij} = \frac{B}{N}\log_2\left(1 + \frac{P_{ij}h_{ij}}{\varphi_0^2 + I_{ij}}\right) \tag{1}$$

where h_{ij} and P_{ij} are respectively the channel gain and allocated power of the $j - th$ sub-carrier on the $i - th$ link, I_{ij} is the sum of interference caused by the remaining links sharing $j - th$ sub-carrier on the $i - th$ link. It can be seen that the reasonable allocation of sub-carriers and power to each link can promote a significant increase in the SNR, and thus increase the data transmission rate on each link.

For the positioning performance, it will be directly affected by the ranging error. According to the literature [16], the CRB (Cramé-Rao Bound) for estimating the ranging delay is equivalent to estimate the inverse of the EFI (Equivalent Fisher Information). We use the following equation represents the ranging error on the link:

$$J_{ei}(N_i, \mathbf{P}_i) = \sum_{j=1}^{N_i} P_{ij}w_{ck}^2 - \left(\sum_{j=1}^{N_i} P_{ij}\mathbf{W}_j\mathbf{b}_j\right)^{\mathbf{T}} \left(\sum_{j=1}^{N_i} P_{ij}\mathbf{W}_j\mathbf{C}_j\mathbf{W}_j\right)^{-1} \left(\sum_{j=1}^{N_i} P_{ij}\mathbf{W}_j\mathbf{b}_j\right) \tag{2}$$

where N_i represents the set of sub-carriers allocated on the $i - th$ link, and $\mathbf{P}_i = [P_{i1}, P_{i2}, \cdots, P_{iN_i}]$ represents the power allocated for each sub-carrier on the $i - th$ link. It can be seen that the number of sub-carriers and power allocation will directly affect the CRB of the ranging accuracy.

Therefore, let the data transmission rate on each link as the primary optimization goal, and let the overall ranging accuracy of the system as the secondary optimization goal, the resource allocation model for joint communication and positioning services can be established as follows:

$$\arg \max_{\delta} \{\min \{J_{ei}(N_i, \mathbf{P}_i), i = l_1, l_2, \ldots, l_n\}\}$$

$$s.t.$$

$$
\begin{array}{lll}
\text{AC1}: & R_i \geq t_i & \forall i \in L \\
\text{AC2}: & N_i \cap N_j = \varnothing & j \notin l(i) \\
\text{AC3}: & \sum\limits_{i=l_1}^{l_n} \sum\limits_{j=1}^{N_i} P_{ij} \leq P_{total} & \\
\text{AC4}: & P_{ij} \geq 0 & \forall i \in l, j \in N_i
\end{array}
\tag{3}
$$

where the objective function adopts the maximization criterion of minimization, that is, take the EFI of the ranging delay on the link with the worst ranging performance in the system as the evaluation of ranging performance, δ is the best resource allocation strategy. AC1 indicates that the primary optimization goal is to meet the data transmission rate on the $i - th$ link. AC2 means that the links in different link groups cannot use the same resources, otherwise interference will occur. AC3 and AC4 limit the power.

3   JCP-OSOP Resource Allocation Strategy

Since the F-OFDM technology is used in the integrated communication and positioning network, the system frequency band can be divided into multiple sub-bands according to the link grouping situation, so that the sub-band resources can be flexibly configured for communication and positioning. Therefore, the optimal sub-carrier allocation strategy in (3) can be converted to the optimal communication sub-band and ranging sub-band division strategy, and the power allocation strategy on each sub-carrier can be converted to the optimal power allocation strategy on sub-band. Due to the high complexity of solving (3), we can decompose it into sub-carrier allocation model and power allocation model.

3.1   The Sub-carrier Allocation Model

Assuming that the power allocated on each sub-carrier is equal, the communication sub-band is divided to meet the data transmission rate of each link firstly. For one single link, the communication sub-band division model is as follows:

$$\arg \min N_{l(i)_c}$$

$$s.t.$$

$$
\begin{array}{lll}
\text{AC1}: & R_i \geq t_i & \forall i \in l(i) \\
\text{AC2}: & N_{i_c} \cap N_{j_c} = \varnothing & j \notin l(i) \\
\text{AC3}: & P_{ij} = \frac{P_{total}}{N} & \forall i \in l(i), j \in N_{l(i)}
\end{array}
\tag{4}
$$

where N_{i_c} represents the set of sub-carriers used for communication on the $i - th$ link.

Then, we establish the ranging sub-band division model as follows:

$$\arg \max_{\delta_N} \left\{ \min \left\{ J_{e_{l(i)}} (N_{l(i)_R}, \mathbf{P}), i = l_1, l_2, \ldots, l_n \right\} \right\}$$

s.t.

$$
\begin{aligned}
\text{AC1} &: N_{i_R} \cap N_{j_R} = \varnothing & j \notin l(i) \\
\text{AC2} &: N_{i_C} \cup N_{i_R} = N_i & i \in l(i) \\
\text{AC3} &: P_{ij} = \frac{P_{total}}{N} & \forall i \in l(i), j \in N_{l(i)}
\end{aligned}
\tag{5}
$$

where N_{i_R} represents the sub-carriers set used for ranging on the $i - th$ link. AC2 means that the sub-carriers used for communication and ranging on the link cannot exceed the allocated sub-carriers. The minimum sub-carrier set $N_{l(i)_R}$, which is allocated to the ranging sub-band $l(i)_R$ can be obtained by solving the above model.

Due to the sub-band division, the sub-carriers contained in each sub-band must be continuous. Therefore, we use the sub-carrier block composed of multiple consecutive sub-carriers as the minimum allocation unit. In order to reduce the complexity of the allocation, we propose an allocation algorithm based on sub-carrier splitting, which allocates the complete sub-carrier block grouping set to the link group. We assume that there are N sub-carriers in the system, and each sub-carrier block contains n sub-carriers, so the number of sub-carrier blocks is $N_{rb} = \frac{N}{n}$, and we use the set $Sub = \{sub_1, sub_2, \ldots, sub_{N_{rb}}\}$ to represent sub-carrier blocks. Then according to coloring the conflict graph, we get the link group set $LG = \{l(1), l(2), \ldots, l(m)\}$. The specific algorithm flow is as follows:

Algorithm 1. Complete Sub-carrier Block Grouping Set Generation Algorithm.

Step 1:

   Define $G = \{2^0, 2^1, \cdots, 2^{N_{rb}-2}\}$, add $m-1$ combination of the elements in G to get the set of decimal numbers $Q_D = \{(q_D)_1, (q_D)_2, \cdots, (q_D)_{C_{N_{rb}-1}^{m-1}}\}$;

Step 2:

   Convert each element in Q_D into the binary number and fill in other bits with 0 to make the length C to get the set of binary number numbers $Q_B = \{(q_B)_1, (q_B)_1, \cdots, (q_B)_{C_{N_r-1}^{m-1}}\}$;

Step 3:

   Let $i = 1$, insert the binary sequence of the $i-th$ element in Q_B into Sub in order, so that a grouping situation can be obtained, which can be stored in the grouping set T^N as the $i - th$ element;

Step 4:

   If $i \leq C_{N_r-1}^{m-1}$ is true, let $i = i + 1$ and jump to Step 2; Otherwise, the final grouping set T^N can be obtained, and the algorithm ends.

We can solve (4) directly to get the minimum sub-carrier block grouping set allocated to the communication sub-band $l(i)_C$ when the communication needs are met, and then get the specific sub-carrier set $N_{l(i)_C}$. Since (5) is a 0–1 integer programming problem, we use an iterative solution algorithm based on Hungary to solve it. The specific steps are as follows:

Algorithm 2. Optimal Ranging Sub-band Division Algorithm Based on Hungary Iterative Algorithm.

Step 1:
   Initialize: $j = 1$;
Step 2:
   Traverse each element in the link group set LG, and record the each link group in the number set G^L;
Step 3:
   Traverse each element in T_j^N, according to the grouping of sub-carriers, record the sub-carriers number of each sub-carrier block in set G_j^N;
Step 4:
   Traverse each element in G^L and G_j^N, and calculate the metric matrix **TP** according to equation (2);
Step 5:
   Take **TP** as the coefficient matrix, find the optimal sub-carrier grouping and optimal matching strategy by the Hungarian algorithm, then store it in **Je** ;
Step 6:
   If $j > |T^N|$, skip and execute Step 7; otherwise, $j = j + 1$, skip and execute Step 2;
Step 7:
   Find the global maximum value in **Je**, according to the column index to get the best sub-carrier grouping and the best sub-carrier block and link matching.

3.2   The Power Allocation Model

After the sub-carrier allocation, we first allocate power resources for the sub-carriers on the communication sub-band to meet the data transmission rate on the link. The power allocation optimization model on one communication sub-band is as follows:

$$\arg\min P_{l(i)_C} = \sum_{j=1}^{N_{l(i)_C}} P_{l(i)_C j}$$

$$s.t. \tag{6}$$

$$AC1: \quad R_i \geq t_i \qquad \forall i \in l(i)$$
$$AC2: \quad P_{l(i)_C j} \geq 0 \quad \forall i \in L, j \in N_{l(i)_C}$$

the minimum transmit power required on the communication sub-band can be obtained by solving it.

Then, taking the two-path channel as an example, (2) can be converted to:

$$J_{e_{l(i)}}(\mathbf{P}_{l(i)_R}) = \frac{2\alpha_1^2 T}{N_0}\left(-\frac{\omega_c^2 T}{2E_T}\mathbf{P}_{l(i)_R}{}^T\mathbf{H}^T\mathbf{H}\mathbf{P}_{l(i)_R} + \mathbf{w}^T\mathbf{P}_{l(i)_R}\right) \tag{7}$$

We establish the sub-carrier power allocation optimization model on the ranging sub-band as follows:

$$\arg \max_{\delta_P} \left\{ \min \left\{ J_{e_{l(i)}}(\mathbf{P}_{l(i)_R}), i = l_1, l_2, \ldots, l_n \right\} \right\}$$

s.t.

$$\text{AC1}: \quad \sum_{i=l_1}^{l_n} P_{l(i)_R} \leq P_{total} - \sum_{i=l_1}^{l_n} P_{l(i)_C}$$

$$\text{AC2}: \quad P_{l(i)_R j} \geq 0 \qquad \forall i \in L, j \in N_{l(i)_R}$$

(8)

where AC1 indicates that the power consumed on all ranging sub-bands and communication sub-bands cannot exceed the total power in the system. By solving the above model, we can get the power on the ranging sub-band, and then get the best power allocation strategy. (6) is a typical nonlinear constrained optimization model, which can be solved by Lagrangian multiplier method. (8) is a quadratic programming problem. Since $\mathbf{H^T H}$ is a positive semi-definite matrix, and the constraints are all linear, (8) as a quadratic programming problem can be converted to a standard convex optimization model and then solved using the interior point method.

4 Performance Evaluation

4.1 Simulation Setup

The simulation parameter settings are shown in Table 1. In order to study the optimality of the JCP-OSOP resource allocation strategy, the ASAP (Average Sub-carrier Allocation and Average Power Allocation) resource allocation strategy in the literature, ASOP (Average Sub-carrier Allocation and Optimized Power Allocation) resource allocation strategy, and the OSAP (Optimized Sub-carrier Allocation and Average Power Allocation) resource allocation strategy in [17] have been simulated and compared.

Table 1. Simulation Parameters

5G band number	n78	Number of sub-carrier blocks	64
Center frequency (MHz)	3450	Number of sub-carriers in sub-carrier block	24
Bandwidth (MHz)	100	Total transmit power W	10
Number of nodes	6	Transmission rate threshold (increasing)	$t_1 - t_8$
Sub-carrier spacing (KHz)	60	Protocol interference model	Two hops

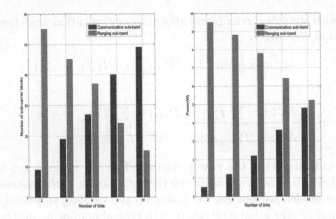

Fig. 2. Sub-carrier block and power allocation changes with network scale (SNR = 10 dB)

4.2   Results and Discussions

As can be seen in Fig. 2, when the resources in the system and the transmission rate threshold are constant, as the network scale becomes larger, the number of links in the system increases. In order to meet the data transmission rate on the links, the number of sub-carriers and power resources occupied by the communication sub-band will gradually increase, while the ranging sub-band will be the opposite.

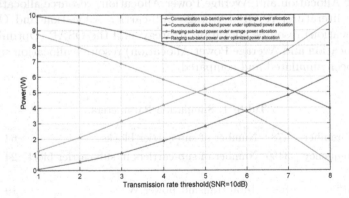

Fig. 3. Comparison results of average power distribution and optimal power distribution on the communication sub-band and the ranging sub-band

It can be seen from Fig. 3 that with the increase of the link transmission rate threshold in the system, since meeting the transmission rate of each link is our primary goal and the total power is constant, the power allocated to the communication sub-band continues to increase, the power of the ranging sub-band continues to decrease. At the same time, under the power optimization allocation strategy, the actual power required by the communication sub-band is much smaller than the pre-allocated power in the average power allocation strategy, and the actual power allocated to the ranging sub-band is much greater than the pre-allocated power in the average power allocation strategy, which will greatly improve the ranging performance of the system.

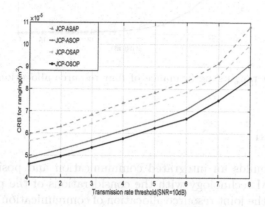

Fig. 4. The system ranging performance of four resource allocation strategies varies with the transmission rate

As shown in Fig. 4, due to the increase of the transmission rate threshold on each link, the resources occupied by the communication sub-bands increase, and the resources occupied by the ranging sub-band decrease, resulting in the ranging performance of the four resource allocation strategies deteriorates. Comparing JCP-ASAP and JCP-ASOP, JCP-OSAP and JCP-OSOP resource allocation strategies, we can see that the latter's ranging performance is greatly improved, which is consistent with the conclusion in Fig. 3. In a comprehensive comparison, the JCP-OSOP resource allocation strategy proposed in this paper has the best ranging performance, which can increase by 20% on average compared to the JCP-ASAP allocation strategy.

In addition, as the SNR in Fig. 5 increases, the resources required for the communication sub-bands are reduced when the data transmission rate on the link are met, and the resources used for the ranging sub-bands in the system increase, and noise has a smaller impact on ranging performance. Therefore, regardless of the resource allocation strategies, the ranging performance is gradually improved. Further, the JCP-OSOP resource allocation strategy proposed in this paper has the best system ranging performance.

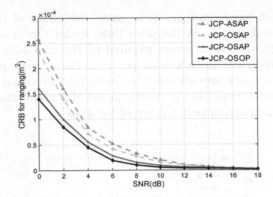

Fig. 5. The system ranging performance of four resource allocation strategies varies with SNR

5 Conclusion

This paper first builds an integrated communication and positioning network based on F-OFDM technology with the considerations of the protocol interference, and models the joint resource allocation of communication and positioning. Next, an iterative Hungary algorithm based on sub-carrier splitting is proposed to solve the sub-carrier allocation problem under the F-OFDM sub-band, then the power allocation problem is solved mathematically, the JCP-OSOP allocation strategy is obtained finally. Through simulation and analysis, the resource allocation strategy proposed in this paper can provide communication and ranging services at the same time, and has better ranging performance than JCP-ASAP, JCP-ASOP and JCP-OSAP resource allocation strategies.

References

1. Agiwal, M., Roy, A., Saxena, N.: Next generation 5G wireless networks: a comprehensive survey. IEEE Commun. Surv. Tutor. **18**(3), 1–1 (2016)
2. Liu, Z., Da, I.W., Win, M.Z.: Mercury: an infrastructure-free system for network localization and navigation. IEEE Trans. Mobile Comput. **PP**(99), 1 (2018)
3. Xiong, J., Sundaresan, K., Jamieson, K.: ToneTrack: leveraging frequency-agile radios for time-based indoor wireless localization. In: International Conference on Mobile Computing and Networking. ACM (2015)
4. He, Z., Ma, Y., Tafazolli, R.: Improved high resolution TOA estimation for OFDM-WLAN based indoor ranging. IEEE Wirel. Commun. Lett. **2**(2), 163–166 (2013)
5. Dai, W., Lindsey, W.C., Win, M.Z.: Accuracy of OFDM ranging systems in the presence of processing impairments. In: IEEE International Conference on Ubiquitous Wireless Broadband. IEEE (2017)
6. Dai, L., Wang, Z., Wang, J., et al.: Positioning with OFDM signals for the next-generation GNSS. IEEE Trans. Consum. Electron. **56**(2), 374–379 (2010)

7. Wymeersch, H., Maran, S., Gifford, W.M., Win, M.Z.: A machine learning approach to ranging error mitigation for UWB localization. IEEE Trans. Commun. **60**(6), 1719–1728 (2012)
8. Dardari, D., Conti, A., Ferner, U.J., Giorgetti, A., Win, M.Z.: Ranging with ultra-wide bandwidth signals in multipath environments. Proc. IEEE **97**(2), 404–426 (2009)
9. Tianheng, Wang, et al.: On OFDM ranging accurary in multipath channels. IEEE Syst. J. **8**(1), 104–114 (2014)
10. Zhang, T., Molisch, A.F., Shen, Y., et al.: Joint power and bandwidth allocation in wireless cooperative localization networks. IEEE Trans. Wirel. Commun. **15**(10), 6527–6540 (2016)
11. Qin, C., Song, L., Zhang, T., et al.: Joint power and spectrum optimization in wireless localization networks. In: Proceedings of the IEEE International Conference on Communication Workshop, pp. 859–864. IEEE (2015)
12. Yaacoub, E.: A survey on uplink resource allocation in OFDMA wireless networks. IEEE Commun. Surv. Tutor. **14**(2), 322–337 (2012)
13. Sumathi, K., Valarmathi, M.L.: Resource allocation in multiuser OFDM systems-a survey. In: Third International Conference on Computing Communication and Networking Technologies. IEEE (2012)
14. Abdoli, J., Jia, M., Ma, J.: Filtered OFDM: a new waveform for future wireless systems. In: 2015 IEEE 16th International Workshop on Signal Processing Advances in Wireless Communications (SPAWC). IEEE (2015)
15. Jacobs, I.M., Wozencraft, J.M.: Principles of communication engineering (1965)
16. Wang, T., et al.: On OFDM ranging accuracy in multipath channels. IEEE Syst. J. **8**(1), 104–114 (2014)
17. Lu, X., Ni, Q., Li, W., et al.: Dynamic user grouping and joint resource allocation with multi-cell cooperation for uplink virtual MIMO systems. IEEE Trans. Wirel. Commun. **PP**(99), 1 (2017)

UAV Formation Using a Dynamic Task Assignment Algorithm with Cooperative Combat

Ying Wang[1(✉)], Yonggang Li[1], Zhichao Zheng[1], Longjiang Li[2], and Xing Zhang[3]

[1] School of Communication and Information Engineering,
Chongqing University of Posts and Telecommunications,
Chongqing 400065, People's Republic of China
1062543772@qq.com
[2] School of Information and Communication Engineering,
University of Electronic Science and Technology of China,
Chengdu 611731, Sichuan, People's Republic of China
[3] State Grid Chongqing Electric Power Company Information
Communication Branch, Chongqing 401120, People's Republic of China

Abstract. As the research model of Unmanned Aerial Vehicle (UAV) formation, it is required to complete the combat tasks such as covering reconnaissance, processing battlefield information and striking targets. A cooperative combat model is proposed to assign combat tasks and plan flight paths for Unmanned Aerial Vehicles (UAVs) in scenarios that need to perform differentiated missions. In view of the UAV function's failure and fail to complete the assigned task, a dynamic task adjustment algorithm is proposed to distinguish the mission stages. Based on the task allocation scheme, the hierarchical adjustment strategy is developed, which take the adjustment time, communication cost and adjustment load into consideration respectively, so as to reduce the adjustment cost and improve the combat capability of the UAV formation on the basis of ensuring the completion of the combat task. Taking the combat formation of UAV as the simulation model, the results show that the proposed algorithm can reduce the adjustment cost and improve the combat capability of the formation on the premise that the formation can complete the combat task.

Keywords: Unmanned Aerial Vehicle (UAV) · Path planning · Cooperative task assignment · Dynamic task adjustment

Project supported by the National Defense Pre-Research Quick Support Foundation of China (no. 80911010302), the Key Project of State Grid Chongqing Electric Power Company under Grant 2021YuDianKeJi23.

1   Introduction

In modern war, UAV has been applied in various aspects, which has great research significance and development space. The research model of UAV formation is of great significance. The UAV formation does not carry out a single type of mission, but a variety of tasks at the same time. Therefore, in this paper, the UAV formation is taken as the research model. UAV formation of needs to have strong cooperative combat capability. The cooperative combat of UAV formation has an important influence on winning the war.

The formation combat mode of cooperative reconnaissance, rapid decision-making and precise strike can improve the success rate of combat in the process of combat, and the assignment of combat tasks and the planning of the flight track of each UAV are key issues in the study of formation combat. According to specific combat tasks, task assignment is carried out for the formation [1–5]. Literature [6] proposed a task allocation strategy based on the Hungarian algorithm. In recent years, scholars at home and abroad have carried out a large number of studies on UAV cluster path planning, including the spatial representation of UAV track, planning objectives and constraints [7–10]. At present, the research on combat formation is relatively simple and does not involve the task allocation that needs to perform multiple tasks. To sum up, this paper puts forward a cooperative combat model, which realizes the cooperation and coordination among UAVs by using task allocation and flight path planning methods.

The rest of this paper is arranged as follows. Section 2 establishes the cooperative combat model of UAV formation. Section 3 introduces dynamic task adjustment. Section 4 provides the simulation results and analysis. Finally, Sect. 5 concludes the paper.

2   Cooperative Combat Modeling

2.1   UAV Formation Scene Research

In the process of combat, according to the needs of the task, the UAV has a variety of functional modules, which form a comprehensive formation with a variety of capabilities, and carry out combat operations to the target in the combat area, in order to meet the strategic needs. By referring to the idea of the circular process of observation, positioning, decision and action (OODA) in modern combat theory, each function of UAV is modeled into corresponding function nodes, that is, reconnaissance node S, decision node D and strike node I. The reconnaissance node S can detect and collect battlefield information in the target area. Decision node D can process the reconnaissance information and command the attack node. Strike node I accepts the instruction of the decision node to strike the target.

In the course of combat, the combat mission is described as follows: in a bounded area A, m heterogeneous UAVs are performing tasks, which require full coverage reconnaissance of A, and they can quickly process reconnaissance information and carry out fire strike tasks.

2.2 Cooperative Task Assignment

Decision-Making Stage. First of all, the division of combat area should be solved in the cooperative task assignment of UAV formation. The decision UAV is responsible for information processing of corresponding region, which is the basis of cooperative operation. The problem of region division can be transformed into the problem of responsible regions of decision nodes. The number of sub-regions can be divided according to the number of decision nodes, i.e. $N_a = N_D$. During region division, D nodes are assigned to each sub-region to obtain N_a subtask sets, and each subtask set contains a D node, that is, $\{T\} = \{T_1, T_2, ..., T_{N_a}\} = \{\{D_1\}, \{D_2\}, ..., \{D_{N_D}\}\}$.

The Reconnaissance Stage. For the reconnaissance stage, it is the purpose of task allocation to minimize the time required to fully cover the reconnaissance combat area. Parallel search is a commonly used cover reconnaissance strategy. The distance between the two parallel lines is defined as the width of the region [12]. In order to ensure the minimum number of turns in the covering reconnaissance process, the reconnaissance node moves along the vertical direction of the minimum width of the region.

For region i, the mission area is divided into $n_{sc,i}$ scan lines along the reconnaissance direction, and S moves along the scan lines. The number of reconnaissance nodes in region i, $n_{Ts,i}$, is determined according to the size of the minimum width L_{min}^i.

$$n_{sc,i} = L_{min}^i / 2R_s \tag{1}$$

$$n_{Ts,i} = N_s \times L_{min}^i / \sum_{i=1}^{N_a} L_{min}^i \tag{2}$$

$$N_S = \sum_{i=1}^{N_a} n_{Ts,i} \tag{3}$$

Considering the multi-functional properties of UAVs, when assigning S nodes to each task set, S nodes belonging to the same UAV as D are first assigned to the sub-task set where the UAVs belong to, and then the remaining S nodes are assigned to each sub-task set according to $n_T s$. When you update the task set, you get the task set with the node number: $\{T\} = \{\{D_1, S_1, ..., S_{n_{Ts,1}}\}, \{D_2, S_2, ..., S_{n_{Ts,2}}\}, ..., \{D_{N_D}, S_1, ..., S_{n_{Ts,N_D}}\}\}$.

The Striking Stage. Attacking node I_i^T in each sub-region executes the attack mission in its region. For targets that appear randomly in space, according to the area of the sub-region, the task assignment in the attacking stage assign nodes to each subtask set. The number of nodes in region i is:

$$n_{Ti,i} = N_I \times S_{ar,i} / S_{ar} \tag{4}$$

where, N_I is the total number of strike nodes in the formation, S_{ar} is the total area of the combat region, and $S_{ar,i}$ is the area of sub-region i.

Node I belonging to the same UAV as D or S is also assigned to the corresponding task set when node D or node S is assigned, while the remaining I

nodes are assigned to each task set according to Eq. (4). The task assignment results of the three types of functional nodes are as follows:

$$\{T\} = \{\{D_1, S_1, ..., S_{n_{Ts},1}, I_1, ..., I_{n_{TI},1}\}, ..., \{D_{N_D}, S_1, ..., S_{n_{Ts},N_D}, I_1, ..., I_{n_{TI},N_D}\}\} \tag{5}$$

2.3 Path Planning Methods

The Track of Node S. According to Eq. (5), the track of UAV in each task set is planned. Firstly, the track of the reconnaissance UAV is planned, that is, the scan line of the mission area responsible for each S node is determined. The goal of track planning of node S is to minimize the time t_{cover} required to complete a single coverage of reconnaissance area A. The communication probability between the two closest UAVs U_n and U_m in the adjacent area should always be greater than the communication probability threshold P_{tv}. The path planning problem of reconnaissance UAV can be described as follows:

$$mint_{\text{cover}} = \min(\max(t_{AS,1}, t_{AS,2}, ..., t_{AS,N_a})),$$
$$t_{AS,i} = \max(t_{S,1}, t_{S,2}, ..., t_{S,n_{Ts,i}}), i \in \{1, ..., N_a\}$$
$$t_{S,j} = \sum_{k=1}^{n_{sc,ij}} l_{ijk}/v + (n_{sc,ij} - 1) \times t_{sw}, j \in \{1, ..., n_{Ts,i}\}$$
$$s.t. \begin{cases} n_{sc,i} = \sum_{j=1}^{n_{Ts,i}} n_{sc,ij} \\ P_{U_n,U_m,t} \geq P_{tv} \end{cases} \tag{6}$$

where, t_{cover} is equal to the maximum of the time taken for each sub-region to complete coverage reconnaissance; $t_{AS,i}$ represents the time required for sub-region i to complete a coverage reconnaissance, which is equal to the maximum $t_{s,j}$ of the time taken by each UAV in this region to complete the mission. $t_{s,j}$ includes reconnaissance mission area scan line time and turn time; $n_{Ts,i}$ is the number of reconnaissance nodes in region i; $n_{sc,ij}$ is the number of scan lines in the area in which UAV j is responsible in region i; l_{ij} is the length of scan lines in the task area k of UAV j; v is the speed of UAV movement; t_{SW} is the time required for UAV to turn.

The Track of Node D. In combat, node D moves along with node S in the region in order to process reconnaissance information. D node belongs to the same UAV as S node, and its flight path is the same as the corresponding S node. The other D nodes are determined by the track of S node in the region.

The communication topology between S nodes can be represented by a weighted directed graph, and the communication distance is represented as the weight of the edge. The communication between UAVs is related to the distance. In region i, the information interaction topology in the directed graph at time t can be represented as a directed minimum spanning tree $T_{Si,t}$ with minimum sum of edge weights $w_{Ts,i}$.

$$w_{Ts,i} = \min \sum_{(u,v) \in T_{Si,t}} w(u,v) \tag{7}$$

According to the minimum spanning tree $T_{Si,t}$, the two S nodes with the farthest distance among all nodes can be obtained. The central position of the connecting line of these two nodes is the position of node D at time t.

According to the track planning of S node, the position of S node at every moment can be obtained. In order to facilitate the study, the position of S node is extracted by taking Δt as the time interval to obtain the position of D node at the same time. Connect all the time positions of D node in time sequence, and finally get the track of D node.

The Track of Node I. The flight path planning of the attack UAV is carried out, and the flight path is the same as that of the corresponding S or D node, which belongs to the I node of an UAV. The rest of the I nodes move around the D nodes. In order to ensure that D and I can communicate with each other and quickly attack the target, the location of node I should be between node D and each node S, and the distance to D should be $d_{S-D}/2$. d_{S-D} is the distance between S node and D node. When the number of I nodes is less than S node, the position of I node is determined from far to near.

3   Dynamic Task Adjustment

3.1   Failure Analysis of Various Nodes

Reconnaissance Node S has Failed. In the dynamic task adjustment strategy, the S node satisfying the optimal adjustment is adjusted from the original task set $\{T_{S,N_P}\}$ to the task set $\{T_{S,P}\}$ with S failure. The ratio of the minimum width of the region to the number of reconnaissance nodes in the region is defined as reconnaissance load C_S, that is

$$C_S = L_{\min}/n_{Ts} \tag{8}$$

Let $C_{S,p}$ and C_{S,N_P} represent the reconnaissance load of $T_{S,P}$ and T_{S,N_P} after adjustment respectively, and the reconnaissance load of $T_{S,P}$ must be greater than or equal to the reconnaissance load of T_{S,N_P}, i.e. $C_{S,p} \geq C_{S,N_P}$. The S node satisfying the optimal adjustment should have the minimum task adjustment time t_p, that is to say, t_p is equal to the minimum task adjustment time required by each adjustment scheme, as shown in Eq. (9).

$$\begin{cases} t_P = \min\{t_{P,N_i}\}, \\ t_{P,N_i} = \max(t_{S,p_i} + \alpha t_{M_i}, t_{S,N_i}). \\ s.t. C_{S,p} \geq C_{S,N_P} \end{cases} \tag{9}$$

where, t_{P,N_i} represents the adjustment time required by the i adjustment scheme, t_{S,p_i} and t_{S,N_i} respectively represent the maximum time taken by $\{T_{S,P}\}$ and $\{T_{S,N_I}\}$ to complete the coverage task after adjustment. t_{M_i} represents the time taken for S node to move from $\{T_{S,N_I}\}$ to $\{T_{S,P}\}$; α is the weight coefficient, representing the influence degree of adjustment time in the combat mission.

When the optimal adjustment results are obtained, the task set is updated. $\{T_{S,P}\}$ and $\{T_{S,N_P}\}$. Track planning is carried out for S nodes of $\{T_{S,N_P}\}$ and $\{T_{S,P}\}$ according to the path planning method.

Strike Node I has Failed. After the failure of node I, the task adjustment strategy is the same as the S node failure strategy, which adjusts the I node satisfying the optimal adjustment from the original task set $\{T_{I,N_P}\}$ to the task set $\{T_{I,P}\}$ when the node fails. Strike load C_I is defined as the ratio of the area to the number of nodes I, that is:

$$C_I = S_{ar}/n_{TI} \tag{10}$$

The optimally adjusted I node should meet the following conditions:

$$\begin{cases} C_{Adj} = \min\{C_{I,P_i}\}, \\ C_{I,P_i} = C_{I,p} + \beta t_{p_i}. \\ s.t. C_{I,p} > C_{I,N_p} \end{cases} \tag{11}$$

The result of the optimal adjustment is to choose the scheme with the minimum C_{I,P_i} of the adjustment load. Where, C_{I,P_i} represents the adjustment load of adjustment scheme i, and t_{p_i} represents the time taken for node I to move from $\{T_{I,N_i}\}$ to $\{T_{I,P}\}$. β is the weight coefficient, which represents the influence of adjustment time on the task. After adjustment, the strike load $C_{I,P}$ of the I node failure task set $\{T_{I,P}\}$ should be larger than the strike load C_{I,N_P} of the original task set $\{T_{I,N_P}\}$.

When the optimal adjustment results are obtained, the task sets $\{T_{I,P}\}$ and $\{T_{I,N_P}\}$ are updated. Then, the path planning is carried out for the I node of $\{T_{I,N_P}\}$ and $\{T_{I,P}\}$ according to the path planning method in this paper.

Decision Node D has Failed. In this paper, for the established network model, a method of network robustness analysis based on function chain is proposed [12]. The network robustness is evaluated by the comprehensive operational capability of all function chains. After the failure of node D, the ability to deal with reconnaissance information is reduced, and the robustness of the network is also reduced. In the dynamic task adjustment, other task sets $\{T_{D,N_i}\}$ are selected for S node and I node in $\{T_{D,P}\}$ with the aim of improving robustness, so that these nodes can maintain communication with the nodes in $\{T_{D,N_i}\}$. The communication cost of S node or I node to $\{T_{D,N_i}\}$ is calculated, and the $\{T_{D,N_i}\}$ with the lowest communication cost w_{k,T_i} is selected as the new task set of S or I. Communication cost w_{k,T_i} is measured by distance cost d_{k,T_i} and error cost f_{k,T_i} of the link. Distance cost of link d_{k,T_i} refers to the distance between U_{Ti} and U_k, which is closest to UAV U_k in task set $\{T_{D,N_i}\}$. Link error cost f_{k,T_i} refers to the length of the communication link between U_k and U_{Ti}, and UAV with decision-making function in $\{T_{D,N_i}\}$ after the establishment of communication.

Normalize d_{k,T_i} and f_{k,T_i}:

$$D_{k,T_i} = \frac{d_{k,T_i} - d_{T_{\min}}}{d_{T_{\max}} - d_{T_{\min}}} \tag{12}$$

$$F_{k,T} = \frac{f_{k,T_i}}{f_{T_{\max}}} \tag{13}$$

where, $d_{T_{\min}}$ and $d_{T_{\max}}$ respectively represent the minimum and maximum distance between UAVs in $\{T_{D,N_i}\}$ after U_k is added to $\{T_{D,N_i}\}$, and $f_{T_{\max}}$ represents the maximum length of communication link in $\{T_{D,N_i}\}$.

The communication cost w_{k,T_i} from UAV U_k to task set $\{T_{D,N_i}\}$ is denoted as:

$$w_{k,T_i} = \gamma D_{k,T_i} + \mu F_{k,T} \tag{14}$$

wherein, γ and μ are weight coefficients, satisfying $\gamma + \mu = 1$.

Traverse the S node in $\{T_{D,P}\}$ to get the adjustment result of each S node; Traverse the I node in $\{T_{D,P}\}$ to get the adjustment result of each I node. Update the task set.

3.2 Dynamic Task Adjustment Algorithm

The corresponding adjustment strategies are designed for the failure of three functional nodes. First, determine whether there is a failure of node S. If so, determine whether there is a failure of node I after adjustment. Finally, determine whether there is a failure of node D. The dynamic task adjustment algorithm is given as follows (Table 1):

Table 1. Dynamic task adjustment algorithm

Algorithm: Dynamic task adjustment algorithm

---

1. Judge whether there is S node failure. If yes, go to 2. If no, go to 5;

2. Calculate $C_{S,i}$ to determine whether it needs to be adjusted from other task sets. If yes, go to 4. If no, go to 3;

3. Re-plan the track of S node in $\{T_{S,P}\}$ and move to 5;

4. The subtask set $\{T_{S,N_i}\}$ satisfying $C_{S,P} \geq C_{S,N_i}$ was obtained, and t_{P,N_i} was calculated. Select the smallest S node of t_{P,N_i} and adjust it to $\{T_{S,P}\}$. Plan the track of S node in $\{T_{S,P}\}$ and $\{T_{S,N_P}\}$;

5. Judge whether there is an I node failure. If yes, go to 6. If no, go to 9;

6. Calculate $C_{I,i}$ to determine whether it needs to be adjusted from other task sets. If so, go to 8; if not, go to 7;

7. Re-plan the path of node I in $\{T_{I,P}\}$ and go to 9;

8. Where $\{T_{I,N_i}\}$ satisfies $C_{I,P} > C_{I,N_P}$, calculate C_{I,P_i}. Select the smallest I node in C_{I,P_i} and adjust it into $\{T_{I,P}\}$. The path of node I in planning $\{T_{I,P}\}$ and $\{T_{I,N_P}\}$;

9. Determine whether there is a D node failure. If so, go to 10; if not, go to 12;

10. Extract UAV coordinates with Δt as an interval, calculate w_{k,T_i}, and select the one withthe least communication cost $\{T_{D,N_i}\}$;

11. Repeat Step 10 to traverse node$S$ and node I in $\{T_{D,P}\}$. Get the adjustment result of each node and go to 12;

12. End.

---

4   Simulation Results and Analysis

4.1   Simulation Scenario and Task Parameter Settings

Before task assignment, the functions of each UAV are determined and the function nodes are numbered. The corresponding relationship between UAV and function nodes is shown in Table 2. The 20 columns represent 20 UAVs, each with one or more functions. The UAV function is modeled into three types of nodes, and the three rows respectively represent the three functional nodes of reconnaissance, decision and attacking. Nodes are numbered, and 0 means that the UAV does not have this function. There are 40 nodes in total, and the number of S, D and I is 15, 6 and 19, respectively.

Table 2. Setting of UAV formation simulation parameters

Function	UAV number																			
	1	2	3	4	5	6	7	8	9	10	11	12	13	14	15	16	17	18	19	20
S	1	2	0	3	4	5	0	0	6	7	8	9	10	11	0	12	0	13	14	15
D	0	16	0	0	0	0	17	18	0	0	19	0	0	0	20	0	21	0	0	0
I	22	23	24	25	26	27	28	0	29	30	31	32	33	34	35	36	37	38	39	40

The combat area A studied in this paper is A rectangular area of $40\,km \times 30\,km$, whose vertices are arranged counterclockwise as $\{v1, v2, v3, v4\}$, and the corresponding two-dimensional coordinates are $(0,0)$,$(40 \times 10^3, 0)$,$(40 \times 10^3, 30 \times 10^3)$,$(0, 30 \times 10^3)$. According to the number of D nodes, A is partitioned, and 6 sub-regions are obtained. Set the reconnaissance radius of UAV as R_s=1000 m, and calculate the minimum width of each sub-region as 13.3 km, 13.3 km, 13.4 km, 13.3 km, 13.3 km and 13.4 km respectively. According to Eq. (1) and Eq. (2), S nodes are assigned to each region. The number of I nodes in each sub-region is determined by calculating the area of each sub-region. The results of initialization task assignment are shown in Table 3.

Table 3. Result of task assignment

Set of tasks	S	D	I	UAV number
1	2,3,4	16	23,25,26	2,4,5
2	1,5	17	22,27,28	1,6,7
3	6,7,9	18	29,30,32	8,9,10,12
4	8,10	19	24,31,33	3,11,13
5	11,12	20	34,35,36	14,15,16
6	13,14,15	21	37,38,39,40	17,18,19,20

Track planning is carried out for node S. The 6 sub-areas divided by a black dotted line in Fig. 1(a) are arranged counterclockwise from the lower left corner to correspond to each task set. The reconnaissance area for each S node in each mission set is separated by a gray dotted line. In general, the cruising speed of small and medium-sized UAVs is between 16.7–33.3 m/s. Therefore, it is assumed that the speed of UAVs is set as $v = 20$ m/s, and the time taken for a single full coverage of S node is 3470 s.

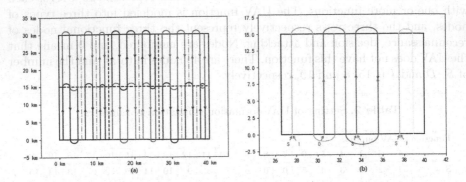

Fig. 1. (a) S node track planning diagram, (b) Node track planning diagram in subtask set $\{T_3\}$ (Color figure online)

Figure 1(b) depicts the path planning of all nodes in subtask set $\{T_3\}$. There are four UAVs in this sub-task set. The three S nodes and the three I nodes belong to the same UAV respectively, and the other UAV only has the decision-making function. Green, red and yellow represent the tracks of S node, D node and I node, respectively.

4.2 Dynamic Task Adjustment Simulation

Node I usually belongs to the same UAV as node S or node D. Therefore, in the simulation study, this paper only considers the failure of S nodes and D nodes.

Table 4. Dynamic task adjustment results

Set of task	1	2	3	4	5	6	Adjustment scheme
S node number	2,3,4	1,5	6,7,9	8,10	11,12	13,14	
	2,3	1,5	6,7,9	8,10	11,12	4,13	$S4 : \{T_1\} \rightarrow \{T_6\}$
	2,3	5,6	7,9	8,10	11,12	4,13	$S6 : \{T_3\} \rightarrow \{T_2\}$
	2,3	5,6	7,9	8,10	11	4,13	
	2,3	5,6	7,9	8	10	4,13	$S10 : \{T_4\} \rightarrow \{T_5\}$

S Node Failure. In the single full coverage of region A by node S, randomly attack five UAVs with node S, whose numbers are 20, 19, 1, 16, 14, and the corresponding nodes of node S are 15, 14, 1, 12, 11. For the convenience of the research, the time when all the UAVs start from the initial position at the same time is 0, then the corresponding failure time of the five UAVs is 234 s, 724 s, 1937 s, 2336 s and 3028 s respectively. If $\alpha = 0.2$, the adjustment result is shown in Table 4.

Figure 2(a) describes the real-time coverage of UAV formation to the combat area in a single coverage reconnaissance. The vertical axis is the covered area. Only the area covered for the first time is considered, and the area of repeated scanning is not calculated. In the random adjustment scheme, other S nodes in the same task set will complete the reconnaissance task of the subtask set after S node fails.

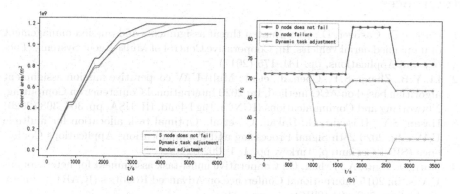

Fig. 2. (a) Real-time area coverage, (b) Real-time changes in network robustness

According to the simulation results, the time to complete a single full coverage scan in the three cases is 3470 s, 4362 s and 5421 s respectively. In the case of S node isn't failure, the time of full coverage scanning is the shortest. After the failure of S node, the formation is adjusted through the dynamic task adjustment scheme, which is compared with the random adjustment, so as to reduce the time that taken to complete the single full coverage.

D Node Failure. Randomly attack two UAVs with D nodes. The UAVs are numbered 2 and 15, and the corresponding D nodes are numbered 16 and 20. The failure times were 927 s and 1780 s. Let $\gamma = 0.3$ and $\mu = 0.7$. Take $\Delta t = 150$ s as the time interval to extract the position of each UAV at each moment, and the network robustness corresponding to each moment is shown in Fig. 2(b). After the failure of node D, the robustness of the damaged network is improved by 13.7103 on average through dynamic task adjustment, and the adjustment effect is obvious.

5   Conclusion

According to the operational requirements, a cooperative combat model is designed, including cooperative task assignment and path planning. The main work of collaborative task assignment is to divide the sub-task set, track planning is to plan the track of each UAV according to the sub-task set. In order to reduce the impact of function failure on the combat capability, a dynamic task adjustment algorithm is designed for the function failure in the battle process. After the UAV formation function failure, the mission adjustment can not only ensure the completion of the mission but also improve the combat capability. The simulation results demonstrate the effectiveness of the modeling for UAV formation combat.

References

1. Garcia, E., Casbeer, D.: Coordinated threat assignments and mission management of unmanned aerial vehicles. In: Cooperative Control of Multi-Agent Systems: Theory and Applications, pp. 141–175 (2017)
2. Li, Y.B., Zhang, S.T., Chen, J., et al.: Multi-UAV cooperative mission assignment algorithm based on ACO method. In: 2020 International Conference on Computing, Networking and Communications (ICNC), Big Island, HI, USA, pp. 304–308 (2020)
3. Hasan, S.Y., Hüseyin, G., Hakan, Ç., et al.: Optimal task allocation for multiple UAVs. In: 2020 28th Signal Processing and Communications Applications Conference (SIU), Gaziantep, Turkey, pp. 1–4 (2020)
4. Nuri, O., Ugur, A., Erhan, O.: Cooperative multi-task assignment for heterogonous UAVs. In: 2015 International Conference on Advanced Robotics (ICAR), Istanbul, Turkey, pp. 599–604 (2015)
5. Fatemeh, A., Mohammad, Z., Abolfazl, R., et al.: A coalition formation approach to coordinated task allocation in heterogeneous UAV networks. In: 2018 Annual American Control Conference (ACC), Milwaukee, WI, USA, pp. 5968–5975 (2018)
6. Liu, Z.J., He, M., Ma, Z.Y., et al.: UAV formation control method based on distributed consistency. Comput. Eng. Appl. **56**(23), 146–152 (2020)
7. Yang, X., Wang, R., Zhang, T.: Review of unmanned aerial vehicle swarm path planning based on intelligent optimization. Control Theor. Appl. **37**(11), 2291–2302 (2020)
8. Wang, Z., Liu, L., Long, T., et al.: Trajectory planning for multi-UAVs sequential convex programming. Acta Aeronaut. et Astronaut. Sin. **37**(10), 3149–3158 (2016)
9. Zhang, X., Duan, L.J.: Fast deployment of UAV networks for optimal wireless coverage. IEEE Trans. Mobile Comput. **18**(3), 588–601 (2018)
10. Lang, R., Wang, J.L., Chen, J., et al.: Energy-efficient multi-UAV coverage deployment in UAV networks: a game-theoretic framework. China Commun. **15**(10), 194–209 (2018)
11. Zhang, X., Moore, C., Newman, M.E.J.: Random graph models for dynamic networks. Eur. Phys. J. B. **90**(10), 1–14 (2017)
12. Li, J., Jiang, J., Yang, K., et al.: Research on functional robustness of heterogeneous combat networks. IEEE Syst. J. **13**, 1–9 (2018)

Complex System Optimization in Edge Computing

Research and Implementation of Multi-Agent UAV System Simulation Platform Based on JADE

Zhichao Zheng[✉], Yonggang Li, and Ying Wang

School of Communication and Information Engineering,
Chonqing University of Posts and Telecommunications,
Chongqing 400065, People's Republic of China
996179491@qq.com

Abstract. The unmanned aerial vehicle (UAV) system under the information condition is a complex adaptive system. The position, speed, efficiency value and other attributes of each unit in the UAV system are changing at high speed all the time. The dynamic equation of a single UAV's motion trajectory and operation is non-linear. The traditional modeling method is difficult to accurately reproduce the complex behavior of the system, so it is very important to put forward a modeling method suitable for the UAV system to reflect the contingency and emergence of the complex system. Therefore, a visualized multi-agent UAV system simulation platform is designed and implemented, regards UAV as an independent agent and the Java agent development framework (JADE) was used to simplify the development of multi-agent. Considering that the agent also needs a motion and obstacle avoidance model, a set of rules system is defined for the agent and an obstacle avoidance algorithm based on fuzzy control is implemented. Judging from the test results, the autonomous obstacle avoidance function can already be realized when the formation is merged, and the trajectory can be planned automatically when scanning the area. At the same time, this paper also provides a reference model for modeling and simulation for task reliability assessment.

Keywords: UAV · Agent · Avoidance algorithm · Task reliability

1 Introduction

The concept of agent is derived from distributed artificial intelligence. People call the computing entities that reside in an environment with the characteristics of autonomy, flexibility, sociality, and responsiveness as agents [1]. With

This work is supported by the National Defense Pre-Research Quick Support Foundation of China (no. 80911010302).

H. Gao et al. (Eds.): ChinaCom 2021, LNICST 433, pp. 115–128, 2022.
https://doi.org/10.1007/978-3-030-99200-2_10

the development of science and technology, agent has attracted much attention and is widely used in various fields [2]. The structure of agent is mainly divided into deliberate agent, reactive agent, Belief-Desire-Intention model (BDI) model and hybrid agent, deliberate agent is also called thinking agent [3]. The reactive agent [4] has no representation and reasoning behavior of the environment, but only works through stimulus response behavior. In the 1990s, the BDI model [5] was proposed by Rao and Georgeff. It was derived from the three words Belief, Desire, and Intention. There are two agent-based modeling theories in existing research: Distributed Artificial Intelligent (DAI) theory and Complex Adaptive System (CAS) theory [6]. The goal of DAI research is to establish a huge complex system composed of multiple subsystems, and each subsystem cooperates with each other to solve specific problems. CAS is a system in which adaptive subjects interact, evolve together, and emerge layer by layer, mainly reflected in the interaction and interactivity between agent and agent. This paper adopts the form of CAS, which can highlight the autonomy, adaptability and intelligence of agent. JADE is a software framework [7] that implements an efficient agent platform and supports the development of multi-agent systems. Literature [8] explains the principle of JADE, introduces the basic method of designing a performance test platform based on JADE, and provides a technical approach to realize the performance test platform. Literature [9] discusses mobile agents and their application in a heterogeneous multi-agent system in the form of a group of collaborative drones and UAV autonomous robots. Literature [10] proposes a multi-agent-based extended travel support system model and the method proposed in literature [11] involves the use of a multi-agent system, which solves the problem of using JADE to calculate the optimal multi-objective control. These literatures focus on the control of agent. However, visualization is also important for multi-agent systems

In this paper, a visualized multi-agent UAV system simulation platform is designed and implemented. In the form of CAS, the program model of agent is constructed with the help of object-oriented programming method to realize the multi-agent system. In order to realize the formation flight and area scanning functions of the agent, the rule system of the agent in the equipment system environment must be determined first, and the movement model and obstacle avoidance model must be realized. Secondly, according to the model, UAV formation and obstacle avoidance functions are realized, and a more efficient area scanning function is also realized. Finally, the GUI is used to display the task reliability curve.

2   Rule System

Rule system is an important part of UAV agent. According to different UAV agents, different rule system is established. Agent performs probability matching according to the weights in the rule base. Firstly, appropriate rules are selected and executed according to task objectives, internal states and environmental parameters. Finally, the rule system is evaluated and modified according to the execution results

The rules in the rule base form corresponding effector instructions according to sensor perception results and internal state monitoring. This paper mainly implements the agent motion model and agent obstacle avoidance model in UAVs. The structure of the agent's rule system is shown in Fig. 1.

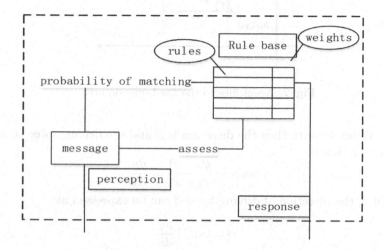

Fig. 1. Agent's rule system

2.1 Agent Mobility Models

Agent moves in order to accomplish tasks, such as agent requires to reach a specified destination, and the follower dynamically adjusts its position according to the position of the leader. Therefore, it is important to discuss the movement of the agent, and to understand how the agent dynamically adjusts its position to reach the desired position. As shown in Fig. 2, an UAV (agent) receives the command from the superior at the point of the global coordinate system $A(x1, y1)$ and moves quickly to the target point $P(x2, y2)$ at the initial speed $\overrightarrow{v}_{\alpha 1}$, which is the direction of movement and has a driving force $\overrightarrow{F}$ to point to the target point.

Then, the kinematics model of UAV moving in the specified direction is:

$$v_x = \overrightarrow{v}_{\alpha 1}\cos\alpha \tag{1}$$

$$v_y = \overrightarrow{v}_{\alpha 1}\sin\alpha \tag{2}$$

where, v_x is the velocity component of the initial velocity $\overrightarrow{v}_{\alpha 1}$ along the x axis, and v_y is the velocity component of the initial velocity $\overrightarrow{v}_{\alpha 1}$ along the y axis. When a UAV receives an order from a superior and needs to reach a certain destination, in order to reach the combat area with the fastest speed, it needs a force pointing to the target point. If the fixed target point and the coordinates

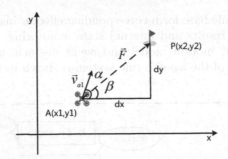

Fig. 2. Agent flies to the fixed mission area

of the UAV are known, then the direction is β and the driving force is, and its dynamic equation is:

$$\tan \beta = \frac{y2 - y1}{x2 - x1} = \frac{dy}{dx} \tag{3}$$

In Eq. 3, the direction of driving force β can be expressed as:

$$\beta = \tan^{-1} \frac{dy}{dx} \tag{4}$$

Given that the weight of the UAV is m kg, the acceleration of the UAV under the action of the driving force $\vec{F}$ can be expressed as:

$$\vec{a} = \frac{\vec{F}}{m} \tag{5}$$

The acceleration along the x and y axes can be expressed as:

$$\vec{a}_x = \vec{a} \cos \beta \tag{6}$$

$$\vec{a}_y = \vec{a} \sin \beta \tag{7}$$

Then the ideal position of UAV at the next moment can be expressed as:

$$x = x1 + \vec{v}_x \Delta t + \frac{1}{2} \vec{a}_x \Delta t^2 \tag{8}$$

In Eq. 8, Δt is the sampling period. The ideal position (x, y) that the agent needs to move in the next step can be obtained through the above formula. By constantly updating the position and initial speed, the UAV can reach the destination at the fastest speed and complete the task.

As shown in Fig. 3, the followers fly to the leaders at different time, and the leaders constantly communicates with the followers to exchange data. The leader sends real-time location information to the followers, and the followers parse the location information sent by the leaders into a fixed target movement, then the followers update the location and speed in real-time through internal calculations. Finally, the task of moving the followers to the moving target is completed.

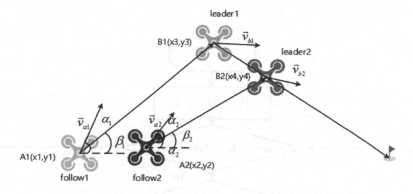

Fig. 3. The follower flies to the leader at different times

2.2 Agent Obstacle Avoidance Algorithm

Agent obstacle avoidance models are divided into two types, one is between agent and static obstacles, the other is between agent and moving objects. The essence of the collision avoidance algorithm lies in the agent can safely reach the designated destination and complete the task in the simulation process without overlapping with the non-self-dangerous area. The current UAV obstacle avoidance algorithm [14] has high time complexity and mostly target static obstacles. This paper proposes an improved speed collision avoidance algorithm based on literature [15,16], which is shown in Fig. 4.

In order to ensure that the agent can complete the task safely, based on the agent's mobile model, the speed of the agent is continuously modified through the collision avoidance algorithm so that the extension direction of the speed does not intersect with the velocity danger zone (VDZ), that means to select suitable speed at intervals for the agent which is outside the VDZ caused by the obstacle. If the direction of the agent's speed is outside the VDZ, the speed is directly used to point to the agent's target position, and the agent can safely move to the target point, that is, the agent chooses a speed that will not collide in the future to move. But the speed obstacle avoidance algorithm, each time it traverses all agents within the communication distance and obtains the best speed will make the algorithm more complicated. In order to satisfy that each agent does not collide with static obstacles or dynamic obstacles in the process of moving, and complete tasks in its own safe area, a speed collision avoidance algorithm based on fuzzy control is proposed. As shown in Fig. 5, in a two-dimensional plane, there is a static obstacle A, whose reference point position is $P_a(x1, y1)$, and whose radius is R_a. In addition, there is an Agent B whose reference point position is $P_b(x2, y2)$, radius is R_b and initial velocity is $\vec{v}_b$.

Assuming that agent slowly approaches the static obstacle so that agent B has exactly an intersection point with obstacle A, which can be expressed as:

$$\sqrt{(x1 - x2)^2 + (y1 - y2)^2} = R_a + R_b \tag{9}$$

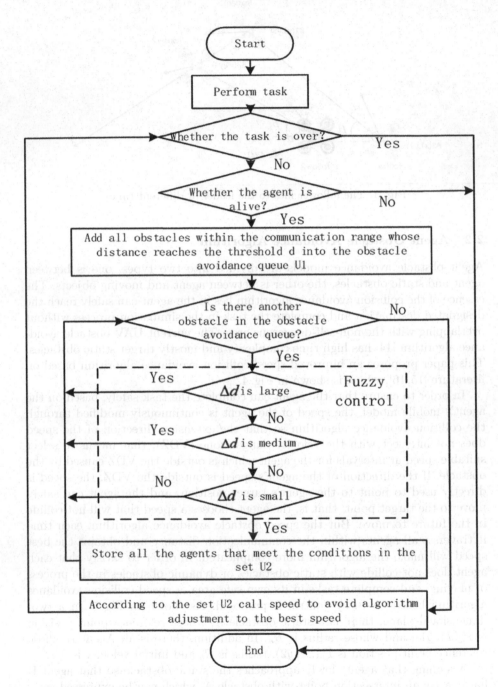

Fig. 4. Speed collision avoidance algorithm based on fuzzy control

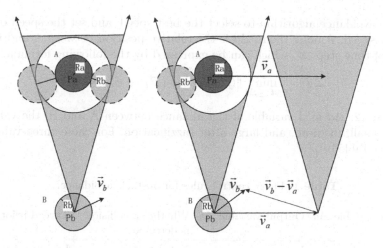

Fig. 5. Agent collides with a 'static' or 'dynamic' object

Set up the equations of agent B and obstacle A:

$$\begin{cases} (x-x1)^2 + (y-y1)^2 = R_a^2 \\ (x-x2)^2 + (y-y2)^2 = R_b^2 \end{cases} \tag{10}$$

By finding the number n of solutions of the equation system, the following conclusions can be obtained.

$$n = \begin{cases} 0, \Delta < 0 \\ 1, \Delta = 0 \\ 2, \Delta > 0 \end{cases} \tag{11}$$

In Eq. 11, when $\Delta = 0$, the agent and the obstacle just have an intersection point, and there is no collision. When $\Delta > 0$, the agent has collided with the obstacle. When $\Delta < 0$, the agent does not collide with the obstacle.

According to Eq. 11, to make the agent avoid obstacles successfully, it only needs to satisfy $n \leq 1$ at all times, that is, $\Delta \leq 0$. As shown in Fig. 5, taking the center P_b of agent B as the starting point to make a tangent to obstacle A, and two rays will be generated. When the direction of the agent's speed is between the two rays, then at some point in the future, agent B will collide with obstacle A. This possible collision speed is expressed as VDZ, that is, at some point in the future, the agent will collide with the obstacle. The number of solutions will be greater than 1. At this time, agent B needs to continuously detect obstacles within the communication range, and adjust the speed autonomously and coordinately, so that agent B can successfully avoid obstacles. When the value of Δd is large, no action is performed and it just continues to detect. When the value of Δd is medium, some simple actions should be performed, such as ignition of a small thrust engine or a warning message to other agents. When the value of Δd is small, add A to B's obstacle avoidance set U2, use the traditional speed obstacle avoidance algorithm to select the best speed, and set the speed of B as the best speed. Among them, Δd is the relative position between A and B in a period of time step Δt, which can be expressed by the following formula:

$$\Delta d = \min_{0 \leq t_s \leq \Delta t} \| (pb - pa) + (\vec{v}_b - \vec{v}_a) \times t_s \| \tag{12}$$

In Eq. 12, Δd as a variable of the distance between A and B, the values of Δd are small, medium, and large after fuzzification. For these three values, as shown in Table 1.

Table 1. Fuzzy control rules for obstacle avoidance.

Input variable Δd	Output variable Δd	Whether a collision is detected	Thread priority
Δd(large)	Long	Not detected	Low priority
Δd(medium)	Medium	Not detected	Medium priority
Δd(small)	Short	Detected	High priority

In the multi-agent UAV system simulation platform, agents and agents couple with each other, form formations, and have a superior-subordinate relationship. In addition to avoiding collisions between agents and the environment, obstacle avoidance functions are also required between agents and agents. As shown in Fig. 5, it contains two moving entities, which can be regarded as two agents or a situation of an agent and a dynamic obstacle. Since both A and B are moving entities, obstacle A is selected as the reference object, and the relative speed $\vec{V}$ of agent B relative to obstacle A is obtained.

$$\vec{V} = \vec{V}_b - \vec{V}_a \tag{13}$$

The obstacle avoidance set U2 of the current agent is updated in real time, and then the traditional obstacle avoidance algorithm is called to obtain the best obstacle avoidance speed, and finally the agent moves to the ideal position. As shown in Fig. 6, there will be a VDZ whether it is an agent, a static obstacle or a dynamic obstacle. If the agent wants to avoid collisions and reach the specified position, it needs to avoid the speed danger zone, choose the speed direction outside the speed danger zone, and continuously update the speed in each sampling period to make it reach the target safely point.

3   Simulation Results

3.1   UAVs Formation Simulation Results

As shown in Fig. 7, when the agent has a multi-component cluster, each formation will select a leader through negotiation. Under the leadership of the leader, when the agents reach the mission area, if the leaders can communicate with

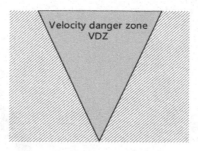

Fig. 6. Collision zone and safety zone

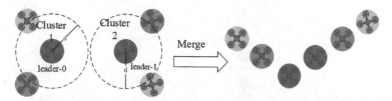

Fig. 7. Two groups of UAVs formation

each other, they will reselect the leader to form a new formation. In the end, the task of all agents is to reach the task area and merge into a cluster.

UAVs can fly in formation through clustering and fusion. As shown in Fig. 8, a total of 10 UVAs are created, and the distance between UAVs is 6 m. The initial formation of the UAVs formation is Fig. 8(a) and the angle in the formation is set to 30°. Figure 8(b) is formation 1, the formation angle is 180° and the distance between UAVs is 6 m. Figure 8(c) is formation 2, the formation angle is 45° and the distance between UAVs is 90 m. Figure 8(d) is formation 3, the formation angle is 120° and the distance between UAVs is 6 m. The test results show that the UAVs formation adjustment function is normal and can meet expectations.

3.2   Collision Avoidance Test Simulation Results

Through the agent obstacle avoidance model, the UAV can avoid obstacles with static and dynamic objects. As shown in Fig. 9, by creating 10 UAVs and setting the flight mode of the drones to formation flying, it is used to test whether the collision avoidance model between UAVs is correct. As shown in Fig. 9(a), when UAV follower-2 goes to its desired position in the formation, there is fol-lower-8 in the trajectory, Fig. 9(b) shows the position where the follower-2 successfully bypassed the follower-8, and the two did not collide. Follower-8 avoids the trajectory of follower-2 by adjusting its position. The test results show that the collision avoidance algorithm proposed in this paper is suitable for this system and can meet expectations.

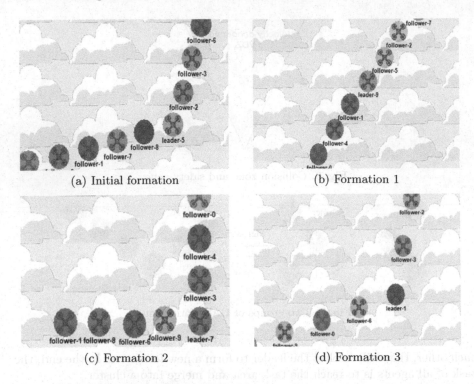

(a) Initial formation   (b) Formation 1

(c) Formation 2   (d) Formation 3

Fig. 8. UAVs formation flying

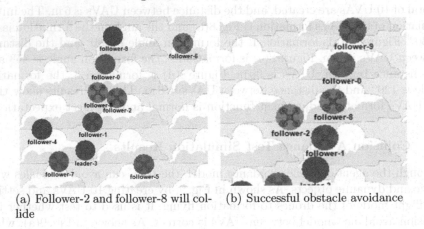

(a) Follower-2 and follower-8 will col-   (b) Successful obstacle avoidance
lide

Fig. 9. Collision avoidance test

3.3   Dynamic Task Reliability Assessment

Simulation Scenario Hypothesis and Task Reliability Definition. The
scene is set as $32 \times 32 \, \text{km}^2$, and the maximum number of UAVs is 32. Each UAV
will perform the straight line cruise scanning task. In the simulation scenario,

the number of aircraft is taken as the constraint condition, and the constraint interval is 1 to 32 aircraft. Under this constraint, the maximum flying speed of the UAV is 15 m/s, and it can cover the scanning area with a diameter of 2 km. The same number of UAVs are simulated to complete the assigned tasks, and the less time-consuming the UAVs are, the higher their reliability will be.

In the large-area detection task, the task completion degree is defined as the task reliability with the goal of completing the scanning task, which can be expressed as:

$$R = \frac{S_h}{S_{all}} \tag{14}$$

In Eq. 14, S_h is the scanned area, and S_{all} is the total area of the scanned area. In this paper, R > 82% indicates that the scan task is completed, but due to the existence of equipment failure, the task may not be completed during the execution of the algorithm.

Comparison Between the Traditional Recovery Mode and Intelligent Recovery Mode. In the traditional recovery mode, the UAV scans a given area. In the initial creation stage, the UAV receives the area scanning task from the superior, and after parsing the task information, it starts the area scanning task immediately after reaching the designated starting point of the task. As shown in Fig. 10(a), with the simulation progresses, a fault is manually injected to randomly strike a UAV which is forced to go offline. Because one of the UAVs has broken down, the scanning task assigned to it cannot be completed in time. The other 31 UAVs continue to perform straight-line cruise scanning tasks, as shown in Fig. 10(b).

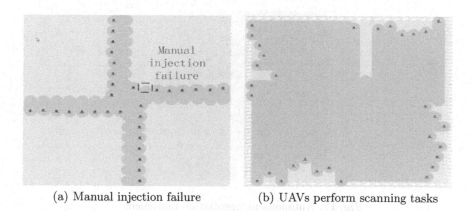

(a) Manual injection failure   (b) UAVs perform scanning tasks

Fig. 10. Traditional recovery mode

After 31 UAVs have completed their straight-line cruise tasks, the superior checks whether the scanning task is completed. If there is an area which is not

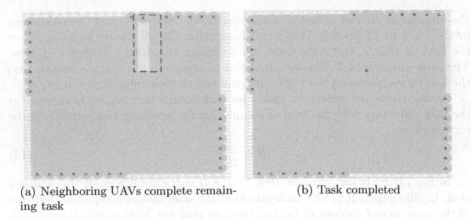

(a) Neighboring UAVs complete remain-   (b) Task completed
ing task

Fig. 11. Traditional recovery mode

accessible to scan, the superior will re-issue the task and let the neighboring
UAV continue to complete the task, as shown in Fig. 11(a).

In the intelligent recovery mode, after fault injection, because the intelligent
recovery UAV is highly autonomous, it can sense its surroundings in real time,
monitor neighbor UAV, communicate with other UAV, collect various information, and finally make a decision, and decide what to do next on the cruise task.
As shown in Fig. 12, the current UAV detects the downline of the neighbor UAV,
and the UAV make decisions through consultation, fill the inner cruise path, and
move the uncompleted task area to the edge.

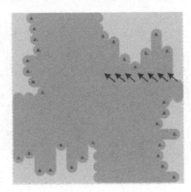

Fig. 12. Autonomous negotiation movement

Finally, the task reliability curves of the traditional recovery mode and the
intelligent recovery mode can be obtained. In Fig. 13, the x-axis represents the
system time, which is calculated from the creation stage, and the y-axis represents the task reliability R. In the intelligent recovery mode, when the system

time reaches about 1600 s, the UAVs have completed the task of regional scanning.

It can be seen that after 1300 s, in the intelligent recovery mode, the task reliability is higher than the traditional recovery mode, and the task can be completed faster, so the intelligent recovery algorithm is better than the traditional algorithm.

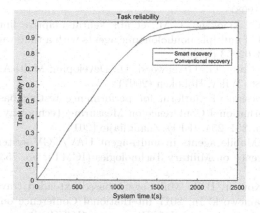

Fig. 13. Task reliability

4   Conclusion

As a complex system, the multi-agent UAV system has many uncertainties. This paper first established motion model and obstacle avoidance model of the UAV, and simulated the functions of UAV movement, obstacle avoidance and area scanning. Then, the user can view the real-time task status through the interactive interface. Finally, the test verification shows that each functional module of the visual multi-agent UAV system simulation platform based on JADE can run nor-mally, meet various performance indicators, and realize the reliability evaluation of dynamic tasks. It can be seen from the test results that the multi-agent UAV system simulation platform can realize functions such as autonomous obstacle avoidance, formation and formation integration, and automatic negotiation between UAVs. Meanwhile, it also provides a reference model for modeling and simulation for task reliability evaluation.

References

1. Jennings, N.: On agent-based software engineering. Artif. Intell. **117**(2), 277–296 (2000)

2. Liu, J.: Operation command decision modeling and simulation research based on agent technique. In: 2019 International Conference on Information Technology and Computer Application (ITCA), pp. 203–206. IEEE, Guangzhou (2019)
3. Jennings, N.R.: On agent-based software engineering. Artif. Intell. **2**(5), 99–110 (2016)
4. Rauff, J.V.: Multi-Agent Systems: An Introduction to Distributed Artificial Intelligence. Addison-Wesley, Boston (1999)
5. Rao, A., Georgeff, M.P.: Decision procedures for BDI logics. J. Logic Comput. **8**(3), 293–343 (1998)
6. Zhou Kaibo, W., Xiaokang, G.M., et al.: Neural-adaptive finite-time formation tracking control of multiple nonholonomic agents with a time-varying target. IEEE Access **8**, 62943–62953 (2020)
7. Bellifemine, F., Caire, G., Greenwood, D.: Developing Multi-Agent Systems with JADE, pp. 1–286. Wiley, Hoboken (2007)
8. Baiquan, X.: Design of platform for performance testing based on JADE. In: 2014 Sixth International Conference on Measuring Technology and Mechatronics Automation, pp. 251–254. IEEE, Zhangjiajie (2014)
9. Obdržálek, Z.: Mobile agents in multi-agent UAV/UGV system. In: 2017 International Conference on Military Technologies (ICMT), pp. 753–759. IEEE, Brno (2017)
10. Sreeja, M.U., Kovoor, B.C.: Multi agent based extended travel support system using JADE framework. In: 2017 International Conference on Energy, Communication, Data Analytics and Soft Computing (ICECDS), pp. 3561–3564. IEEE, Brno (2017)
11. Estrada, D.F., Lee, K.Y.: Multi-agent system implementation in JADE environment for power plant control. In: 2013 IEEE Power and Energy Society General Meeting, pp. 1–5. IEEE, Brno (2017)
12. Abeywickrama, H.V., Jayawickrama, B.A., He, Y., Dutkiewicz, E.: Algorithm for energy efficient inter-UAV collision avoidance. In: 17th International Symposium on Communications and Information Technologies (ISCIT), pp. 1–5. IEEE, Cairns (2017)
13. Yoo, C., Cho, A., Park, B., Kang, Y., Shim, S., Lee, I.: Collision avoidance of Smart UAV in multiple intruders. In: 12th International Conference on Control, Automation and Systems, pp. 443–447. IEEE, Jeju (2012)
14. Ling, L., Niu, Y., Zhu, H.: Lyapunov method-based collision avoidance for UAVs. In: The 27th Chinese Control and Decision Conference (2015 CCDC), pp. 4716–4720. IEEE, Qingdao (2015)
15. Van den Berg, J., Lin, M., Manocha, D.: Reciprocal velocity obstacles for real-time multi-agent navigation. In: 2008 IEEE International Conference on Robotics and Automation, pp. 19–23. IEEE, Pasadena (2008)
16. Rezaee, H., Abdollahi, F., Menhaj, M.B.: Model-free fuzzy leader-follower formation control of fixed wing UAVs. In: 13th Iranian Conference on Fuzzy Systems (IFSC), pp. 1–5. IEEE, Qazvin (2013)

A Complex Neural Network Adaptive Beamforming for Multi-channel Speech Enhancement in Time Domain

Tao Jiang[1]([✉]), Hongqing Liu[1], Yi Zhou[1], and Lu Gan[2]

[1] School of Communication and Information Engineering,
Chongqing University of Posts and Telecommunications, Chongqing, China
s190101065@stu.cqupt.edu.cn
[2] College of Engineering, Design and Physical Science, Brunel University,
London UB8 3PH, U.K.

Abstract. This paper presents a novel end-to-end multi-channel speech enhancement using complex time-domain operations. To that end, in time-domain, Hilbert transform is utilized to construct a complex time-domain analytic signal as the training inputs of the neural network. The proposed network system is composed of complex adaptive complex neural network beamforming and complex fully convolutional network (CNAB-CFCN). The real and imaginary parts (RI) of the clean speech analytic signal are used as training targets of the CNAB-CFCN network, and the weights of the CNAB-CFCN network are updated by calculating the scale invariant signal-to-distortion ratio (SI-SDR) loss function of the enhanced RI and clean RI. It is fundamentally different from the complex frequency domain single channel approach. The experimental results show that the proposed method demonstrates a significant improvement in end-to-end multi-channel speech enhancement scenarios.

Index Terms: End-to-end · Multi-channel · Speech enhancement · Complex operations

1 Introduction

The purpose of the speech enhancement algorithms is to suppress the background noise and to improve the quality and intelligibility of speech [1]. Recent studies show that deep learning based single-channel speech enhancement methods have achieved a great success, for example, the convolutional recurrent network (CRN) in [2] and dual-signal transformation LSTM network (DTLN) in [3] demonstrate promising results. The further studies indicate that these methods can also be applied to multi-channel speech enhancement. Due to the availability of multiple microphones, multi-channel signals contain spatial information, which can improve the system performance over single-channel speech enhancement, if utilized properly.

© ICST Institute for Computer Sciences, Social Informatics and Telecommunications Engineering 2022
Published by Springer Nature Switzerland AG 2022. All Rights Reserved
H. Gao et al. (Eds.): ChinaCom 2021, LNICST 433, pp. 129–139, 2022.
https://doi.org/10.1007/978-3-030-99200-2_11

In supervised learning, the techniques of estimating the time-frequency mask have become popular in both multi-channel and single-channel scenarios. In [2], intra-channel and inter-channel features are used as the input of the model to estimate phase sensitive mask (PSM) [4,5] and then applied to the reference channel of dual-channel speech. However, this method ignores the phase information and directly uses the phase from the noisy signal to reconstruct the enhanced speech, which may result in phase distortion of the enhanced speech. Several methods that utilize phase information have been proposed [5,6], but they still operate in the real number domain. This limits the upper limit of speech enhancement when the phase information estimation is not accurate enough.

To overcome the lack of properly utilizing phase information, deep complex U-net [7] that combines the advantages of deep complex network and U-net [8] is developed to process spectrogram with complex values to further improve the performance of speech enhancement. In [9], a complex number network is designed to simulate complex value operations, termed as deep complex convolution recurrent network (DCCRN), where both CNN and RNN structures handle the complex-valued operations. In DCCRN, the real and imaginary (RI) parts of the complex STFT spectrogram of the mixture are used as the input to the network, and both the amplitude and phase of the spectrogram can be reconstructed by the estimated RI. However, this is a single channel based approach and it is in frequency domain. In 2020, Wang [10] proposed a complex spectral mapping combined with minimum variance distortion-less response (MVDR) beamforming [11] for multi-channel speech enhancement approach. The enhanced signal spectrum is predicted in the neural network, and by calculating the covariance matrix of the signal and noise, the beamforming filter coefficients are produced. Without computing the covariance matrix, neural network adaptive beamforming (NAB) directly learns beamforming filters from noisy data, which avoids estimating the direction of arrival (DOA) [12], and the results demonstrate that NAB outperforms the traditional beamforming methods such as MVDR.

In this work, we propose an end-to-end multi-channel speech enhancement using complex operations that implicitly explore the phase information. To that aim, we first develop a complex neural network adaptive beamforming (CNAB) to predict complex time domain beamforming filter coefficients. It is worth noting that the coefficient will be updated according to the changes of the noisy dataset during the training process, which is different from the fixed filter in [13,14]. After that, the obtained complex beamforming filter coefficients by the CNAB are convolved with the input of each channel. The resulting signal is now single channel and to further process the signal, we develop a second time-domain complex network, called complex full convolutional network (CFCN), to predict the complex time-domain information of the enhanced speech. The proposed network is called CNAB-CFCN and results show that the proposed network demonstrates a superior performance over the current networks.

A. System flowchart

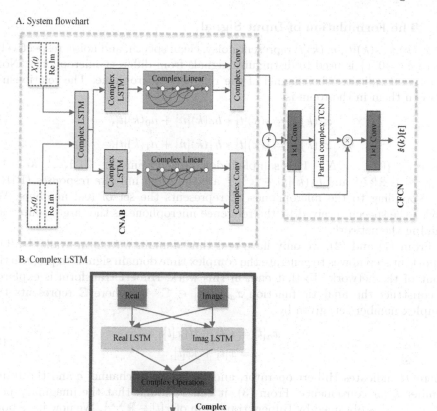

Fig. 1. (A) System flowchart. The input of each channel contains RI components. The entire system is composed of CNAB and CFCN. (B) Complex LSTM. The operation process of RI feature in complex LSTM.

2   Complex Neural Network Adaptive Beamforming

The proposed end-to-end time-domain multi-channel speech enhancement using complex value operation network framework is depicted in Fig. 1. It is of interest to point out that the flowchart provides a dual-channel description, but the extension to multi-channel is straightforward. It consists of complex neural network adaptive beamforming and a complex fully convolutional network (CNAB-CFCN). The input is complex time domain waveform, and the CNAB-CFCN model is updated by calculating scale invariant signal-to-distortion ratio (SI-SDR) loss [15].

2.1   The Formulation of Input Signal

Let $x_c(k)[t]$, $s(k)[t]$, $n_c(k)[t]$ represent noisy, clean speech, and noise, respectively, where $c \in \{0,1\}$ is used to distinguish signals from different microphones. Note that $c = 0$ indicates the channel of the reference microphone. The relationship between them in the room is

$$x_0(k)[t] = s(k)[t] * h_0(k)[n] + n_0(k)[t], \tag{1}$$

$$x_1(k)[t] = s(k)[t] * h_1(k)[n] + n_1(k)[t], \tag{2}$$

where $t \in \{0,1,\ldots,N-1\}$ is sample index in each frame $k \in \{0,1,\ldots,M-1\}$, $h_0(k)[n] \in \mathbb{R}^{K \times 1}$ and $h_1(k)[n] \in \mathbb{R}^{K \times 1}$ are the room impulse responses (RIRs) corresponding to the microphones, $\mathbb{R}$ represents the set of real numbers. We choose the speech received by the reference microphone as the target source for training the network.

From (1) and (2), we only have real time domain waveform available. The important step now is to generate the complex time domain signals to prepare the input of the network. To that end, in this work, Hilbert transform is explored to construct the analytic function $x_{ac}(k)[t] \in \mathbb{C}^{N \times 1}$, where $\mathbb{C}$ represents the complex number set, given by

$$\begin{aligned} x_a[t] &= x[t] + \mathcal{H}(x[t]) \\ &= x[t] + j\hat{x}[t], \end{aligned} \tag{3}$$

where $\mathcal{H}$ indicates Hilbert operator, and we omit the channel c and the frame number k for convenience. From (3), it can be found that the imaginary part $\hat{x}[t] \in \mathbb{R}^{N \times 1}$ is obtained by Hilbert transform of $x[t] \in \mathbb{R}^{N \times 1}$. We now have both the real and imaginary time domain waveforms for the proposed network.

2.2   Complex Adaptive Spatial Filtering

The dual-channel noisy speech signal is subjected to Hilbert transform to obtain the sequence of real and imaginary parts as the input of CNAB model architecture. The purpose of CNAB is to estimate the beamforming filters, which also includes real and imaginary parts. The complex convolution of the beamforming filter coefficients and the input RI is

$$\begin{aligned} y_a[t] &= (conv_r(Re(x_a[t])) - conv_i(Im(x_a[t]))) \\ &\quad + j(conv_r(Im(x_a[t]) + conv_i(Re(x_a[t])), \end{aligned} \tag{4}$$

where $conv$ denotes the convolution operation, and the subscripts r and i are the real and imaginary parts of the CNAB, respectively, $y_a[t]$ is the output of the complex convolution in one channel, and $Re(\cdot)$ and $Im(\cdot)$ respectively takes the real and imaginary part of a complex signal.

Summing the results of different channels yields the final output

$$N_{out} = y_{a0}[t] + y_{a1}[t], \tag{5}$$

where $y_{a0}[t]$ and $y_{a1}[t]$ respectively represent the beamforming results of two channels, and N_{out} is the final output of CNAB.

2.3   CNAB-CFCN Architecture

In our CNAB-CFCN framework, we all use the rules of complex number operations. In Fig. 1, for the CNAB part, complex LSTM is utilized to estimate the filter coefficients, where the rule of complex LSTM is provided in Fig. 1(B). The first layer of the CNAB is a complex number LSTM, termed as complex shared-LSTM, which takes complex time domain waveforms generated by Hilbert transform as input. The next layer has two separated complex LSTMs, which process two corresponding futures of each channel, called complex splitted-LSTM. Finally, the beamforming filters are produced by complex linear activations, and the enhanced speech features are estimated through complex convolutions. The specific operations of complex LSTM is

$$L_r = LSTM_r(Re(x_a[t])) - LSTM_i(Im(x_a[t])), \tag{6}$$

$$L_i = LSTM_r(Re(x_a[t])) + LSTM_i(Re(x_a[t])), \tag{7}$$

$$L_{out} = L_r + jL_i, \tag{8}$$

where $LSTM_r$ and $LSTM_i$ are two ordinary LSTM networks, representing the real part and imaginary part of complex LSTM, respectively, $L_r \in \mathbb{R}^{N \times 1}$ and $L_i \in \mathbb{R}^{N \times 1}$ are feature mappings of real part and imaginary part. The feature mapping output $L_{out} \in \mathbb{C}^{N \times 1}$ by a complex LSTM is still a complex feature.

The output of CNAB now is a single channel time domain waveform and to further improve the system performance, we develop another complex network, called complex fully convolutional network (CFCN). In Fig. 1, for the CFCN part, 1×1 *conv* first separately processes the real and imaginary parts of N_{out}, and then the output features are stacked together. After that, in partial complex TCN, we repeat X 1-D convolution blocks with the dilated convolution factor $d = \{1, 2, 4, \ldots, 2^{X-1}\}$ for R times, and the size of kernel is P. Note that only the last 1-D convolution blocks is a complex network. The operation rules are as follows.

$$\begin{aligned} C_{out} &= (conv_r(M_r) - conv_i(M_i)) \\ &\quad + j(conv_r(M_i) + conv_i(M_r)), \end{aligned} \tag{9}$$

where $conv_r$ represents the feature mapping function corresponding to the real convolution layer, and $conv_i$ represents the feature mapping function corresponding to the imaginary convolution layer, M is the feature map output by the upper layer of the network, and $C_{out} \in \mathbb{C}^{T \times 1}$ is the output of the CFCN, which is also the final output of the whole network.

2.4   Loss Function

We train CNAB-CFCN to estimate the real and imaginary parts of clean speech from noisy speech, and the weighted complex SI-SDR as a loss function is utilized to train our model, given by

$$\text{SI-SDR} = (1 - \lambda) \times 10 \log_{10} \frac{\|\beta_r \times Re(s_a[t])\|^2}{\|\beta_r \times Re(s_a[t]) - \hat{R}_t\|^2}$$

$$+ \lambda \times 10 \log_{10} \frac{\|\beta_i \times Im(s_a[t])\|^2}{\|\beta_i \times Im(s_a[t]) - \hat{I}_t\|^2}, \tag{10}$$

$$\beta_r = \frac{\hat{R}_t^T Re(s_a(t))}{\|s_a[t]\|^2} = \arg\min_{\beta_r} \|\beta_r \times Re(s_a(t)) - \hat{R}_t\|^2, \tag{11}$$

$$\beta_i = \frac{\hat{I}_t^T Im(s_a(t))}{\|s_a[t]\|^2} = \arg\min_{\beta_i} \|\beta_i \times Im(s_a(t)) - \hat{I}_t\|^2, \tag{12}$$

where $\hat{R}_t \in \mathbb{R}^{N \times 1}$ and $\hat{I}_t \in \mathbb{R}^{N \times 1}$ indicate the real and imaginary parts of each frame estimated by the CNAB-CFCN model, $s_a[t]$ is the analytical signal of the clean speech from the reference channel, λ is a weighting constant in the range of $[0, 1]$. When $\lambda = 0$, the network only uses the real part information to update the network parameters, whereas when $\lambda = 1$ means that the network uses the imaginary part information to update the parameters. The experiments indicate that $\lambda = 0.5$ is a good empirical hyperparameter in this work.

Table 1. CNAB-CFCN model parameter configuration, where B is the batch size and L is the length of the feature mapping.

Layer name	Input size	Hyperparameters
Complex input	(B, 2, 16000) × 2	–
Complex sh-LSTM	(B, 2, 100, 160)	(160, 512)
Complex sp-LSTM × 2	(B, 2, 512)	(512, 256)
Complex linear	(B, 2, 256)	(256, 25)
N_{out}	(B, 2, 16000)	–
1 × 1 Conv	(B, 16000) × 2	(1, 256), 40, 20
Layernorm	(B, 2, 256, L)	BatchNorm2d
Complex 1 × 1 Conv	(B, 2, 256, L)	(1, 1), (5, 2), (2, 1)
1-D Conv × 23	(B, 256, L)	...
Complex 1-D Conv	(B, 128, L)	...
1 × 1 Conv	(B, 256, L)	(256, 512), 1, 1

3   Experimental Setup

To generate dual-channel speech, image method [16] was used. The azimuth of the target source is on the same horizontal line as the two microphones and close to the reference microphone. We define this angle as 0°. The noise direction angle is uniformly distributed between 0 and 90° at a 15° interval. The space between the two microphones is 3 cm and the clean speech and noise are 1 m away from the microphone. The simulated reverberation room size is 10 × 7 × 3 m. The dual-channel clean and dual-channel noise generated by the above configuration are randomly mixed from −5 to 10 dB, and the signal-to-noise ratio (SNR) interval is 1 dB in the training set and the verification set. Note that noise and clean are also randomly mixed at different directions. The SNR of speech-noise mixtures include {−5, 0, 5, 10, 20} dB in the test set.

3.1 Production of Dual-Channel Dataset

In this section, we use the dataset provided by deep noise suppression (DNS) challenge [17] to train and evaluate our speech enhancement model, where all audio clips are 16 kHz. We use the script provided by the DNS to generate 75 h audio clips, and the size of each clip is 6 s long. In total, we generate 40400 clips for training set and 4600 clips for validation set. In addition, we generate 3500 clips for testing, and the speech and noise did not appear in the training set.

Table 2. Number of complex 1-D convolution blocks.

No.	0	3	6	9	12
PESQ	3.257	3.291	2.501	2.497	2.506

3.2 Experimental Setting

In this study, we divide 16 kHz speech signal with a duration of 6 s into 1 s segments, and each segment contains 16000 sampling points. The Hilbert transform is performed on the input speech segments to create complex time-domain signals. Table 1 summarizes the parameter configurations of the model used, where the input feature is dual-channel, and sh-LSTM and sp-LSTM are short for shared-LSTM and split-LSTM, respectively. The format of hyperparameter *LSTM* is *input* and *output channels*, and the format of *Conv* is *input* and *output channels*, *kernelsize*, and *stride*. The structure of 1-D convolution blocks refers to [18]. A complex 1-D convolution block is composed of two 1-D convolution blocks, representing real 1-D Conv and imaginary 1-D conv, respectively. The input channel size of the complex 1-D convolution block structure is half of 1-D convolution blocks. To verify the effect of the number of complex 1-D convolution blocks, in Table 2, the PESQs of the proposed approach versus number of complex Conv are provided. It is found that increasing the number of complex 1-D Conv does not always improve the performance of the model, and at the same time, it will increase the amount of model parameters by using more complex Convs. Therefore, in current work, three complex 1-D Conv are utilized and remaining Convs are still real.

Table 3. PESQ and STOI (%) on the simulated DNS dataset.

Model	Mics	Para (M)	PESQ					STOI (%)				
			−5 dB	0 dB	5 dB	10 dB	20 dB	−5 dB	0 dB	5 dB	10 dB	20 dB
Noisy	–	–	1.37	1.67	1.98	2.32	2.99	66.91	75.57	83.26	89.32	96.43
C-TasNet	1	5.1	2.39	2.75	3.02	3.27	3.65	81.38	88.95	93.17	95.48	98.32
MFMVDR	1	5.3	2.45	2.83	3.08	3.35	3.70	84.73	90.52	93.34	96.25	98.43
CRN-i	2	0.08	1.56	1.89	2.28	2.63	3.24	69.51	78.68	86.27	91.77	97.36
CRN-ii	2	17.6	1.61	1.96	2.33	2.66	3.30	71.06	79.85	87.15	92.30	97.93
Prop.	2	9.2	**2.75**	**3.07**	**3.32**	**3.52**	**3.81**	**89.22**	**93.29**	**95.74**	**97.27**	**98.81**

3.3   Training Baseline and Training Results

For comparisons, we reproduce CRN [2], Conv-TasNet (C-TasNet) [18], and MFMVDR [19] based on our dataset for training and testing. Conv-TasNet is a single-channel speech enhancement, and one channel of our noisy datasets is used for training and testing. MFMVDR is trained according to the open source project provided in [19]. Since MFMVDR is also a single-channel speech enhancement, the datasets are the same as C-TasNet. The parameters of CRN-i are selected according to [2], while CRN-ii is the parameter we tuned based on the dataset with reference to [20]. The experimental results are provided in Table 3, where the best results are highlighted with bold numbers. Compared with the single-channel speech enhancement MFMVDR, our proposed method has a significant improvement, which shows the benefits of multiple channels. Compared with the dual-channel CRN model, the proposed method also demonstrates a superior performance, regardless the sizes of the CRN model.

Table 4. The influence of interference sources in different directions on PESQ.

Method	Interference direction					
	15°	30°	45°	60°	75°	90°
Noisy	2.07	2.01	1.98	1.97	1.96	1.95
MVDR	2.03	2.02	2.03	2.12	2.21	2.27
Prop.	**2.97**	**3.00**	**3.15**	**3.39**	**4.11**	**3.37**

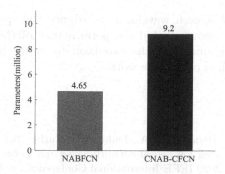

Fig. 2. The number of trainable parameters (unit: million).

Table 5. The influence of complex time domain network on speech quality in terms of PESQ and STOI.

Model	Metrics	SNR				
		−5 dB	0 dB	5 dB	10 dB	20 dB
NABFCN	PESQ	2.65	2.97	3.21	3.42	3.74
	STOI (%)	87.44	92.14	94.98	96.77	98.62
CNAB-CFCN	PESQ	**2.75**	**3.07**	**3.32**	**3.52**	**3.81**
	STOI (%)	**89.22**	**93.29**	**95.74**	**97.27**	**98.81**

In Table 4, we also analyze the effect of interference sources on the performance of CNAB-CFCN model at different azimuths. It indicates that the denoising performance is different at different directions, which agrees with the concept of the traditional beamforming. However, compared with traditional beamforming MVDR, the proposed neural beamforming indeed produces a better performance.

Finally, we study the benefits brought by complex operations. The NABFCN model is the same structure as the proposed CNAB-CFCN, but with real operations. The model sizes of NABFCN and CNAB-CFCN are shown in Fig. 2. Under the same scenarios, in Table 5, it is seen that the proposed complex network outperforms its corresponding real one across all the input SNRs. Due to the complex operations, the CNAB-CFCN also increases the model size over the NABFCN, which needs further compression in real-time applications.

4   Conclusions

In this study, we propose a novel complex time domain means to perform an end-to-end multi-channel speech enhancement network, termed as CNAB-CFCN. The Hilbert transform is explored to generate complex time-domain waveforms. With the introductions of complex operation rules, the proposed model outperforms its corresponding real network, and other single- and dual-channel networks. In addition, the loss function SI-SDR considers both the real part and imaginary part of the speech waveform to balance the speech quality. In the experiments, we only demonstrated the performance of the proposed network in the case of dual-channel, but the extension to more than two channels is straightforward, which is our future work.

References

1. Xia, Y., Braun, S., Reddy, C.K.A., Dubey, H., Cutler, R., Tashev, I.: Weighted speech distortion losses for neural-network-based real-time speech enhancement. In: ICASSP 2020 - 2020 IEEE International Conference on Acoustics, Speech and Signal Processing (ICASSP), pp. 871–875. IEEE (2020)
2. Tan, K., Zhang, X., Wang, D.L.: Real-time speech enhancement using an efficient convolutional recurrent network for dual-microphone mobile phones in close-talk scenarios. In: ICASSP 2019 - 2019 IEEE International Conference on Acoustics, Speech and Signal Processing (ICASSP), pp. 5751–5755. IEEE (2019)
3. Westhausen, N.L., Meyer, B.T.: Dual-signal transformation LSTM network for real-time noise suppression. arXiv preprint arXiv:2005.07551 (2020)
4. Erdogan, H., Hershey, J.R., Watanabe, S., Roux, J.L.: Phase-sensitive and recognition-boosted speech separation using deep recurrent neural networks. In: 2015 IEEE International Conference on Acoustics, Speech and Signal Processing (ICASSP), pp. 708–712. IEEE (2015)
5. Wang, Y., Wang, D.L.: A deep neural network for time-domain signal reconstruction. In: 2015 IEEE International Conference on Acoustics, Speech and Signal Processing (ICASSP), pp. 4390–4394. IEEE (2015)
6. Liu, Y., Zhang, H., Zhang, X., Yang, L.: Supervised speech enhancement with real spectrum approximation. In: ICASSP 2019 - 2019 IEEE International Conference on Acoustics, Speech and Signal Processing (ICASSP), pp. 5746–5750. IEEE (2019)
7. Choi, H.-S., Kim, J.-H., Huh, J., Kim, A., Ha, J.-W., Lee, K.: Phase-aware speech enhancement with deep complex U-net. In: International Conference on Learning Representations (2018)
8. Ronneberger, O., Fischer, P., Brox, T.: U-net: convolutional networks for biomedical image segmentation. In: Navab, N., Hornegger, J., Wells, W.M., Frangi, A.F. (eds.) MICCAI 2015. LNCS, vol. 9351, pp. 234–241. Springer, Cham (2015). https://doi.org/10.1007/978-3-319-24574-4_28
9. Hu, Y., et al.: DCCRN: deep complex convolution recurrent network for phase-aware speech enhancement. arXiv preprint arXiv:2008.00264 (2020)
10. Wang, Z.-Q., Wang, P., Wang, D.L.: Complex spectral mapping for single-and multi-channel speech enhancement and robust ASR. IEEE/ACM Trans. Audio Speech Lang. Process. **28**, 1778–1787 (2020)
11. Zhang, J., Chepuri, S.P., Hendriks, R.C., Heusdens, R.: Microphone subset selection for MVDR beamformer based noise reduction. IEEE/ACM Trans. Audio Speech Lang. Process. **26**(3), 550–563 (2017)
12. Benesty, J., Chen, J., Huang, Y.: Microphone Array Signal Processing, vol. 1. Springer, Heidelberg (2008). https://doi.org/10.1007/978-3-540-78612-2
13. Hoshen, Y., Weiss, R.J., Wilson, K.W.: Speech acoustic modeling from raw multi-channel waveforms. In: 2015 IEEE International Conference on Acoustics, Speech and Signal Processing (ICASSP), pp. 4624–4628. IEEE (2015)

14. Sainath, T.N., Weiss, R.J., Wilson, K.W., Narayanan, A., Bacchiani, M., et al.: Speaker location and microphone spacing invariant acoustic modeling from raw multichannel waveforms. In: 2015 IEEE Workshop on Automatic Speech Recognition and Understanding (ASRU), pp. 30–36. IEEE (2015)
15. Kolbæk, M., Tan, Z.-H., Jensen, S.H., Jensen, J.: On loss functions for supervised monaural time-domain speech enhancement. IEEE/ACM Trans. Audio Speech Lang. Process. **28**, 825–838 (2020)
16. Allen, J.B., Berkley, D.A.: Image method for efficiently simulating small-room acoustics. J. Acoust. Soc. Am. **65**(4), 943–950 (1979)
17. Reddy, C.K.A., et al.: The INTERSPEECH 2020 deep noise suppression challenge: datasets, subjective testing framework, and challenge results. arXiv preprint arXiv:2005.13981 (2020)
18. Luo, Y., Mesgarani, N.: Conv-TasNet: surpassing ideal time-frequency magnitude masking for speech separation. IEEE/ACM Trans. Audio Speech Lang. Process. **27**(8), 1256–1266 (2019)
19. Tammen, M., Doclo, S.: Deep multi-frame MVDR filtering for single-microphone speech enhancement. arXiv preprint arXiv:2011.10345 (2020)
20. Tan, K., Wang, D.L.: A convolutional recurrent neural network for real-time speech enhancement. In: Interspeech, pp. 3229–3233 (2018)

Robust Transmission Design
for IRS-Aided MISO Network
with Reflection Coefficient Mismatch

Ran Yang[1,2]([✉]) [iD], Ning Wei[2] [iD], Zheng Dong[1] [iD], Hongji Xu[1] [iD], and Ju Liu[1] [iD]

[1] School of Information Science and Engineering, Shandong University,
Qingdao, China
yangran6710@outlook.com, {zhengdong,hongjixu,juliu}@sdu.edu.cn
[2] National Key Laboratory of Science and Technology on Communications,
University of Electronic Science and Technology of China, Chengdu, China
wn@uestc.edu.cn

Abstract. Intelligent reflection surface (IRS) has been recognized as a revolutionary technology to achieve spectrum and energy efficient wireless communications due to its capability to reconfigure the propagation channels. However, due to the limited cost and space of each reflection element, it is difficult to accurately adjust the reflection coefficients of the passive elements. In this paper, we propose a worst-case robust reflection coefficient design for an IRS-aided single-user multiple-input single-output (SU-MISO) system where one IRS is deployed to enhance the received signal quality. Based on the fact of imperfect adjustment of reflection coefficients, our goal is to minimize the transmission power subject to the signal-noise ratio (SNR) constraint on the receiver end and the unit-modulus constraints on the reflection coefficients. The resulting optimization problem is non-convex and in general hard to solve. To tackle this problem, we adopt the linear approximation and alternating optimization (AO) methods to convert the original optimization problem into a sequence of convex subproblems that could be efficiently solved. We then extend our work to a practical situation where only limited phase shifts at each element are available. Numerical results demonstrate the robustness of the transmission scheme and show that high resolution for phase shifts is not an essential condition to approach the ideal performance.

Keywords: Intelligent reflection surface (IRS) · Robust design ·
Convex optimization · Reflection coefficient error · Discrete phase shift

This work was supported in part by the National Natural Science Foundation of China under Grants 61901245, 62071275, 91938202, and 61871070, and the Natural Science Foundation of Shandong Province of China under Grant ZR2020MF139, and the Fundamental Research Funds of Shandong University under Grant 61170079614095.

H. Gao et al. (Eds.): ChinaCom 2021, LNICST 433, pp. 140–150, 2022.
https://doi.org/10.1007/978-3-030-99200-2_12

1   Introduction

Although the 5th generation (5G) cellular systems are still under deployment globally, both academia and industry have begun to seek next-generation solutions that are faster, smarter, and greener. Thanks to the technological breakthrough in meta-surfaces, the intelligent reflection surface (IRS) has drawn wide attention recently as a promising technology to achieve smart and reconfigurable wireless environments in the physical layer [1–3]. Specifically, the IRS is a planar composed of a large number of passive elements, each of which could independently adjust the amplitude and phase of the incident signals, thus collaboratively improve the performance of communication systems by upgrading the propagation environments of wireless signals [4].

Note that the IRS is significantly different from the existing technologies such as the amplify-and-forward (AF) relaying networks and the conventional massive multiple-input multiple-output (MIMO) systems [5]. Firstly, different from AFs that actively amplify and forward the received signals, the IRS is generally passive, which incurs a negligible power consumption. Secondly, the IRS is also different from the active antennas in massive MIMO due to their different architectures (passive versus active) and operating mechanisms (reflect versus transmit), resulting in a relatively low cost. Besides, the IRS is able to create a virtual line-of-sight (LoS) channel to bypass obstacles, which is meaningful for applications of millimeter wave (mmWave).

Due to the aforementioned advantages, the IRS has been considered as a key enabler for future wireless communications. The acquisition of channel state information (CSI) of auxiliary links [6–9], and the transmission/beamforming vector design [10–13] are two major design issues about IRS-aided communications. However, the above-mentioned works are all based on the perfect adjustment of reflection coefficients, which is impractical due to the complexity and cost constraints on reflection elements. In addition, prior works mainly consider continuous phase shifts at the IRS, which is also difficult to realize.

We note that there are several works focusing on the discrete phase optimization [14, 15], but very complex processing methods are required. To address this problem, we investigate the worst-case robust transmission design problem based on the inaccurate reflection coefficients, which could also be interpreted as phase mismatch due to instability of meta-surfaces. More specifically, our goal is to minimize the total transmission power subject to the signal-noise ratio (SNR) constraint on the receiver end and the unit-modulus constraints on the reflection coefficients while ensuring no outage happens under all possible reflection coefficient error realizations. Numerical simulations were conducted which verified the robustness of the proposed transmission scheme with the obtained reflection phases being set to the nearest discrete values to evaluate the practical robust performance of the IRS-aided communication. Our main contributions can be summarized as follows:

– The original optimization problem is non-convex and generally hard to solve. By employing the linear approximation and alternating optimization (AO)

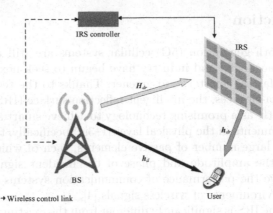

Fig. 1. IRS-aided communication system.

methods, we successfully convert the original problem into a sequence of convex subproblems that could be efficiently solved by using the CVX toolbox.

– We showed that a very high resolution for phase shifts is not necessary to achieve the ideal performance.

2 The System Model

2.1 Transmission Model

We consider an IRS-aided single-user multiple-input single-output (SU-MISO) system where the base station (BS) and the IRS are equipped with N antennas and M reflection elements, respectively, as depicted in Fig. 1. The channel coefficients from the BS to the user, from the IRS to the user, and from the BS to the IRS are denoted by $h_d \in \mathbb{C}^{N \times 1}$, $h_{dr} \in \mathbb{C}^{M \times 1}$, and $H_{dr} \in \mathbb{C}^{M \times N}$, respectively. Let $\theta = [e^{j\theta_1}, \ldots, e^{j\theta_M}]$ denote the phase shifts and a diagonal matrix $\Theta = \mathrm{diag}(\beta_1 e^{j\theta_1}, \ldots, \beta_M e^{j\theta_M})$ denote the corresponding reflection coefficient matrix. The reflection amplitude and phase shift of the nth element at the IRS are given by $\beta_n \in [0,1]$ and $\theta_n \in [0, 2\pi)$, respectively. In order to maximize the signal reflection power, we set $\beta_n = 1, n \in \mathcal{M}$, where $\mathcal{M} = [1, \ldots, M]$, and then investigate the phase shifts optimization. Practically, it is difficult to realize continuous phase shifts due to cost and hardware limitations. Motivated by this fact, we first focus on the optimization of continuous phase shifts and then extend our work to the practical cases where each element at the IRS only has finite levels for phase shifts. In this paper, we consider the linear precoding at the BS, with $f \in \mathbb{C}^{N \times 1}$ denoting the precoding vector. The transmission power is denoted by $\|f\|_2^2$. Then, the received signal at the user is given by

$$y = (h_d^{\mathrm{H}} + h_{dr}^{\mathrm{H}} \Theta H_{dr}) f s + n_g \tag{1}$$

where s is the transmitted information carrying symbol with zero mean and unit variance, i.e., $s \sim \mathcal{CN}(0,1)$, and n_g denotes additive white Gaussian noise (AWGN) at the receiver with zero mean and variance σ^2, i.e., $n_g \sim \mathcal{CN}(0,\sigma^2)$. Note that $\boldsymbol{h}_{dr}^{\mathrm{H}} \boldsymbol{\Theta} \boldsymbol{H}_{dr} = \boldsymbol{v}^{\mathrm{H}} \boldsymbol{G}$, where the cascaded channel $\boldsymbol{G} = \mathrm{diag}(\boldsymbol{h}_{dr}^{\mathrm{H}}) \boldsymbol{H}_{dr}$, the passive beamforming $\boldsymbol{v} = [v_1, \ldots, v_M]^{\mathrm{H}}$, the nth entry $v_n = e^{j\theta_n}, n \in \mathcal{M}$. Accordingly, the SNR of the received signal is given by

$$\mathrm{SNR} = \frac{|(\boldsymbol{h}_d^{\mathrm{H}} + \boldsymbol{v}^{\mathrm{H}} \boldsymbol{G})\boldsymbol{f}|^2}{\sigma^2}. \tag{2}$$

2.2 Problem Formulation

As mentioned above, many of the current IRS-aided systems concentrate on the transmission design and the acquisition of CSI corresponding to the cascaded links. However, most of them are based on the assumption of perfect adjustment of reflection coefficients, which is too idealistic due to the limited space and cost of the passive elements. The performance of the IRS is highly affected by the accuracy of the reflection coefficient adjustment and thus this problem should be addressed. In this paper, we adopt the bounded error model to investigate the worst-case robust transmission design of the IRS-aided communication, i.e., the practical passive beamforming could be rewritten as

$$\hat{\boldsymbol{v}} = \boldsymbol{v} + \triangle \boldsymbol{v}$$

where $\|\triangle \boldsymbol{v}\|_2 \leq \epsilon$. Then, the worst-case optimization problem can be formulated as

$$\min_{\boldsymbol{f},\boldsymbol{v}} \quad \|\boldsymbol{f}\|_2^2 \tag{3a}$$

$$\mathrm{s.t.} \quad \frac{|(\boldsymbol{h}_d^{\mathrm{H}} + \hat{\boldsymbol{v}}^{\mathrm{H}} \boldsymbol{G})\boldsymbol{f}|^2}{\sigma^2} \geq \gamma, \tag{3b}$$

$$\|\triangle \boldsymbol{v}\|_2 \leq \epsilon, \tag{3c}$$

$$|v_m|^2 = 1, \forall m \in \mathcal{M}. \tag{3d}$$

The constraint (3b) is the minimum SNR target for the user; (3c) is the bounded model for coefficient error; (3d) corresponds to the unit-modulus requirements for the reflection coefficients.

The above problem is non-convex due to the following reasons: *(a)* The constraint (3b) is a quadratic infinite inequality, the solution set of which is non-convex; *(b)* The precoding $\boldsymbol{f}$ and the passive beamforming $\boldsymbol{v}$ are coupled, which is non-convex and hard to optimize simultaneously; *(c)* The unit-modulus constraints (3d) are strongly non-convex and intractable.

3 Robust Optimization Solutions

In this section, we aim to propose the solutions to Problem (3) which is non-convex due to the reasons mentioned in Sect. 2. Following the non-convexity analysis, we first deal with the quadratic inequality in (3b) by linear approximation and then uncouple the variables by the alternating optimization (AO) method, thus convert the primal problem into a sequence of convex subproblems that could be efficiently solved.

3.1 Problem Transformation

We first handle the quadratic inequality in (3b) by using the linear approximation method.

Lemma 1. *Letting $f^{(n)}$ and $v^{(n)}$ be the optimal solution obtained at the nth iteration, the linear lower bound of $|(h_d^{\mathrm{H}} + \hat{v}^{\mathrm{H}}G)f|^2$ at $(f^{(n)}, v^{(n)})$ could be denoted by*

$$\triangle v^{\mathrm{H}} X \triangle v + 2\,\mathrm{Re}\{x^{\mathrm{H}}\triangle v\} + c \tag{4}$$

where

$$
\begin{aligned}
X &= Gff^{\mathrm{H},(n)}G^{\mathrm{H}} + Gf^{(n)}f^{\mathrm{H}}G^{\mathrm{H}} - Gf^{(n)}f^{\mathrm{H},(n)}G^{\mathrm{H}}, \\
x &= -Gf^{(n)}f^{\mathrm{H},(n)}h_d - Gf^{(n)}f^{\mathrm{H},(n)}G^{\mathrm{H}}v^{(n)} \\
&\quad + Gf^{(n)}f^{\mathrm{H}}h_d + Gf^{(n)}f^{\mathrm{H}}G^{\mathrm{H}}v + Gff^{\mathrm{H},(n)}h_d \\
&\quad + Gff^{\mathrm{H},(n)}G^{\mathrm{H}}v^{(n)}, \\
c &= 2\,\mathrm{Re}\{f^{\mathrm{H}}(h_dh_d^{\mathrm{H}} + G^{\mathrm{H}}vh_d^{\mathrm{H}} + h_dv^{\mathrm{H},(n)}G \\
&\quad + G^{\mathrm{H}}vv^{\mathrm{H},(n)}G)f^{(n)}\} - f^{\mathrm{H},(n)}(h_dh_d^{\mathrm{H}} + G^{\mathrm{H}}v^{(n)}h_d^{\mathrm{H}} \\
&\quad + h_dv^{\mathrm{H},(n)}G + G^{\mathrm{H}}v^{(n)}v^{\mathrm{H},(n)}G)f^{(n)}.
\end{aligned}
$$

∎

Proof: Let a be a complex scalar variable. Then, the first-order Taylor expansion is expressed as

$$|a|^2 \geq a^{(n),*}a + a^*a^{(n)} - a^{(n),*}a^{(n)}.$$

The linear approximation inequality holds for any fixed $a^{(n)}$, where $a^{(n)}$ is a constant scalar. Then we replace a and $a^{(n)}$ with $(h_d^{\mathrm{H}} + (v^{\mathrm{H}} + \triangle v^{\mathrm{H}})G)f$ and $(h_d^{\mathrm{H}} + (v^{\mathrm{H},(n)} + \triangle v^{\mathrm{H}})G)f^{(n)}$, respectively. The proof is completed.   □

With the lower bound (4) of $|(h_d^{\mathrm{H}} + \hat{v}^{\mathrm{H}}G)f|^2$, we have

$$\triangle v^{\mathrm{H}} X \triangle v + 2\mathrm{Re}\{x^{\mathrm{H}}\triangle v\} + c - \gamma\sigma^2 \geq 0. \tag{5}$$

Note that (5) contains random variable $\triangle v$ and it is not known to be tractable. A possible way is to derive a bound for the error constraint for any $\|\triangle v\|_2 \leq \epsilon$ and then perform optimizations based on the bound. Here, similar to [16], we directly derive the exact equivalent condition from (5) by employing the S-lemma [17]. Thus, we have the equivalent constraint of (5) formulated as

$$\begin{bmatrix} X + wI & x \\ x^{\mathrm{H}} & c - \gamma\sigma_u^2 - w\epsilon^2 \end{bmatrix} \succeq 0,\ w \geq 0 \tag{6}$$

where I is an identity matrix and w is a slack variable. Thus we have the reformulated optimization problem

$$\min_{f,v,w} \quad \|f\|_2^2 \tag{7a}$$

$$\text{s.t.} \quad \begin{bmatrix} X + wI & x \\ x^H & c - \gamma\sigma^2 - w\epsilon^2 \end{bmatrix} \succeq 0, \tag{7b}$$

$$|v_m|^2 = 1, \forall m \in \mathcal{M}, \tag{7c}$$

$$w \geq 0. \tag{7d}$$

According to the non-convexity analysis, Problem (7) is still non-convex due to the coupled precoding at the BS and passive beamforming at the IRS, which is hard to optimize simultaneously. Thus, in the next section, we propose to use the AO method to handle the coupling optimization problem by iteratively updating precoding with the passive beamforming fixed and optimizing passive beamforming with the precoding fixed, respectively.

3.2 The Alternating Optimization Method

We first fix the passive beamforming coefficient and optimize the precoding vector. With reflection coefficients fixed, the unit-modulus constraints (7c) could be ignored. We then have the subproblem of Problem (7) formulated as

$$\min_{f,w} \quad \|f\|_2^2 \tag{8a}$$

$$\text{s.t.} \quad \begin{bmatrix} X + wI & x \\ x^H & c - \gamma\sigma^2 - w\epsilon^2 \end{bmatrix} \succeq 0, \tag{8b}$$

$$w \geq 0. \tag{8c}$$

Problem (8) is a semidefinite program (SDP) problem and could be solved by employing the CVX toolbox [18,19]. Note that in the above problem reflection coefficients are constant numbers, which satisfy $v = v^{(n)}$.

We then consider the subproblem about the passive beamforming. With the precoding fixed, Problem (7) is reduced to a feasibility-check problem. We introduce a slack variable $\eta \geq 0$ to improve the convergence speed, which could be interpreted as 'SNR residual'; see [11] for details. Thus (5) could be rewritten as

$$\triangle v^H X \triangle v + 2\text{Re}\{x^H \triangle v\} + c - \gamma\sigma^2 - \eta \geq 0. \tag{9}$$

The subproblem about the passive beamforming of Problem (7) could be formulated by

$$\max_{\eta,v,w} \quad \eta \tag{10a}$$

$$\text{s.t.} \quad \begin{bmatrix} X + wI & x \\ x^H & c - \gamma\sigma^2 - w\epsilon^2 - \eta \end{bmatrix} \succeq 0, \tag{10b}$$

$$w \geq 0, \eta \geq 0, \tag{10c}$$

$$|v_m|^2 = 1, \forall m \in \mathcal{M}. \tag{10d}$$

Algorithm 1. Penalty convex-concave procedure optimization for IRS

1: **Initialize:** set $i = 0$, and initialize $\boldsymbol{v}^{[0]}$, $p > 1$.
2: **repeat**
3:    **if** $i < I_{\max}$ **then**
4:        Update $\boldsymbol{v}^{[i+1]}$ from Problem (11);
5:        $\lambda^{[i+1]} = \min\{p\lambda^{[i]}, \lambda_{\max}\}$;
6:        $i = i + 1$;
7:    **else**
8:        Initialize with a new random $\boldsymbol{v}^{[0]}$, and set $i = 0$.
9:    **end if**
10: **until** $\|\boldsymbol{b}\|_2 \leq \chi$ and $\|\boldsymbol{v}^{[i]} - \boldsymbol{v}^{[i-1]}\|_2 \leq z$
11: **Output** $\boldsymbol{v}^{(n+1)} = \boldsymbol{v}^{[i]}$.

It is worth noting that Problem (10) is still non-convex due to the unit-modulus constraints (10d). Here we adopt penalty convex-concave procedure (CCP) to handle the non-convex part [10]. Following the penalty framework, our first step is to relax the unit-modulus constraints and then implement penalty on the target function according to the degree of violation of the constraints. The intensity of penalty increases as the number of iterations goes on. The modulus of each reflection coefficient is compelled to be 1 when the penalty is sufficiently large.

Specifically, $|v_m|^2 = 1, \forall m \in \mathcal{M}$ could be rewritten as $1 \leq |v_m|^2 \leq 1, \forall m \in \mathcal{M}$. Note that the constraints $|v_m|^2 \geq 1, \forall m \in \mathcal{M}$ are non-convex, which are again handled by linear approximation, i.e., $|v_m^{[i]}|^2 - 2\mathrm{Re}(v_m^* v_m^{[i]}) \leq -1, \forall m \in \mathcal{M}$, where $v_m^{[i]}$ is the optimal solution obtained at the ith iteration. Here, we let the modulus of $v_m, \forall m \in \mathcal{M}$ violate the unit-modulus constraints and impose slack variables $\boldsymbol{b} = [b_1, \dots, b_{2M}]^{\mathrm{T}} \geq 0$ over the relaxed constraints, i.e., $|v_m^{[i]}|^2 - 2\mathrm{Re}(v_m^* v_m^{[i]}) \leq b_m - 1, |v_m|^2 \leq 1 + b_{M+m}, \forall m \in \mathcal{M}$. The slack variables could be interpreted as the penalty term for optimization variables that violate the unit-modulus constraints. Thus, we have the reformulated problem as follows

$$\max_{\eta, v, w, b} \quad \eta - \lambda^{[i]} \|\boldsymbol{b}\|_2 \tag{11a}$$

$$\text{s.t.} \quad \begin{bmatrix} \boldsymbol{X} + w\boldsymbol{I} & \boldsymbol{x} \\ \boldsymbol{x}^{\mathrm{H}} & c - \gamma\sigma^2 - w\epsilon^2 - \eta \end{bmatrix} \succeq 0, \tag{11b}$$

$$w \geq 0, \eta \geq 0, \boldsymbol{b} \geq 0, \tag{11c}$$

$$|v_m^{[i]}|^2 - 2\mathrm{Re}(v_m^* v_m^{[i]}) \leq b_m - 1, \forall m \in \mathcal{M}, \tag{11d}$$

$$|v_m|^2 \leq 1 + b_{M+m}, \forall m \in \mathcal{M}. \tag{11e}$$

The above problem is an SDP and could be solved by the CVX tool, and the algorithm is summarized in Algorithm 1.

Note that the feasible solution directly obtained from Problem (11) may not be the solution of Problem (10) due to the limited intensity of penalty and iterations. The feasibility is guaranteed by imposing a strong penalty over the target function and a large number of iterations. The variable $\lambda^{[i]}$ is the

---
Algorithm 2. Alternating optimization

1: **Initialize:** initialize $v^{(0)}$ and $f^{(0)}$, set $n = 0$
2: **repeat**
3:    Update $f^{(n+1)}$ from Problem (8);
4:    Update $v^{(n+1)}$ from Problem (10);
5:    $n = n + 1$;
6: **until** The objective value (3) converges.

---

regularization factor to measure the influence of the penalty factor $\|b\|_2$; $\|b\|_2 \leq \chi$ and $\|v^{[i]} - v^{[i-1]}\|_2 \leq z$ control the convergence of Algorithm 1, see [10] for more details. Note that in Problems (10) and (11) the precoding vector is constant, which satisfies $f = f^{(n)}$. We iteratively solve Problem (8) and Problem (10), thus the reformulated Problem (7) could be efficiently solved. The alternating optimization algorithm is summarized in Algorithm 2. The convergence of AO is guaranteed because the feasible solution to Problem (10) is also feasible to Problem (8), thus the value of target function (3), i.e., the total transmission power is non-increasing. Therefore, AO is guaranteed to converge.

It is worth noting that only limited quantization bits are deployed into each element at the IRS practically. Hence, only a finite number of phase shift levels are available at the passive elements. Specifically, let q denote quantization bits for phase shifts at each element, and the discrete values are generated by uniformly quantizing the continuous phase shift set $[0, 2\pi)$. Thus, the set of discrete phase shift values is given by

$$\mathcal{F}_d = \{0, \triangle\theta, \ldots, (L-1)\triangle\theta\} \tag{12}$$

where $\triangle\theta = 2\pi/L, L = 2^q$. We then set the phases obtained from AO to the nearest discrete values to evaluate the practical performance of IRS-aided communication.

4   Simulation Results

In this section, numerical simulations are performed to verify the robustness of the proposed algorithm and evaluate the performance of the practical IRS-aided communication systems.

We consider a uniform linear array (ULA) at the BS equipped with $N = 6$ antennas and a uniform rectangular array (URA) at the IRS equipped with $M = 16$ elements. A rectangular coordinate is adopted to locate the system, i.e., the BS is located at $(0\,\text{m}, 0\,\text{m})$, the IRS is located at $(50\,\text{m}, 10\,\text{m})$, and the user is located at $(70\,\text{m}, 0\,\text{m})$, respectively. The distance-dependent path loss model is given by $PL = -30 - 10\alpha\log_{10}(d)$ dB, where α is the path loss exponent and d is the distance in meters. We adopt Ricean channels to model h_d, h_{dr}, and H_{dr}. All the Ricean factors are set to be 5. The LoS parts are given by the product of the steering vectors of the transmitter and receiver. The non-LoS parts follow Rayleigh distribution. The adjustment error bound of IRS reflection coefficients

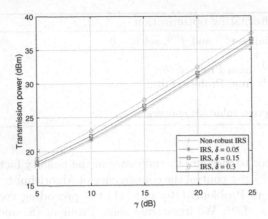

Fig. 2. Transmission power versus target SNR γ.

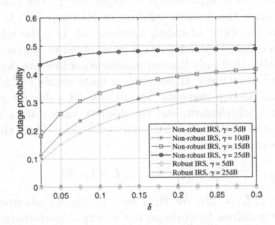

Fig. 3. Outage probability versus coefficient uncertainty δ.

is defined as $\epsilon = \delta\|v\|_2$, where δ could be interpreted as the amount of error. The power of AWGN is set to be -80 dBm.

Figure 2 shows the transmission power versus the target SNR of the user. It is observed that the total power increases with the target SNR and the coefficient error δ. The transmission power of the robust design is higher than that of the non-robust design as expected. The additional power can be interpreted as the price to overcome the performance degradation caused by the mismatch. Note that even when the uncertainty δ in reflection coefficients is highly increased, the problem is still feasible at the cost of more energy consumption.

Figure 3 plots the outage probability of the non-robust transmission design. The outage probability is defined as the probability that the target SNR of the user is not satisfied. It is observed that the outage probability increases with the coefficient error δ and the target SNR γ when we ignore the error of reflection

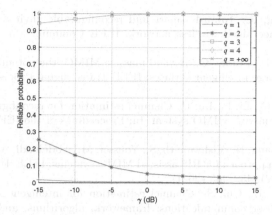

Fig. 4. Reliable probability versus target SNR γ under $\delta = 0.05$.

coefficients. However, when the robust transmission scheme is adopted, we can guarantee that no outage occurs.

In Fig. 4, we extend our work to the practical case where only finite phase levels are available. The reliable probability is defined as the probability that the SNR of the received signal is higher than γ. It is observed that when the number of quantization bits is small, i.e., 1-bit or 2-bit, the IRS-aided communication suffers severe performance degradation. However, the 3-bit IRS is arguably able to approach the ideal communication quality.

5   Conclusion

We have investigated the robust transmission design problem based on the inaccurate adjustment of the reflection coefficients at the IRS. The non-convex parts of the obtained optimization problem were successfully tackled by using the linear approximation and AO methods and thus could be efficiently solved. Numerical results verified the robustness of the transmission scheme and showed that high resolution for phase shifts is not mandatory to approach the ideal IRS performance.

References

1. Wu, Q., Zhang, S., Zheng, B., You, C., Zhang, R.: Intelligent reflecting surface-aided wireless communications: a tutorial. IEEE Trans. Commun. **69**(5), 3313–3351 (2021)
2. Di Renzo, M., Debbah, M., Phan-Huy, D.-T., et al.: Smart radio environments empowered by reconfigurable AI meta-surfaces: an idea whose time has come. EURASIP J. Wirel. Commn. **2019**(1), 1–20 (2019)
3. Liaskos, C., Nie, S., Tsioliaridou, A., Pitsillides, A., Ioannidis, S., Akyildiz, I.: A new wireless communication paradigm through software-controlled metasurfaces. IEEE Commun. Mag. **56**(9), 162–169 (2018)

4. Wu, Q., Zhang, R.: Towards smart and reconfigurable environment: Intelligent reflecting surface aided wireless network. IEEE Commun. Mag. **58**(1), 106–112 (2020)
5. Hu, S., Rusek, F., Edfors, O.: Beyond massive MIMO: the potential of data transmission with large intelligent surfaces. IEEE Trans. Signal Process. **66**(10), 2746–2758 (2018)
6. Zhang, J., Qi, C., Li, P., Lu, P.: Channel estimation for reconfigurable intelligent surface aided massive MIMO system. In: Proceedings of the IEEE SPAWC, May 2020 (2020)
7. Shtaiwi, E., Zhang, H., Vishwanath, S., Youssef, M., Abdelhadi, A., Han, Z.: Channel estimation approach for RIS assisted MIMO systems. IEEE Trans. Cogn. Commun. Netw. **7**(2), 452–465 (2021)
8. Wang, Z., Liu, L., Cui, S.: Channel estimation for intelligent reflecting surface assisted multiuser communications: framework, algorithms, and analysis. IEEE Trans. Wireless Commun. **19**(10), 6607–6620 (2020)
9. Wang, P., Fang, J., Duan, H., Li, H.: Compressed channel estimation for intelligent reflecting surface-assisted millimeter wave systems. IEEE Signal Process. Lett. **27**, 905–909 (2020)
10. Zhou, G., Pan, C., Ren, H., Wang, K., Nallanathan, A.: A framework of robust transmission design for IRS-aided MISO communications with imperfect cascaded channels. IEEE Trans. Signal Process. **68**, 5092–5106 (2020)
11. Wu, Q., Zhang, R.: Intelligent reflecting surface enhanced wireless network via joint active and passive beamforming. IEEE Trans. Wireless Commun. **18**(11), 5394–5409 (2019)
12. Yang, H., Xiong, Z., Zhao, J., Niyato, D., Xiao, L., Wu, Q.: Deep reinforcement learning-based intelligent reflecting surface for secure wireless communications. IEEE Trans. Wireless Commun. **20**(1), 375–388 (2021)
13. Huang, C., Zappone, A., Alexandropoulos, G.C., Debbah, M., Yuen, C.: Reconfigurable intelligent surfaces for energy efficiency in wireless communication. IEEE Trans. Wireless Commun. **18**(8), 4157–4170 (2019)
14. Wu, Q., Zhang, R.: Beamforming optimization for wireless network aided by intelligent reflecting surface with discrete phase shifts. IEEE Trans. Commun. **68**(3), 1838–1851 (2020)
15. Abeywickrama, S., Zhang, R., Wu, Q., Yuen, C.: Intelligent reflecting surface: practical phase shift model and beamforming optimization. IEEE Trans. Commun. **68**(9), 5849–5863 (2020)
16. Zheng, G., Wong, K.-K., Ottersten, B.: Robust cognitive beamforming with bounded channel uncertainties. IEEE Trans. Signal Process. **57**(12), 4871–4881 (2009)
17. Boyd, S., Vandenberghe, L., Faybusovich, L.: Convex optimization. IEEE Trans. Automat. Contr. **51**(11), 1859 (2006)
18. Grant, M., Boyd, S.: CVX: Matlab software for disciplined convex programming, version 2.0 beta, September 2013 (2013). http://cvxr.com/cvx
19. Grant, M., Boyd, S.: Graph implementations for nonsmooth convex programs. In: Blondel, V.D., Boyd, S.P., Kimura, H. (eds.) Recent Advances in Learning and Control. LNCIS, vol. 371, pp. 95–110. Springer, London (2008). https://doi.org/10.1007/978-1-84800-155-8_7

Network Communication Enhancement

Distributed Deep Reinforcement Learning Based Mode Selection and Resource Allocation for VR Transmission in Edge Networks

Jie Luo[1(✉)], Bei Liu[2], Hui Gao[3], and Xin Su[2]

[1] China Academy of Telecommunications Technology, Beijing, China
nsluojie1016@163.com
[2] Tsinghua University, Beijing, China
[3] Beijing University of Posts and Telecommunications, Beijing, China

Abstract. Wireless virtual reality (VR) is expected to be one of the most pivotal applications in 5G and beyond, which provides an immersive experience and will greatly renovate the way people communicate. However, the challenges of VR service transmission to provide high quality of experience (QoE) and a huge data rate remain unsolved. In this paper, we formulate an optimization of the mode selection and resource allocation to maximize the QoE of VR users, aiming at the optimal transmission of VR service based on the cloud-edge-end architecture. Moreover, a distributed game theory based deep reinforcement learning (DGTB-DRL) algorithm is proposed to solve the problem, which can achieve a Nash equilibrium (NE) rapidly. The simulation results demonstrate that the proposed method can achieve better performance in terms of training efficiency, QoE utility values.

Keywords: Virtual reality · Reinforcement learning · Resource allocation · Mobile edge network

1 Introduction

Based on the existing three service scenarios in wireless communication, enhanced mobile broadband (eMBB), massive machine type communications (mMTC) and ultra-reliable low latency communications (uRLLC), virtual reality (VR) develops rapidly and is considered to be one of the most promising application in the next generation, which will greatly renovate the way people communicate [1]. With the growth of multiple access for various scenarios, video applications represented by VR are required with an exponential increase of data rate, which brings up challenges. It is foreseeable that the system capacity will desire to be multiple 1000. Additionally, VR services are obliged to provide low latency and high quality of experience (QoE) to prevent users from feeling dizzy.

This work was supported by National Key R&D Program of China (2020YFB1806702).

H. Gao et al. (Eds.): ChinaCom 2021, LNICST 433, pp. 153–167, 2022.
https://doi.org/10.1007/978-3-030-99200-2_13

Benefit from mobile edge computing (MEC) which has been considered as an essential network architecture for future wireless networks [2], many works focused on reducing the amount of data transmission by delivering computation and communication resources to the network edge. The authors in [3] introduced MEC into the internet of things (IoT) network and proposed a deep reinforcement learning (DRL) based scheme to optimize the communication and computing resource allocation. To apply MEC in vehicle networks, [4] proposed to utilize DRL to find the policies of computation offloading and resource allocation in stochastic traffic and other uncertain communication conditions.

Additionally, there are dozens of works that concentrate on QoE optimization of cross-layer transmission in the wireless network. The author in [5] proposed an artificial intelligence (AI) aided joint bit rate selection and radio resource allocation scheme in fog-computing based radio access networks (F-RANs) based on multi-agent hierarchy DRL. [6] came up with an online learning method to solve the fast device-to-device (D2D) clustering and mode selection joint problem in both large-scale and small-scale scenarios, while [7] proposed a deep learning approach with an adaptive VR framework to conquer association, offloading and caching problem in real-time VR rendering tasks. To optimize VR content delivery while meeting high transmission rate requirements, [8] formed a Lagrangian dual decomposition approach to solve multiple dimensional knapsack problem, which realized a communications-caching-computing tradeoff for mobile VR devices.

However, there is little research to develop a distributed DRL method based on game theory to solve the joint optimization issue. As the matter of fact, the distributed learning methods have better flexibility and adaptability than the centralized ones [7]. Taking the adaptation capability and the scalability into account, in this paper, we focus on the transmission of VR service based on the cloud-edge-end architecture and formulate a joint optimization of the mode selection and resource allocation to maximize the QoE of all VR users. Furthermore, we propose a distributed game theory based DRL (DGTB-DRL) algorithm to conquer the above highly complex problem, which can achieve a Nash equilibrium (NE) within less time. At last, the simulation results show the superiority of the proposed method in terms of training efficiency, QoE utility values.

The remainder of this paper is organized as follows. Section 2 and Sect. 3 introduce the system model and present the problem formulation. The proposed algorithm is presented in Sect. 4. In Sect. 5, simulation results of the proposed method are presented and analyzed. Finally, the conclusion is given in Sect. 6.

2 System Model

2.1 Network Model

In this paper, we consider a mobile edge network composed of Y cloud nodes, M edge nodes and U user equipments as a cloud-edge-end collaborative framework, as shown in Fig. 1. The original VR game resources are centrally managed by the cloud server on the cloud nodes. Equipped with edge servers, edge nodes served as access points are distributedly deployed in the VR environment on a large scale. End nodes are defined as VR headsets worn by real users.

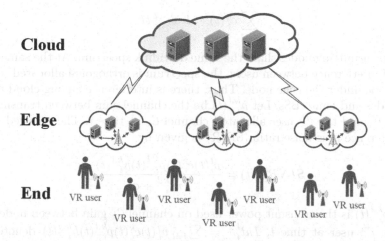

Fig. 1. Cloud-edge-end VR system model.

Considering the different quality of service requirements and the computation capabilities of different users, the VR user can choose three modes to transmit the contents, (1) cloud mode, (2) edge mode, and (3) local mode. The cloud nodes transmit the contents directly in cloud mode while the edge nodes serve users simply in edge mode. Specially, there are F femto base stations (BS) for short distance transmission in local mode. Therefore, we denotes the set of all transmitting nodes by $\mathcal{TN} = \{pn_1, \ldots, pn_y, mn_1, \ldots, mn_m, fn_1, \ldots, fn_f\}$, with the set of the transmitting nodes' indices $\mathcal{N} = \{0, 1, \ldots, L-1\}$, where $L = Y + M + F$.

2.2 Communication Model

Assume that different transmitting nodes may associate with different sets of end nodes with K shared orthogonal channels. A binary mode selection indicator is given by $\mathcal{V} = \{v_i^l(t), i \in \mathcal{U}, l \in \mathcal{N}\}$, where $\mathcal{U} = \{1, \ldots, U\}$. And we have $v_i^l(t) \in \{0, 1\}$, where $v_i^l(t) = 1$ indicates that the ith user selects the $\mathcal{TN}_l$ as its transmitting node and $v_i^l(t) = 0$ otherwise. We assume that each user can only choose one mode at any time, the following constraint should be met,

$$\sum_{l=0}^{L-1} v_i^l(t) \leq 1, \forall i \in \mathcal{U}. \tag{1}$$

Meanwhile, assume that spectrum resource is divided into $\mathcal{K} = \{1, \ldots, K\}$ channels for users. We denote a binary channel allocation indicator by $\mathcal{C} = \{c_i^k(t), i \in \mathcal{U}, k \in \mathcal{K}\}$. We have $c_i^k(t) \in \{0, 1\}$, where $c_i^k(t) = 1$ indicates that the

i^{th} user select channel $\mathcal{C}_k$ at time t. Considering the limitation of simultaneous transmission, we assume that each user can only choose one channel at any time, the following constraint should be met,

$$\sum_{k=1}^{K} c_i^k(t) \leq 1, \forall i \in \mathcal{U}. \tag{2}$$

All transmitting nodes share the same downlink spectrum. At the same time, to avoid interference between users, the spectrum is orthogonal allocated to each client node under the edge node. Thus, there is interference among cloud nodes, edge nodes and femto BSs. Let $h_i^{k,l}(t)$ be the channel gain between transmitting node $\mathcal{TN}_l$ and the ith user allocated channel $\mathcal{C}_k$ at time t. The received signal to interference plus noise ratio (SINR) is given by

$$SINR_i^{k,l}(t) = \frac{v_i^l(t)c_i^k(t)p_i^{k,l}(t)h_i^{k,l}(t)}{Int_i^{k,l} + \sigma^2}, \tag{3}$$

where $p_i^{k,l}(t)$ is the transmit power used on channel $\mathcal{C}_k$ gain between node $\mathcal{TN}_l$ and the ith user at time t, $Int_i^{k,l} = \sum_{j \neq l} v_i^j(t)c_i^k(t)p_i^{k,l}(t)h_i^{k,j}(t)$ denotes the interference and σ^2 denotes the noise power.

As a result, the downlink data rate of the ith user from node $\mathcal{TN}_l$ on channel $\mathcal{C}_k$ can be calculated by

$$r_i^{k,l}(t) = W \log_2 \left(1 + SINR_i^{k,l}(t)\right), \tag{4}$$

where W denoted the channel bandwidth.

For convenience, we focus on the chosen channel (the rate is abbreviated as $r_i^l(t)$) and assume that $q_i^l(t)$ denotes the total bits of transmission, the transmission time is given by

$$TR_i^l(t) = \frac{q_i^l(t)}{r_i^l(t)}, \forall i, l. \tag{5}$$

Therefore, we have the total downlink data rate of the ith user which is expressed by

$$r_i(t) = \sum_{l=0}^{L-1} \sum_{k=1}^{K} r_i^{k,l}(t) = \sum_{l=0}^{L-1} \sum_{k=1}^{K} W \log_2 \left(1 + SINR_i^{k,l}(t)\right). \tag{6}$$

2.3   Computing Model

In this system, the frame calculation of video rendering is also indispensable for VR users. The computing task requested by users is scheduled by the servers on the transmitting nodes. At time t, $C_i^l(t)$ denotes the computational resource which is assigned to the ith user from node $\mathcal{TN}_l$. Thus, the time consumed for computing tasks is given by

$$TC_i^l(t) = \frac{D_i^l(t)}{C_i^l(t)\beta}, \forall i, l, \tag{7}$$

where $D_i^l(t)$ denotes the data size at time t, β is the computation capacity of the server per CPU cycle. Let C_{sum} denote the total computational resource, the following constraint should be met,

$$\sum_{i=1}^{U} \sum_{l=0}^{L-1} C_i^l(t) \leq C_{sum}. \tag{8}$$

2.4   Quality of Experience Model

Most existing QoE model building methods rely on the prior assumption that the QoE score and quality of service (QoS) parameters have specific mathematical expressions. A commonly used model for QoE prediction is given by a rational model with a logarithmic function, which is defined by network-level parameters (packet loss rate) and application-level parameters (i.e., send bitrate, frame rate). Based on [9], the time of video stalling should be kept at a low level to enhance the user's experience. We convert logarithm to linear weighting and make a reasonable migration assumption. Similar to [10], the time-average stalling probability is defined as

$$\lim_{t \to \infty} \frac{1}{t} \sum_{\tau=0}^{t-1} \frac{1}{LK} \sum_{l=0}^{L-1} \frac{TC_i^l(\tau)}{TR_i^l(\tau)}, \tag{9}$$

and the time average bitrate is calculated by

$$\lim_{t \to \infty} \frac{1}{t} \sum_{\tau=0}^{t-1} \frac{1}{LK} \sum_{l=0}^{L-1} \sum_{k=1}^{K} r_i^{k,l}(\tau). \tag{10}$$

In order to jointly consider the above factors, we define the average QoE for the ith user at processing period t as

$$QoE_i(t) = \frac{1}{LK} \sum_{l=0}^{L-1} \sum_{k=1}^{K} \left[\omega_1 r_i^{k,l}(\tau) - \omega_2 \frac{TC_i^l(\tau)}{TR_i^l(\tau)} \right], \tag{11}$$

where ω_1 and ω_2 are non-negative weights representing the relative importance of QoE.

3   Problem Formulation

Consider that all end nodes are desired to get the maximum transmission rate from transmitting nodes while keeping a quite low latency. We assume that the SINR of the ith user $SINR_i(t)$ should not be less than the minimum QoS threshold ξ_i,

$$SINR_i(t) = \sum_{l=0}^{L-1} \sum_{k=1}^{K} SINR_i^{k,l}(t) \geq \xi_i, \tag{12}$$

Furthermore, taking transmission cost into consideration, we define λ_l as the unit price of the $\mathcal{TN}_l$ transmit power. Distinctly, λ_l has a negative correlation with ω_1. Thus, we make $\omega_2 = \lambda_l$.

Similarly, the ith user obtains the achieved profit which is given by ρ_i. We have $\omega_1 = \rho_i$.

Our goal is to develop an effective joint mode selection and resource allocation scheme to maximize the QoE of the VR users. The optimization problem can be formulated as follows,

$$P0 : \max_{\mathcal{V},\mathcal{C}} QoE_i(t)$$

$$s.t. C1 : \sum_{l=0}^{L-1} v_i^l(t) \leq 1, \sum_{k=1}^{K} c_i^k(t) \leq 1, \forall i \in \mathcal{U}$$

$$C2 : \sum_{i=1}^{U} \sum_{l=0}^{L-1} C_i^l(t) \leq C_{sum} \tag{13}$$

$$C3 : \sum_{l=0}^{L-1} \sum_{k=1}^{K} SINR_i^{k,l}(t) \geq \xi_i, \forall i \in \mathcal{U}$$

where the first constraint $C1$ denotes the mode and the channel limitation. $C2$ limits the computation resource of computing servers, and the third constraint $C3$ means that each user should achieve the minimum QoS threshold.

Sequentially, mode-selection action taken by users may consume cost, the reward of the ith user should be given as the QoE utility minus the action-selection cost Ψ_i, that is,

$$R_i(t) = QoE_i(t) - \Psi_i \tag{14}$$

where $\Psi_i \geq 0$ acts as a punishment for the negative reward. To achieve the minimum QoS of all users, the punishment should be set large enough. Besides, the joint optimization problem is to maximize the long-term reward. We define the long-term reward Φ_i as the weighted sum of the instantaneous rewards over a finite period T. Hence, we can transfer P0 to P1 as

$$P1 : \max_{\mathcal{V},\mathcal{C}} \sum_{t=0}^{T-1} \gamma^t R_i(t)$$

$$s.t. C1 : \sum_{l=0}^{L-1} v_i^l(t) \leq 1, \sum_{k=1}^{K} c_i^k(t) \leq 1, \forall i \in \mathcal{U}$$

$$C2 : \sum_{i=1}^{U} \sum_{l=0}^{L-1} C_i^l(t) \leq C_{sum} \tag{15}$$

$$C3 : \sum_{l=0}^{L-1} \sum_{k=1}^{K} SINR_i^{k,l}(t) \geq \xi_i, \forall i \in \mathcal{U}$$

where $\gamma \in [0, 1)$ denotes the discount rate to determine the weight of the future reward. When $\gamma = 0$, we only focus on the immediate reward. $\gamma \le 1$ means that future rewards are smaller than the rewards in the earlier periods.

4   DGTB-DRL for the Optimization Problem

In particular, it is worth noting that the action space of the issue increases exponentially with the growth of the number of transmitting nodes and channels. Owing to the non-convex and combinatorial characteristics, it is a great challenge to find a globally optimal strategy for the joint mode selection and resource allocation problem. Besides, the traditional centralized learning method may need global information of all nodes to achieve the optimal solution. Hence, we develop a distributed game theory based DRL (DGTB-DRL) method for the optimization problem in the mobile edge network.

Suppose that all users do not know the network environment and the quality of transmitting nodes. At any time t, the reward of each user relies on the current state of the network environment and the action of other users. Thus, the learning game meets Markov property [11]. We formulate the joint optimization problem as a regular Markov decision process (MDP). The corresponding quadruple $\langle \mathcal{S}, \mathcal{A}_i, \mathcal{P}, \mathcal{R}_i \rangle$ is defined as follows.

1. *State space $\mathcal{S}$:* For the sake of convenience, let $s(t)$ denote the state of link quality at time t,

$$s(t) = \{s_1(t), s_2(t), \ldots, s_U(t)\}, \tag{16}$$

where $s_i(t) \in \{0, 1\}$ is a binary indicator. $s_i(t) = 1$ means that the ith user satisfies the minimum QoS threshold ξ_i and $s_i(t) = 0$ otherwise. It is critical that the number of possible states is as large as 2^U.

2. *Action space $\mathcal{A}_i$:* Considering the uniqueness of chosen transmitting node and channel for each user, we define the action space as

$$a_i^l(t) = \{v_i^l(t), c_i^l(t)\}, \tag{17}$$

where $v_i^l(t) \in \{0, 1\}$ and $c_i^k(t) \in \{0, 1\}$ are binary indicators and $\boldsymbol{v}_i^l(t) \in \{v_i^0(t), \ldots, v_i^{(L-1)}(t)\}$, $\boldsymbol{c}_i^k(t) \in \{c_i^1(t), \ldots, c_i^K(t)\}$.

3. *State transition probability $\mathcal{P}$:* The state is transferred by taking the action. We have state transition probability as follows:

$$P_{ss'}(\vec{a}) = P\left[\mathcal{S}_{t+1} = s' \mid \mathcal{S}_t = s, \mathcal{A}_t = a\right], \tag{18}$$

where $\vec{a} = (a_1, \ldots, a_U)^T$ is the joint action of all users.

4. *Reward function $\mathcal{R}_i$:* When the ith user makes a decision and takes the action $a_i^l(t)$, it receives an immediate reward which is referred to (14). We have $\boldsymbol{R}_i(\boldsymbol{t}) \in \{R_1(t), \ldots, R_U(t)\}$. Assume that the whole system is stationary, the

policy π_i is defined as a time-invariant mapping $\mathcal{S} \to \mathcal{A}_i$. According to formulation (15), the long-term reward can be formulated as follows.

$$\mathcal{R}_i(s, \pi_i, \pi_{-i}) = \sum_{t=0}^{T-1} \gamma^t R_i(s(t), \pi_i(t), \pi_{-i}(t) \mid s(0)), \tag{19}$$

where $\pi_{-i} = (\pi_1, \ldots, \pi_{i-1}, \pi_{i+1}, \ldots, \pi_U)$ is the policy of the rest $U - 1$ agents.

Notice that the reward of each user depends not only on its decisions but also on the decisions of other users. Therefore, the centralized optimization for such complex problems is unsuitable. To conquer the above shortcoming, we propose a distributed learning method. When the network changes dynamically, the system needs to rerun the entire scheme to reach the Nash equilibrium (NE) with traditional methods. Using the DGTB-DRL method proposed in this paper, each user can predict the utility generated by each action from multiple states with little consumption of computation and latency.

Hence, we formulate a stochastic game $\langle \mathcal{U}, \{\mathcal{A}_i\}_{i \in \mathcal{U}}, \{R_i\}_{i \in \mathcal{U}} \rangle$ that is involved in all users, where the utility function of the game is equivalent to Eq. (19) for convenience. Besides, $(\vec{a}_i, \vec{a}_{-i}) \in \mathcal{A}_i$ is defined as the feasible solution space of this game where $\vec{a}_{-i}$ are actions of the other users.

To solve the proposed problem, we used the pure-strategy NE theorem, that is:

$$\mathcal{R}_i\left(s, \vec{a}_i^*, \vec{a}_{-i}^*\right) \geq \mathcal{R}_i\left(s, \vec{a}_i, \vec{a}_{-i}^*\right), \tag{20}$$

The game reaches its NE state if and only if the inequalities above stay true.

Next, we resort to the DRL method to obtain an NE strategy, where we adopt the double-dueling DQN model to interact with the dynamic environment. Figure 2 shows the procedure of the proposed method.

To solve the problem of overestimation in typical DQN, the online network and the target network are linked up to calculate the target instead of using the target network alone, which is the essence of double DQN [12]. The target is expressed by

$$y_i = R_i + \gamma \hat{Q}_i\left(s', \arg\max_{a_i \in \mathcal{A}_i} Q_i\left(s', a_i; \theta\right); \theta^-\right). \tag{21}$$

Moreover, the dueling architecture [13] is introduced to improve the training efficiency of DQN by estimating a part of the value of actions, which can be expressed by

$$Q(s, a) = A(s, a) + V(s) \tag{22}$$

The overall algorithm is illustrated in Algorithm 1.

According to the increasing utility and the existence of its upper bound, the purpose game will achieve convergence eventually within finite steps. Therefore, the purpose DGTB-DRL algorithm can guarantee to achieve a pure strategy NE.

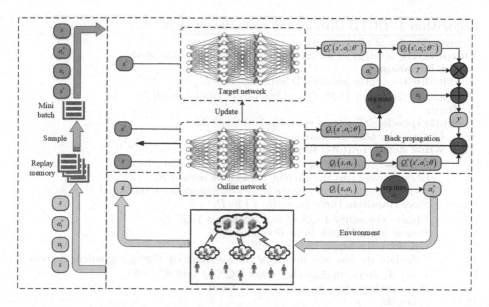

Fig. 2. Deep reinforcement learning with double-dueling DQN model.

5 Simulation Results and Discussions

In this section, simulations are executed to evaluate the performance of the proposed algorithm which shows the advantages of the proposed method in terms of training efficiency and system utility. Similar to [14], we consider a mixed VR scenario consisting of 2 cloud nodes, 8 edge nodes 30 end nodes whose correlated transmission radiuses are 500 m, 100 m and 30 m, respectively. We conduct the simulation on a 64-bit computer with 16 GB of RAM and Intel i7 1.8 GHz. The other simulation parameters of the proposed solution are defined in Table 1.

Table 1. Simulation parameters.

Parameter	Value
Channel bandwidth W	180 MHz
Orthogonal channels K	20
Transmit power	40, 30, 20 dBm
Path loss model of cloud/edge node	$34 + 40\log(d)$
Spectral noise power density N_0	-174 dBm/Hz
Minimum QoS threshold ξ_i	5 dB
Action-selection cost Ψ_i	0.001
Non-negative weight ω_1	0.5
Non-negative weight ω_2	0.0005

Algorithm 1. DGTB-DRL for the optimization problem

Input: The list of allowed actions taken by all users.
Output: The strategy of all users that meets the QoS threshold.
1: **Initialization**
2: Initialize the replay memory $\mathcal{D}$ with capacity L_{rp}.
3: Initialize the online DQN Q and the target DQN $\hat{Q}$ with $\theta^- = \theta$.
4: **Run:**
5: **while** episode $\leq T_1$ (total episodes in a trial) **do**
6:    Observe the network state s.
7:    **while** step $\leq T_2$ (total steps in an episode) **do**
8:       Each user selects an action a_i using ϵ-greedy policy from $\hat{Q}$.
9:       Each user obtains the current immediate reward R_i .
10:      Each user gets the new state s' by communications and sets $s \leftarrow s'$.
11:      Store transition tuple (s, a_i, R_i, s') in $\mathcal{D}$.
12:      Update the online DQN Q and the target DQN $\hat{Q}$.
13:      Sample a mini-batch from $\mathcal{D}$ randomly.
14:      Calculate the target based on (21).
15:      Calculate the loss and minimize the loss function through gradient descent
16:      Every T_0 steps, update the target DQN $\hat{Q}$ with $\theta^- = \theta$
17:      **if** the system state is $s = (1, \ldots, 1)$ **then**
18:         Break.
19:      **end if**
20:   **end while**
21: **end while**
22: **Return:** The optimal allocation scheme, mode selection strategy and the total
    reward.

In our experiments, the deep neural networks in DGTB-DRL are composed of an input layer, 3 hidden layers and an output layer using the ReLU function for activation function. We initialize 500 episodes and 500 steps for a trial, 8 for mini-batch size and 500 for replay memory $\mathcal{D}$. The ϵ-greedy policy is utilized linearly with ϵ from 0 to 0.9.

5.1   Evaluation of Different Learning Parameters

Firstly, the proposed method is evaluated with various learning rates η. Figure 3 demonstrates the training efficiency with varying η. At the early phase of the process, training steps are huge in all trials. As the number of episodes grows up, training steps tend to converge with narrow fluctuation. Furthermore, as η decreases, the speed of convergence increases that shows all users satisfy the QoS threshold through exploration and exploitation. However, when $\eta = 0.0001$, the learning convergence becomes unstable at the end of the process. Considering comprehensive performance, η is hence chosen to be 0.01.

Next, we evaluate the performance of the proposed method with different optimization strategies. As shown in Fig. 4, as well as learning rates, training steps are very large in all cases at the early phase. As the number of episodes

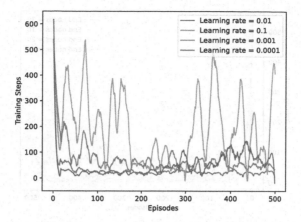

Fig. 3. Training steps of different learning rates η.

grows up, the RMSProp optimizer and Adagrad optimizer show better convergence than Adam. Consequently, we choose the Adagrad optimization strategy in our DGTB-DRL method.

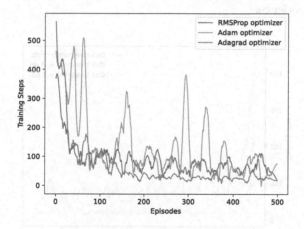

Fig. 4. Training steps of different optimization strategies.

5.2   Simulation of Different Scenarios

Secondly, simulations of various scenarios are executed to verify the proposed method. The number of end nodes is varied in this experiment. Figure 5 indicates that the efficiency of the DGTB-DRL method is decreasing with the growth of the number of users. Concretely, when $U = 10$ the speed of convergence is faster than $U = 20$ and $U = 30$, which means that as U increases, it takes more time for agents to get their optimal action vector and achieve the NE due to the increase of interference.

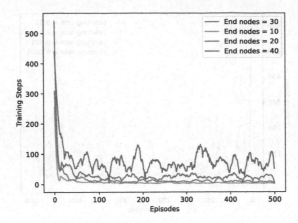

Fig. 5. Training steps with various numbers of end nodes.

Moreover, Fig. 6 demonstrates the performance of the DGTB-DRL method with different minimum QoS thresholds of users ξ_i. When $\xi_i = 0\,\text{dB}$ and $\xi_i = -5\,\text{dB}$, the curves of learning efficiency perform well. However, with the growth of QoS requirements, more steps need to be tried for the proposed method to meet users' experience.

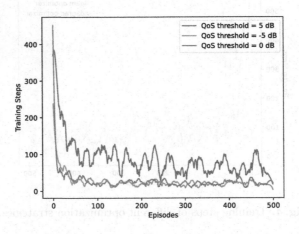

Fig. 6. Training steps with different minimum QoS thresholds of users.

5.3 Performance of Different Algorithms

Ultimately, the performance of training efficiency and total QoE utility value of the DGTB-DRL method are compared with several mainstream optimization algorithms which are presented in Figs. 7 and 8. For the convenience of comparison, the curves of training steps in Fig. 7 are fitted. The proposed DGTB-DRL algorithm shows better performance in terms of training speed and convergence. Moreover, the simple genetic algorithm (SGA) as the classic method to solve the optimization problem is also considered with varying numbers of users. On account of computational complexity, the experiment of SGA terminates at 30. As the number of users increases, the total QoE utility increases in all optimization methods. Specifically, three learning methods perform better than SGA and achieve approximately the same QoE utility. However, the DQN method and Q-learning method obtain less QoE utility when users of the system get higher, which shows the superiority of the proposed DGTB-DRL method.

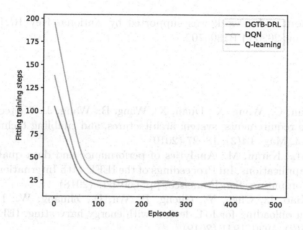

Fig. 7. Training steps of different learning methods.

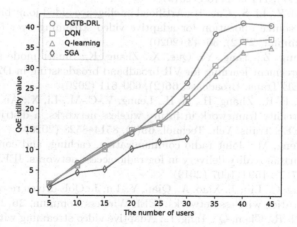

Fig. 8. Total QoE utility of different optimization methods.

6   Conclusion

In this paper, we focus on the transmission of VR services based on the cloud-edge-end architecture towards future generation communication. Foremost, we formulate a joint optimization of the mode selection and resource allocation to maximize the QoE of VR users. Then we propose the DGTB-DRL algorithm to conquer the high complexity problem, which can achieve a Nash equilibrium (NE) and maintain the adaptation capability and scalability. The numerical results show the superiority of the proposed method compared with prevalent schemes in terms of training efficiency, QoE utility values.

Acknowledgment. This work was supported by National Key R&D Program of China under Grant 2020YFB1806702.

References

1. Zong, B., Fan, C., Wang, X., Duan, X., Wang, B., Wang, J.: 6G technologies: key drivers, core requirements, system architectures, and enabling technologies. IEEE Veh. Technol. Mag. **14**(3), 18–27 (2019)
2. Truong, H.-L., Karan, M.: Analytics of performance and data quality for mobile edge cloud applications. In: Proceedings of the IEEE 11th International Conference on Cloud Computing, San Francisco, pp. 660–667 (2018)
3. Min, M., Xiao, L., Chen, Y., Cheng, P., Wu, D., Zhuang, W.: Learning-based computation offloading for IoT devices with energy harvesting. IEEE Trans. Veh. Technol. **68**(2), 1930–1941 (2019)
4. Liu, Y., Yu, H., Xie, S., Zhang, Y.: Deep reinforcement learning for offloading and resource allocation in vehicle edge computing and networks. IEEE Trans. Veh. Technol. **68**(11), 11158–11168 (2019)
5. Chen, J., Wei, Z., Li, S., Cao, B.: Artificial intelligence aided joint bit rate selection and radio resource allocation for adaptive video streaming over F-RANs. IEEE Wireless Commun. **27**(2), 36–43 (2020)
6. Feng, L., Yang, Z., Yang, Y., Que, X., Zhang, K.: Smart mode selection using online reinforcement learning for VR broadband broadcasting in D2D assisted 5G hetNets. IEEE Trans. Broadcast. **66**(2), 600–611 (2020)
7. Guo, F., Yu, F.-R., Zhang, H., Ji, H., Leung, V.-C.-M., Li, X.: An adaptive wireless virtual reality framework in future wireless networks: a distributed learning approach. IEEE Trans. Veh. Technol. **69**(8), 8514–8528 (2020)
8. Dang, T., Peng, M.: Joint radio communication, caching, and computing design for mobile virtual reality delivery in fog radio access networks. IEEE J. Sel. Areas Commun. **37**(7), 1594–1607 (2019)
9. Tao, X., Jiang, C., Liu, J., Xiao, A., Qian, Y., Lu, J.: QoE driven resource allocation in next generation wireless networks. IEEE Wireless Commun. **26**(2), 78–85 (2019)
10. Luo, J., Yu, F.R., Chen, Q., Tang, L.: Adaptive video streaming with edge caching and video transcoding over software-defined mobile networks: a deep reinforcement learning approach. IEEE Trans. Wireless Commun. **19**(3), 1577–1592 (2019)
11. Neyman, A., Sorin, S.: Stochastic Games and Applications. Kluwer, Dordrecht (2003)

12. Hasselt, H., Guez, A., Silver, D.: Deep reinforcement learning with double Q-learning. In: Proceedings of the 30th AAAI Conference on Artificial Intelligence (2016)
13. Wang, Z., Freitas, N., Lanctot, M.: Dueling network architectures for deep reinforcement learning (2015). arXiv:1511.06581
14. Zhao, N., Liang, Y., Niyato, D., Pei, Y., Wu, M., Jiang, Y.: Deep reinforcement learning for user association and resource allocation in heterogeneous cellular networks. IEEE Trans. Wireless Commun. **18**(11), 5141–5152 (2019)

Joint Optimization of D2D-Enabled Heterogeneous Network Based on Delay and Reliability Constraints

Dengsong Yang[1], Baili Ni[1], Haidong Wang[2], and Baoxiang Wei[2(✉)]

[1] Jiangsu Expressway Company Limited, Nanjing, China
[2] College of Computer and Information, Hohai University, Nanjing, China
2819886353@qq.com

Abstract. The D2D-enabled heterogeneous network services have strict requirements on communication delay and quality. D2D users can share spectrum resources with cellular users, which can effectively improve the spectrum efficiency. But it can cause serious co-channel interference in dense user scenarios. In order to satisfy the highly dynamic patterns of the data traffic and wireless communication environment, a joint optimization algorithm of D2D-enabled heterogeneous network resources is proposed while meeting the delay and quality requirements. Specifically, a heterogeneous network is established where users are clustered. We maximize system energy efficiency under the premise of meeting the user delay and reliability requirements. In addition, we use the Lyapunov algorithm to optimize the dynamic allocation of wireless resources. The results of the experiment show that the proposed algorithm can improve the spectrum efficiency, reduce the D2D power consumption and reduce the co-channel interference, while ensuring the reliability and time delay requirements of both D2D users and cellular users.

Keywords: D2D · Resources allocation · Lyapunov optimization

1 Introduction

In traditional cellular networks, communication is divided into uplink and downlink. Users can't communicate directly bypassing the base station. With the rapid development of social economy and the advent of the information age, wireless spectrum resources are becoming increasingly tense.

Device-to-device (D2D) communication is a technology under the control of a cellular network system, which enables two terminal users to communicate with each other directly without a base station. D2D communication is a good solution to the spectrum crisis, which provides high data rate and reduces user delay and transmission

Supported by Major Technological Projects of Jiangsu Province Transportation Department (2019Z07).

power. Therefore, it is necessary to study cellular communication and the heterogeneous network of terminal direct communication.

In the heterogeneous network, D2D users' reuse of cellular user (CUE) spectrum resources will produce complex co-channel interference, which will reduce the communication quality of users. Besides, D2D communication is only applicable to short-range communication scenarios. It is necessary to study the network resource allocation to reduce the co-channel interference between users, to improve the spectrum utilization and system performance. At present, some work about resource allocation of heterogeneous network focuses on resource allocation of D2D user side. And, other studies on the joint optimization of D2D user and cellular user resources usually only take the performance of D2D user or cellular user as the optimization goal, ignoring the overall performance optimization. In addition, these studies don't consider the characteristics of random arrival of user data traffic in the network. [1] investigated the problem of network resource allocation by reusing downlink spectrum resources of cellular users. Han et al. considered the dynamic change of data traffic in network by establishing user queue model [2, 3].

Aiming at the above problems, in the scenario where D2D user and cellular user data flow arrive randomly, we optimize spectrum reuse and power consumption of D2D user side and cellular user side. Under the constraints of delay and reliability of D2D users and cellular users, we maximize the energy efficiency of the overall system and ensure the data queue stability of D2D users and cellular users.

2 System Model

Considering that multiple D2D users in a single cell reuse downlink spectrum resources of cellular users, a system model is established as shown in Fig. 1.

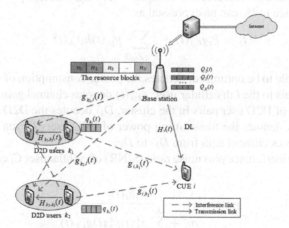

Fig. 1. Resource joint optimization interference model in heterogeneous network

We assume that both base stations and users are equipped with a single omnidirectional antenna. The system includes a base station located in the regional center.

We denote the set of N cellular users by $C = \{C_1, C_2, \cdots, C_N\}$, and denote by $B = \{B_1, B_2, \cdots, B_N\}$ the set of resource blocks that matched by D2D user pair. In addition, we denote K D2D user pairs by $D = \{D_1, D_2, \cdots, D_K\}$, and each D2D user pair contain a sender and a receiver. It is stipulated that one D2D user pair can reuse multiple cellular user spectrum resources, and at the same time, cellular user spectrum resources can be reused by multiple D2D users in different clusters. We assume all users' geographical location obey Poisson distribution, and the base station can obtain the location information of all users.

In order to reduce co-channel interference between adjacent D2D user pairs, we group D2D user pairs into different regions based on their geographic location, which is called clustering. The part about clustering will be introduced in detail later.

2.1 Interference Model

The system considers not only the data queue of D2D users, but also the cache queue of cellular users caused by dynamic data traffic. The power consumption of signals sent by base station to cellular users is no longer a constant, but a variable to be optimized. When D2D user pair D_k reuses the downlink spectrum resources of cellular user C_i, cellular user C_i will be interfered by the sender of D2D user pair D_k. The receiver of D_k will be interfered by the base station and other D2D users. The interference to cellular user C_i can be expressed as

$$I_i = \sum_{k=1}^{K} x_k^i(t) p_k^i(t) g_{i,k}(t) \tag{1}$$

where $P_k^i(t)$ denotes the transmitting power of D_k on spectrum resource block B_i, $g_{i,k}(t)$ denotes channel gain from D2D user pair D_k transmitter to cellular user C_i. When D_k reuse downlink spectrum resources of C_i, $x_k^i(t) = 1$, otherwise $x_k^i(t) = 0$.

The interference to D_k can be expressed as

$$I_k^i = P_i g_{k,i}(t) + \sum_{k' \in Z_b/k} p_{k'}^i(t) h_{k',k}(t) \tag{2}$$

where P_i, a variable to be optimized, denotes the power consumption of the base station transmitting signals to the i th cellular user, $g_{k,i}(t)$ denotes channel gain of D_k receiver, Z_b is a collection of D2D user pairs in the cluster, $D_{k'}$ denotes the D2D users in cluster Z_b except D_k, $p_{k'}^i$ denotes the transmitting power of D2D users $D_{k'}$ on resource block B_i, and $h_{k',k}$ denotes channel gain from $D_{k'}$ to D_k.

The signal to interference plus noise ratio (SINR) of cellular user C_i can be expressed as

$$\gamma_i(t) = \frac{P_i h_i(t)}{\sigma^2 + \sum_{k=1}^{K} x_k^i(t) p_k^i(t) g_{i,k}(t)} \tag{3}$$

where $h_i(t)$ denotes channel gain from the base station to cellular user C_i, σ^2 denotes power of Gaussian white noise. Assuming the SINR threshold of cellular user is γ_{iTH}, so communication quality requirement for cellular user can be expressed as $\gamma_i(t) \geq \gamma_{iTH}$.

When D_k reuse C_i, the SINR of D_k receiver can be expressed as

$$\gamma_k^i(t) = \frac{p_k^i(t)h_{k,k}(t)}{\sigma^2 + P_i g_{k,i}(t) + \sum_{k' \in Z/k} p_{k'}^i(t)h_{k',k}(t)} \tag{4}$$

where $h_{k,k}(t)$ denotes channel gain from D_k transmitter to D_k receiver. D_k and $D_{k'}$ belong to the collection Z, which includes all users in a cluster. The content of cluster will be introduced in the following section. We assume the SINR threshold of D2D user pair is γ_{kTH}^i. Then, reliability requirements for D2D user pair can be expressed as $\gamma_k^i(t) \geq \gamma_{kTH}^i$.

2.2 Delay and Reliability Model

The queuing delay of task traffic and communication reliability that can be measured by the length of user queue are mainly considered in the heterogeneous scenario [1].

Each D2D user pair maintains a cache queue at the transmitter, defining the queue length vector as $q(t) = [q_1(t), \cdots, q_K(t)]$, where $q_k(t)$ denotes the queue length of D2D user pair D_k. Similarly, we define the cell user queue length vector as $Q(t) = [Q_1(t), \cdots, Q_N(t)]$, where $Q_i(t)$ denotes the queue length of the i th cellular user.

According to Little Law in Queuing Theory [4], in the heterogeneous network system in this paper, the delay is proportional to the system data queue length, and inversely proportional to the data arrival rate. Therefore, the data queue length can be used as a measure of delay. According to the above conclusions, we can set an upper bound d_k for the average queuing delay of each D2D user pair and an upper bound D_i for the average queuing delay of each cellular user.

In addition to queuing delay, queue length is also related to the reliability requirement of communication, so the reliability index can be measured by the violation of queue length boundary. We impose probability to restrain the queue length of each D2D user pair, as

$$\lim_{T \to \infty} \frac{1}{T} \sum_{t=1}^{T} \Pr(q_k(t) \geq l_k) \leq \varepsilon_k \tag{5}$$

where l_k denotes maximum allowable queue length for D2D users, ε_k denotes a tolerable violation probability. However, this probability constraint is difficult to deal with, so the nonlinear constraint is replaced by the linear equivalent constraint, which can be expressed as

$$\lim_{T \to \infty} \frac{1}{T} \sum_{t=1}^{T} E[1\{q_k(t) \geq l_k\}] \leq \varepsilon_k \tag{6}$$

In a similar way, we impose probability to restrain the queue length of each cellular user, as

$$\lim_{T \to \infty} \frac{1}{T} \sum_{t=1}^{T} \Pr(Q_i(t) \geq L_i) \leq E_i \tag{7}$$

where L_i denotes the maximum allowable queue length for cellular user, and E_i denotes a tolerable violation probability. Replace nonlinear constraint of cellular users with linear equivalent constraint, which can be expressed as

$$\lim_{T \to \infty} \frac{1}{T} \sum_{t=1}^{T} E[1\{Q_i(t) \geq L_i\}] \leq E_i \tag{8}$$

Through the above series of constraints on user queue length, the delay and reliability constraints of D2D users and cellular users can be satisfied.

2.3 Optimization

Our goal is to optimize the control of transmission power of the D2D users, the power consumption transmitted by the base station to the cellular user and the spectrum resource matching at the base station, in order to maximize system energy efficiency under the premise of meeting the user delay and reliability requirements. We characterize the system performance by energy efficiency, which is defined as ratio of system throughput to system power consumption.

The spectrum resource block allocation matrix is $X(t) = [x_k^i(t)]$, and the D2D users power matrix is $p(t) = [p_k^i(t)]$, the power matrix of the signal transmitted by the base station to the cellular user is $P(t) = [P_i(t)]$. In this way, our joint optimization problem can be summarized as

$$\max_{X(t), p(t), P(t)} \frac{\sum_{k=1}^{K} r_k(t) + \sum_{i=1}^{N} R_i(t)}{\sum_{k=1}^{K} y_k(t) p_k^{\max} + P^{\max}} \tag{9}$$

meeting:

$$x_k^i(t) \in \{0, 1\}, \forall t, k \in K, i \in N \tag{9a}$$

$$\sum_{i \in N} x_k^i(t) \leq n_k, \forall k \in K \tag{9b}$$

$$\lim_{T \to \infty} \frac{1}{T} \sum_{t=1}^{T} E[q_k(t)] \leq \bar{\lambda}_k d_k, \forall k \in K \tag{9c}$$

$$\lim_{T \to \infty} \frac{1}{T} \sum_{t=1}^{T} E[Q_i(t)] \leq \bar{\lambda}_i D_i, \forall i \in N \tag{9d}$$

$$\lim_{T \to \infty} \frac{1}{T} \sum_{t=1}^{T} E[1\{q_k(t) \geq l_k\}] \leq \varepsilon_k, \forall k \in K \tag{9e}$$

$$\lim_{T \to \infty} \frac{1}{T} \sum_{t=1}^{T} E[1\{Q_i(t) \geq L_i\}] \leq E_i, \forall i \in N \tag{9f}$$

$$\sum_{n\in N} x_k^i(t) p_k^i(t) = p_k^{\max}, \forall t, k \in K \tag{9g}$$

$$p_k^i(t) \geq 0, \forall t, k \in K, i \in N \tag{9h}$$

$$\sum_{i=1}^{N} P_i(t) = P^{\max} \tag{9i}$$

$$P_i(t) \geq 0, i \in N \tag{9j}$$

$$\gamma_k^i(t) \geq \gamma_{kTH}^i \tag{9k}$$

$$\gamma_i(t) \geq \gamma_{iTH} \tag{9l}$$

where $r_k(t)$ denotes transmission rate of D2D user pair D_k, $R_i(t)$ denotes transmission rate of cellular user C_i. $y_k(t)$ is an indicator variable. If D_k matches to the spectrum resource block, $y_k(t) = 0$; otherwise, $y_k(t) = 1$. $\bar{\lambda}_k$ and $\bar{\lambda}_i$ denote the average task data arrival of D_k and C_i respectively. In order to facilitate representation, the energy efficiency expression (9) is denoted by EE.

3   Dynamic Allocation of Network Resources

3.1   Spectrum Resource Matching Algorithm Based on Cluster

In the scenario of the dense distribution of D2D users in the region, we can group D2D users into different regions to reduce the co-frequency interference between adjacent D2D users, according to their geographical location. The region is named a cluster. Then, it is stipulated that the D2D users in the same cluster can't reuse the same spectrum resource block, so that there will be no co-frequency interference between the D2D users in the same cluster.

D2D Users Clustering Algorithm
We use a dynamic region formation strategy to divide a collection of regions into non-overlapping regions, as

$$\bigcup_{\forall z \in Z} K_z = K \tag{10}$$

$$K_z \cap K_{z'} = \emptyset, \forall z \neq z', z, z' \in Z \tag{11}$$

Gaussian distance similarity is based on the distance between D_k and $D_{k'}$, which can be expressed as

$$g_{kk'}(i) = \exp(\frac{-\|g_k - g_{k'}\|_2}{2\sigma_s^2}) \tag{12}$$

174   D. Yang et al.

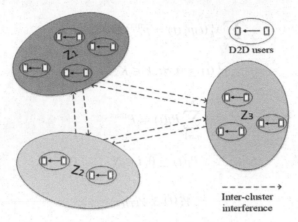

Fig. 2. Clustering mechanism

where g_k and $g_{k'}$ denote two-dimensional geographic coordinates of D_k and $D_{k'}$, parameter σ_s controls influence range of area size. The relationship among D2D users after clustering is shown in Fig. 2. Since the intra-cluster D2D users strictly obey the principle of orthogonal frequency division multiplexing, the interference between users in the cluster is eliminated, and there is only inter-cluster interference.

We note that the solution of the problem depends on the resource block matching and power optimization scheme. And resource block matching and power allocation are coupled. The solution to the problem (9) requires global information, such as $X(t)$ and $P(t)$ in the network. But it's difficult to achieve. Therefore, we use the semi-distributed method to decouple the resource block matching and the power allocation, according to the Lyapunov optimization theory [5].

Spectrum Resource Matching Algorithm

Based on the clustering results, D2D users and the spectrum resource blocks of cellular users are matched bilaterally according to the mutual preference relationship. Since the heterogeneous network resource allocation in the paper involves the power optimization of base station transmitting to cellular users, it is necessary to propose a new utility function according to the model.

We use the framework of matching theory to solve the problem [6–10]. The matching $K_z \to N$ of each region is essentially a bilateral matching problem of two disjoint users, namely, the matching of D2D users set K_z and resource block set N. We assume that each user in K_z matches one or more users in N, and each user in N matches one user in K_z at most. In the proposed matching, D2D users with longer queues need more resource blocks to transmit data from the queue buffer.

The matching framework defines the utility that reflects the preferences of D2D users and resource blocks. The utility of D_k matched to B_i can be expressed as

$$U_k(t) = -\sum_{i \in N} \Gamma_{ki}(t) \tag{13}$$

Accordingly, the utility of B_i matched to D_k can be expressed as

$$U_{ki}(t) = -\Gamma_{ki}(t) \tag{14}$$

$\Gamma_{ki}(t)$ will be introduced in detail in the following section.

When optimizing resource block selection to maintain queue stability, each D2D user maximizes its utility. Therefore, D2D users sorts the subset of the resource block, namely, $P \subseteq N$ and $P' \subseteq N$. We can define the preference of D_k as

$$P >_k P' \Leftrightarrow U_k(P) > U_k(P') \tag{15}$$

On the other hand, each resource block will sort D_k and $D_{k'}$ to maximize its utility value, which can be expressed as

$$k >_n k' \Leftrightarrow U_{ki} > U_{k'i} \tag{16}$$

Using the above formulas, we obtain an algorithm that can be used for appropriate matching. The matching result is that the set of matched resource blocks for D_k is N_k, and $\cup N_k = N, k \in K_z$.

3.2 Power Optimization Based on Lyapunov Method

Lyapunov optimization theory can provide online solutions to meet the optimization objectives. For the dynamic network in the paper, it can dynamically adapt to changes. With the help of virtual queue method, the original stochastic optimization problem is transformed into equivalent solvable problem. Then we can use the Lyapunov optimization framework to solve the problem [11].

Optimization Problem Transformation
By using Lyapunov optimization, the time-average inequality constraint is transformed into a virtual queue. If the stability of the virtual queue is maintained, the time constraint can be satisfied. Therefore, virtual queues $J(t)$ and $F(t)$ for D2D users are defined by constraints (9c) and (9e) respectively, as follows

$$J_k(t + 1) = \max\{J_k(t) + q_k(t + 1) - \bar{\lambda}_k \cdot d_k, 0\} \tag{17}$$

$$F_k(t + 1) = \max\{F_k(t) + 1\{q_k(t + 1) \geq l_k\} - e_k, 0\} \tag{18}$$

If the corresponding virtual queue is stable, in other words, $\lim_{t\to\infty} E[|J_k(t)|]/T = 0$ and $\lim_{t\to\infty} E[|F_k(t)|]/T = 0$, the time average inequality constraint is satisfied.

Similarity, virtual queues $B(t)$ and $M(t)$ for cellular users are defined by constraints (9c) and (9e) respectively, as follows

$$B_i(t + 1) = \max\{B_i(t) + Q_i(t + 1) - \bar{\lambda}_i \cdot D_i, 0\} \tag{19}$$

$$M_i(t + 1) = \max\{M_i(t) + 1\{Q_i(t + 1) \geq l_i\} - E_i, 0\} \tag{20}$$

In this way, the original optimization problem is equivalent to maximize the system energy efficiency in the condition that the average rate of virtual queue is stable.

D2D Network Resource Dynamic Allocation Algorithm

Using Lyapunov Optimization Theory, combination of virtual queues can be expressed as $y(t) = [J(t), F(t), B(t), M(t)]$. Define the Lyapunov function $L(y(t))$ as

$$L(y(t)) = \frac{1}{2}\{\|J(t)\|^2 + \|F(t)\|^2 + \|B(t)\|^2 + \|M(t)\|^2\} \tag{21}$$

We define the Lyapunov drift function as

$$\Delta(L(t)) = E[L(y(t+1)) - L(y(t))]$$

$$= E[\sum_{k=1}^{K}\sum_{i=1}^{N}(\frac{J_k(t+1)^2}{2} - \frac{J_k(t)^2}{2} + \frac{F_k(t+1)^2}{2} - \frac{F_k(t)^2}{2}$$

$$+ \frac{B_i(t+1)^2}{2} - \frac{B_i(t)^2}{2} + \frac{M_i(t+1)^2}{2} - \frac{M_i(t)^2}{2})] \tag{22}$$

By expanding $J(t), F(t), B(t), M(t)$ and taking into account $\max\{x, 0\}^2 \leq x^2$, we can get

$$\Delta(L(t)) \leq E[\sum_{k=1}^{K}\sum_{i=1}^{N}(-\tau(J_k(t) + F_k(t) + 2q_k(t) + 2\lambda_k(t))r_k^i(t)$$

$$- \tau(B_i(t) + M_i(t) + 2Q_i(t) + 2\lambda_i(t))R_i(t))] + C \tag{23}$$

where C can be regarded as a constant part, which can be ignored in optimization problem. The stability of virtual queues can be guaranteed by minimizing the nonconstant part. Our goal is to maximize the system energy efficiency, namely $\max\{EE\}$ or $\min\{-EE\}$.

Therefore, if the sum of the negative value of the Lyapunov drift function and the energy efficiency function is minimized, the constraint conditions can be satisfied and the optimization objective can be realized. The optimal solution can be obtained by implementing the following polynomial

$$\min_{X(t),p(t),P(t)} imize \sum_{k=1}^{K}\sum_{i=1}^{N}\Gamma_k^i(t) \tag{24}$$

where

$$\Gamma_k^i(t) = -\tau(J_k(t) + F_k(t) + 2q_k(t) + 2\lambda_k(t))r_k^i(t)$$

$$- \tau(B_i(t) + M_i(t) + 2Q_i(t) + 2\lambda_i(t))R_i(t) - EE \tag{25}$$

After subset B_k of the spectrum resource block is assigned to the D2D users in cluster z_b, the D2D user pair $D_k \in K_{z_b}$ optimizes the transmission power of the allocated spectrum resource blocks. D2D users optimize the power allocation locally, and the base station optimizes the power consumption transmitted to cellular users. Therefore, local power allocation based on matching algorithm can be written as a convex optimization problem, as follows

$$\min_{p_k^i, P_i} \{ \sum_{k=1}^{K} \sum_{i=1}^{N} (-\tau(J_k(t) + F_k(t) + 2q_k(t) + 2\lambda_k(t))r_k^i(t)$$

$$- \tau(B_i(t) + M_i(t) + 2Q_i(t) + 2\lambda_i(t))R_i(t)) - EE \} \qquad (26)$$

meeting the following constraints

$$\gamma_k^i \geq \gamma_{kTH}^i, \forall k \in K_{z_b}, i \in B_k \qquad (26a)$$

$$\gamma_i \geq \gamma_{iTH}, \forall k \in K_{z_b}, i \in B_k \qquad (26b)$$

$$\sum_{i \in B_k} p_k^i(t) = p_k^{\max}, \forall t, k \in K \qquad (26c)$$

$$\sum_{i=1}^{N} P_i(t) = P^{\max} \qquad (26d)$$

where K_{z_b} denotes the set of D2D users in cluster z_b, B_k denotes the set of matched spectrum resource blocks for D_k. The power optimization solution can be obtained by solving the convex optimization problem in each time slot. The paper uses MATLAB convex optimization tool CVX to solve the convex optimization problem [12].

4 Simulation Results and Analysis

4.1 Simulation Configuration

D2D users and cellular users are distributed in the rectangular area of 400 m × 400 m and they obey Poisson distribution. The loss is a free-space propagation model $L_0 = 32.45 + 20 \log(f) + 20 \log(d)$. The carrier frequency used by cellular users is 5.8 GHz. We set the number of cellular users $N = 10$, which is the same as the number of resource blocks. The number of D2D users is $K = 15$. For each D2D user pair, the maximum transmit power is $P_k^{\max} = 20$ dBm. Base station transmit power is $P^{\max} = 20$ dBm. D2D users and cellular users packet size is fixed, $N_k = N_i = 6400$ bit. The packet arrival rate is independent and identically distributed in different time, which follows Poisson distribution. The average packet arrival rate of D2D users is $\bar{\zeta}_k = 100$, and the average packet arrival rate of cellular users is $\bar{\zeta}_i = 10$. The bandwidth of spectrum resource block is $\omega = 180$ kHz, the time slot length is $\tau = 5$ ms. D2D users' delay constraint threshold is $d_k = 10$ ms, cellular users' delay constraint threshold is $D_i = 20$ ms. The number of clusters is b.

4.2   Results and Analysis

For the convenience of expression, the average distribution scheme of base station transmitting power is denoted as APRA, and the joint optimization scheme proposed in the paper is denoted as OPRA.

Energy Efficiency

System energy efficiency reflects the amount of transmitted data per unit time per unit power. It's very meaningful to improve energy efficiency in the era of green communication. Figure 3 shows the complementary cumulative distribution function (CCDF) of system energy efficiency in OPRA and APRA. It can be seen that in the process of increasing the number of clusters from 3 to 5, the energy efficiency of the two systems is continuously improving. With the increase in the number of clusters, the number of spectrum resource blocks that can be matched to each D2D user pair will increase. Then, the transmission rate and the system energy efficiency of D2D users will be significantly improved, while cellular users are less affected by co-channel interference.

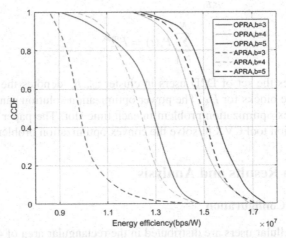

Fig. 3. System energy efficiency CCDF of OPRA and APRA

By comparison, the energy efficiency performance of OPRA is better than that of APRA. It's due to the decrease of the co-frequency interference among users. After obtaining the matching results of spectrum resource blocks, optimizing the transmit power of the base station can improve data transmission rate of D2D users and reduce the co-channel interference of the downlink signal of the base station to the D2D users. Meanwhile, optimization of power consumption for D2D users also reduces interference to cellular users.

User Delay in Heterogeneous Networks

Figure 4 shows the CCDF of D2D users and cellular users delay. No matter the number of clusters, the delay performance of D2D users and cellular users is OPRA better than APRA. With the increase of the number of clusters, the delay performance of D2D users improves greatly. When the number of clusters increases, the number of D2D users in a cluster will decrease accordingly, the number of spectrum resource blocks matched to each D2D user pair will increase, the transmission rate will increase, and the delay will decrease. With the increase of the number of clusters, the delay of cellular users increases. It's because the more clusters, the greater the co-frequency interference of D2D users to cellular users is, and the delay will increase.

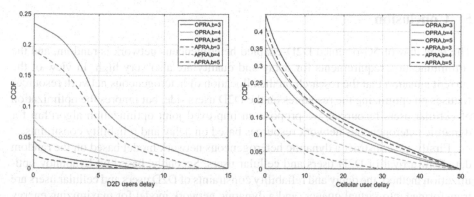

Fig. 4. Time delay CCDF of D2D users and cellular users

It can be seen that there is a trade-off between the delay performance of D2D users and that of cellular users. The decrease of D2D users delay will lead to the increase of cellular users delay, and vice versa.

User Reliability of Heterogeneous Network

Because we set an upper limit threshold for the cache queue, the situation that exceeds this threshold is considered unreliable, and the situation that doesn't exceed the threshold is considered reliable. Figure 5 shows the comparison of D2D users and cellular users reliability in different clustering cases.

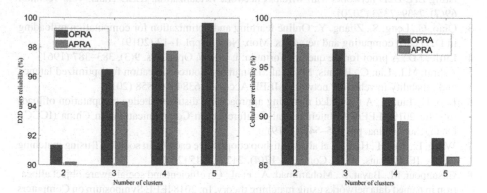

Fig. 5. Reliability of D2D users and cellular users in some scenarios

With the increase of clusters, the reliability of D2D users increases, meanwhile the reliability of cellular users decrease. The reasons are the same as before, which are given in delay section. We note that OPRA is more reliable than APRA for both D2D users and cellular users. Because OPRA outperforms APRA in terms of energy efficiency, which means that the transmission capability of OPRA is better than APRA at the same power consumption.

5 Conclusion

The data flow of 5G-oriented D2D-enabled heterogeneous network is random, and the communication requirements for delay and quality are also very high. In view of the current situation that the research on the allocation of heterogeneous network resources focuses on optimizing the resources on the D2D users side, but ignores the optimization of cellular users resources, we propose an improved joint optimization algorithm for dynamic heterogeneous network resources based on delay and reliability constraints.

Firstly, we construct a dynamic heterogeneous network model based on the random data flow arrival of D2D users and cellular users. Secondly, based on Lyapunov optimization method, the delay and reliability constraints of D2D users and cellular users are transformed into virtual queues, and a dynamic network model for maximizing energy efficiency of heterogeneous network system is established. Finally, we propose an energy efficiency maximization algorithm for heterogeneous network system. Compared with the average allocation scheme of base station transmitting power, the algorithm has improved the overall system energy efficiency and the performance of users delay and reliability.

References

1. Ashraf, M.I., Liu, C.-F., Bennis, M., Saad, W., Hong, C.S.: Dynamic resource allocation for optimized latency and reliability in vehicular networks. IEEE Access **6**, 63843–63858 (2018)
2. Han, Y., Tao, X., Zhang, X., Jia, S.: Delay-aware resource management for multi-service coexisting LTE-D2D networks with wireless network virtualization. IEEE Trans. Veh. Technol. **69**(7), 7339–7353 (2020)
3. Qiao, G., Leng, S., Zhang, Y.: Online learning and optimization for computation offloading in D2D edge computing and networks. Mob. Netw. Appl. 1–12 (2019)
4. Little, J.D.: A proof for the queuing formula: L= λ W. Oper. Res. **9**(3), 383–387 (1961)
5. Ashraf, M.I., Liu, C., Bennis, M., et al.: Dynamic resource allocation for optimized latency and reliability in vehicular networks. IEEE Access **6**, 63843–63858 (2018)
6. Bao, H., Liu, Y.: A two-sided matching approach for distributed edge computation offloading. In: 2019 IEEE/CIC International Conference on Communications in China (ICCC), Changchun, China, pp. 535–540 (2019)
7. Wang, L., Wu, H., Han, Z., et al.: Multi-hop cooperative caching in social IoT using matching theory. IEEE Trans. Wirel. Commun. **17**(4), 2127–2145 (2018)
8. Ehsanpour, M., Bayat, S., Mohammad, A., et al.: On efficient and social-aware object allocation in named data networks using matching theory. In: 2018 IEEE Symposium on Computers and Communications (ISCC), pp. 298–303 (2018)

9. Hmila, M., Manuel, F., Miguel, R.: Matching-theory-based resource allocation for underlay device to multi-device communications. In: 2018 14th International Conference on Wireless and Mobile Computing Networking and Communications (WiMob), pp. 28–35 (2018)
10. Li, J., Liu, M., Lu, J., et al.: On social-aware content caching for D2D-enabled cellular networks with matching theory. Internet Things J. IEEE **6**(1), 297–310 (2019)
11. Neely, M.J.: Stochastic Network Optimization with Application to Communication and Queueing Systems. Morgan & Claypool, San Rafael (2010)
12. Grant, M., Boyd, S.: CVX: Matlab software for disciplined convex programming. Wordpress and MathJax (2014)

Channel Allocation for Medical Extra-WBAN Communications in Hybrid LiFi-WiFi Networks

Novignon C. Acakpo-Addra[1]([✉]), Dapeng Wu[1], and Andrews A. Okine[1,2]

[1] School of Communication and Information Engineering,
Chongqing University of Posts and Telecommunications, Chongqing 400065, China
`no.vignon@hotmail.fr`, `wudp@cqupt.edu.cn`
[2] Chongqing Key Lab of Mobile Communications Technology,
Chongqing University of Posts and Telecommunications, Chongqing 400065, China

Abstract. Electronic health (e-health) systems based on optical wireless communication (OWC) provide a means of meeting the low latency requirements of medical applications, while ensuring little or no interference to sensitive devices. Notwithstanding, an optical wireless link is susceptible to temporal obstructions and a reliable radio frequency (RF) link may still be required. Against this backdrop, hybrid radio-optical extra-body area networks are considered viable solutions towards the attainment of pervasive healthcare in the Internet of Things (IoT) era. In practice, these networks will more often than not be used for both medical and non-medical applications, which will increase the competition for limited channel resources. Thus, in this paper, we propose a channel allocation framework for hybrid LiFi and WiFi networks, with the objective of safeguarding the quality of service (QoS) of medical applications. The scheme allocates channels for medical applications first, and then shares the remaining channels in a proportionally fair manner. Simulation results validate the effectiveness of the proposed solution in minimizing the waiting time of the delay-constrained medical packets.

Keywords: Channel allocation · Optical wireless communications · Remote health monitoring · Wireless body area networks · Wireless sensor networks

1 Introduction

To overcome the shortcomings of legacy healthcare systems, information and communication technologies (ICTs) are being introduced to enable new healthcare services such as remote health monitoring. Upcoming electronic health (e-health) systems will achieve pervasive healthcare, by providing quality and cost-effective services to patients irrespective of their location using ICTs, enabling the efficient monitoring of patients with chronic conditions and provision of proactive treatment [1]. Key to the realization of ubiquitous e-health systems is the application of wireless body area networks (WBANs).

A WBAN that is deployed on a patient typically consists of a gateway and heterogeneous medical sensors for monitoring physiological signals such as body temperature, respiratory rate, and blood pressure [2]. Real-time communication among medical sensors, the gateways or coordinator nodes (CNs), and the access points (APs) is critical to ensure that the sensed medical data is sent to the appropriate medical center on time for processing and analysis. This will allow for timely treatment decisions to be taken by the health professionals [3]. The unique requirements of medical WBANs poses a number of technical challenges to be overcome including ensuring reliable, low latency, and energy-efficient communications [4]. Yet, medical sensor nodes in a WBAN may have different data rate and latency requirements. Another daunting task in WBAN networking is maintaining uninterrupted communications, in view of physical obstructions and signal interference from nearby wireless networks. Owing to the non-stationary nature of their surrounding networks, adaptive strategies will have to be devised to meet the quality of service (QoS) targets of WBAN transmissions over extra-body links.

Due to the interference caused by radio frequency (RF) technologies to sensitive devices, the non-interfering optical wireless technologies are becoming attractive for application in healthcare systems [5]. Besides avoiding interference, optical wireless communication (OWC) offers tremendous data rates on the back of license-free high bandwidth channels. However, link reliability is of more concern, due to the possibility of misalignment between transceivers, and the vulnerability of optical wireless links to blockages. In [6], the authors focused on channel characterization for the uplink (UL) communication between the CN on a patient's body and an optical wireless AP in a typical hospital room, taking into account the impact of local body movements, user's mobility, and body shadowing. An all-optical wireless network was designed in [7] for off-body sensor communications using infrared (IR) in the UL and visible light for the downlink (DL). As an alternative to RF-based body area networks, an optical body area network (OBAN) was proposed in [7] to capture vital signals from several patients by means of visible light communication (VLC) and orthogonal codes. In terms of remote health monitoring, OWC and RF access networks can be integrated to achieve reliable high rate extra-body communications with minimal or infrequent interference to devices. For instance, the authors in [8] developed a dual hop communication system, in which an optical wireless transceiver acts as relay node, by converting optical signals to RF signals and transmitting them to a distant laboratory via RF links. Envisioning the hospital of the future, the authors in [9] presented a hybrid wireless network that can be reconfigured for both optical and radio-based extra-body wireless communications.

WBANs located at the home or the workplace may have to integrate with the existing wireless networks such as WiFi or LiFi, and coexist with their associated user devices (UDs). This suggest the need for channel reservation for medical applications, to ensure that the activities of UDs do not jeopardize their QoS. However, to the best of our knowledge, none of the preceding works on OWC networks for e-health tackled the issue of channel allocation for the extra-body tier. Hence, in this paper, we address the channel allocation problem in hybrid LiFi and WiFi networks for extra-body communications, in consideration of traditional UDs normally associated with local area networks. The rest of the paper is summarized as follows. Section 2 describes the system model of the hybrid LiFi and WiFi extra-body area network. In Sect. 3, we present the channel allocation strategy for the network. The performance of the proposed scheme is evaluated in Sect. 4 and the conclusions are drawn in Sect. 5.

2   System Model

The hybrid network consists of a LiFi AP coexisting with a WiFi AP in an indoor setting, to provide enhanced wireless coverage and capacity for the wireless devices. Both LiFi and WiFi offer bi-directional communication for UDs equipped with a multimode transceiver. UDs receive visible light or radio signals in the DL and transmit infrared or radio signals in the UL. Additionally, there are coexisting WBANs consisting of medical sensors and a CN as shown in Fig. 1. Data from the medical sensors are collected by the CN and transmitted via a LiFi or WiFi AP. In this sense, an extra-WBAN link refers to a wireless link that connects an WBAN and a wireless AP via the CN. The AP controller routes data between the Internet and indoor network, and selects an AP for DL transmission to a UD within coverage area of that AP. In addition, the controller manages the channel resource allocation and packet scheduling for UDs and extra-WBAN communications simultaneously.

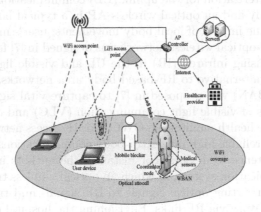

Fig. 1. System model of hybrid LiFi and WiFi access network.

In our system model, each WBAN contains heterogeneous sensors for collecting different medical data including electrocardiogram (ECG), electroencephalogram (EEG), electromyogram (EMG), blood glucose level, blood pressure, and respiratory rate. Based on the type of data, medical packets have different delay constraints. For instance, the ECG, the EEG, and the EMG packets require a maximum delay of 250 ms, whereas medical packets with information on blood glucose, blood pressure, and respiratory rate have delay requirements of 20 ms, 750 ms, and 600 ms respectively [10]. Fulfilling the delay requirements of medical packets is critical for the early detection of complications and the timely intervention by healthcare practitioners. Besides, if the waiting time of a medical packet in the queue exceeds its delay requirement it will be dropped by the CN. We classify the data packets into two main categories, based on the nature of information carried. Packets sent from the WBANs are referred to as medical packets, whereas the packets that emanate from traditional UDs, such as smartphones and laptops, are known here as general packets. The medical packets are further grouped according to their delay requirements, where C is the number of classes. In this work, we consider four classes of medical packets ($C = 4$). A class refers to a group of medical packets with the same delay requirements, such as blood glucose ($c = 1$), ECG/EEG/EMG ($c = 2$), respiratory rate ($c = 3$), and blood pressure ($c = 4$).

We denote the general and medical packets by g and m respectively. Packet arrivals follow a Poisson distribution, in which λ_g represents the aggregate arrival rate of general packets and λ_m denotes the aggregate arrival rate of medical packets. Moreover, τ_c denote the delay requirement of class c medical packets. Supposing there are $|W|$ WBANs and $|U|$ UDs in the hybrid network and each device has its own queue for packet transmissions, the number of UL queues for medical data and general data transmissions are W and U respectively. There are L channels for UL transmissions in the LiFi network, and the WiFi network has R UL channels. Part of the UL channels resources in each network is allocated to the UDs and part for extra-WBAN transmissions. Packets from a WBAN or UD could be served by the LiFi or WiFi UL resources, depending on the condition of optical wireless channel and the transmission mode. To this end, a portion of the UL channels in the WiFi network is allocated for extra-WBAN medical data transmissions and part for the general packets transmissions. The wireless infrared connections are prone to link interruptions, during which a change of transmission mode to RF may be required to maintain data transmission.

3   A Proposed Channel Allocation Strategy

In this section, the strategy for allocating the UL channels resources is presented, taking into account the occurrence of optical wireless link blockages and the need for link switching.

Let $E[D_w]$ be the expected transmission delay of medical packets of the wth CN and $\hat{D}_m$ be the delay threshold for meeting the QoS requirements of any medical packet. Then, we should always aim at fulfilling the following requirement:

$$E[D_w] \leq \hat{D}_m \tag{1}$$

Given an $M/M/1$ queuing system with a first-in, first-out (FIFO) policy [11], average waiting delay D can be computed as

$$D = \frac{1}{\mu - \lambda}, \qquad (2)$$

where μ is the service rate of the queue and λ is the arrival rate of packets to the queue. Thus, to satisfy Eq. 2, the service rate must meet the following constraint for a LiFi UL:

$$\hat{\mu}_w^{IR} \geq \frac{1}{\hat{D}_m + \lambda_w}, \qquad (3)$$

where $\hat{\mu}_w^{IR}$ is the minimum LiFi service rate offered to the UL queue of the wth CN and is obtained by

$$\hat{\mu}_w^{IR} = \frac{X_{IR}\hat{N}_w^{IR}}{\hat{\kappa}_m}, \qquad (4)$$

where κ_m and X_{IR} are the expected size of medical packets and the LiFi per-channel transmission rate respectively. $\hat{N}_w^{IR}$ is the minimum UL channel resources allocated to the wth CN for medical packet transmissions. Combining Eqs. 3 and 4, $\hat{N}_w^{IR}$ can be expressed as

$$\hat{N}_w^{IR} = \frac{\kappa_m}{X_{IR}} \left[\frac{1}{\hat{D}_m} + \lambda_w \right], \qquad (5)$$

Due to the criticality of extra-WBAN transmissions, we derive the amount of LiFi UL channel resources allocated for the wth CN, when transmitting over LiFi network, as

$$N_w^{IR} \simeq \hat{N}_w^{IR} + (L - A)(\lambda_w/\lambda_m'), \qquad A \leq L \qquad (6)$$

where λ_w is the arrival rate of UL medical packets at the wth CN and λ_m' is the aggregate arrival rate of UL medical packets of the CNs in LiFi network. $A = \sum_{w=1}^{W} \hat{N}_w^{IR} + \sum_{u=1}^{U} N_u^{IR}$, where N_u^{IR} is the number of LiFi UL channel resources allocated to the uth UD in LiFi network and is given by

$$N_u^{IR} \simeq \left[L - \sum_{w=1}^{W} \hat{N}_w^{IR} \right] \frac{\kappa_g \lambda_u}{\kappa_g \lambda_g + \kappa_m \lambda_m} \qquad (7)$$

By integrating the LiFi network into an existing wireless local area network (WLAN) such as WiFi network, packets from CNs that cannot be served by the LiFi network, due to blockage or coverage limitations, can be served by the WiFi after a switching delay. To this end, the number of UL channel resources reserved for medical packet transmissions of the wth CN over WiFi network, is given by

$$N_w^{RF} \simeq \hat{N}_w^{RF} + (R - H) \left(\frac{T_w'}{\sum_{w=1}^{W} T_w'} \right), \qquad H \leq R \qquad (8)$$

where T_w' is the cumulative connection time of the wth CN to the WiFi network and $\hat{N}_w^{RF}$ is the minimum amount of UL channel resources reserved for the medical packet transmissions of the wth CN over WiFi network. $H = \sum_{w=1}^{W} \hat{N}_w^{RF} + \sum_{u=1}^{U} N_u^{RF}$, where N_u^{RF} is the number of WiFi UL channel resources allocated to the uth UD and is obtained as

$$N_u^{IR} \simeq \left[R - \sum_{w=1}^{W} \hat{N}_w^{RF} \right] \frac{\kappa_g \lambda_u T_u'}{\kappa_g \lambda_g \sum_{u=1}^{U} T_u' + \kappa_m \lambda_m \sum_{w=1}^{W} T_w'}, \tag{9}$$

where T_u' is the total connection time of the uth UD to the WiFi network. $\hat{N}_u^{RF}$ is the minimum WiFi UL channel resources allocated to the device and is given by

$$\hat{N}_w^{RF} = \frac{\kappa_m}{X_{RF}} \left[\frac{1}{\hat{D}_m} + \lambda_w \right], \tag{10}$$

where X_{RF} is the per-channel transmission rate of WiFi network.

4 Simulation and Results Discussion

In the simulation scenario, there are two WBANs within the network area. One of the WBANs is stationary in the LiFi attocell and is described as WBAN-S. The other WBAN, which is denoted as WBAN-M, is mobile between the LiFi and the WiFi only coverage areas. In addition, a number of UDs that can connect to either LiFi or WiFi are integrated into the simulation process. The UL queue of the CNs of each WBAN is modeled and simulated using the $M/M/1$ queuing model, in which the packet arrival process is Poisson distributed and the service time is exponentially distributed. We also factor the blockage of the optical wireless link via the Bernoulli distribution, using the link blockage probability as parameter. The switching delay encountered during a change of UL transmission mode from LiFi to WiFi is considered to be a normal random variable with a mean value of 100 ms and a variance of 20 ms. Besides, four (4) classes of medical packets with equal probabilities of arrival, and delay requirements as in Sect. 2, are used for the simulation. For the general packet transmissions, we assume the interactive non-real-time service class, which includes Web browsing and voice messaging.

We compare our proposed channel allocation scheme with a heuristic method which is a modified version of a general allocation scheme given in [12]. In the heuristic method, channel allocation for the wth WBAN in LiFi is obtained by

$$N_w^{IR} \simeq L \left[\frac{\kappa_m \lambda_w}{\kappa_g \lambda_g (1 - \rho) + \kappa_m \lambda_m \rho} \right], \tag{11}$$

where $\rho = E[\hat{D}_m]/(E[\hat{D}_m] + \hat{D}_g)$ is a normalizing factor, accounting for the relatively stringent delay requirements of medical packets. $E[\hat{D}_m]$ is the expected delay requirements of an arriving medical packet and $\hat{D}_g$ is the delay threshold

Table 1. Simulation parameters

Parameter	Value
Number of user devices (UDs)	3
UL packet arrival rate for UD	1 packet/s
UL packet arrival rate for CN	0.5–5 packets/s
Number of LiFi UL channels L	50
Number of WiFi UL channels R	20
LiFi per-channel transmission rate X_{IR}	2 Mbps
WiFi per-channel transmission rate X_{RF}	0.5 Mbps
Probability of UD in LiFi coverage area	0.3
Probability of mobile WBAN in LiFi coverage area	0.2
Probability of optical wireless link blockage	0.2
Mean value of switching delay	100 ms
Variance of the switching delay	20 ms
Expected size of medical packets κ_m	50 Kbits
Expected size of general packets κ_g	500 Kbits
Expected delay constraint of medical packets $E[\hat{D}_m]$	405 ms
Delay threshold of general packets $\hat{D}_g$	1 s
Number of medical packets for M/M/1 simulation	500
Number of iterations	1000

for meeting the QoS requirements of any general packet. Based on the heuristic method, channel allocation in WiFi for the wth WBAN is also given as

$$N_w^{RF} \simeq R \left[\frac{\kappa_m \lambda_w}{\kappa_g \lambda_g (1 - \rho) + \kappa_m \lambda_m \rho} \right], \tag{12}$$

The average UL queue length and the average response time of the system are used as performance metrics in the analysis. These metrics are measured against the arrival rates of medical and general packets, considering the outcomes for both the stationary WBAN and the mobile WBAN (Table 1).

Figure 2 investigates the impact of arrival rates of the two type of packets (medical and general) on the response times of medical packets in the queueing system. Our proposed channel allocation strategy is shown to be more effective than the heuristic method, in terms of minimizing the response time of medical packets. When the medical packet arrival rate increases, the heuristic method allocates more channels to WBAN UL with the goal of enhancing channel utilization. Meanwhile, our proposed method aims specifically at meeting the delay requirements of medical packets by allocating more UL channels for medical data transmissions. This correctly offsets any additional waiting time and its impact on response time. Hence, it is striking to note in Fig. 2a that the response time performance of our scheme remains relatively stable, irrespective of the arrival

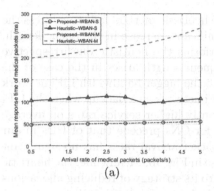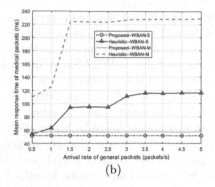

Fig. 2. Mean response of medical packets against the (a) arrival rate of medical packets at the gateway (b) per-UD general packets arrival rate

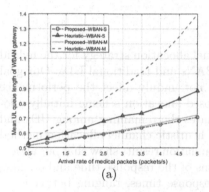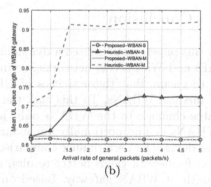

Fig. 3. Mean UL queue length of WBAN gateway against the (a) arrival rate of medical packets at the gateway (b) per-UD general packets arrival rate

rate of medical packets or the movement of WBAN. However, the stationary WBAN attains a lower response time than that of the mobile WBAN for both allocation methods, since it is connected to the higher service rate LiFi network for a longer period of time. Medical packets are given priority in our proposed allocation scheme. Therefore, any increase in the general packet arrival rate has minimal impact on the allocations for medical packet transmissions, as indicated in Fig. 2b. The heuristic method does not prioritize medical packets and, thus, decreases the allocations for medical data transmissions in favor of general packets, with a view of better channel utilization.

We proceed to examine the effects of arrival rates of the two type of packets (medical and general) on the mean queue length of the CN uplink in Fig. 3. By relying on our proposed allocation strategy, an increase in medical packet arrival rate has little impact on the UL queue length. The proposed method allocates more channels for WBAN UL transmissions in response to higher arrival rates of medical packets. This results in faster service rates and the reduction of significant waiting periods, resulting in smaller queue lengths than expected.

Due to a minimal change in service rate, the heuristic method fails to address the additional waiting periods, leading to a significant increase in the queue length. From Fig. 3a, our proposed method outperforms the heuristic one irrespective of the mobility of WBANs. A shorter UL queue length arises in the case of the stationary WBAN, because it is hooked onto the bigger bandwidth LiFi network for a longer period and only falls on the WiFi network as a backup when the LiFi link is obstructed. Thus, it attains significantly higher service rates on average than the mobile WBAN. Since allocations for CNs precede that of UDs in our proposed scheme, any changes in general packet arrival rates has no significant impact on the UL queue length, as is the case in Fig. 3b. Meanwhile, the heuristic method causes longer queue lengths, owing to its strategy of reducing allocations for medical data transmissions when the arrival rate of general packets increases, resulting in lower service rates and longer response times.

5   Conclusion

In this paper, we presented a dynamic channel allocation strategy for the hybrid LiFi and WiFi access networks serving WBANs and traditional UDs. Our goal is to ensure that the activities of UDs do not unduly affect the performance of medical WBANs. To this end, the proposed allocation scheme initially allocates channels to the WBANs only, and shares the remaining among UDs and WBANs in a proportionally fair manner. A heuristic channel allocation method was devised for performance comparison. Simulation results show that our allocation framework yields better results, in terms of the response time and queuing length at WBAN gateway. Based on its response times, ranging between 38–61 ms, the proposed scheme has clearly proven its ability to meet the stringent QoS targets of medical packets. Nevertheless, the packet dropping probability will be investigated in our future work.

References

1. Yi, C., Cai, J.: Delay-dependent priority-aware transmission scheduling for e-health networks: a mechanism design approach. IEEE Trans. Veh. Technol. **68**(7), 6997–7010 (2019). https://doi.org/10.1109/tvt.2019.2916496
2. Haddad, O., Khalighi, M.A.: Enabling communication technologies for medical wireless body-area networks. In: 2019 Global LIFI Congress (GLC). IEEE (2019). https://doi.org/10.1109/glc.2019.8864122
3. Pramanik, P.K.D., Nayyar, A., Pareek, G.: WBAN: driving e-healthcare beyond telemedicine to remote health monitoring. In: Telemedicine Technologies, pp. 89–119. Elsevier (2019). https://doi.org/10.1016/b978-0-12-816948-3.00007-6
4. Fang, G., Dutkiewicz, E., Huq, M.A., Vesilo, R., Yang, Y.: Medical body area networks: opportunities, challenges and practices. In: 2011 11th International Symposium on Communications and Information Technologies (ISCIT). IEEE (2011). https://doi.org/10.1109/iscit.2011.6089699

5. Julien-Vergonjanne, A., Sahuguède, S., Chevalier, L.: Optical wireless body area networks for healthcare applications. In: Uysal, M., Capsoni, C., Ghassemlooy, Z., Boucouvalas, A., Udvary, E. (eds.) Optical Wireless Communications. SCT, pp. 569–587. Springer, Cham (2016). https://doi.org/10.1007/978-3-319-30201-0_26
6. Haddad, O., Khalighi, A., Zvanovec, S.: Channel characterization for optical extra-WBAN links considering local and global user mobility. In: Dingel, B.B., Tsukamoto, K., Mikroulis, S. (eds.) Broadband Access Communication Technologies XIV. SPIE (2020). https://doi.org/10.1117/12.2544901
7. Hoang, T.B., Sahuguede, S., Julien-Vergonjanne, A.: Optical wireless network design for off-body-sensor based monitoring. Wirel. Commun. Mob. Comput. **2019**, 1–13 (2019). https://doi.org/10.1155/2019/5473923
8. Vats, A., Aggarwal, M., Ahuja, S., Vashisth, S.: Hybrid VLC-RF system for real time health care applications. In: Bhattacharya, I., Chakrabarti, S., Reehal, H.S., Lakshminarayanan, V. (eds.) Advances in Optical Science and Engineering. SPP, vol. 194, pp. 347–353. Springer, Singapore (2017). https://doi.org/10.1007/978-981-10-3908-9_42
9. Ahmed, I., Kumpuniemi, T., Katz, M.: A hybrid optical-radio wireless network concept for the hospital of the future. In: Sugimoto, C., Farhadi, H., Hämäläinen, M. (eds.) BODYNETS 2018. EICC, pp. 157–170. Springer, Cham (2020). https://doi.org/10.1007/978-3-030-29897-5_13
10. Yi, C., Cai, J.: A truthful mechanism for scheduling delay-constrained wireless transmissions in IoT-based healthcare networks. IEEE Trans. Wirel. Commun. **18**(2), 912–925 (2019). https://doi.org/10.1109/twc.2018.2886255
11. Bouloukakis, G., Moscholios, I., Georgantas, N., Issarny, V.: Simulation-based queueing models for performance analysis of IoT applications. In: 2018 11th International Symposium on Communication Systems, Networks and Digital Signal Processing (CSNDSP). IEEE (2018). https://doi.org/10.1109/csndsp.2018.8471798
12. Chowdhury, M.Z., Uddin, M.S., Jang, Y.M.: Dynamic channel allocation for class-based QoS provisioning and call admission in visible light communication. Arab. J. Sci. Eng. **39**(2), 1007–1016 (2013). https://doi.org/10.1007/s13369-013-0680-4

Cache Allocation Scheme in Information-Centric Satellite-Terrestrial Integrated Networks

Jie Duan$^{(\boxtimes)}$, Xianjing Hu, Hao Liu, and Zhihong Zhang

School of Communication and Information Engineering,
Chongqing University of Posts and Telecommunications, Chongqing 400065, China
duanjie@cqupt.edu.cn

Abstract. With the development of satellite communication technologies, satellite-terrestrial integrated network (STIN) is often used for content delivery services as satellites are with wide-area coverage. In terrestrial networks, in-network caching has been proved to be an effective method to improve network performance in terms of throughput and delay, so information-centric networking (ICN) will be deployed in STIN architecture (STI^{2}CN). However, the current cache research scenarios in satellite networks usually ignore cache cooperation among nodes, resulting in lower cache hit rate. Secondly, the cache configuration is mainly implemented in the coverage area, which does not consider satellites mobility and dynamic network topology, so that network transmission efficiency is decreased. Thus, we formulate cache content allocation problem as user average hop count minimization problem, which is solved by matching algorithm to obtain exchange stability between satellites and contents. In addition, a proactive pushing scheme is proposed to solve satellites mobility problem. The cached contents will be pushed to the subsequent satellite covering the area in advance to further improve content retrieval efficiency. Simulation results show that the proposed scheme can significantly reduce user average hop count and improve cache hit rate.

Keywords: Satellite-terrestrial integrated network ·
Information-centric networking · Cache allocation · Proactive pushing

1   Introduction

As the popularization of mobile communication devices, such as smartphones and tablet PCs, making user requirements for pervasive network access any-

This work was supported by National Natural Science Foundation of China (NO. 61701058), Foundation of key Laboratory in National Defense Science and Technology for Equipment pre-research in the 13-th Five-year Plan (NO. 61422090301), Science and Technology Research Program of Chongqing Municipal Education Commission (NO. KJQN201800633).

time anywhere become more intense. However, the current network infrastructures that provide Internet services are made up of terrestrial devices, which are prone to be damaged in disasters to interrupt communication. In addition, it is inconvenient to deploy network infrastructures in the remote and isolated areas such as deserts, seas and forests. To address the above problems, a new network architecture that integrates satellites into the terrestrial network is proposed—satellite-terrestrial integrated network (STIN) [1–3]. In STIN, satellites can provide services in extreme environments without terrestrial infrastructures, which makes it a strong supplement to traditional terrestrial networks [4,5].

The current Internet traffic is increasing exponentially. From the perspective of content retrieval, users just concern about contents themselves, not the servers or hosts where contents are located. In this case, applying the traditional TCP/IP communication mechanism to STIN will have problems such as low content delivery efficiency, poor mobility and scalability [6]. Information-centric networking (ICN) [7,8] has emerged as one of the promising candidates for the architecture of the future Internet, which has two major characteristics: in-network caching and routing-by-name. These characteristics enable each node to cache passing-by data and reduce redundant transmissions. Therefore, establishing an information-based satellite-terrestrial integrated information-centric network (STI^{2}CN) can solve a series of problems mentioned above. In-network caching has drawn lots of attention in satellite networks and proved its effectiveness in both improving transmission efficiency and reducing delay [9–15].

The critical characteristics of integrating satellites into ICN architecture were analyzed in [9] and [10], which provide a preliminary direction for exploring caching research in satellite communication systems. Recently, ICN architecture was applied in satellite-assisted emergency communications scenarios to solve mobility and link interruption problems, while effectively reducing end-to-end delay [11]. Literature [12] proposed caching strategy named SatCache in satellite networks to maximize cache hit ratio on the interest profile of users. In fact, with the rapid development of technologies such as on-board processing and storage capabilities, [13] first proposed the caching on-satellite to further improve network performance. However, a two-layer caching model based on genetic algorithm heavily depends on the initialization parameters. In [14], a scheme based on simulated annealing optimization algorithm was proposed in 5G-satellite backhaul networks. Although [13] and [14] studied the cache allocation algorithm for on-board cache, the research scenarios were limited to single-satellite nodes. Thus, a novel caching strategy based on matching game to minimize content access delay was proposed in [15]. But the work mainly focuses on optimal caching content placement at a certain moment, and not considering satellite network topology time-varying characteristics.

All these above have made contributions to solve cache in satellites. However, the research scenarios of existing cache schemes are limited to a single satellite node, non-cooperation among satellites causes cache redundancy. Secondly, the current algorithms have high complexity and slow convergence in a large-scale satellite network. Motivated by the above observations, the paper considers user preferences and cache content cooperation, and formulates cache content allocation problem as user average hop count minimization problem. Since the formulated optimization problem contains 0–1 variables and it cannot be solved conveniently, we propose a matching-based cache allocation algorithm with fast convergence and low complexity, and finally obtains exchange stability between satellites and contents. In addition, the mobility of LEO satellites incurs users' requests not being responded to in time, which will increase user delay. Therefore, in the case of in-network caching, a proactive pushing scheme is proposed. The cached contents will be pushed to the subsequent satellite covering the area in advance to further improve network transmission efficiency.

The rest of the paper is organized as follows. Section 2 introduces our STI^{2}CN scenarios. In Sect. 3, the cache allocation problem is described and formulated. Then, we propose a matching-based cache allocation algorithm and proactive pushing scheme. Performance analysis are shown in Sect. 4. At last, conclusions are drawn in Sect. 5.

2   Network Scenarios

In this paper, when it is difficult for the conventional terrestrial network to achieve coverage and communication in some remote areas, STI^{2}CN providers Internet access services for users. As shown in Fig. 1, STI^{2}CN mainly comprises two parts: the terrestrial network and satellite network. The terrestrial network is the network currently in use, which is mainly composed of users, ground stations and content servers. It can provide high data rates to users, but the network coverage in rural and remote areas is limited. The satellite network is a strong supplement to terrestrial network, which consists of geostationary earth orbit (GEO) and low earth orbit (LEO) satellites. GEO satellites are relatively stationary with the ground, while exhibiting the highest propagation delay due to the long distance. Thus, GEO satellites are considered as controller to in charge of routing strategy calculation, mobility management and so on. By comparison, LEO satellites are close to the ground and have a short propagation delay, which mainly realize in-network caching and provide wireless access to users for obtaining long-distance global communications. However, due to the high-speed movements of LEO satellites, data transmission in such dynamic environment suffers frequently periodic interruptions, leading to lack of persistent end-to-end paths. Therefore, users and LEO satellites are constantly handover to ensure network connection. Normally, when the LEO satellite moves to another area, users' incomplete data transmissions can be migrated under the guidance of the GEO satellites. The following will briefly illustrate the process of user communication and content pushing with Fig. 1 as an example.

Firstly, users obtain contents as follows: when a user sends a request for a file, the ground station will search for the file in its local cache. If ground station already caches the file, it will respond to the request directly. Otherwise, the request will be sent to LEO satellites. Then, LEO satellites check whether the desired file is currently stored in the cache. If cache hits, LEO satellites will

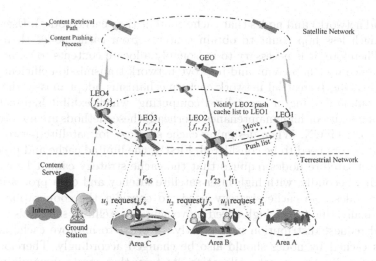

Fig. 1. Network scenario.

return the data to user along the requested path. Otherwise, they pass the file request to the next hop towards the destination of the requested file. Compared to space network, user delay in ground network is negligible, so it will not be discussed in detail. The red dashed arrows in Fig. 1 represent content retrieval path. The first case is that user u_1 acquires the file f_1 from the access satellite LEO2, and content retrieval path is r_{11}. In case of user u_2 requesting file f_3, it will obtain the requested file through inter-satellite links(ISL) from LEO3, content retrieval path is r_{23}. The user u_3 gained file f_6 from content server in the last case, content retrieval path is r_{36}.

In addition, data transmission will often be interrupted due to the mobility of LEO satellite nodes. Generally speaking, the file is waiting to be issued in the caching of the access satellite via ISL. However, ground-to-satellites links (GSL) may be insufficient in the satellite communication, resulting in the link being interrupted before file is not completely delivered, and the subsequent satellite covering the area will transmit file for user. If the file is not contained in the caching of the subsequent satellite, which will request the file from other satellites and cause longer delay. Hence, the cached contents will be pushed to the subsequent satellite in advance. The specific pushing process is shown by the green solid arrows in Fig. 1. GEO detects that LEO2 is about to leave coverage area B, GEO notifies LEO2 to push the cache list to LEO1, and LEO1 sends interest packet to LEO2 to request cached contents.

3 Dynamic Cache Allocation Scheme

3.1 Problem Description

For users, the desired contents can be obtained from content sources, which include source nodes that publish contents (i.e., satellite nodes or content servers in ground network) and nodes that cache content (i.e., cache nodes). Users experience much less hop count to obtain contents from cache nodes than source nodes. Therefore, it is necessary to reasonably allocate contents to cache nodes, which reduce user hop count and improve network transmission efficiency.

The existing terrestrial network cache mechanisms adopt mostly the design ideas of online caching or centralized computing, which exhibit high computational complexity or high maintenance overhead, these methods are not obviously suitable for STI^{2}CN. When designing cache strategy for satellite networks, the following factors need to be considered. Firstly, the limited cache and processing capacity of satellite nodes requires that the cache strategy cannot be excessive complexity. Secondly, with high inter-satellite latency and data processing rate of cache nodes, the cache strategy needs to consider the cooperation among nodes. Finally, due to the high-speed movement of satellites, satellite's coverage area and request distribution vary greatly. In order to improve cache hit rate, contents cached by nodes should also be changed accordingly. Therefore, it is necessary to propose a cache allocation strategy that meets dynamic network topology and lower overhead.

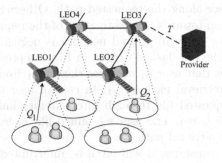

Fig. 2. Example problem description.

As shown in Fig. 2, Q_1 and Q_2 represent the request rate for LEO1 and LEO3, respectively. T is hop count from content provider (i.e., nodes that publish contents) to LEO3. Requests can be forwarded to content provider If caches miss in nodes. It is preferable for users to directly fetch contents from cache nodes rather than content provider which experience longer delay. The cost of satellite nodes cache deployment is node cache capacity, and profit is hop count for users to obtain contents. The smaller profit, the smaller hop count.

We assume that LEO1, LEO2 and LEO3 satellites deploy caches, and the cache capacity of each satellite is C, cache allocation results are shown in Table 1. The single node cache means that only one satellite is deployed cache, and users in other regions can obtain contents from this satellite node. The profit of cache deployment LEO2 = 1 is the best, namely hop count is minimal. Supposing that there are multi-satellite nodes deploying cache, and cache cooperation among satellite nodes is considered. Obviously, the hop count of cache deployment

Table 1. Cache allocation results.

Classification	Cache Deployment	Cost	Profit
Single node cache	LEO1 = 1	C	$2Q_2$
	LEO2 = 1	C	$Q_1 + Q_2$
	LEO3 = 1	C	$2Q_1$
Cooperative cache	LEO1 = 1, LEO2 = 1	$2C$	Q_2
	LEO1 = 1, LEO3 = 1	$2C$	0
	LEO2 = 1, LEO3 = 1	$2C$	Q_1
Non-cooperative cache	**LEO1 = 1, LEO2 = 1, LEO3 = 1**	$3C$	0

LEO1 = 1, LEO3 = 1 is the smallest. Compared with cooperative cache, the last non-cooperative cache leads to larger cache redundancy and waste cache space. In fact, the cache capacity is very limited and it is not practical to deploy large cache space in nodes. Thus, cooperative cache allocation gains better results than the other two cache allocation schemes from network revenue perspective.

3.2   Problem Formulation

Cooperative cache means that the probability of caching the same file among nodes obeys certain rules. By modeling cooperative cache, contents with higher popularity can be cached in nodes closer to users, and contents cached in the network is more diversified to improve transmission performance.

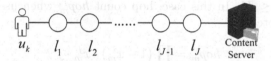

Fig. 3. Path topology for user to obtain contents.

When establishing the cooperative relationship model between nodes, cooperation parameter is proposed to reflect the probability that nodes cache the same file. The path topology for any user to obtain contents is shown in Fig. 3, supposing $L = \{l_j \mid 1 \leq j \leq J\}$ indicates that J LEO satellites are connected between user u_k and content server. $F = \{f_m \mid 1 \leq m \leq M\}$ means M files published by satellite nodes or content server. The size of file f_m is denoted by b_m in bytes. As described in Sect. 2, when a user sends request, the target file may be obtained at the access satellite l_1. Or the request will be forwarded to other satellites, namely $l_2, \cdots, l_j, \cdots, l_J$. Otherwise, the request will eventually reach content server. To describe the relationship between file and satellite node, Boolean variable $x_m^j \in \{0, 1\}, \forall m, \forall j$ is used to indicate the cache state of file f_m requested by the user at the satellite node l_j, which can be expressed as

$$x_m^j = \begin{cases} 1, & \text{if file } f_m \text{ is cached at node } l_j \\ 0, & \text{otherwise} \end{cases} \tag{1}$$

According to the path for user to obtain contents, it can be known that all the contents must be stored in the content server. Thus, the probability of caching the same file between nodes should have a certain cooperative relationship. For the cache of the file f_m, when the node close to user caches f_m, the probability that the next-level node caches the same file can be expressed as

$$g_m^{j+1} = \frac{\beta}{1 + (-1)^{x_m^j} \cdot g_m^j} = \begin{cases} \frac{\beta}{1 - g_m^j}, & x_m^j = 1 \\ \frac{\beta}{1 + g_m^j}, & x_m^j = 0 \end{cases} \tag{2}$$

Where, g_m^j denotes the probability that the node close to user caches the file f_m, which can be regarded as the popularity of the file f_m. β is a constant between $[0, 1]$, which is seen as the cooperation degree between nodes.

The popularity of many current network traffic has been proved to follow the Zipf distribution. There are M files in the system, contents are sequenced in a descending order according to their popularity, the requested probability p_m for m_{th} popular content is given as

$$p_m = \frac{1/m^\alpha}{C} \tag{3}$$

Where the range of m is $[1, M]$, $C = \sum_{m \in [1, M]} 1/m^\alpha$. α is Zipf parameter, which is denoted by $Zipf(a)$. If the total request rate is Q, the request rate of m_{th} most popular content can be expressed as $Q_m = Q \cdot p_m$.

For notational convenience, all satellite nodes before hit node l_j are denoted as $L = \{l_i | 0 \le i \le J\}$. In this case, hop count hop_m^j when user request to hit node can be denoted as

$$hop_m^j = \prod_{i<j} \left(1 - x_m^i\right) \cdot x_m^j \cdot j \tag{4}$$

Where x_m^i is the cache state of all nodes before the hit node l_j, and j is hop count from user to hit node l_j. Considering that in practice not only the hit node stores the target file, but also other nodes and server may cache the target file. Thus, the total hop count hop_m^J for user to obtain files from satellite nodes or server can be calculated as

$$hop_m^J = \sum_{j<J+1} \prod_{i<j} \left(1 - x_m^i\right) \cdot x_m^j \cdot j \tag{5}$$

In satellite networks, topology is regularly changing and predictable. Assuming that the maximum hop count for user to obtain contents is H, the probability that user will experience hop count is $\frac{1}{H}$. At this time, the hop count expectation hop_m for user to gain files is represented by

$$hop_m = \frac{1}{H} \sum_{J \in [1,H]} hop_m^J \tag{6}$$

The cooperative cache allocation scheme aims to minimize user average hop count hop, which is formulated as: for given cache capacity in satellites, content popularity and network topology, which contents should be cached in each satellite so that user average hop count is minimized. Thus, the optimization problem can be expressed as

$$\min hop = \sum_{m \in [1,M]} hop_m \cdot p_m \tag{7}$$

$$\text{s.t.} \quad C1: g_m^{j+1} = \frac{\beta}{1 + (-1)^{x_m^j} \cdot g_m^j} \tag{8}$$

$$C2: \sum_{m \in [1,M]} x_m^j \cdot b_m \leq C_j, \forall j \tag{9}$$

$$C3: \sum_{u \in J} Q_u - \sum_{u \in J} Q_{uj} = 0, \text{ if } u = l_i \tag{10}$$

$$C4: \sum_{u \in J} Q_{ju} - \sum_{u \in J} x_m^u = 0, \text{ if } u = l_j \tag{11}$$

$$C5: \sum_{u \in J} Q_{ju} - \sum_{u \in J} Q_{uj} = 0, \text{ otherwise} \tag{12}$$

$$C6: x_m^j \in \{0, 1\}, \forall m, \forall j \tag{13}$$

The decision variable for the optimization objective is files to be cached on every satellite, that is content placement matrix X, and the values of the elements in the matrix X can be obtained by $x_m^j \in \{0, 1\}$. C_j is the cache capacity of each satellite in bytes. C1 ensures cooperative cache relationships between nodes. C2 sets the cache space limit for each satellite node. C3-C5 are flow conservation constraints. C6 limits the storage variables of the satellite nodes to 0 or 1.

3.3 Matching-Based Cache Allocation Algorithm

After modeling the system, it can be concluded that the optimization objective is a problem with high computational complexity because it contains 0–1 variables and it is difficult to find the problem optimal solution in polynomial time. Considering the time-varying characteristics of the satellite network topology and the limited on-board computing power, a low-complexity solution scheme should be explored. The matching theory [16] is a promising approach to perform resource management in satellite networks, which can effectively deal with the allocation problems such as on-board storage resources and cached contents. Thus, a matching-based cache allocation algorithm is proposed. For a node in satellite networks, a certain number of files need to be cached based on storage

space; also, files need to be cached on multiple satellite nodes based on user preferences. Therefore, the problem can be modeled as a many-to-many matching,
in which satellites and files are players in matching games. Based on the above
descriptions, we give the following definitions.

Definition 1: In the many-to-many matching model, a matching μ is function
from the set $L \cup F$ into the set of all subsets of $L \cup F$, $\mu : L \cup F \to L \cup F$, for
every $l_j \in L$ and $f_m \in F$ so that

1) $\mu(f_m) \subseteq L$ and $\mu(l_j) \subseteq F$

2) $\sum_{m \in [1,M]} |\mu(l_j)| \cdot b_m \leq C_j, \; \forall j$

3) $l_j \in \mu(f_m)$ iff $f_m \in \mu(l_j)$

In 1), the matching node corresponding to the file belongs to the satellite, while
the matching node corresponding to the satellite belongs to the file. 2) is the
cache space limit for each satellite node. 3) indicates that file f_m is in the matching set of satellite l_j, if and only if that satellite l_j is also in the matching set of
the file f_m.

Next, we need to build a preference list between the file and satellite. Each
file has different presences for different satellite nodes, and each satellite node
has different preferences for different files similarly. Here these two kinds of
preference values are defined. First, based on the optimization objective, we
define the preference value of a satellite over a file, which should consider the
following principle: the satellite cached files make the user hop count smaller,
the preference value of these files is higher. So, the preference value of satellites
for different files is set as the ascending order of the utility function as

$$\varepsilon_j(m, \mu) = p_m \cdot hop_m^j + \sum_{m' \neq m} p_{m'} hop_{m'} \tag{14}$$

Similarly, we define the preference value of files over satellite nodes, which
can be expressed as

$$\varepsilon_m(j, \mu) = p_m \cdot hop_m^j \tag{15}$$

It can be seen from the above defined preference value that the matching
in this paper contains externalities, which the preference value of each satellite
node for not only depends on the current satellite cache state, but also is affected
by other satellite cache states. In many-to-many matching with externalities, a
stability concept cannot be defined straightforwardly because the gain from a
matching pair depends on the matching results of other players. Inspired by the
definition of exchange stability, swap matching is defined to achieve stability
[17].

Definition 2: In the matching μ, a swap matching is defined as

$$\mu_{mj}^{m'j'} = \{\mu \setminus \{(m, \mu(m)), (m', \mu(m'))\}\} \cup$$
$$\{(m, \{\{\mu(m) \setminus \{j\}\} \cup \{j'\}\}), (m', \{\{\mu(m') \setminus \{j'\}\} \cup \{j\}\})\} \tag{16}$$

where $j \in \mu(m)$, $j' \in \mu(m')$, $j \notin \mu(m')$ and $j' \notin \mu(m)$. In swap matching, two agents in the same set exchange their matches in the opposite set while keeping all other agents' matching state unchanged. It is worth noticing that one of satellite $l_{j'}$ can be a hole which allowing for satellite moving to available vacancies. Based on the operations of swap matching, we will the concept of swap-blocking pairs as follows.

Definition 3: $(l_j, l_{j'})$ is a swap-blocking pairs in the matching μ, such that

$$1) \; \forall i \in \{m, m', j, j'\}, \varepsilon_i\left(\mu_{mj}^{m'j'}\right) \leq \varepsilon_i(\mu)$$

$$2) \; \exists i \in \{m, m', j, j'\}, \varepsilon_i\left(\mu_{mj}^{m'j'}\right) < \varepsilon_i(\mu)$$

In Definition 3, condition 1) implies that the utilities of all involved satellites and files should not be reduced after the exchange operation between the swap-blocking pairs $(l_j, l_{j'})$. Condition 2) indicates that at least one of the players' utilities is increased after the exchange operation between the swap-blocking pair. The matching μ is two-sided exchange-stable if and only if there are no swap-blocking pairs.

Algorithm 1: Proposed Matching-Based Cache Allocation Strategy.

1 *Initialization*
2 Satellites and files are randomly matched with each other subject to constraints (8)-(13), forming a matching state μ.
3 Set iteration count $iter = 0$.
4 *Exchange Matching Process*
5 For each satellite l_j, it searches for another satellite $l_{j'}$ or file's available vacancies O to form a swap-blocking pair.
6 If $(l_j, l_{j'})$ or (l_j, O) forms a swap-blocking pair along with $j \in \mu(m)$, and $j' \in \mu(m')$,
7 Update the current matching state to $\mu_{mj}^{m'j'}$ and reset $iter = 0$.
8 *Else if* there does not exist such a swap-blocking pari,
9 Keep matching state μ unchanged and update $iter = iter + 1$.
10 Repeat Step 5 to 9 until $iter > iter_{max}$.
11 *End Algorithm*

The proposed algorithm is shown in Algorithm 1, which includes two stages: Initialization and exchange matching stage. In the initialization stage, complete the random initial matching scheme in the scene, and set iteration count to 0. In the exchange matching stage, it cyclically searches for swap-blocking pairs and performs exchange operations until there are no swap-blocking pairs for certain consecutive times. At this time, the matching formed after multiple exchange operations is a stable matching. In the algorithm, satellites and files try to find swap-blocking pairs according to the preference value, then the number of exchange operations is $O(J \cdot M)$. During each exchange operation, the complexity caused by the Quick Sort ordering operations is $O\left(JM\log_2\left(JM\right)\right)$. Thus, the complexity of entire caching allocation algorithm can be calculated as $O\left(J^2M^2\log_2\left(JM\right)\right)$.

3.4 Proactive Pushing Scheme

As mentioned above, LEO satellites will make cache decision in the satellite's coverage area based on the matching-based cache allocation algorithm proposed in Sect. 3.3. When file arrives at satellite node, it is directly cached if the node cache space is left; otherwise, the name of the arriving file is matched with the file in the cache decision. If the match is successful, the cached file is replaced according to the cache decision; if the match is unsuccessful, no operation is performed on the arriving file.

However, the link between user and access satellite be interrupted due to the mobility of LEO satellites, the subsequent satellite covering the area will provide service for user. In this case, interest packet cannot be responded in time. For the user, the requested file is cached in the previous satellite, and the files cached by the subsequent satellite are not preferred by the user, which will increase hop count for users to obtain contents. Furthermore, there are tremendous differences in user preferences and the number of content requests in different geographic regions, simply adopting local caching policies will lead to frequent cache updates. Therefore, in the case of in-network caching, we propose a proactive pushing scheme to improve network transmission efficiency.

The basic idea of the proactive pushing scheme here is that when an LEO satellite is about to leave the divided area, GEO satellite will notify the LEO to push the cached list to the subsequent satellite covering the area in advance. With the equal division of the area, each LEO satellite can start the proactive cache pushing process one by one under the coordination of GEO satellite. The proactive pushing algorithm can be described as follows.

Step 1: LEO satellites will make cache decision based on the matching-based cache allocation algorithm proposed in Sect. 3.3, while GEO satellites monitor the movement of LEO satellites.
Step 2: According to the movement speed and trajectory of LEO satellite l_j, GEO judges whether satellite l_j is about to leave the coverage area. If so, continue to Step 3. Otherwise, go to Step 1.
Step 3: GEO informs satellite l_j to generate a cache list CL_j with only the file name based on the cached files, and push it to the subsequent satellite l_{j+1}.
Step 4: The subsequent satellite l_{j+1} determines whether CL_j is consistent with CL_{j+1}. If not, the corresponding interest packets will be generated, and sent to satellite l_j to request cache files that are not in CL_{j+1}. Until satellite l_{j+1} no longer sends interest packets, the step ends.

4   Simulation Results

To demonstrate the effectiveness of the proposed cache allocation scheme, MAT-LAB is used in our simulations. In the simulation environment, the STI^2CN scenario should ensure normal satellite communication and the coverage of communication on the whole earth. Therefore, Iridium system is adopted, which consists of 6 orbital planes with 11 satellites residing in each orbit, 86.4° inclined orbit planes with the altitude of 780 km. Each satellite is linked to the forward and rear satellites of adjacent planes. Total number of 1,000 files is supposed Zipf distribution, where $\alpha \in [0.2, 1.2]$ varies from different file types. The sizes of files are same, $b_m = 1, \forall m$. The cache capacity of each LEO satellite is same, which varies from 0.005 to 0.025. The maximum hop count $H = 15$ and simulation parameters $iter_{max} = 20$ is used.

In addition, three caching schemes are selected as comparison schemes in the simulation process: the first one is NoCache that no caching in satellites and cache capacity is 0, the second is Random that each satellite caches files randomly, the last one is content popularity cache that most popular files are cached in each satellite. During the simulation, the impact on the network performance is observed by changing the size of nodes' cache capacity and the Zipf parameters α. We choose two metrics to evaluate the performance of caching allocation scheme: average hop count and cache hit rate. Average hop count refers to the average hop count that interest packet caches hit or reaches to content source to obtain files. Cache hit rate refers to the percentage of cache hits in the satellite nodes when all the requests arrive at the satellite nodes.

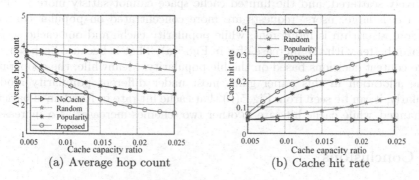

(a) Average hop count   (b) Cache hit rate

Fig. 4. The impact of cache capacity ratio.

Figure 4 shows the average hop count and cache hit rate versus different cache capacity ratio, the following simulation results set the Zipf parameter $\alpha = 0.8$. When the cache capacity is small, the increase in cache capacity enables more requests to be responded in neighboring nodes, and the cache distribution of popular files tends to stabilize as the cache capacity increases further. In Fig. 4(a), with the increasing of cache capacity ratio, average hop count among the three schemes of Random, Popularity and Proposed is diminishing. For popularity cache, each satellite tends to cache most popular files based on the preference of its served users, making it perform better than Random. However, popularity cache considers no cooperation resulting in the cache redundancy between adjacent satellites. As comparison, our scheme further decreases user average hop count by considering cooperation among satellites. Another result in Fig. 4(b) shows that cache hit rate increases as cache capacity ratio increases and eventually plateaus, but our scheme's cache hit rate is always higher than the other two schemes.

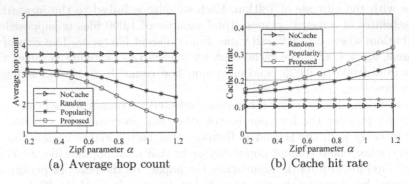

(a) Average hop count   (b) Cache hit rate

Fig. 5. The impact of Zipf parameter α.

Figure 5 shows the average hop count and cache hit rate versus different α, the following simulation results set cache capacity ratio 0.015. The Zipf parameter α reflects the concentration of file requests. When α is small, users' requests are relatively scattered, and the limited cache space cannot satisfy more requests. When α is large, users' requests are more concentrated on popular contents. Random algorithm is indifferent, while popularity cache and our cache scheme perform better with the increase of α in Fig. 5(a). That is because both schemes make content decisions based on the file popularity. Meanwhile, the cooperative cache allocation in this paper gains most under different popularity models. Similarly, it can be seen from Fig. 5(b) that cache hit rate in Random is basically unchanged, while cache hit rate in other two schemes increases as α increases.

5   Conclusions

In this article, we investigate content cache allocation scheme in information-centric satellite-terrestrial integrated networks. Firstly, the cache allocation problem is formulated as user average hop count minimization by considering content popularity and cache cooperation. Then, a matching algorithm with fast convergence and low complexity is used to solve it, and finally obtains exchange stability between satellites and contents. Additionally, a proactive pushing scheme is proposed to accelerate user content retrieval and further improve network performance. The simulation results show that proposed scheme can effectively reduce user average hop count and increase cache hit rate.

References

1. Liu, J., Shi, Y., Fadlullah, Z.M., Kato, N.: Space-air-ground integrated network: a survey. IEEE Commun. Surv. Tutor. **20**(4), 2714–2741 (2018)
2. Xie, R., Tang, Q.: Satellite-terrestrial integrated edge computing networks: architecture, challenges, and open issues. IEEE Netw. **34**(3), 224–231 (2020)
3. Li, J., Xue, K., Liu, J., Zhang, Y.: An ICN/SDN-based network architecture and efficient content retrieval for future satellite-terrestrial integrated networks. IEEE Netw. **34**(1), 188–195 (2019)
4. Bi, Y., et al.: Software defined space-terrestrial integrated networks: architecture, challenges, and solutions. IEEE Netw. **33**(1), 22–28 (2019)
5. Shi, Y., Cao, Y., Liu, J., Kato, N.: A cross-domain SDN architecture for multi-layered space-terrestrial integrated networks. IEEE Netw. **33**(1), 29–35 (2019)
6. Ji, S., Sheng, M., Zhou, D., Bai, W.: Flexible and distributed mobility management for integrated terrestrial-satellite networks: challenges, architectures, and approaches. IEEE Netw. **35**(4), 73–81 (2021)
7. Din, I.U., Hassan, S., Khan, M.K., Guizani, M.: Caching in information-centric networking: strategies, challenges, and future research directions. IEEE Commun. Surv. Tutor. **20**(2), 1443–1474 (2017)
8. Ngaffo, A.N., El Ayeb, W.: Information-centric networking challenges and opportunities in service discovery: a survey. In: 2020 IEEE Eighth International Conference on Communications and Networking (ComNet), pp. 1–8. IEEE (2020)
9. Liu, Z., Zhu, J., Pan, C., Song, G.: Satellite network architecture design based on SDN and ICN technology. In: 2018 8th International Conference on Electronics Information and Emergency Communication (ICEIEC), pp. 124–131. IEEE (2018)
10. Siris, V.A., Ververidis, C.N., Polyzos, G.C.: Information-centric networking (ICN) architectures for integration of satellites into the future internet. In: 2012 IEEE First AESS European Conference on Satellite Telecommunications (ESTEL), pp. 1–6. IEEE (2012)
11. de Cola, T., Gonzalez, G., et al.: Applicability of ICN-based network architectures to satellite-assisted emergency communications. In: 2016 IEEE Global Communications Conference (GLOBECOM), pp. 1–6. IEEE (2016)
12. D'Oro, S., Galluccio, L., Morabito, G., Palazzo, S.: SatCache: a profile-aware caching strategy for information-centric satellite networks. Trans. Emerg. Telecommun. Technol. **25**(4), 436–444 (2014)
13. Wu, H., Li, J., Lu, H., Hong, P.: A two-layer caching model for content delivery services in satellite-terrestrial networks. In: 2016 IEEE global communications conference (GLOBECOM), pp. 1–6. IEEE (2016)
14. Feng, Y., Wang, W., Liu, S., Cui, G., Zhang, Y.: Research on cooperative caching strategy in 5G-satellite backhaul network. In: Yu, Q. (ed.) SINC 2017. CCIS, vol. 803, pp. 236–248. Springer, Singapore (2018). https://doi.org/10.1007/978-981-10-7877-4_21
15. Liu, S., Hu, X., Wang, Y., Cui, G., Wang, W.: Distributed caching based on matching game in LEO satellite constellation networks. IEEE Commun. Lett. **22**(2), 300–303 (2017)
16. Gu, Y., Saad, W., Bennis, M., Debbah, M.: Matching theory for future wireless networks: fundamentals and applications. IEEE Commun. Mag. **53**(5), 52–59 (2015)
17. Zhao, J., Liu, Y., Chai, K.K., Chen, Y.: Many-to-many matching with externalities for device-to-device communications. IEEE Wirel. Commun. Lett. **6**(1), 138–141 (2016)

Signal Processing and Communication Optimization

Resource Allocation in Massive Non-Orthogonal Multiple Access System

Wen Zhang[1($\boxtimes$)], Jie Zeng[2], and Zhong Li[1]

[1] School of Communication and Information Engineering,
Chongqing University of Posts and Telecommunications, Chongqing 400065, China
{s190131003,s190131188}@stu.cqupt.edu.cn
[2] Department of Electronic Engineering, Tsinghua University, Beijing 100084, China
zengjie@tsinghua.edu.cn

Abstract. With the popularity of large-scale communication scenarios, massive multiple access technology has attracted academic attention. In this paper, a resource allocation problem in massive Non-Orthogonal Multiple Access (NOMA) network is studied. The optimization goal is the energy efficiency (EE) of the system. The resource allocation problem includes two parts: subcarrier allocation and power allocation. Firstly, a many-to-one matching algorithm is proposed to assign subcarriers. Then, a power allocation algorithm is designed to optimize the objective function using successive convex approximation. Further, an iterative algorithm for power allocation is studied and the suboptimal solution of the function is obtained. Simulation results show that the proposed resource allocation scheme can effectively improve the EE of the system.

Keywords: Massive NOMA · Resource allocation · Energy efficiency

1 Introduction

With the rapidly development of the Internet, indicators such as EE, delay and reliability of the network have attracted more attention. With the popularization of B5G, massive multiple access is becoming more important due to its ability to improve spectrum efficiency [1,2]. Compared with Code Division Multiple Access (CDMA) used in 3G [3], Orthogonal Frequency Division Multiple Access (OFDMA) used in 4G [4], NOMA, which greatly improves the spectrum efficiency and capacity of the system, can cope with the explosive growth of IIoT applications.

Supported by the National Key R&D Program (No. 2018YFB1801102), the Natural Science Foundation of Beijing (No. L192025), the National Natural Science Foundation of China (No. 62001264), and the China Postdoctoral Science Foundation (No. 2020M680559 and No. 2021T140390).

H. Gao et al. (Eds.): ChinaCom 2021, LNICST 433, pp. 209–219, 2022.
https://doi.org/10.1007/978-3-030-99200-2_17

In NOMA systems, Successive Interference Cancellation (SIC) applied to the receiving end can achieve high spectral efficiency at the cost of increasing the complexity of the receiver [5,6]. Literature [7] uses the inherent characteristics of NOMA to solve the problems of non-orthogonal coordination direct transmission and relay transmission. In order to minimize the interruption probability of relay selection schemes, a two-stage relay selection strategy is proposed [8]. In order to maximize the throughput of the single-carrier MIMO-NOMA system, a suboptimal joint power allocation and precoding design are proposed [9]. In addition, two relay-assisted NOMA schemes were proposed to achieve low-delay and high-reliability for Vehicle-to-Everything (V2X) services [10]. Resource allocation was decentralized in a multi-antenna system with Full-Duplex (FD) base stations, to achieve simultaneous uplink and downlink transmissions [11].

In recent years, massive multiply access has been paid more attention. Literature [1] gives a general overview of massive access wireless communication, and describes the research results and progress of it at the current stage. Literature [12] studies unlicensed Massive-Device Multiple Access (MaDMA) Multiuser Detection (MUD) problem. Literature [13] explains requirements and challenges about massive Machine-Type Communications (MTC) application, and an overview of key technologies to overcome these.

Literature [14] proposes a massive NOMA technology, which is expected to support a large number of IoT devices in cellular networks. In literature [15], the information theoretical upper bound of the overall transmission rate is derived under the massive communication scenario in uplink. It proved the relationship between the optimal number of active devices and coherent time slots, and designed a two-stage practical communication framework. Literature [16] studies the capacity range of multi-input multi-output massive multiple channel, and used the information theory method based on Gallager error index analysis to characterize the finite dimensional region of the channels.

Most of the existing researches are focused on the limitations of massive NOMA systems. So we choose its EE as the optimization objective and optimize the objective function through the proposed resource allocation algorithm. Finally, the effectiveness of the proposed optimization method is proved by simulation.

The remainder of this paper is outlined as follows. In Sect. 2, we describe the system model. Section 3 addresses Recourse Allocation Algorithm, while Sect. 4 demonstrates and analyze the simulation results. Finally, we summarize the conclusions of research in Sect. 5.

2   System Model

In this section, we describe the new downlink massive NOMA network setting consisting of a BS and N UEs, as shown in Fig. 1. Each of the nodes is equipped with a transmit antenna and a receive antenna. The system frequency band is divided into K subcarriers. The signals of different UEs or different packets of an UE can be superposed in one subcarrier to transmit simultaneously. Besides, all the UEs are connected to the BS. The SIC commonly used in NOMA systems is used in the receiver continuation. We assume that the UEs follow independent Poisson point processes (PPPs) with the density of λ_u. The channels of the downlink stage are independent Rayleigh fading channels, and the path loss exponent is α.

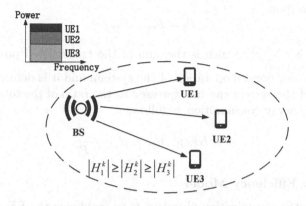

Fig. 1. A missive multiple access downlink system model

2.1   Signaling Model

At the downlink system, we take the nth UE ($n \in 1, 2, ..., N$) as an example to analyze the signals. Suppose that the signal transmitted from the BS to the nth UE on the kth subcarrier is x_n^k, and the power of the signal is p_n^k. Additionally, h_n and ϖ_n denote the small-scale and the large-scale channel coefficients from the BS to the nth UE. Then, the received signal on the nth UE as

$$y_n^k = \sqrt{\varpi_n p_n^k} h_n^k x_n^k + \underbrace{\sum_{s \neq n}^{N} \sqrt{\varpi_s q_s^k} h_s^k x_s^k}_{\text{Interference from other UEs}} + z_n^k, \tag{1}$$

where z_n^k denotes the Additive White Gaussian Noise (AWGN) with mean zero and variance σ^2 at the receiver of the nth UE.

The SIC decoding sequence is determined by the channel coefficient. At the system, we assume that the channel coefficients follow $|H_1^k| \geq |H_2^k| \geq ... \geq |H_N^k|$, where $H_n^k = |h_n^k|^2 \varpi_n$. UEs with poor channel conditions adopt higher transmission power to meet the Quality of Service (QoS), so the decoding sequence is arranged in ascending order according to the size of the channel coefficient. The Signal-to-Interference-plus-Noise Ratio (SINR) of the nth UE is given by

$$SINR_n^k = \frac{\left|h_n^k\right|^2 \varpi_n p_n^k}{\sigma^2 + \sum_{s=1}^{n-1} \left|h_s^k\right|^2 \varpi_s p_s^k}. \tag{2}$$

Thus, the throughput of the nth UE is

$$R = \sum_{k=1}^{K} \sum_{n=1}^{N} W c_n^k \log_2(1 + SINR_n), \tag{3}$$

where c_n^k indicates whether the nth UE is served by the kth subcarrier. Total power of the system is

$$P_t = P_m + P_o, \tag{4}$$

where $P_m = \sum\limits_{k=1}^{K} \sum\limits_{n=1}^{N} p_n^k$, which is the sum of the transmitted power, P_o is the power required for normal operation of the system, and it is determined. Therefore, the EE of the system can be expressed as the ratio of the total throughput to the overall energy consumption as follows

$$max EE(p_1^k, p_2^k, ..., p_n^k) = \frac{R}{P_t}. \tag{5}$$

2.2 Energy Efficiency Model

In this paper, the optimization objective is to maximize the EE of UEs in the system. The formula for the optimization function can be expressed as follows

$$\max EE\left(p_1^k, p_2^k, \ldots, p_n^k\right) = \frac{\sum\limits_{k=1}^{K} \sum\limits_{n=1}^{N} W c_n^k \log_2\left(1 + \dfrac{|h_n^k|^2 \sigma_n p_n^k}{\sigma^2 + \sum\limits_{s=1}^{n-1} |h_s^k|^2 \varpi_s p_s^k}\right)}{\sum\limits_{k=1}^{K} \sum\limits_{n=1}^{N} p_n^k + P_o}. \tag{6}$$

Due to the need for overall power constraints and QoS of UE in missive multiple access systems, constraints C1, C2 and C3 are set as

$$C1 : P_m \leq P_{max}, \tag{7}$$

$$C2 : R_n \geq R_{min}, \forall n, \tag{8}$$

$$C3 : c_n^k \in \{0, 1\}, \forall n, k \tag{9}$$

$$\sum_{n=1}^{N} c_n^k \leq \delta, \forall n,$$

where P_{max} is the upper limit of system transmission power, R_{min} is the minimum value of each UE under the condition that QoS of UE is guaranteed. δ is the upper limit of the number of UE that can be served by a subcarrier. At the same time, in order to ensure the correct decoding of the receiver, the system also needs to meet the SIC decoding threshold, so the constraint C4 is set as

$$C4 : |h_n^k|^2 \varpi_n p_n^k - \sum_{s=1}^{n-1} |h_s^k|^2 \varpi_s p_s^k \geq p_{thr}, \forall n, k, \tag{10}$$

where p_{thr} is the decoding threshold of SIC.

3   Recourse Allocation Algorithm

The optimization objective is a non-convex function, so it is difficult to obtain the global optimal solution by conventional methods. In this paper, the optimization objective is divided into two sub-problems, namely, subcarrier allocation and power allocation. Finally, the suboptimal solution of the objective function is obtained by using iterative algorithm.

3.1   Subcarrier Allocation

The priority of subcarriers is determined according to the channel coefficient, and the priority list of UE is determined according to the EE, their priority list is

$$\{SC\} = \{SC_1, SC_2, ..., SC_k, ..., SC_K\},$$
$$\{UE\} = \{UE_1, UE_2, ..., UE_n, ..., UE_N\}, \tag{11}$$

where $\{SC\}$, $\{UE\}$ represent the total priority list of the subcarriers and the total priority list of the UEs, respectively. $\{SC_k\}$ represents the priority list of all UE corresponding to the kth subcarrier, $\{UE_n\}$ represents the priority list of all subcarriers corresponding to the nth UE. $\{UE^u\}$ represents UEs which have been unmatched. According to the literature [17], we propose a many-to-one matching algorithm to match the subcarriers and the UEs.

Algorithm 1. Many-to-one Matching Algorithm

1: Initializes the priority list of the subcarriers and UEs, initializes the unmatched list of UEs
2: **while** $\{UE^u\} \neq 0$ **do**
3:    Each UE (represented by n) sends matching request to the subcarriers (represented by k) in order of priority from the highest to the lowest in $\{UE^u\}$
4:    **if** The number of UE serviced by k is less than δ **then**
5:       Match n with k, and take n out of $\{UE^u\}$
6:    **else**
7:       Compare n with the UEs which subcarrier k have been matched according to $\{SC_k\}$
8:       **if** There is an UE whose priority is less than n **then**
9:          Replace the lowest priority UE that subcarrier k has matched with n, put the replaced UE into $\{UE^u\}$, removing UE n from $\{UE^u\}$
10:      **else**
11:         Subcarrier rejects the request from n
12:      **end if**
13:   **end if**
14: **end while**

3.2 Power Allocation

Since we have matched the subcarriers to the UEs in the last part, we can treat c_n^k as a definite value. In this paper, the following equation is always true

$$\log_2\left(1 + SINR_n^k\right) = \log_2\left(\sigma^2 + \sum_{s=1}^{n}\left|h_s^k\right|^2 \varpi_s p_s^k\right) - \log_2\left(\sigma^2 + \sum_{s=1}^{n-1}\left|h_s^k\right|^2 \varpi_s p_s^k\right).$$

(12)

We can define it as $\log_2\left(1 + SINR_n^k\right) \triangleq f_n^k(p) - g_n^k(p)$, in the same way, the formula of the throughput can be converted to

$$R = \sum_{k=1}^{K}\sum_{n=1}^{N} W c_n^k \log_2\left(1 + \text{SINR}_n\right) \triangleq F(p) - G(p).$$

(13)

Since $f_n^k(p)$ and $g_n^k(p)$ is convex, $F(p)$, $G(p)$ and P_t are all convex functions according to the properties of convex functions. The objective function can be expressed as

$$\max_{p} EE\left(p_1^k, p_2^k, \ldots, p_n^k\right) = \frac{F(p) - G(p)}{P_t}.$$

(14)

Among all the constraints, $C1$, $C3$ and $C4$ satisfy the definition of a convex function, $C2$ can be represented as $R_{\min} + W g_n^k(p) - W f_n^k(p) \leq 0$, it is the difference form of two convex functions. At this time, the optimization objective is still A non-convex function. We can assume that a maximum EE is ε^*, and the optimization objective can be transformed into B, forming A differential convex function. Therefore, successive convex approximation can be used to calculate the suboptimal solution of the objective function.

$f_n^k(p)$ is a convex function, for any feasible point $p*$, we have

$$f_n^k(p) \geq f_n^k\left(p^*\right) + f_n^k(p)'\big|_{p=p^*}\left(p - p^*\right) \triangleq f_n^k(p)^*,$$

(15)

where

$$f_n^k(p)'\big|_{p=p^*}\left(p - p^*\right) = \frac{\sum_{s=1}^{n}\left|h_s^k\right|^2 \varpi_s\left(p - p^*\right)}{\left(\sigma^2 + \sum_{s=1}^{n}\left|h_s^k\right|^2 \varpi_s p^*\right)\ln 2}.$$

(16)

So we can get

$$F(p) \geq \sum_{k=1}^{K}\sum_{n=1}^{N} W c_n^k f_n^k(p)^* \triangleq F(p)^*.$$

(17)

The optimization goal can be translated as

$$\min_{p}\varepsilon^* P_t + G(p) - F(p)^*$$

$$\text{s.t. } C1, C3, C4$$

(18)

$$C2 : R_{\min} + W g_n^k(p) - W f_n^k(p)^* \leq 0.$$

The suboptimal solution of the function can be obtained by using the convex function solver. We propose an iterative algorithm to obtain the suboptimal solution of the function.

Algorithm 2. Iterative Algorithm for Power Allocation

1: Set iteration index $\alpha = 1$, Set the maximum number of iterations is α_{max}
2: Initializing ε_1^*, randomly taking a viable solution to ε_1^*
3: **while** $\varepsilon_n^* < \varepsilon_{n+1}^*$ and $\alpha < \alpha_{max}$ **do**
4:    Use the standard convex function solver to solve equation (18) and obtain the value of p_n^*
5:    Using the obtained value of p_n^* and equation (6), update the value of ε_n^*
6:    Update iteration index, let $\alpha = \alpha + 1$
7: **end while**
8: Get and output the suboptimal solution of the function.

---

4 Simulation Results

Table 1. parameter setting

Parameters	Value
The system bandwidth	3 MHz
Carrier frequency	2 GHz
The subcarrier bandwidth	15 KHz
Path loss exponent α	3
Noise power spectral density	-174 dBm/Hz
The density of UEs λ_u	200 per/Km^2
The maximum power of BS	46 dBm
The target data rate of UEs	2 bps/Hz
The transmit power P_m	20 dBm
The system operating power P_o	20 dBm

In this section, the proposed resource allocation algorithm in the massive NOMA system is simulated and analyzed, and it is compared with other algorithms. We deploy a BS in an area with a radius of 500 m. The UE model is an independent PPP model with a density of λ_u which is randomly generated in the area, and the other simulation parameters are shown in Table 1.

In Fig. 2, we research EE versus the density of UEs when the subcarrier match with different maximum numbers of superposed signals. It can be clearly seen that with the increase of the density of UEs, the greater the value of δ is, the greater EE will be. The rate of increase in EE is also proportional to the value of δ. However, when the value of δ increases to a certain extent, EE will decrease with the increase of the density of UEs. It is worth mentioning that when the value of δ is different, the corresponding maximum EE will also be different, which is 8.7, 16.5, 19.3, and 23.9, respectively.

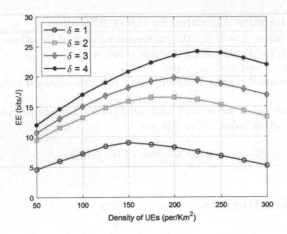

Fig. 2. EE versus the density of UEs with different value of δ

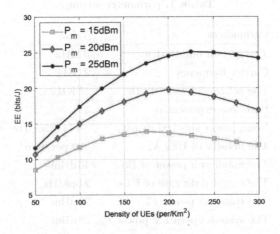

Fig. 3. EE versus the density of UEs with different maximum of the transmit power

In Fig. 3, we compared the relationship between EE and the density of UEs under different maximum transmission power. It can be seen that, with the increase of the density of UEs, the value of EE increases first and then decreases, and the increase speed is proportional to the size of the maximum transmission power. However, when the density of UEs is between 200 and 300, the value of EE corresponding to $P_{m2} = 20\,\text{dBm}$ decreases the fastest.

In Fig. 4 and Fig. 5, we compare different algorithms under different systems. Among them, the subcarrier of OFDMA system can serve three UEs, and there is interference between the UEs, and the optimization goal is to maximize the EE of the system. The subcarriers in the NOMA system can also serve three UEs, but the optimization goal is to maximize the throughput of the system.

Figure 4 studies the EE versus the density if UEs under different algorithms. With the increase of the density of UEs, EE in NOMA system grows at a rate

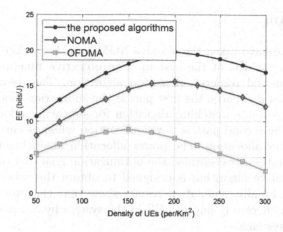

Fig. 4. EE versus the density of UEs in different algorithms

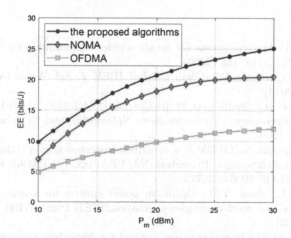

Fig. 5. EE versus the maximum of the transmit power in different algorithms

almost equal to that of the proposed algorithm, but EE in OFDMA system grows at the slowest rate and declines at the fastest. As shown in Fig. 5, we research the EE versus the maximum transmitted power under different algorithms, which shows that EE of three algorithms increase monotonically with the increase of the maximum transmitted power in the range of P_m is 10 dBm to 30 dBm. When P_m is less than 16.25 dBm, the growth rate of EE in NOMA system is the fastest; when P_m is more than 20 dBm, the growth rate of EE in the proposed algorithm is the fastest. We can see that the proposed algorithm greatly improves the EE of the system.

5 Conclusion

In this article, we have taken the massive NOMA as the background, the optimization goal is the EE of the system. The objective function is optimized through the proposed resource allocation algorithm. The resource allocation algorithm includes two parts, the first part is the subcarrier allocation, we have proposed a many-to-one matching algorithm for subcarrier allocation which is user-centered. The second part is power allocation which is carried out on the basis of subcarrier allocation. The power allocation algorithm uses successive convex approximation, transformed the optimization goal to a convex function. And then a iterative algorithm is designed to obtain the suboptimal solution of the function. Finally, simulation results show that the proposed optimization method can effectively improve EE in the system by comparing with other algorithms and systems.

References

1. Wu, Y., et al.: Massive access for future wireless communication systems. Wirel. Commun. **27**(4), 148–156 (2020)
2. Andrews, J.G., et al.: What will 5G be? IEEE J. Sel. Areas Commun. **32**(6), 1065–1082 (2014)
3. Hanzo, L., et al.: Single and Multi-Carrier DS-CDMA: Multi-User Detection, Space-Time Spreading, Synchronisation, Networking and Standards, pp. 35–80. IEEE Xplore (2003)
4. Yin, H., Alamouti, S.: OFDMA: a broadband wireless access technology. In: 2006 IEEE Sarnoff Symposium, Princeton, NJ, USA, pp. 1–4 (2006). https://doi.org/10.1109/SARNOF.2006.4534773
5. Andrews, J.G., Meng, T.H.: Optimum power control for successive interference cancellation with imperfect channel estimation. IEEE Trans. Wirel. Commun. **2**(2), 375–383 (2003)
6. Agrawal, A., et al.: Iterative power control for imperfect successive interference cancellation. IEEE Trans. Wirel. Commun. **4**(3), 878–884 (2005)
7. Kim, J.B., Lee, I.H.: Non-orthogonal multiple access in coordinated direct and relay transmission. IEEE Commun. Lett. **19**(11), 2037–2040 (2015)
8. Ding, Z., Dai, H., Poor, H.V.: Relay selection for cooperative NOMA. IEEE Wirel. Commun. Lett. **5**(4), 416–419 (2016)
9. Hanif, M.F., et al.: A minorization-maximization method for optimizing sum rate in the downlink of non-orthogonal multiple access systems. IEEE Trans. Sig. Process. **64**(1), 76–88 (2015)
10. Liu, G., et al.: Cooperative NOMA broadcasting/multicasting for low-latency and high-reliability 5G cellular V2X communications. IEEE Internet Things J. **6**(5), 7828–7838 (2019)
11. Ng, D., Wu, Y., Schober, R.: Power efficient resource allocation for full-duplex radio distributed antenna networks. IEEE Trans. Wirel. Commun. **15**(4), 2896–2911 (2016)
12. Ding, T., Yuan, X., Liew, S.C.: Sparsity learning-based multiuser detection in grant-free massive-device multiple access. IEEE Trans. Wirel. Commun. **18**(7), 3569–3582 (2019)

13. Dawy, Z., et al.: Toward massive machine type cellular communications. Wirel. Commun. **24**(1), 120–128 (2016)
14. Shirvanimoghaddam, M., Dohler, M., Johnson, S.J.: Massive non-orthogonal multiple access for cellular IoT: potentials and limitations. IEEE Commun. Mag. **55**(9), 55–61 (2017)
15. Yu, W.: On the fundamental limits of massive connectivity. In: 2017 IEEE Information Theory and Applications Workshop (ITA), San Diego, CA, USA, pp. 1–6 (2017). https://doi.org/10.1109/ITA.2017.8023482
16. Wei, F., et al.: On the fundamental limits of MIMO massive multiple access channels. In: ICC 2019-2019 IEEE International Conference on Communications (ICC), Shanghai, China, pp. 1–6 (2016). https://doi.org/10.1109/ICC.2019.8761543
17. Sethuraman, J., Teo, C.P., Qian, L.: Many-to-one stable matching: geometry and fairness. Math. Oper. Res. **31**(3), 581–596 (2006)

Distortionless MVDR Beamformer for Conformal Array GNSS Receiver

Han Li[1], Di He[1(✉)], Xin Chen[1], Jiaqing Qu[2], and Lieen Guo[3(✉)]

[1] Shanghai Key Laboratory of Navigation and Location-Based Services, Shanghai Jiao Tong University, Shanghai, People's Republic of China
{hanzai_1101,dihe,xin.chen}@sjtu.edu.cn
[2] Shanghai Radio Equipment Research Institute, Shanghai, People's Republic of China
[3] School of Mechanical and Electrical Engineering, Nanchang University, Nanchang, People's Republic of China
guolieen@163.com

Abstract. The minimum variance distortionless response (MVDR) beamforming technique and space-time adaptive processing (STAP) have been playing important roles in interference suppression of globe navigation satellite system (GNSS) receiver. However, the demand for conformal arrays is increasing these days and its characteristics will vitiate the traditional MVDR method. And on the other side, traditional MVDR based on STAP will inevitably distort the Beidou signal, which is unacceptable in high precision GNSS applications. To address the above issues, first, a conformal array signal processing model is proposed; and based on that, a distortionless MVDR method is proposed in this study. The simulation results show that the proposed method can not only suppress the interference better than the traditional MVDR, but also guarantee the Beidou signal to be undistorted.

Keywords: Conformal antenna array · Interference suppression · Distortionless STAP

1 Introduction

GNSS is one of the most powerful infrastructures that provide positioning, navigation and timing (PNT) service for users all over the world [1,2]. Currently, there are four operating GNSS, the Global Positioning System (GPS) of the United States, the Global Navigation Satellite System (GLONASS) of Russia, the Galileo of the European Union and the Beidou Navigation Satellite System (BDS) of China. The BDS is a versatile GNSS completely designed and constructed independently by China. The BDS III was officially put into use and provide positioning service to the world in 2020, marking the BDS has formally become a GNSS.

Similar to other existing GNSS, BDS is also designed to withstand a certain level ($10 log_{10} 10230 \approx 40\,\mathrm{dB}$) of radio frequency interference [3], which is

H. Gao et al. (Eds.): ChinaCom 2021, LNICST 433, pp. 220–234, 2022.
https://doi.org/10.1007/978-3-030-99200-2_18

achieved by the direct-sequence spread spectrum (DSSS) technique [4] and is capable of coping with the most common noise. However, the application environment in real life is not always ideal. In addition to the receiver thermal noise, unintentional and deliberate interferences exist under many circumstances and undermine the reliability of PNT. Common unintentional interferences include some specialized communication systems like DME, TACAN, DVB-T, and so on. The deliberate jammer is used in electronic warfare (EW) or to defend one's own privacy, especially through the personal privacy devices (PPDs). Related researches show that a single jamming device can disable GNSS signal reception over a range of several kilometers [5]. To address these situations and enhance the robustness of the GNSS receiver, variety of GNSS interference mitigation countermeasures have been developed in the past decades, such as inertial aiding using IMU, vector tracking technique and filtering technique [3,5]. Among them, the filtering technique is most frequently used in the GNSS receiver, which can be grouped into time, frequency, time-frequency and spatial-time domains. It has been proved in practice that the spatial-time domain filtering is the most powerful interference mitigation method in aforementioned methods [6]. The spatial-time adaptive filtering (STAP) techniques utilize the antenna array combined with FIR filters to adaptively rearrange the received signals at the antenna in a weighted sum version to steer the antenna array response beam pointing to the desired direction and nulls pointing to the undesired interferences, which allow the BDS signals to pass unchanged and suppress the jamming effectively. Obviously, the effectiveness of receiving BDS signals and rejecting interferences is determined by the weights of each delayed tap. Therefore, the core of STAP algorithm is the principle of calculating the weights. Different approaches have been developed to realize the objective, including minimum mean square error (MMSE), MVDR and power inversion (PI), and they work well on suppressing the interference with high power [7–9]. However, the FIR filters introduced in STAP lead to non-linearity phase response which in turn cause the distortion of the desired signals [10,11]. Previous experiments suggest that without special attention on this issue, the positioning error brought by STAP can be tens of meters [11]. Therefore, in high-accuracy GNSS applications, this issue must be taken into consideration. Among some methods proposed to reduce the biases in code and carrier phase [6,10,12], the method proposed in [13] guarantees phase linearity and zero biases in code and carrier phase measurements.

A conformal array is a kind of antenna array that conforms to its bearer's surface which is designed for better aerodynamic and hydrodynamic performance or aesthetic consideration [14]. Therefore, it can be very useful in many applications such as radar or BDS receiver for high-speed aircraft. While most of the traditional interference suppression techniques can not be applied for conformal array directly since they generally assume that all the antennas are omnidirectional element and have identical response pattern. However, due to the non-planar curvature of the bearer surface, directional element model should be adopted and different antenna usually has different orientation, which in turn leads to different response to the same signal on different antenna [15,16]. Furthermore, it means that the element response function can not be separated as a common factor of the array manifold which is a conventional practice used in planar array analysis. Accordingly, the mathematical model for the signal receiving procedure must be modified to take the above non-ideal conditions into consideration. Research about conformal array signal processing mainly focus on pattern synthesis and DOA estimation. Less attention is paid to the interference suppression technique in conformal array. So in this study, an array signal processing model integrating the element response is proposed.

The remainder of this article is organized as the following sections. In Sect. 2, a traditional procedure of suppressing interference with linear array is introduced. Next in Sect. 3, an extended signal processing model applicable to conformal array is given, and based on that a distortionless MVDR method for conformal array is proposed and analyzed. Then in Sect. 4, some simulation results are demonstrated to prove the effectiveness of the proposed method on conformal characteristic and distortionless performance. Finally, Sect. 5 concludes the paper.

2   Interference Suppression Using Antenna Array

In this section, a planar antenna array is considered to illustrate the traditional interference suppression method for GNSS using STAP and MVDR. Meanwhile, a general array signal processing model is established. Before that, the generation of the Beidou B3I signal and barrage jammer is introduced.

2.1   Signal Generation

Beidou B3I Signal. The Beidou B3I signal is composed of ranging code and navigation message modulated on the carrier using BPSK [4], which can be expressed as

$$S_{B3I}(t) = A_{B3I}C_{B3I}(t)D_{B3I}(t)\cos(2\pi ft + \varphi_{B3I}) \tag{1}$$

Among them, A_{B3I} represents the amplitude of the B3I signal. $D_{B3I}(t)$ is the data code of the B3I signal. In this study, it is assumed that only the D1 navigation message is used. $C_{B3I}(t)$ is the ranging code modulated on the data code by direct sequence spread spectrum (DSSS). The chip rate of C_{B3I} is $10.23Mcps$ and the code length is 10230. The C_{B3I} is generated by truncating a Gold code which is the result of truncating and XORing two linear sequences G1 and G2. The G1 and G2 sequences are respectively derived from two 13-bit linear shift registers, and its period is 8191 chips. The code sequence generated by G1 is truncated with the last one chip, making it into a CA sequence with a period of 8190 chips. The CA sequence with a period of 8191 chips is generated by G2. The C_{B3I} with a period of 10230 chips is generated by means of Modulo-2 addition of CA and CB sequences. f represents the carrier frequency of the B3I signal, which is 1268.52 MHz, φ_{B3I} represents the carrier initial phase of the B3I signal [4]. From the above, it is derived that the bandwidth of the B3I signal is 20.46 MHz. The Beidou B3I signal generation process is shown in Fig. 1.

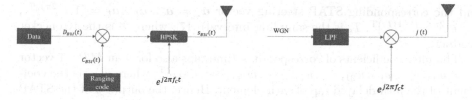

Fig. 1. Generation of the B3I signal.   **Fig. 2.** Generation of the barrage jammer.

Interference. In this study, the interference is considered to be a barrage jammer covering the entire frequency band of B3I. To generate such a jammer. Firstly, a white Gaussian noise (WGN) is generated and passed through a low-pass filter with half of the bandwidth of the Beidou signal as passband bandwidth, and then the filter output signal is multiplied by the complex carrier signal $e^{j2\pi f_c t}$, modulated to the center frequency of B3I. The interference generation process is shown in Fig. 2.

2.2 Signal Receiving Model

Without loss of generality and for the sake of simplicity, an M-element omnidirectional linear array along with one B3I signal and K interference is considered. For a narrowband signal impinges from θ_k, its corresponding steering vector $\boldsymbol{a}$ is

$$\boldsymbol{a}(\theta_k) = [1 \quad e^{-j\phi_k} \dots e^{-j(M-1)\phi_k}]^{\mathrm{T}} \tag{2}$$

where $[\cdot]^{\mathrm{T}}$ represents the transpose of a vector or matrix, ϕ_k is the phase delay due to propagation and can be expressed as

$$\phi_k = \frac{2\pi d_k}{\lambda} \sin\theta_k, \quad 1 \le k \le K \tag{3}$$

where d_k is the coordinate of the k-th antenna. Therefore, the observation vector $\boldsymbol{X}(n)$ of the array with dimension $M \times 1$ can be denoted as

$$\boldsymbol{X}(n) = \boldsymbol{a}(\theta_d)d(n) + \sum_{k=1}^{K} \boldsymbol{a}(\theta_k)j_k(n) + \boldsymbol{D}(n) \tag{4}$$

where $d(n)$ and $j_k(n)$ are the desired signal and the k-th interference, respectively [9]. Thus, $\boldsymbol{a}(\theta_d)$ and $\boldsymbol{a}(\theta_k)$ are the corresponding steering vectors. $\boldsymbol{D}(n)$ is an $M \times 1$ vector representing the internal noise of the array.

Through a standard STAP implementation of P time taps, the final STAP observation vector can be expressed as

$$\boldsymbol{V}(n) = \begin{bmatrix} \boldsymbol{X}(n) \\ \boldsymbol{X}(n-1) \\ \vdots \\ \boldsymbol{X}(n-(P-1)) \end{bmatrix} \tag{5}$$

and the corresponding STAP steering vector $a_1 = a \otimes a_t$, $(a_t = [1, e^{j2\pi f_c T_s},$ $\ldots, e^{j2\pi f_c(P-1)T_s}]^T$, T_s is the sampling interval) [17], where $\otimes$ is the Kronecker product.

The filter coefficients of corresponding time taps also form an $MP \times 1$ vector $w = [w_{11}, \cdots, w_{1P}, w_{21}, \cdots, w_{2P}, \cdots, w_{M1}, \cdots, w_{MP}]^T$, where w_{mp} is the coefficient of the p-th delayed tap of mth element. Hence, the output y of the STAP filter is

$$y = w^H V \tag{6}$$

2.3 Interference Suppression Using MVDR Beamformer

The MVDR beamformer minimizes the output signal power while maintaining the desired signal, which can be expressed in the following formulations [9]

$$\begin{cases} \min_{w} E\{\|y\|^2\} = \min_{w} w^H R_{VV} w \\ w^H a_1(\theta_d) = 1 \end{cases} \tag{7}$$

$E\{\cdot\}$ denotes the mathematical expectation, R_{VV} denotes the space-time covariance matrix defined as $R_{VV} = E\{VV^H\}$. Using Lagrange multipliers, the object function can be expressed as

$$f(w, \lambda) = w^H R_{VV} w + \lambda(w^H a_1 - 1) \tag{8}$$

and then the optimal coefficient vector $w^\star$ can be acquired as

$$w^\star = \frac{R_{VV}^{-1} a_1}{a_1{}^H R_{VV}^{-1} a_1} \tag{9}$$

3 Proposed Method

As mentioned above, the demand on the suppression based on the conformal antenna array is increasing. However, in traditional array signal processing, the radiation pattern function of all array elements are generally considered to be omnidirectional, and in this case, the global polar coordinate system is consistent with the local polar coordinate system of each array element. Therefore, the radiation function of the array element can be put forward as a common factor, which is called the Pattern Product Theorem (PPT), so that the radiation pattern of the array is only related to the geometric structure of the array. However, in the conformal array, firstly a directional antenna element model is more appropriate. Secondly, the different orientations of the array elements lead to the inconsistency between the global polar coordinate system and the local polar coordinate system of each array element, which makes the PPT fails and furthermore causes the error in signal array manifold estimation. So, the traditional beamforming algorithms do not work very well in the conformal antenna array structure.

In this section, a novel MVDR method based on Euler rotation is presented to solve the conformal array beamforming problem. Besides, the traditional STAP procedure is inevitable to introduce distortion on the desired signal due to the deployment of delayed time taps. Based on the conformal array beamforming above, a distortionless constraint is also introduced and analyzed in detail.

3.1 Conformal Array Signal Receiving Model

Here a more general conformal array model containing M elements is considered. In the global coordinate system, the position of each element is $\boldsymbol{p_i} = [x_i, y_i, z_i]^{\mathrm{T}}$, $(i = 1, 2, \cdots, M)$. For this array, first examine the synthesis of its array response pattern. Considering that a far-field narrow-band plane wave with a propagation direction of $\boldsymbol{a} = -[\sin\theta\cos\varphi \quad \sin\theta\sin\varphi \quad \cos\theta]^{\mathrm{T}}$ and wave length of λ_0 impinges on the array, so the corresponding wave vector is

$$k = \frac{2\pi}{\lambda_0}a \tag{10}$$

Among them, θ and φ are the elevation angle and azimuth angle in the global coordinate system. Furthermore, the radiation pattern of this array can be obtained

$$\boldsymbol{F}(\theta, \varphi) = \sum_{m=1}^{M} [w_m^* \boldsymbol{f_m}(\theta, \varphi) e^{-j\boldsymbol{k}^{\mathrm{T}}\boldsymbol{p_m}}] \tag{11}$$

where $\boldsymbol{f_i}(\theta, \varphi)$ is the radiation direction function of the i-th element itself, w_m is the corresponding complex weighting factor. Note that the global polar coordinate system is not consistent with local polar coordinate system of each element. To address this issue and for simplicity and generality, a five-element array with a central element conforms to a cylindrical surface is used for analysis. The specific array structure is shown in Fig. 3 below, all the elements point to the normal line. All elements have the same radiation function $\boldsymbol{f}(\theta, \varphi)$. Take p_i as an example to analyze its contribution to the array pattern. First, (θ, φ) is the polar coordinate corresponding to the global Cartesian coordinate system $O - xyz$, (θ_i, φ_i) is the coordinate in the polar coordinate system corresponding to the local Cartesian coordinate system $O_i - x_i y_i z_i$ of the ith array element. For a signal impinging from (θ, φ) in the global coordinate system, it first needs to be converted to local polar coordinate of each element, which can be achieved with Euler rotation.

It is known that the relationship between rectangular and polar coordinate system is

$$\begin{cases} x = \rho\sin\theta\cos\varphi \\ y = \rho\sin\theta\sin\varphi \\ z = \rho\cos\theta \end{cases} \tag{12}$$

where ρ is the distance between the signal source and the origin O. At the same time a new coordinate $O_i - x_i' y_i' z_i'$ is defined, which is generated by translate $O - xyz$ to make O overlap O_i

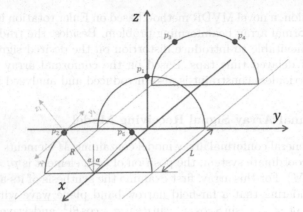

Fig. 3. Cylindrical conformal array structure.

$$\begin{bmatrix} x_i' \\ y_i' \\ z_i' \end{bmatrix} = \begin{bmatrix} x \\ y \\ z \end{bmatrix} - p_i \qquad (13)$$

It is obvious that $O_i - x_i y_i z_i$ can be generated by rotating $O_i - x_i' y_i' z_i'$ around O_i. Therefore, the relationship between the above two coordinates can be derived using Euler rotation [18] as follows

$$\begin{bmatrix} x_i \\ y_i \\ z_i \end{bmatrix} = R_x R_y R_z \begin{bmatrix} x_i' \\ y_i' \\ z_i' \end{bmatrix} \qquad (14)$$

and for the cylindrical array proposed in Fig. 3, only the rotation of $\alpha°$ with $O_i x_i'$ as the rotation axis is needed. Then $R_y R_z = I$, combining with (13) and (14), the following transformation can be obtained

$$\begin{bmatrix} x_i \\ y_i \\ z_i \end{bmatrix} = \begin{bmatrix} 1 & 0 & 0 \\ 0 & \cos\alpha & -\sin\alpha \\ 0 & \sin\alpha & \cos\alpha \end{bmatrix} \begin{bmatrix} x - p_{ix} \\ y - p_{iy} \\ z - p_{iz} \end{bmatrix} = \begin{bmatrix} x - p_{ix} \\ (y - p_{iy})\cos\alpha - (z - p_{iz})\sin\alpha \\ (y - p_{iy})\sin\alpha + (z - p_{iz})\cos\alpha \end{bmatrix} \qquad (15)$$

Besides, the signal is assumed to be far-field, so ρ can be regarded as $+\infty$. Using the definition of *arctan* function, local polar coordinate can be solved as

$$\begin{cases} \theta_i = \lim_{\rho \to \infty} \arctan \dfrac{\sqrt{x_i^2 + y_i^2}}{z_i} \\ \qquad = \arctan \dfrac{\sqrt{(\sin\theta\cos\varphi)^2 + (\sin\theta\sin\varphi\cos\alpha - \cos\theta\sin\alpha)^2}}{\sin\theta\sin\varphi\sin\alpha + \cos\theta\cos\alpha} \\ \varphi_i = \arctan 2(x_i, y_i) \end{cases} \qquad (16)$$

where $arctan2$ function is the four-quadrant inverse tangent. Now the array radiation pattern can be transformed to

$$F(\theta,\varphi) = \sum_{m=1}^{M} [w_m^* f(\theta_m,\varphi_m)e^{-jk^T p_m}] \tag{17}$$

As to the element radiation function, a cosine element model showed in Fig. 4 is used to better depict the radiation characteristic of elements on the convex conformal surface.

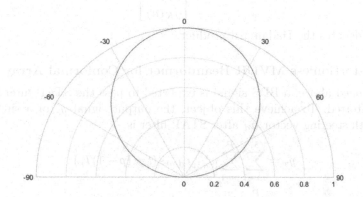

Fig. 4. Radiation pattern of the cosine element.

$$f(\theta,\varphi) = \begin{cases} \cos\theta, & 0 \le \theta \le \frac{\pi}{2} \\ 0, & else \end{cases} \tag{18}$$

Based on the above analysis, a general signal receiving model for an M-element convex conformal array can be obtained. Assume there are K interferences $j_k(t)(k = 1,2,\cdots,K)$, and one B3I signal $d(t)$. For arbitrary signal $s(t)$ with incident angle of (θ,φ) and wave vector k, its steering vector is

$$a(\theta,\varphi) = [1, e^{-jk^T \overline{r_2}}, \cdots, e^{-jk^T \overline{r_M}}]^T \tag{19}$$

where $\overline{r_m} = p_m - p_1$. Using the conversion in (16), the array response for $s(t)$ can be expressed as

$$x_s = \begin{bmatrix} f(\theta_1(\theta,\varphi)) \\ f(\theta_2(\theta,\varphi)) \\ \vdots \\ f(\theta_M(\theta,\varphi)) \end{bmatrix} a(\theta,\varphi)s(n) + D(n) \tag{20}$$

Besides, the array manifold of the incoming signals can be expressed as

$$A_{M\times(K+1)} = [a(\theta_d,\varphi_d) \quad a(\theta_{j_1},\varphi_{j_1}) \quad \cdots \quad a(\theta_{j_K},\varphi_{j_K})] \tag{21}$$

And for the convenience, the conformal factor matrix (CFM) G is defined to incorporate the conformal characteristic

$$G = [g(\theta_d, \varphi_d) \quad g(\theta_{j_1}, \varphi_{j_1}) \cdots g(\theta_{j_K}, \varphi_{j_K})] \tag{22}$$

where $g(\theta, \varphi) = [f(\theta_1(\theta, \varphi)), \cdots, f(\theta_M(\theta, \varphi))]^{\mathrm{T}}$. Finally, the observation vector of the conformal array can be written as

$$X(n) = G \odot A \begin{bmatrix} d(n) \\ j_1(n) \\ \vdots \\ j_K(n) \end{bmatrix} + D(n) \tag{23}$$

where $\odot$ denotes the Hadamard product.

3.2   Distortionless MVDR Beamformer for Conformal Array

As mentioned above, a BDS signal is expected to pass the STAP filter without being distorted. To achieve this object, the output signal y_s of a single BDS signal with steering vector a_d after STAP filter is

$$y_s = \sum_{m=1}^{M} \sum_{p=1}^{P} w_{m,p}^* c_{d,m} s\left(t - (p-1)T_s\right)$$

$$= \sum_{p=1}^{P} w_p{}^{\mathrm{H}} c_d s\left(t - (p-1)T_s\right) \tag{24}$$

where $c_d = g(\theta_d, \varphi_d) \odot a(\theta_d, \varphi_d)$, and $w_p = [w_{1,p}, w_{2,p}, \cdots, w_{M,p}]^{\mathrm{T}}$ denotes the weight vector at mth delayed tap, $[\cdot]^*$ denotes the conjugate operation. Actually, frequency response $H(f)$ of the STAP filter toward signal with steering vector a_d is in the same form [13]

$$H(f) = \sum_{p=1}^{P} w_p{}^{\mathrm{H}} c_d e^{-j2\pi f(p-1)T_s} = \sum_{p=1}^{P} h(p) e^{-j2\pi f(p-1)T_s} \tag{25}$$

It can be found that (25) is also equivalent to a FIR filter given in Fig. 5 with coefficient vector $h(p) = w_p{}^{\mathrm{H}} c_d$ $(p = 1, 2, \cdots, P)$.

Now, substitute the second constraint of the optimization problem (7) with (26) to guarantee the linearity of $H(f)$

$$\begin{cases} P \text{ is odd} \\ w^{\mathrm{H}} c = 1 \end{cases} \tag{26}$$

where $c = \begin{bmatrix} \underbrace{0 \cdots 0}_{M(P-1)/2} & c_d{}^{\mathrm{T}} & \underbrace{0 \cdots 0}_{M(P-1)/2} \end{bmatrix}^{\mathrm{T}}$. Then the optimal weight can be solved using the Lagrange multipliers method introduced in Sect. 2.3 as

$$w^* = \frac{R_{VV}^{-1} c}{c^{\mathrm{H}} R_{VV}^{-1} c} \tag{27}$$

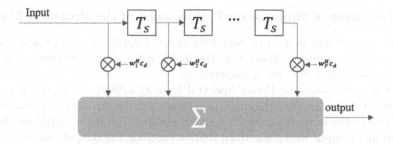

Fig. 5. Structure of the equivalent FIR filter.

The proof of this method applied on the traditional planar array has been detailed analyzed in [13]. Here the a_s in [13] is substituted by c_d which is the Hadamard product of a_s and $g(\theta_d, \varphi_d)$. It is readily verified that this alteration does not affect the form and characteristic of R_{VV}^{-1} and R_J described in [13]. Finally, h can be obtained as

$$
h = \frac{\mu M}{\delta_n^2} \left[\frac{|\rho|^2 \delta_n^2}{MP_J} r_{1,(P+1)/2} \quad \cdots \quad 1 - |\rho|^2 + \frac{|\rho|^2 \delta_n^2}{MP_J} r_{(P+1)/2,(P+1)/2} \right.
$$
$$
\left. \cdots \quad \frac{|\rho|^2 \delta_n^2}{MP_J} r_{P,(P+1)/2} \right]^{\mathrm{T}}
$$

(28)

where δ_n^2, ρ, P_J denotes the noise variance, spatial correlation coefficient [19] between the desired signal and the interference, interference power, respectively. And more importantly, h is conjugate symmetric [13]. Combining that P is odd, the FIR filter coefficients meet the requirement of linear-phase system [20] with a constant bias $\frac{(P-1)T_s}{2}$. This bias does not affect the positioning accuracy because it remains the same for all incoming Beidou signals. Therefore, the proposed method is linear phased.

4   Simulation Results

In the simulation experiments, the array structure is the same as shown in Fig. 3 with $P = 5$, $R = 0.1\,\mathrm{m}$, $\alpha = 30°$, and $L = 0.1025\,\mathrm{m}$, and signals are generated based on the signal model given in Sect. 2, or by the procedure shown in Fig. 1 and Fig. 2. The received signals are down converted to intermediate frequency $f_{IF} = 46.52\,\mathrm{MHz}$ and then sampled with sampling rate $f_s = 62\,\mathrm{MHz}$, and sent to the TDLs where $T_s = 1/f_s$. The thermal noise power is $p_N = N_0 B$, where $B = 20.46\,\mathrm{MHz}$ is the bandwidth of Beidou B3I signal, $N_0 = k_B T_0$ is the noise power density, and k_B is the Boltzmann constant and T_0 is set to 290 K. The Beidou signal power used in the following simulations is set to SNR = $-30\,\mathrm{dB}$, and the jammer-noise ratio is set to JNR = 70 dB.

4.1 Interference Suppression Performance of the Proposed Method

The first experiment is used to demonstrate the interference suppression performance of the proposed method. One Beidou signal and one interference impinge from $(30°, 30°)$ and $(45°, 90°)$, respectively.

In Fig. 6, it shows the Power Spectral Density (PSD) of the signal received on p_1 and that of the output signal. It is clear that the strong barrage jammer is dampened to the noise floor. From another perspective, if the signal received on any antenna is input into a standard Beidou receiver, the output would definitely be a garbled because the bit error rate (BER) is about 50%. While after the interference suppression, even though the desired signal mixed in the output signal is still overwhelmed by thermal noise, it can be decode by the standard Beidou receiver for the DSSS characteristic.

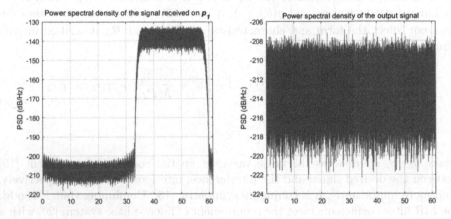

Fig. 6. PSD estimation comparision. The horizontal axis is the frequency in MHz.

From (17), the gain of the array at f is shown in Fig. 7. Obviously, a fairly deep null about -121 dB is formed in the direction of the interference. While in the direction of the Beidou signal, there is about only -0.04 dB attenuation. And the SINR is improved from about -100 dB to about -26 dB, which can guarantee the receiver to decode correctly.

4.2 SINR Performance Comparison with the Traditional MVDR

Here a comparision experiment is adopted to prove that extra considerations on the antenna direction in the proposed method can help it outperform the traditional MVDR. In this simulation, the interference direction is fixed at $(45°, 90°)$, and the azimuth of the Beidou signal is fixed to $30°$. Suppose the elevation angle varies from $0°$ to $90°$, and the output signal SINR is plotted in Fig. 8.

As shown in Fig. 8, the output SINR of the traditional MVDR degrades faster than the proposed method as the altitude angle increases where the largest deficit

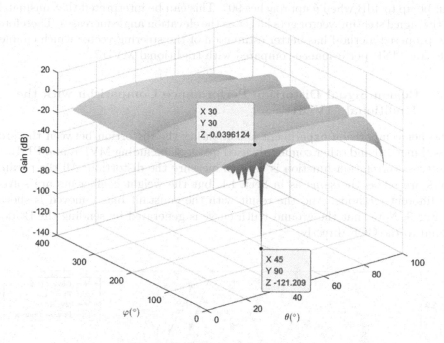

Fig. 7. Performance of array gain.

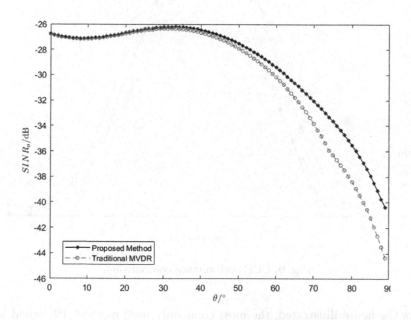

Fig. 8. SINR performance comparision.

can be up to 4dB when θ approaches 90°. This can be interpreted that mismatch of the signal steering vector gets larger as the elevation angle increases. Therefore, the proposed method has better estimation of the steering vector which enables a better SINR performance compared with traditional MVDR.

4.3   Beidou Signal Distortion Performance Comparison with the Traditional MVDR

Another comparision experiment is comparing the distortion between the proposed method and other commonly used methods including MVDR and PI. Here the cross-correlation function is used to measure the distortion. All the parameters are set as the same as in Sect. 4.1, but the weight coefficients are solved in different methods. And the result with the constant bias removed is shown in Fig. 9. Note that the ground truth curve is generated by sending the Beidou signal to the CCF directly.

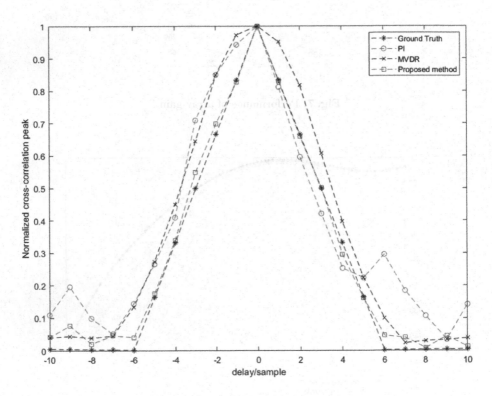

Fig. 9. CCF performance comparision.

As the figure illustrated, the most commonly used method, PI, would leads to a distortion in the cross-correlation peak significantly, undermine the symmetry in cross-correlation peak, and the MVDR method would broaden the cross-correlation peak. The above phenomena in turn result in error in code phase estimation. On the contrary, the cross-correlation peak of the proposed method is quite close to the ground truth, indicating the desired signal is barely distorted.

5 Conclusion

A convex conformal array signal processing model is proposed in this study, and based on that a distortionless MVDR beamformer is introduced. The effectiveness of the proposed method is validated by theoretical analysis and numerical simulations. The results show that the proposed conformal array signal processing model can significantly improve the estimation of the steering vector and enhance the interference suppression performance. What's more, the results also show that the additional distortionless constraints can ensure the linearity of the STAP frequency response, maintain the Beidou signal and furthermore ensure the correct decoding of the standard receiver.

Acknowledgment. This research work was supported by the National Natural Science Foundation of China under Grant Nos. 61971278 and 61771308, the Joint Foundation of the Eighth Research Institute of China Aerospace Science and Technology corporation and Shanghai Jiao Tong University under Grant No. USCAST2020-26.

References

1. Xie, G., et al.: Principles of GPS and Receiver Design, vol. 7, pp. 61–63. Publishing House of Electronics Industry, Beijing (2009)
2. Borre, K., Akos, D.M., Bertelsen, N., Rinder, P., Jensen, S.H.: A Software-Defined GPS and Galileo Receiver: A Single-Frequency Approach. Springer, Boston (2007). https://doi.org/10.1007/978-0-8176-4540-3
3. Gao, G.X., Sgammini, M., Lu, M., Kubo, N.: Protecting GNSS receivers from jamming and interference. Proc. IEEE **104**(6), 1327–1338 (2016)
4. BeiDou navigation satellite system signal in space interface control document. http://www.beidou.gov.cn/xt/gfxz/201802/P020180209623601401189.pdf
5. Ioannides, R.T., Pany, T., Gibbons, G.: Known vulnerabilities of global navigation satellite systems, status, and potential mitigation techniques. Proc. IEEE **104**(6), 1174–1194 (2016)
6. Fante, R.L., Vaccaro, J.J.: Wideband cancellation of interference in a GPS receive array. IEEE Trans. Aerosp. Electron. Syst. **36**(2), 549–564 (2000)
7. Compton, R.: The power-inversion adaptive array: concept and performance. IEEE Trans. Aerosp. Electron. Syst. **6**, 803–814 (1979)
8. Van Veen, B., Buckley, K.M.: Beamforming techniques for spatial filtering. In: Digital Signal Processing Handbook, p. 61-1 (1997)
9. Van Trees, H.L.: Optimum Array Processing: Part IV of Detection, Estimation, and Modulation Theory. Wiley, Hoboken (2004)
10. Marathe, T., Daneshmand, S., Lachapelle, G.: Assessment of measurement distortions in GNSS antenna array space-time processing. Int. J. Antennas Propag. **2016**, 1–18 (2016)

11. O'Brien, A.J., Gupta, I.J.: Mitigation of adaptive antenna induced bias errors in GNSS receivers. IEEE Trans. Aerosp. Electron. Syst. **47**(1), 524–538 (2011)
12. Shuangxun, L., Zhu, C., Kan, H., Hongyin, X.: A compensating approach for signal distortion introduced by STAP. In: 2006 International Conference on Communication Technology, pp. 1–4. IEEE (2006)
13. Dai, X., Nie, J., Chen, F., Ou, G.: Distortionless space-time adaptive processor based on MVDR beamformer for GNSS receiver. IET Radar Sonar Navig. **11**(10), 1488–1494 (2017)
14. Hansen, R.: Conformal antenna array design handbook. Technical report, Naval Air Systems Command, Washington DC (1981)
15. Rasekh, M., Seydnejad, S.R.: Design of an adaptive wideband beamforming algorithm for conformal arrays. IEEE Commun. Lett. **18**(11), 1955–1958 (2014)
16. Zou, L., Laseby, J., He, Z.: Beamformer for cylindrical conformal array of non-isotropic antennas. Adv. Electr. Comput. Eng. **11**(1), 39–42 (2011)
17. Daneshmand, S.: GNSS interference mitigation using antenna array processing (2013)
18. Burger, H.A.: Use of Euler-rotation angles for generating antenna patterns. IEEE Antennas Propag. Mag. **37**(2), 56–63 (1995)
19. Lin, H.C.: Spatial correlations in adaptive arrays. IEEE Trans. Antennas Propag. **30**(2), 212–223 (1982)
20. Oppenheim, A.V.: Discrete-Time Signal Processing. Pearson Education India (1999)

HEVC Rate Control Optimization Algorithm Based on Video Characteristics

Qiang Li and Jun Nie[✉] iD

School of Communication and Information Engineering, Chongqing University of Posts and
Telecommunications, Chongqing, China
liqiang@cqupt.edu.cn, 724413152@qq.com

Abstract. Based on the relationship between the frame-level and the large coding
unit (LCU)-level in the high-efficiency video coding (HEVC) rate control scheme,
this paper proposes novel optimization algorithms for bit allocation at the frame-
level and improves algorithms for code rate control in LCU level for the space-time
domain. For the frame-level, the statistical characteristics of the whole source are
considered, and the information entropy of the coded source is added to the R-
lambda model. An optimization model of bit allocation considering the current
coded frame is used to guide the bit allocation in the frame layer. This algorithm
improves the bit accuracy in the frame level. For the LCU-level, the Hadamard
transform algorithm detects the energy distribution region and motion region. It
constructs a new complexity from the energy value and the predicted residual
value of the image so that the target bits of the LCU layer can be reasonably
adjusted and accurately assigned. The coding structure in LCU-level achieves a
more accurate update of the model parameters. The experimental results show that
the algorithm in this paper reduces the relative error of bit rate by 0.017% and
0.016% on average and improves the rate-distortion performance by 2.7% and
2.6% on average compared with the adaptive ratio bit allocation algorithm under
low-delay B configuration and P configuration.

Keywords: HEVC · Rate control · Space-time domain · Information entropy

1 Introduction

With the rapid development of computers, the Internet, intelligent terminals in today's
era, people's video demands are increasing. It also causes the increasing amount of data.
The International Telecommunication Union ITU-T and the International Organization
for Standardization ISO/IEC proposed the high-efficiency video coding (HEVC) standard [1] in 2013 to improve the coding efficiency of video compression and meet the
needs of high-definition and ultra-high-definition video compression in practical applications. In practical applications such as surveillance and video conferencing, ensuring
the quality of encoded video with limited bandwidth resources is the problem to solve by
bit rate control technology. Therefore, bit rate control technology is an essential process

© ICST Institute for Computer Sciences, Social Informatics and Telecommunications Engineering 2022
Published by Springer Nature Switzerland AG 2022. All Rights Reserved
H. Gao et al. (Eds.): ChinaCom 2021, LNICST 433, pp. 235–249, 2022.
https://doi.org/10.1007/978-3-030-99200-2_19

in video compression coding. Code rate control is a decision problem in video compression. How to allocate the code rate efficiently and how to convert the target allocated code rate into the actual encoded bit rate are the two most important problems need to be solved by code rate control techniques. These two problems also affect the final coding efficiency and observation effect. In the evolving video compression technology, the bit rate control part is also ceaselessly updated. The bit rate control algorithm has been a hot research topic in video compression technology. Therefore, the study of bit rate control algorithm in video compression technology has significant practical value.

The rate control algorithm allocates the target number of bits to the group of pictures (GOP) level, image level, a large coding unit (LCU)-level through a particular strategy and then calculates the Lagrange multiplier (λ) by the target number of bits, Update the model parameters. In the process of formulating the high-efficiency video coding standard HEVC/H.265, a series of continuously improving bit rate allocation schemes have also been successfully formed. The main influential ones are JCTVC-H0213 [1] and JCTVC-K0103 [2]. The JCTVC-H0213 proposal is the first-rate control algorithm proposed by HEVC. This algorithm has certain defects, mainly due to the allocation model's unreasonable setting, which leads to a large gap between the code rate generated after the final encoding and the expected rate, and the video. The quality after compression is not very good either. Because of the JCTVC-H0213 rate control algorithm's shortcomings, the rate implementation model in the JCTVC-K0103 control algorithm is modified, and a rate implementation model based on the R-λ-QP model is proposed, which significantly improves the insufficiency of previous rate control algorithm. K0103 is also not the optimal algorithm. Because of its shortcomings, the follow-up meeting offered JCTVC-M0257 [3], JCTVC-M0036 [4], and other proposals. In recent years, many rate control algorithms based on the R-λ model have emerged. The work [5] found that there is a more robust correspondence between R and the λ, and proposed a novel λ-domain rate control algorithm, which has been implemented in the latest video coding standard. The authors [6] proposed image-level and basic unit-level bit allocation algorithms based on basic RD optimization theory, which makes full use of video content to guide bit allocation. The work [7] took into account the characteristics of inter-frame correlation. A precise H.265/HEVC frame-level bit allocation algorithm is developed, which improves coding efficiency. In [8], considering the recursive rate-distortion model with dependence between frames caused by motion compensation, the algorithm treats frame-level rate allocation as a convex optimization problem.[9] used the recursive Taylor expansion equation to solve the constraint equation, and the optimized R-D model performs target bit allocation on the CTU layer. [10] proposed an improved R-λ to establish a rate control model based on joint time-space information and HVS characteristics. In this model, mutual time-space information based on gradient information is used to guide bit allocation at the frame and CTU level, where the time coefficient is adaptively corrected. The work [11] proposed a gradient-based R-lambda model for intra-frame rate control, which can effectively measure the frame content complexity and enhance traditional R-lambda methods' performance. And the work developed LCU-level bit allocation method.

Although there are many work designs on rate control algorithms, most studies have not considered the impact of the frame levels target bit on the bit allocation of the LCU-level and the coding characteristics of HEVC itself. This paper proposes an optimized Frame-level and LCU-level rate control algorithm. The code rate control algorithm further improves the coding performance. Unlike the traditional way of frame-level bit weight allocation, it also considers the amount of information in the aggregation feature of the gray distribution in the current frame. The purpose is to make the bit allocation of the current frame fully consider the current source's overall information measurement. Besides, based on the space-time theory of the LCU-level, a new linear bit allocation weight calculation method for the LCU-level is proposed. Experimental results show that the proposed rate control algorithm further improves the coding performance. In summary, the main contributions of this article are as follows:

In this article, a rate control algorithm incorporating the current coded frame's information entropy into the R-D model is proposed. Therefore, compared with the traditional frame-level bit allocation algorithm, the proposed rate control algorithm can make the actual rate closer to the target rate and improve the encoding quality to a certain extent.

The new LCU-level bitrate control algorithm takes account of the Hadamard transform algorithm, and takes more fully into account each frame's texture characteristics, time-domain prediction information, and the video content characteristics. Besides, the algorithm designs two parts to control them.

The rest of this article is organized as follows. The second section introduces the relevant technical principles, and the third section discusses the rate control scheme at the frame and LCU-level. Next, the experimental results and conclusions are presented in Sect. 4 and Sect. 5, respectively.

2 Related Technical Principles

Like other rate control schemes, the R-λ rate control model is mainly divided into two parts, the one is bit allocation, the other is the calculation of the λ and the quantization parameter QP. Literature [5] established the exponential relationship between the code rate and the λ, and its model is:

$$\lambda = -\frac{\partial D}{\partial R} = CK \cdot R^{-K-1} \triangleq \alpha R^\beta \tag{1}$$

Among them, D is the encoding distortion. C and K are model parameters related to the video sequence. α and β are model parameters related to the characteristics of the video content. R represents the coding bit, the unit is bpp (bit per pixel). If the target bit of a particular frame or a certain LCU is T and the number of pixels is N, the calculation equation is:

$$\text{bpp} = \frac{T}{N} \tag{2}$$

Among them α and β are model parameters, which will be updated as each LCU or frame is encoded. Then QP can be determined by the empirical equation [5].

$$QP = 4.2005 \ln \lambda + 13.7122 \tag{3}$$

Bit allocation will be implemented in the GOP level, frame-level, and basic coding unit level. First, calculate the target number of bits for each picture.

$$R_{PicAvg} = \frac{R_{tar}}{f} \tag{4}$$

Suppose the number of encoded pictures is $N_{coded}N_{coded}N_{coded} \ N_{coded}$, the number of bits used by these pictures is R_{coded}, the number of frames in the current GOP is N_{GOP}, SW is the size of the sliding window for smooth bit allocation, which is used to make bit consumption changes, and the quality of encoded pictures more smooth. The bit allocation of the GOP level is:

$$T_{AvgPic} = \frac{R_{PicAvg} \times (N_{coded} + SW) - R_{coded}}{SW} \tag{5}$$

$$T_{GOP} = T_{AvgPic} \cdot N_{GOP} \tag{6}$$

We hope to reach the target bit rate after the SW frame. If the SW frame can consume bits in each frame T_{AvgPic}, the above equation can be rewritten as:

$$T_{AvgPic} = R_{PicAvg} + \frac{R_{PicAvg} \cdot N_{coded} - R_{coded}}{SW} \tag{7}$$

The first part of equation represents the target bit rate, and the second part describes the buffer state.

Then there is the bit allocation at the picture level. Suppose the number of bits used in the current GOP is $Coded_{GOP}$, ω is the bit allocation weight of each picture, so the target bit rate of the recent frame is:

$$T_{CurrPic} = \frac{T_{GOP} - Coded_{GOP}}{\sum\limits_{NotCodedPictures} \omega_i} \cdot \omega_{CurrPic} \tag{8}$$

The bit allocation of the LCU-level is considered in the proposal that a fundamental unit contains each LCU, and the following equation determines the target number of bits:

$$T_{CurrLCU} = \frac{T_{CurrPic} - Bit_{header} - Coded_{Pic}}{\sum\limits_{NotCodedLCUs} \omega_i} \cdot \omega_{CurrLCU} \tag{9}$$

Among them Bit_{header} is the estimated value of all header information bits, which is estimated by the actual number of header information bits of the coded picture in the same layer.

3 Proposed Rate Control Model

3.1 Frame-Level Bit Allocation Model

In this section, this paper first proposes a frame-level bit allocation model based on information entropy. The amount of information is used to measure an event's uncertainty, and image entropy refers to the average amount of information in the image. The greater the probability of an event, and the smaller the uncertainty, the smaller the amount of information it carries; conversely, the greater the entropy, the richer the information. To describe that the amount of information in different messages is different, we use the mathematical expectation of self-information to the amount of information, called information entropy, $\chi_i (i = 1, 2, \cdots, n)$ means that an information source sends out n symbolic messages, $p(\chi_i)$ means the probability of occurrence of different symbolic messages. The calculation equation is shown in (10):

$$H(X) = E[I(\chi_i)] = E\left(\log_2 \frac{1}{p(\chi_i)}\right) = -\sum_{i=1}^{n} p(\chi_i) \times \log_2 p(\chi_i) \qquad (10)$$

Among them $I(\chi_i)$ is the self-information of the symbolic message χ_i, $p(\chi_i)$ represents the probability of the symbolic message χ_i appearing, and n represents the total number of symbolic messages.

Because the image information entropy can characterize the overall characteristics of the source and describe the local details of the image, it can well reflect the amount of information contained in the aggregation features of the grayscale distribution in the image. Image information entropy can well reflect the complexity of the image. The calculation equation of image information entropy is as follows:

$$EI = -\sum_{\chi=0}^{N-1} p(\chi) \times \log_2[p(\chi)] \qquad (11)$$

EI is the image's information entropy, $p(\chi)$ is the proportion of pixels with gray value χ in the image, and N is the total number of image gray levels. Each pixel gray level is quantized with 8 bits, and the total number of gray image levels is 256 gray levels.

This algorithm considers the frame-level fixed weight allocation model's information entropy in the original R-lambda model rate control algorithm. Unlike the traditional linear allocation model, the higher the frame level's information entropy, the more allocated bits. The current encoded frame has different bit demands on the frame level, and the information entropy can represent the value of the information. Equation (12) accumulates all the weights and information entropy within a GOP and finds the multiplier between them, ensuring that the weights and information entropy are guaranteed to be of the same order of magnitude.

$$Af = \sum_{AllPictures} \omega_i / \sum_{AllPictures} EI_i \qquad (12)$$

The bit allocation algorithm takes into account the information entropy of the currently encoded frame. The calculation equations for the bit allocation weight ω'_{pic} and total bit weight ω_{total} of the new frame-level can be obtained as follows:

$$\omega'_{pic} = Af \times EI_i + \omega_{original_i} \qquad (13)$$

$$\omega_{total} = \sum_{NotCodedPictures} \omega_i + \sum_{NotCodedPictures} (EI_i \times Af) \qquad (14)$$

Among them EI_i is the information entropy of the current frame image, $\omega_{original_i}$ representing the weight of the current frame image in the rate control algorithm based on the R-λ model, and $\sum_{NotCodedPictures} \omega_i$ assigns the sum of all uncoded images in the current image group.

Because the frame-level bit allocation weight does not fully consider each frame's content's characteristics and allocates the target bits of the frame-level more reasonably, this section proposes a novel optimized frame-level target bit allocation algorithm. The layer bit allocation weight ω'_{pic} and the new total bit weight ω_{total} perform high-precision bit allocation to the frame. Therefore, equation for the allocation of target bits at the frame-level in this algorithm is shown in Eq. (15):

$$T_{CurrPic} = \frac{T_{GOP} - Coded_{GOP}}{\omega_{total}} \cdot \omega'_{pic} \tag{15}$$

3.2 LCU-Level Bit Allocation Model

Hadamard transform is often used in image and video processing to calculate the residual signal's SATD (sum of absolute transformed difference). SATD calculates the sum of each element's absolute value after Hadamard transforms the residual signal. Suppose the square matrix of residual signal is X, then SATD is:

$$SATD = \sum_M \sum_M |HXH| \tag{16}$$

Where M is the size of the square matrix, and H is the $M \times M$ normalized Hadamard matrix.

Since the SATD value reflects some extent, the energy magnitude of the residual signal in the frequency domain and the size of the coded output stream, the energy of the LCU can be statistically derived by calculating the SATD value of each sub-block of the LCU. The energy magnitude of each LCU in a frame is allocated accordingly to improve the quality of video coding.

First, split each LCU into 8×8 sub-blocks and using the Hadamard matrix $H_{8 \times 8}$. Use Eq. (16) to calculate each sub-block's SATD value and then using Eq. (17) to calculate the SATD value D'_1 of each LCU.

$$D'_1 = \sum_{i=1}^{m} SATD_i \tag{17}$$

Among them, m is the number of LCU sub-blocks; $SATD_i$ is the SATD value of the i-th LCU sub-block.

Then, using Eq. (18) to calculate the average SATD value D'_2 of a frame.

$$D'_2 = \frac{1}{n} \sum_{j=1}^{n} SATD_j \tag{18}$$

Among them, n is the number of LCUs in a frame; $SATD_j$ is the SATD value of the j-th LCU in the current frame.

Finally, according to D_1' of each LCU and D_2' of the current frame, Eq. (19) calculates the energy proportion factor θ_1 that can reflect each LCU in the frame. The larger the value of θ_1, the larger the energy ratio of the LCU, and the more bits should be allocated.

$$\theta_1 = \frac{D_1'}{D_2'} \tag{19}$$

Inter-frame prediction refers to the method which uses the video time domain's correlation to predict the pixels of the current image based on adjacent coded image pixels in order to remove redundant information in the video time domain. Since video sequences usually include solid temporal correlation, the prediction residuals are generally flat. That is, many residual values are close to "0". The residual signal is used as the input of the subsequent module for transformation, quantization, and entropy coding, realizing the video signal's high-efficiency compression. The residual value of the current LCU can be obtained by estimating the actual reconstruction value of the LCU corresponding to the previous coded position, which can well reflect the relevant information between the video frames. The SATD algorithm is used to count the predicted residual value R_1' of each LCU in a frame, as shown in Eq. (20):

$$R_1' = \sum_{i=1}^{m} SATD_i' \tag{20}$$

Where m is the number of sub-pictures in each LCU. $SATD_i'$ is the residual prediction value of one sub-picture in each LCU. The average prediction residual R_2' in the current frame is counted using LCU as the basic unit, as shown in Eq. (21):

$$R_2' = \frac{1}{k} \sum_{j=1}^{k} SATD_j' \tag{21}$$

Where k is the number of LCUs in a frame and $SATD_j'$ is the predicted residual value of the j-th LCU in the current frame.

According to the residual prediction value R_1' of the current LCU and the average prediction residual value R_2' of the current frame, Eq. (22) calculates the residual prediction factor θ_2 of the LCU block.

$$\theta_2 = \frac{R_1'}{R_2'} \tag{22}$$

3.3 New Complexity Based on Time and Space

Because of the shortcomings of the rate control algorithm based on the R-λ model in HEVC, this section weights the energy proportion factor and the residual prediction factor of the LCU-level to obtain a new type of complexity NC. It can effectively distinguish the motion area, and perform a good fitting based on the actual complexity of the

LCU, that is a linear relationship. After adopting this weighted combination method, this layer's target bits can be allocated more reasonably. More target bits can be distributed to areas in HEVC with inaccurate predictions, high complexity, or intense motion, and vice versa. To better detect the texture area and the residual prediction area, combining Eq. (15) and Eq. (19), the calculation equation NC is shown in Eq. (23):

$$NC = a \times \eta_1 + (1 - a) \times \eta_2 \tag{23}$$

a is a weighting coefficient, and the coefficient value is greater than 0 and less than 1. In this paper, by performing feature quantization on all LCU energy values and predicted residual values in a frame, the coefficients normalized by the linear function are used as the weights of η_1 and η_2 convenient for the indicators of energy ratio and residual factor. Perform comparison and weighting, and improve the accuracy of the algorithm to a certain extent. Its feature quantification equation is as follows:

$$a_{\eta_1} = \frac{D'_1 - D'_{MIN}}{D'_{MAX} - D'_{MIN}} \tag{24}$$

$$a_{\eta_2} = \frac{R'_1 - R'_{MIN}}{R'_{MAX} - R'_{MIN}} \tag{25}$$

Among them, $D'_{MIN}, D'_{MAX}, R'_{MIN}$, and R'_{MAX} are the minimum and maximum values of all LCU energy values and the minimum and maximum values of residual prediction values in a frame, respectively. The feature quantization weighting coefficient a is calculated by Eqs. (24) and (25), as shown in Eq. (26):

$$a = \frac{a_{\eta_1}}{a_{\eta_1} + a_{\eta_2}} \tag{26}$$

3.4 LCU-Level Bit Allocation Equation

For the B frame or P frame, the target bit allocation algorithm that takes the time and space domain of the LCU-level into consideration is adopted, that is, after the new weight NC is obtained by (23), and use it as the allocation weight to allocate target bits to the LCU-level.

According to the weight of the current LCU, through Eq. (27), the initial target bit allocation for the LCU is performed using the target remaining bit number of the current frame.

$$T_{LCU} = (1 - b) \times \frac{T_{CurrPic} - Bit_{header} - Coded_{Pic}}{\sum\limits_{NotCodedLCUs} \omega_i} \cdot \omega_{CurrLCU} + b \times \frac{T_{CurrPic} - Bit_{header}}{\sum\limits_{AllLCUs} NC_i} \cdot NC_{CurrLCU} \tag{27}$$

Among them, $T_{CurrPic}$ represents the current frame's target bit; Bit_{header} is the estimated number of bits of the header information; $\omega_{CurrLCU}$ represents the adaptive bit allocation weight of the original platform LCU; $NC_{CurrLCU}$ represents the bit allocation weight of each LCU of the proposed algorithm. b is a weighting coefficient, and its value is greater than 0 and less than 1. In this paper, several test sequences were tested for coding performance to determine the value of b. The effect of different values of b on RD performance and code rate control accuracy was calculated. Finally, the value of b was set to 0.4.

According to the buffer status, use Eq. (28) to dynamically adjust the target bit T_{LCU} initially allocated by the current LCU to obtain the final target bit number $T_{CurrLCU}$.

$$T_{CurrLCU} = T_{LCU} - (\text{totalWeight} - B_{left})/\text{realInfluenceLUC} + 0.5 \qquad (28)$$

Equation B_{left} represents the actual remaining bits in the current frame, *totalWeight* represents the sum of bits required by the remaining LCU including the current LCU, *realInfluenceLCU* represents the actual smoothing window size.

3.5 Model Parameter Update

The video sequence's RD characteristic is significant for bit allocation and rate control in the video encoding process. With the changes in the encoded frame's complexity, the motion area, and the reference frame structure, the R-λ of each encoded frame, The λ model parameters (α and β) are not the same. Its rate distortion relationship is not available until the next frame is encoded. The frame's R-λ model parameters to be encoded in [5, 6] are estimated using the closest encoded frame information that belongs to the same level and is updated. The method only uses the actual coded bits and Lagrange multipliers of the coded frame. The convergence speed and accuracy of its parameter update are not ideal, which affects the rate control's RD performance. Using the frame-level and LCU-level bit allocation algorithm proposed in this article can make the target bit allocation more accurate and reasonable. As the encoding quality of the previous encoded CTU frame becomes higher, more valid reference information can be obtained for the encoded frame, including the actual coded bit R_{real}, the Lagrange multiplier λ_{real}, and the coding distortion D_{real}. The updated R-D model parameters can be directly derived from the R-λ function and R-D function of Eq. (1) and Eq. (29):

$$D(R) = CR^{-K} \qquad (29)$$

$$C_{new} = \frac{D_{real}}{R_{real}^{-\lambda_{real} \times R_{real}/D_{real}}} \qquad (30)$$

$$K_{new} = \frac{\lambda_{real} \times R_{real}}{D_{real}} \qquad (31)$$

Then, the R-λ model parameters are expressed as:

$$\alpha_{new} = C_{new} \cdot K_{new} \qquad (32)$$

$$\beta_{new} = -K_{new} - 1 \qquad (33)$$

The α_{new} and β_{new} calculated by Eqs. (32) and (33) will be more accurate because the next GOP after encoding can obtain the R-lambda model parameters so that the parameter update can make full use of the information of the relevant coding unit, which will get better RD performance and higher bit rate accuracy.

4   Experimental Result

The HM-16.7 default algorithm without rate control is used as the benchmark scheme to prove the proposed algorithm's effectiveness. The algorithms in [5] and [6] have been applied to the software HM as a comparison solution. All algorithms are implemented on the HEVC reference platform HM-16.7, and the target bit rate of each sequence is under the condition of a fixed QP (QP = 22, 27, 32, 37). After following the general test under the platform of the HM-16.7 reference software, get. Sixteen video sequences with different content and characteristics were tested using the default low-latency B and P configurations. The encoding results include actual bit rate, PSNR, BD-Rate, and RCE calculations thoroughly compare the encoding performance. Table 1 shows the test sequences of Type B, C, D, and E attributes, including resolution, frame rate, and the number of coded frames.

Table 1. Basic information of the test sequence

Sequences	Number of sequences	Resolution	Frame count	Frame rate
Class B	5	1920 × 1080	240&500&600	24&50&60
Class C	4	832 × 480	300&500&600	30&50&60
Class D	4	416 × 240	300&500&600	30&50&60
Class E	3	1280 × 720	600	60

4.1   R-D Performance

RD performance is an important indicator to measure the video coding system. BDBR represents the bit rate saving percentage of the comparison scheme relative to the original method under the same video coding quality. A positive value indicates that the compression performance becomes worse, and a negative value indicates that the compression performance becomes better. HM16.7 without code rate control is the baseline solution, and HM16.7 with code rate control algorithm is the test solution. This article tested RC [5], RC [6], and the proposed algorithm under LDP and LDB configuration. BD-Rate, the test results are shown in Table 2.

Under normal circumstances, compared with the encoder without rate control, RD performance of the encoder with rate control will be reduced due to the encoding bit restriction on GOP, frame, and LCU. The average BDBR of RC [5] and RC [6] under LDP and LDB configurations are 1.7% and 2.7%, 0.1%, and 0.7%, respectively. Although the coding performance of some test sequences has been improved, the overall performance of these two algorithms the rate-distortion performance is not as good as the benchmark scheme; the average BDBR of the algorithm in the LDP and LDB configuration is − 2.5% and −2.0%, respectively, and the RD performance of all test sequences is better than that of [5] and [6]. Table 2 allows to derive the average BDBR values for each type of test sequence. This paper's algorithm can improve the C-type sequence's performance the most. The texture of this type of sequence is relatively complex, and the movement is fierce. Because the algorithm in this paper can detect the texture complexity and motion area of the image area well, the energy proportion factor and residual prediction factor proposed can better reflect the sequence's texture characteristics and motion characteristics. More bits are allocated to areas with more intense motion, which improves the coding quality of the video, and provides a better reference frame for the next GOP, thereby improving the RD performance of the entire coding sequence.

Table 2. Performance comparison of BD-Rate (%) in LDP and LDB configuration

Class	Sequence	RC [7]		RC [8]		Proposed	
		LDP	LDB	LDP	LDB	LDP	LDB
B	Cactus	−2.7	−1.1	−2.2	−1.2	−5.5	−5.0
	BasketballDrive	6.8	8.4	3.2	3.5	1.1	1.3
	BQTerrace	3.9	7.1	4.5	7.1	−1.0	1.5
	Kimono	11.3	11.5	6.6	6.5	4.5	4.4
	ParkScene	2.4	2.7	1.7	2.1	−2.5	-2.2
	Average	4.3	5.7	2.8	3.6	−0.7	0.0
C	BasketballDrill	−5.9	−5.9	−5.9	−5.5	−8.4	−8.4
	BQMall	−1.6	−1.4	−1.8	−1.7	−3.9	−4.1
	PartyScene	−1.3	−1.2	−2.3	−2.4	−5.0	−5.5
	RaceHorsesC	1.5	2.7	−1.4	−0.2	−2.8	−1.8
	Average	−1.8	−1.5	−2.8	−2.5	−5.0	−5.0
D	BasketballPass	−2.7	−2.5	−4.8	−4.9	−6.9	−7.1
	BlowingBubbles	0.9	0.6	−1.7	−2.1	−4.3	−4.8
	BQSquare	5.4	7.9	1.1	2.3	0.6	1.4
	RaceHorses	1.2	1.9	0.6	0.8	−0.4	−0.2
	Average	1.2	2.0	−1.2	−1.0	−2.8	−2.7
E	FourPeople	2.3	2.8	1.2	1.7	−3.2	−2.5
	Johnny	5.9	8.1	3.0	4.9	1.5	3.7
	KristenAndSara	0.4	0.9	−0.8	0.0	−4.0	−3.4
	Average	2.9	4.0	1.1	2.2	−1.9	−0.7
Total average		1.7	2.7	0.1	0.7	−2.5	−2.0

Figure 1 and Fig. 2 are the comparison diagrams of Bits cost and Y-PSNR of each frame obtained by encoding the class B sequence BQTerrace with the target code rate of 841.254 kbps the algorithm of this paper and RC [6] under the LDP configuration. The actual output bit rate and Y-PSNR are 841.278 kbps and 31.3942 dB. The actual output bit rate and Y-PSNR of RC [6] are 841.343 kbps and 31.1997 dB. Figure 3 and Fig. 4 are the RD curves of the Basketballpass sequence and the Cactus sequence. It can be seen from Figs. 1 and 2 that for the non-homogeneous BQTerrace sequence, the algorithm in this paper has better rate control performance. Most of the coded frames can get a higher Y with a relatively small number of bits. -PSNR, which improves the average PSNR of the sequence and increases the smoothness. Among them, there is a very obvious shot switch from frame 150 to frame 250. This paper's algorithm can detect the image texture complexity and motion areas better than RC [6] to obtain a very high coding quality. Figures 3 and 4 reflect the non-homogeneous and violent motion sequence with complex background and object rotation. This paper's algorithm is used to adjust and accurately allocate the number of bits in the frame and the LCU-level. It has better RD performance under the condition of high bit rate and high bit rate.

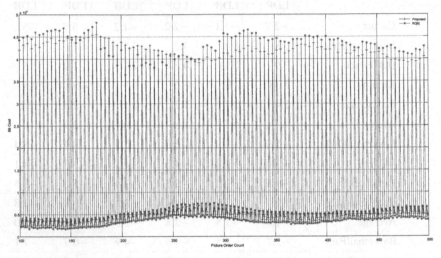

Fig. 1. Bits cost comparison chart of BQTerrace sequence

4.2 Rate Control Performance

The purpose of code rate control is to make the encoder's actual output code rate equal to the target code rate as much as possible while minimizing the coding distortion of the sequence. Therefore, the bit rate accuracy is another crucial performance index measured by the bit rate's relative error (R_{Err}). The inaccuracy of the code rate is defined as:

$$R_{Err} = \frac{|R_{actual} - R_{target}|}{R_{target}} \times 100\% \tag{34}$$

R_{actual} and R_{target} represent the actual code rate and target code rate. The trimmer R_{Err} is, the higher the rate control accuracy is. In the experiment, each test sequence is set with four target bit rates from low to high. Due to space limitations, the average value of the four QPs is used as the result. As shown in Table 3, it can be observed that this

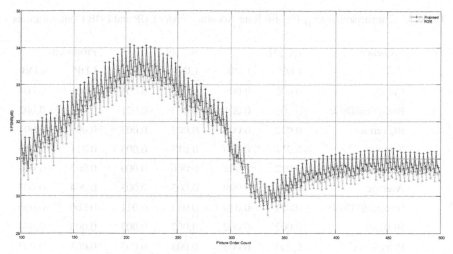

Fig. 2. Y-PSNR comparison chart of BQTerrace sequence

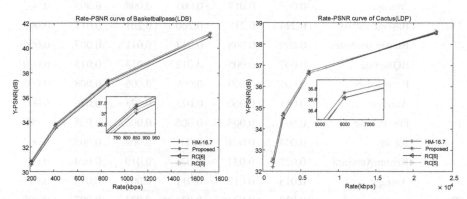

Fig. 3. RD curve of Basketballpass sequence   **Fig. 4.** RD curve of Cactus sequence

paper can achieve more accurate rate control accuracy than RC [5] and RC [6]. Under the configuration of LDP and LDB, the algorithm's average in this paper is 0.007% and 0.006%, respectively. Among them, the relative error of the average code rate of Class B, Class C, and Class D is better than that of the comparison method, and the accuracy reaches 0.004%, 0.003%, and 0.010%. The rate control accuracy of Class D is 0.063% higher than that of RC [6]. It is worth noting that the algorithm in this paper achieves a considerable improvement in RD performance and improves the rate control accuracy.

Table 3. Comparison of R_{Err} (%) Bit Rate Accuracy under LDP and LDB Configurations

Class	Sequence	RC [5]		RC [6]		Proposed	
		LDP	LDB	LDP	LDB	LDP	LDB
B	Cactus	0.002	0.005	0.002	0.006	0.001	0.001
	BasketballDrive	0.003	0.001	0.003	0.001	0.002	0.002
	BQTerrace	0.002	0.002	0.003	0.003	0.003	0.001
	Kimono	0.011	0.009	0.013	0.003	0.011	0.011
	ParkScene	0.022	0.028	0.006	0.009	0.003	0.001
	Average	0.008	0.009	0.005	0.005	0.004	0.003
C	BasketballDrill	0.021	0.029	0.012	0.011	0.006	0.005
	BQMall	0.002	0.002	0.002	0.002	0.003	0.004
	PartyScene	0.004	0.003	0.004	0.003	0.001	0.003
	RaceHorsesC	0.003	0.035	0.002	0.015	0.002	0.001
	Average	0.007	0.017	0.005	0.008	0.003	0.003
D	BasketballPass	0.224	0.218	0.260	0.261	0.011	0.005
	BlowingBubbles	0.009	0.008	0.013	0.011	0.007	0.006
	BQSquare	0.007	0.005	0.012	0.015	0.015	0.019
	RaceHorses	0.005	0.009	0.008	0.006	0.008	0.006
	Average	0.061	0.060	0.073	0.073	0.010	0.009
E	FourPeople	0.007	0.003	0.006	0.006	0.005	0.007
	Johnny	0.004	0.001	0.001	0.002	0.005	0.002
	KristenAndSara	0.027	0.031	0.025	0.019	0.024	0.023
	Average	0.013	0.012	0.011	0.009	0.011	0.011
Total average		0.022	0.024	0.023	0.023	0.007	0.006

5 Conclusion

This article aims to improve the visual quality and coding performance, so a novel frame-level precision bit allocation optimization algorithm and LCU-level rate control improvement algorithm for image complexity are proposed. The frame level bit allocation is performed based on the average amount of information in the image by the new type of weight bit allocation algorithm, and then the Hadamard transform algorithm is performed to detect the texture complexity area and the residual prediction area, besides, take the image's complexity and the unique coding features of HEVC into account to allocate the bits of the LCU-level reasonably. This paper's proposed algorithm has excellent benefits compared with other advanced bit allocation algorithms and can achieve more stable rate control accuracy, more minor bit fluctuations, and better RD performance.

References

1. Choi, H., Nam, J., Yoo, J., et al.: Rate control based on unified RQ model for HEVC. ITU-T SG16 Contribution, JCTVC-H0213, 1–13 (2012)
2. Li, B., Li, H., Li, L., et al.: Rate control by R-lambda model for HEVC. ITU-T SG16 Contribution, JCTVC-K0103, 1–5 (2012)
3. Karczewicz, M., Wang, X.: Intra frame rate control based on SATD. Document JCTVC-M0257, Incheon, Korea, April 2013
4. Li, B., Li, H., Li, L.: Adaptive bit allocation for R-lambda model rate control in HM. Document JCTVC-M0036, Incheon, Korea, April 2013
5. Li, B., Li, H., Li, L., Zhang, J.: λ domain rate control algorithm for high efficiency video coding. IEEE Trans. Image Process. **23**(9), 3841–3854 (2014)
6. Li, L., Li, B., Li, H., Chen, C.W.: λ-domain optimal bit allocation algorithm for high efficiency video coding. IEEE Trans. Circuits Syst. Video Technol. **28**(1), 130–142 (2018)
7. Liu, M., Ren, P., Xiang, Z.: Frame-level bit allocation for hierarchical coding of H.265/HEVC considering dependent rate-distortion characteristics. Sig. Image Video Process. **10**(8), 1457–1463 (2016)
8. Fiengo, A., Chierchia, G., Cagnazzo, M., Pesquet-Popescu, B.: Rate allocation in predictive video coding using a convex optimization framework. IEEE Trans. Image Process. **26**(1), 479–489 (2017)
9. Lu, X., et al.: Improved Bit Allocation Using Distortion for the CTU-Level Rate Control in HEVC (2018)
10. Zhao, Z., Xiong, S., Sun, W., He, X., Zhang, F.: An improved R-λ rate control model based on joint spatial-temporal domain information and HVS characteristics. Multimed. Tools Appl. **80**(1), 345–366 (2020). https://doi.org/10.1007/s11042-020-09721-9
11. Wang, M., Ngan, K.N., Li, H.: An efficient frame-content based intra frame rate control for high efficiency video coding. IEEE Sig. Process. Lett. **22**(7), 896–900 (2015)

Performance Analysis of Radar Communication Shared Signal Based on OFDM

Zeyu Liu[ID], Ying Zhang[✉], and Xinmin Luo

The Department of Telecommunications, Xi'an Jiaotong University, Xi'an, China
983279987@qq.com,{yzhang627,luoxm}@mail.xjtu.edu.cn

Abstract. Orthogonal Frequency Division Multiplexing (OFDM) signal is proved to be an appropriate shared signal in Integrated Sensing and Communication (ISAC). In this paper, an OFDM waveform with radar and communication functions is proposed, which is obtained by the traditional OFDM communication signal pulse processing. In order to measure the radar resolution performance of the shared signal, the effects of the number of subcarriers N_c and OFDM symbols N_s on the performance of radar ambiguity function (AF) are analyzed. Meanwhile, the influence of cyclic prefix (CP) on range ambiguity function is also considered. The results show that the inherent range resolution and inherent velocity resolution of shared signal are inversely proportional to the signal bandwidth and pulse width, respectively; The sidelobe characteristics of the velocity ambiguity function and range ambiguity function can be improved by adding the number of N_c and N_s, respectively. In addition, using zero padding (ZP) can eliminate the influence of CP on the range ambiguity function.

Keywords: OFDM signal · Integrated Sensing and Communication · Ambiguity function.

1 Introduction

With the rapid evolution of wireless communication technology, the explosive growth of wireless communication equipment leads to the increasing tension of wireless spectrum resources. In the field of military application, faced with the increasingly complex electromagnetic environment and the threat of new weapons and equipment, the demand for a new type of combat platform with high integration, miniaturization and parallel coordination becomes more urgent. The integrated sensing and communication (ISAC) is an effective way to solve the above problems. With the development of radar technology, the boundary between radar detection and communication is more and more blurred. From the perspective of the working frequency band, signal waveform, system structure

H. Gao et al. (Eds.): ChinaCom 2021, LNICST 433, pp. 250–263, 2022.
https://doi.org/10.1007/978-3-030-99200-2_20

and information processing of radar and communication, it is obvious that the two can share each other which have been separated before [1,2].

ISAC has become one of the core technologies of future 6G research, it will be applied in some cases, such as high-accuracy localization and tracking, augmented human sense, etc. [3], and signal design is one of the core issues of ISAC. High spectrum utilization and good anti-multipath performance make OFDM signal an appropriate choice for integration waveform design of both target detection and communication transmission [4]. From the perspective of waveform design, N. Levanon introduced the OFDM technology into the radar field for the first time and proposed the multi-carrier phase coded signal [5]. C. Sturm proposed the continuous wave transmitting mode with split transceiver and receiver to realize the integration of OFDM radar communication, and discussed the design of system parameters in the Industrial Scientific Medical (ISM) Band vehicle scenario [6]. The fully shared signal of target detection and information transmission was realized by using the random stepping frequency between OFDM signal pulses to transmit data in [7]. The radar communication integrated signal of single-symbol OFDM was designed, but its communication rate was too low [8]. From the perspective of ambiguity function of OFDM integrated waveform, S. Sen proposed an adaptive technique to design the spectrum of OFDM waveform which can improve the radars wideband ambiguity function (WAF) [9]. Liu Yongjun studied the ambiguity function of OFDM radar communication shared signal and reduced the sensitivity of the ambiguity function to communication modulation information through communication premodulation [10]. In reference [11], the performance of ambiguity function of multi-symbol OFDM radar communication shared signal was analyzed, but it didn't adopt oversampling and CP was not considered either.

In this paper, OFDM radar communication shared signal is realized by pulsing the OFDM signal of traditional communication, and the ambiguity function of shared signal without CP is deduced theoretically, which is simulated after adding CP by a fast algorithm. The effects of the number of subcarriers, the number of OFDM symbols and the CP on ambiguity function are analyzed in the end.

2 Signal Model

The traditional OFDM radar signal is shown in Fig. 1. The transmitted signal only sends one OFDM symbol in each pulse repetition interval (PRI), and usually doesn't consider the transmitted communication information, so it does not contain the CP. If using this signal as a radar communication shared signal, it has low communication rate and synchronization is also difficult. This paper takes the OFDM radar communication shared signal as shown in Fig. 2, which sends multiple OFDM symbols in a radar PRI. Compared with the traditional OFDM radar signal, the shared signal increases the communication rate under the same bandwidth, and it is easier to synchronize [10].

In addition, in order to overcome multipath interference in communication, the CP is usually added to OFDM symbols to suppress inter-symbol Interference

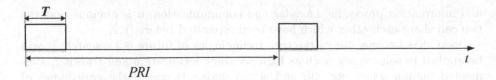

Fig. 1. The traditional OFDM radar transmitted signal

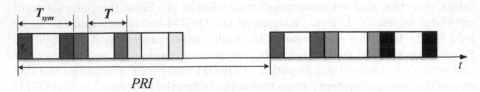

Fig. 2. OFDM radar communication shared signal

(ISI) and inter-carrier Interference (ICI), and the same form is used in radar communication shared signals. The structure of single OFDM symbol is also shown in Fig. 2, which consists of the basic OFDM symbol duration T and the cyclic prefix T_G, the full OFDM symbol time satisfies $T_{\text{sym}} = T + T_G$.

Assuming that the radar continuously sends N_s OFDM symbols in each PRI, and each OFDM symbol has N_c subcarriers, the baseband form of the OFDM radar communication shared signal in the first PRI (without adding the cyclic prefix) can be expressed as (For the sake of simulation, the subscript of $a_{n,m}$ starts at 1)

$$s(t) = \sum_{m=1}^{N_s} \sum_{n=1}^{N_c} a_{n,m} \exp(j2\pi f_n t) \text{rect}\left[\frac{t - (m-1)T}{T}\right] \quad (1)$$

where $a_{n,m}$ represents the communication data modulated on the nth subcarrier of the mth OFDM symbol. In propagation scenarios, with multipath propagation, Doppler and fading, only low-order modulation schemes, like BPSK and QPSK, can be successfully implemented [6], so choosing the QPSK modulation in this paper and $a_{n,m} = e^{j\phi_{n,m}}$, the subcarrier frequency satisfies $f_n = (n-1)\,\Delta f$. In order to ensure the orthogonality between subcarriers, the subcarrier interval satisfies $\Delta f = 1/T$, $\text{rect}\,(t/T) = \begin{cases} 1 & ,0 < t < T \\ 0 & , otherwise \end{cases}$.

3   Ambiguity Function

As an important tool for radar waveform design and analysis, ambiguity function can describe the characteristics of waveform and corresponding matched filter. By analyzing the ambiguity function of radar transmitted waveform, the resolution, measurement accuracy and ambiguity of radar system can be obtained when the optimal matched filter is used [12]. The ambiguity function has many definitions. In this paper, the ambiguity function defined from the output of the matched filter is adopted. Its expression is as follows

$$\chi(\tau, f_d) = \int_{-\infty}^{+\infty} s(t)s^*(t-\tau)e^{j2\pi f_d t}dt \tag{2}$$

where $s(t)$ is the transmitted signal of radar, τ is the relative delay between two targets, f_d is the relative Doppler frequency shift, and $s^*(t-\tau)$ is the conjugate of transmitted signal delay.

Substituting Eq. (1) and the conjugate of its delay into Eq. (2), we can get

$$\chi(\tau, f_d) = \int_{-\infty}^{+\infty} \sum_{m=1}^{N_s} \sum_{n=1}^{N_c} a_{n,m} \exp(j2\pi f_n t) \text{rect}\left[\frac{t-(m-1)T}{T}\right] \times$$

$$\sum_{p=1}^{N_s} \sum_{q=1}^{N_c} a_{q,p}^* \exp[-j2\pi f_q(t-\tau)] \text{rect}\left[\frac{t-\tau-(p-1)T}{T}\right] \exp(j2\pi f_d t)dt$$

$$= \sum_{n=1}^{N_c} \sum_{q=1}^{N_c} \exp(j2\pi f_q \tau) \sum_{m=1}^{N_s} \sum_{p=1}^{N_s} a_{n,m} a_{q,p}^* \int_{-\infty}^{+\infty} \exp[j2\pi(f_n - f_q + f_d)t] \times$$

$$\text{rect}\left[\frac{t-(m-1)T}{T}\right] \text{rect}\left[\frac{t-\tau-(p-1)T}{T}\right] dt \tag{3}$$

Next we derive the closed-form expression of $\chi(\tau, f_d)$:

i. when $|\tau| \geq N_s T$, since the delay exceeds the signal duration in the first PRI, $\chi(\tau, f_d) = 0$, $|\tau|$ is the absolute value of the delay;

ii. when $-N_s T < \tau < 0$, it is shown in Fig. 3. Let's make $\left|\left[\frac{\tau}{T}\right]\right| = k$, where $[\cdot]$ is the least integer function, and $|\cdot|$ represents taking the absolute value, then the ambiguity function can be calculated as

$$\chi(\tau, f_d) = (-kT - \tau) \sum_{m=1}^{N_s-k-1} \sum_{n=1}^{N_c} \sum_{q=1}^{N_c} a_{n,m} a_{q,m+k+1}^* \times$$

$$\exp(j2\pi f_q \tau) \exp\left[j2\pi M \frac{\tau + (2m+k)T}{2}\right] \text{sinc}[M(-kT - \tau)]$$

$$+ [(k+1)T + \tau] \sum_{m=1}^{N_s-k} \sum_{n=1}^{N_c} \sum_{q=1}^{N_c} a_{n,m} a_{q,m+k}^* \times$$

$$\exp(j2\pi f_q \tau) \exp\left[j2\pi M \frac{\tau + (2m+k-1)T}{2}\right] \text{sinc}\{M[(k+1)T + \tau]\} \tag{4}$$

with $M = \Delta f(n-q) + f_d$, $f_q = (q-1)\Delta f$, $\text{sinc}(x) = \frac{\sin(\pi x)}{\pi x}$.

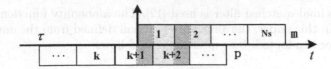

Fig. 3. When the delay satisfies $-N_sT < \tau < 0$

iii. when $0 < \tau < N_sT$, it is shown in Fig. 4. Let's make $\left[\frac{\tau}{T}\right] = k$, where $[\cdot]$ is the least integer function, then the ambiguity function can be calculated as

$$
\begin{aligned}
\chi\left(\tau, f_d\right) = & (-kT + \tau) \sum_{p=1}^{N_s-k-1} \sum_{n=1}^{N_c} \sum_{q=1}^{N_c} a_{n,p+k+1} a_{q,p}^* \times \\
& \exp(j2\pi f_q \tau) \exp\left[j2\pi M \frac{\tau + (2p+k)T}{2}\right] \operatorname{sinc}[M(-kT + \tau)] \\
& + [(k+1)T - \tau] \sum_{p=1}^{N_s-k} \sum_{n=1}^{N_c} \sum_{q=1}^{N_c} a_{n,p+k} a_{q,p}^* \times \\
& \exp(j2\pi f_q \tau) \exp\left[j2\pi M \frac{\tau + (2p+k-1)T}{2}\right] \operatorname{sinc}\{M[(k+1)T - \tau]\}
\end{aligned}
$$

(5)

with $M = \Delta f(n-q) + f_d$, $f_q = (q-1)\Delta f$, $\operatorname{sinc}(x) = \frac{\sin(\pi x)}{\pi x}$.

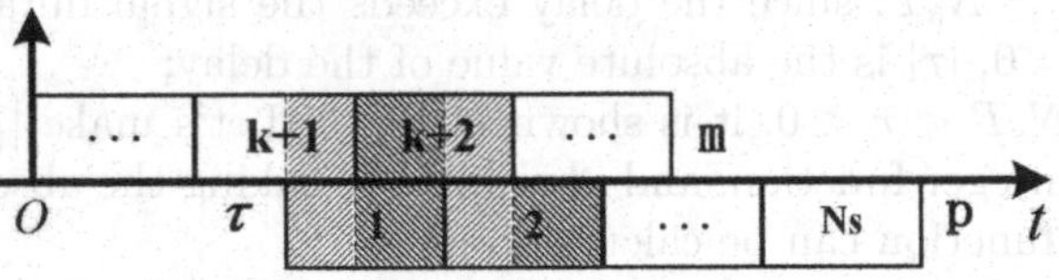

Fig. 4. When the delay satisfies $0 < \tau < N_sT$

3.1 Range Ambiguity Function

When the Doppler frequency shift satisfies $f_d = 0$, the range ambiguity function of the signal can be obtained from the Eq. (2)

$$
\chi(\tau, 0) = \int_{-\infty}^{+\infty} s(t) s^*(t - \tau) dt
$$

(6)

It can be seen from Eq. (6) that the range ambiguity function is the complex correlation function of the signal $s(t)$, and also the output response of the signal through its matched filter $s^*(-t)$.

Substituting $f_d = 0$ into Eq. (4), now $-N_s T < \tau < 0$ and $\left|\left[\frac{\tau}{T}\right]\right| = k$, then it can be obtained

$$
\begin{aligned}
\chi(\tau, 0) =& (-kT - \tau) \sum_{m=1}^{N_s-k-1} \sum_{n=1}^{N_c} \sum_{q=1}^{N_c} a_{n,m} a_{q,m+k+1}^* \times \\
& \exp(j2\pi f_q \tau) \exp\left[j2\pi M' \frac{\tau + (2m+k)T}{2}\right] \operatorname{sinc}[M'(-kT - \tau)] \\
& + [(k+1)T + \tau] \sum_{m=1}^{N_s-k} \sum_{n=1}^{N_c} \sum_{q=1}^{N_c} a_{n,m} a_{q,m+k}^* \times \\
& \exp(j2\pi f_q \tau) \exp\left[j2\pi M' \frac{\tau + (2m+k-1)T}{2}\right] \operatorname{sinc}\{M'[(k+1)T + \tau]\}
\end{aligned}
\tag{7}
$$

with $M' = \Delta f(n - q)$, $f_q = (q-1)\Delta f$, $\operatorname{sinc}(x) = \frac{\sin(\pi x)}{\pi x}$.

In order to simplify the Eq. (7), $R_{m,p}^+(\tau)$ and $R_{m,p}^-(\tau)$ respectively denote the correlation function of single OFDM symbol with index m and single OFDM symbol with index p when the delay is $0 \le \tau \le T$ and $-T \le \tau \le 0$ (as shown in Fig. 5). The expressions are as follows

$$
R_{m,p}^+(\tau) = (T - \tau) \sum_{n=1}^{N_c} \sum_{q=1}^{N_c} a_{n,m} a_{q,p}^* \exp(j2\pi f_q \tau) \exp(j2\pi M' \frac{T+\tau}{2}) \operatorname{sinc}[M'(T - \tau)]
\tag{8}
$$

$$
R_{m,p}^-(\tau) = (T + \tau) \sum_{n-1}^{N_c} \sum_{q-1}^{N_c} a_{n,m} a_{q,p}^* \exp(j2\pi f_q \tau) \exp(j2\pi M' \frac{T+\tau}{2}) \operatorname{sinc}[M'(T + \tau)]
\tag{9}
$$

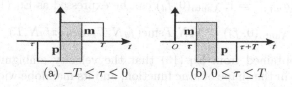

(a) $-T \le \tau \le 0$ (b) $0 \le \tau \le T$

Fig. 5. Integral of single symbol correlation function

By comparing and observing Eqs. (7) and (8) and (9), let's make $\tau' = \tau + (k+1)T$, $\tau'' = \tau + kT$, and easy to know $0 < \tau' < T$, $-T < \tau'' < 0$, then Eq. (7) can be rewritten as

$$
\chi(\tau, 0) = \sum_{m=1}^{N_s-k-1} R_{m,m+k+1}^+(\tau') + \sum_{m=1}^{N_s-k} R_{m,m+k}^-(\tau'')
\tag{10}
$$

It can be seen from Eq. (10) that the range ambiguity function of OFDM radar communication shared signal is composed of auto-correlation function and cross-correlation function of each symbol. For the case of $0 < \tau < N_sT$, according to the symmetry of ambiguity function, the same conclusion can be obtained.

3.2 Velocity Ambiguity Function

When the relative delay satisfies $\tau = 0$, the velocity ambiguity function of the signal can be obtained from the Eq. (2)

$$\chi(0, f_d) = \int_{-\infty}^{+\infty} s(t)s^*(t)e^{j2\pi f_d t}dt \tag{11}$$

Substituting $\tau = 0$ into Eq. (3), and we can get

$$\begin{aligned}
\chi(0, f_d) &= \sum_{m=p=1}^{N_s} \sum_{n=1}^{N_c} \sum_{q=1}^{N_c} a_{n,m} a_{q,m}^* \int_{(m-1)T}^{mT} e^{j2\pi(f_n - f_q + f_d)t}dt \\
&= T \sum_{m=1}^{N_s} \sum_{n=q=1}^{N_c} |a_{n,m}|^2 \operatorname{sinc}(f_d T) \exp[j\pi f_d(2m-1)T] \\
&\quad + T \sum_{m=1}^{N_s} \sum_{\substack{n=1 \\ n \neq q}}^{N_c} \sum_{q=1}^{N_c} a_{n,m} a_{q,m}^* \operatorname{sinc}(MT) \exp[j\pi M(2m-1)T] \\
&= \chi_{\text{auto}}(0, f_d) + \chi_{\text{cross}}(0, f_d)
\end{aligned} \tag{12}$$

with $M = \Delta f(n-q) + f_d$, $\operatorname{sinc}(x) = \frac{\sin(\pi x)}{\pi x}$.

In Eq. (12), $\chi_{\text{auto}}(0, f_d)$ is the main part of the velocity ambiguity function corresponding $n = q$, while $\chi_{\text{cross}}(0, f_d)$ is adjacent-channel interference corresponding $n \neq q$.

Because of $|a_{n,m}|^2 = 1$, $\chi_{\text{auto}}(0, f_d)$ can be expressed as Eq. (13) further

$$\chi_{\text{auto}}(0, f_d) = N_s N_c T \operatorname{sinc}(f_d N_s T) \exp(j\pi f_d N_s T) \tag{13}$$

It can be obtained from Eq. (13) that the velocity ambiguity function of shared signal is in the shape of *sinc* function, and its mainlobe width is inversely proportional to the pulse width.

3.3 Evaluation Indicator

In order to evaluate the performance of the ambiguity function, the Peak Sidelobes Ratio (PSLR) and Integrated Sidelobes Ratio (ISLR) are used to measure the sidelobe characteristics of the range ambiguity function and the velocity ambiguity function [10]. The width of the mainlobe (the first zero point region) is used to measure the range resolution and velocity resolution.

PSLR is defined as the ratio of the highest sidelobe peak P_s to the mainlobe peak P_m:

$$\text{PSLR} = 20\lg\frac{P_s}{P_m}(dB) \tag{14}$$

ISLR is defined as the ratio of sidelobe energy E_s to mainlobe energy E_m:

$$\text{ISLR} = 20\lg\frac{E_s}{E_m}(dB) \tag{15}$$

3.4 Fast Algorithm for Ambiguity Function

It is not easy to express the specific mathematical equation of the shared signal when the cyclic prefix is added, so a fast algorithm of ambiguity function is used for simulation. Equation (2) can be rewritten as

$$\chi(\tau, f_d) = \int_{-\infty}^{+\infty} [s(t)e^{j2\pi f_d t}]s^*(t - \tau)dt \tag{16}$$

By the definition of signal convolution $x(t) * h(t) = \int_{-\infty}^{+\infty} x(\tau)h(t - \tau)d\tau$, swap τ and t to get

$$x(\tau) * h(\tau) = \int_{-\infty}^{+\infty} x(t)h(\tau - t)dt \tag{17}$$

By comparing Eqs. (16) and (17), the ambiguity function can be written as the convolution of two signals

$$\chi(\tau, f_d) = [s(\tau)e^{j2\pi f_d \tau}] * s^*(-\tau) \tag{18}$$

According to the time domain convolution theorem, the convolution of two signals in the time domain is equal to the product of the Fourier transform of two signals in the frequency domain, so it can be realized by FFT-IFFT algorithm in computer. Using this algorithm, the program running speed is much higher than simulating the Eq. (2) directly. The fast algorithm process is shown in Fig. 6.

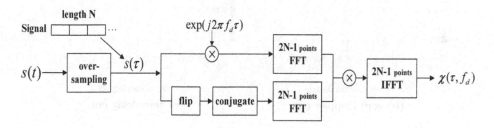

Fig. 6. Fast algorithm for ambiguity function

4   Simulation and Analysis

Simulation parameters are set as follows [11]: QPSK modulation is adopted in the communication modulation mode, the number of OFDM symbols $N_s = 10$, the number of subcarriers $N_c = 16$, the basic OFDM symbol duration $T = 1\,\text{ms}$, the subcarrier interval $\Delta f = 1\,\text{MHz}$, the length of cyclic prefix $T_G = \frac{1}{4}T$, the complete OFDM symbol duration $T_{sym} = T_G + T = 1.25\,\text{ms}$, and the oversampling factor $L = 16$. The simulation results are shown in Fig. 7 (The delay normalization factor is T_{sym} and the Doppler normalization factor is Δf).

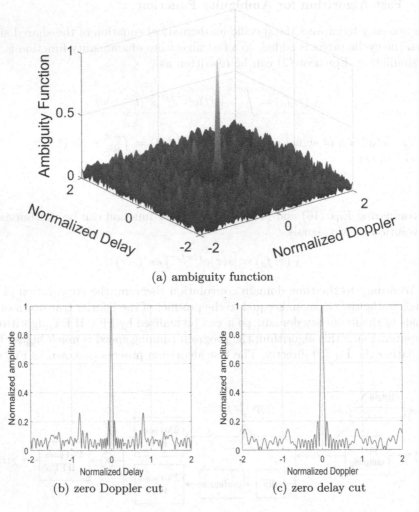

(a) ambiguity function

(b) zero Doppler cut

(c) zero delay cut

Fig. 7. Ambiguity function and its cut of OFDM shared signal

As can be seen from Fig. 7(a), the ambiguity function of OFDM radar communication shared signal has a "thumbtack" shape with single central peak. Figure 7(b) is the simulation result of range ambiguity function, which is the zero Doppler cut of Fig. 7(a). Figure 7(c) is the simulation result of velocity ambiguity function, which is the zero delay cut of Fig. 7(a). As can be seen from Fig. 7(b) and Fig. 7(c), a narrow central peak implies high range and Doppler resolution, so it has high accuracy of range and speed measurement. On the other hand, there are some fluctuations in the sidelobe region of OFDM radar communication shared signal, it may affect the detection of the target.

4.1   Influence of the Number of Subcarriers

Fixing the number of OFDM symbols $N_s = 10$, and the number of subcarriers $N_c = 8, 16, 32, 64$ are used for simulation comparison to explore the influence of the number of subcarriers on the range and velocity ambiguity function, and the statistical average value of each indicator is obtained after 1000 Monte Carlo methods. The simulation results are shown in Fig. 8 and Table 1.

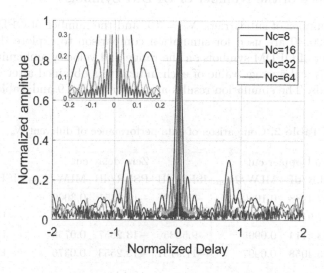

Fig. 8. Zero Doppler cuts of different N_c when $N_s = 10$

For the zero Doppler cut, by observing Fig. 8 and the data in Table 1, it can be obtained obviously that the mainlobe width is also reduced by two times when the N_c is doubled, and PSLR and ISLR don't change significantly with the increase of N_c. In addition, by changing the normalized factor, the mainlobe width of the zero Doppler cut is approximately inversely proportional to the bandwidth of the shared signal, that is when using $B = N_c * \Delta f$ as the normalized factor, we can get that mainlobe width is equal to 1. As for zero delay cut, the shape is similar to Fig. 7(c), so it will not be shown here. By observing the data in Table 1, PSLR and mainlobe width don't change obviously with the increase of N_c, while ISLR decreases gradually with the increase of N_c. To sum up, the increase of the N_c improves the inherent range resolution of the shared signal which is approximately inversely proportional to the bandwidth, but has little effect on the sidelobe characteristics of the range ambiguity function. Meanwhile, the sidelobe characteristic of velocity ambiguity function is improved to some extent.

Table 1. Comparison of cuts performance of different N_c

Nc	Zero Doppler cut			Zero delay cut		
	PSLR/dB	MLW/T_sym	ISLR/dB	PSLR/dB	MLW/T_sym	ISLR/dB
8	−12.4296	0.2090	−6.8986	−13.0512	0.1543	−7.9760
16	−12.9021	0.1006	−6.5161	−13.2568	0.1514	−10.9989
32	−12.9766	0.0497	−6.2470	−13.2572	0.1498	−14.0773
64	−13.1038	0.0250	−6.2715	−13.0440	0.1487	−16.8074

4.2   Influence of the Number of OFDM Symbols

Fixing the number of subcarriers $N_c = 16$, and the number of OFDM symbols $N_s = 5, 10, 20, 40$ are used for simulation comparison to explore the influence of the number of OFDM symbols on the range and velocity ambiguity function, and the statistical average value of each indicator is obtained after 1000 Monte Carlo methods. The simulation results are shown in Fig. 9 and Table 2.

Table 2. Comparison of cuts performance of different N_s

Ns	Zero Doppler cut			Zero delay cut		
	PSLR/dB	MLW/T_sym	ISLR/dB	PSLR/dB	MLW/T_sym	ISLR/dB
5	−12.4788	0.1080	−3.2469	−12.9851	0.3009	−11.0098
10	−12.9140	0.1004	−6.5039	−13.2766	0.1514	−11.1299
20	−13.2114	0.0995	−9.5385	−13.2657	0.0758	−11.3472
40	−13.4058	0.0997	−11.9731	−13.2553	0.0376	−11.6498

For zero delay cut, by observing Fig. 9 and the data in Table 2, it can be obtained obviously that the mainlobe width is also reduced by two times when the N_s is doubled, and PSLR and ISLR don't change significantly with the increase of N_s. In addition, by changing the normalized factor, the mainlobe width of the zero delay cut is approximately inversely proportional to the pulse width of the shared signal, that is when using $taup = N_s * T_\text{sym}$ as the normalized factor, we can get that mainlobe width is equal to 1. As for the zero Doppler cut, the shape is similar to Fig. 7(b), so it will not be shown here. By observing the data in Table 2, PSLR and mainlobe width don't change obviously with the increase of N_s, while ISLR decreases gradually with the increase of N_s. To sum up, the increase of the N_s improves the inherent velocity resolution of the shared signal which is approximately inversely proportional to the pulse width, but has little effect on the sidelobe characteristics of the velocity ambiguity function. Meanwhile, the sidelobe characteristic of range ambiguity function is improved to some extent.

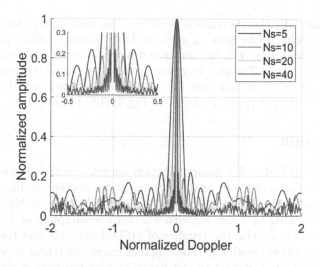

Fig. 9. Zero delay cuts of different N_s when $N_c = 16$

4.3   Influence of Cyclic Prefix

When delay is equal to basic OFDM symbol duration namely $\tau = T$, the integral of the ambiguity function is shown in Fig. 10.

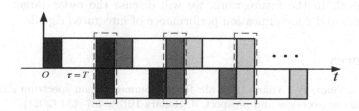

Fig. 10. When the delay satisfies $\tau = T$

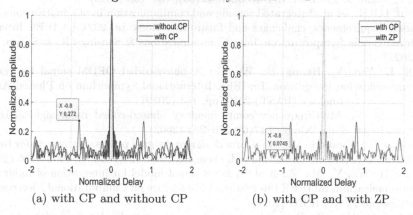

(a) with CP and without CP	(b) with CP and with ZP

Fig. 11. The influence of CP and ZP on the ambiguity function($N_s = 10$, $N_c = 16$, $T_G = \frac{1}{4}T$)

Due to the CP, the correlation between the two signals will be exactly equal at the cyclic prefix part of each OFDM symbol, which causes a symmetric pseudo-peak that will be introduced into the range ambiguity function. As shown in Fig. 11(a), The zero Doppler cut shows a symmetric pseudo-peak at $\tau = \pm T/T_{\mathrm{sym}} = \pm 0.8$. In this case, ZP can be chosen to replace CP to eliminate the appearance of false peak, as shown in Fig. 11(b) obviously.

5 Conclusion

In this paper, OFDM radar communication shared signal is realized by pulsing the OFDM signal of traditional communication, and the ambiguity function of shared signal without CP is deduced theoretically. The ambiguity function with CP is simulated by a fast algorithm under oversampling and the effects of the number of subcarriers, the number of OFDM symbols and the CP on the ambiguity function are analyzed. The inherent range resolution which is approximately inversely proportional to the bandwidth of the shared signal and the sidelobe characteristics of the velocity ambiguity function can be improved by adding N_c, and the inherent velocity resolution which is approximately inversely proportional to the pulse width of the shared signal and the sidelobe characteristics of the range ambiguity function can be improved by adding N_s. In addition, using ZP can eliminate the influence of CP on the range ambiguity function. This provides a reference for parameter design of OFDM radar communication shared signal. In the future work, we will discuss the radar target detection performance and communication performance of integrated signals.

References

1. Liu, F., Yuan, W., Yuan, J., et al.: Radar-communication spectrum sharing and integration: overview and prospect. J. Radars **10**(3), 467–484 (2021)
2. Xiao, B., Huo, K., Liu, Y.: Development and prospect of radar and communication integration. J. Electron. Inf. Technol. **41**(03), 739–750 (2019)
3. Tan, D.P.K., et al.: Integrated sensing and communication in 6G: motivations, use cases, requirements, challenges and future directions. In: 2021 1st IEEE International Online Symposium on Joint Communications & Sensing (JC & S), pp. 1–6 (2021)
4. Qi, L., Yao, Y., Huang, B., Wu, G.: A phase-coded OFDM signal for radar-communication integration. In: IEEE International Symposium on Phased Array System & Technology (PAST) 2019, pp. 1–4 (2019)
5. Levanon, N.: Multifrequency complementary phased-coded radar signals. IEEE Proc. Radar Sonar Navig. **147**(6), 276–284 (2002)
6. Sturm, C., Wiesbeck, W.: Waveform design and signal processing aspects for fusion of wireless communications and radar sensing. Proc. IEEE **59**(4), 1902–1906 (2011)
7. Lou, H., Wu, Y., Ma, Z., et al.: A novel signal model for integration of radar and communication. In: IEEE International Conference on Computational Electromagnetics, pp. 14–16. IEEE (2017)
8. Zhang, C., Gao, Y.: Research on signal processing technology in radar and communication integrated system based on OFDM. Radio Eng. **47**(3), 19–22 (2017)

9. Sen, S., Nehorai, A.: Adaptive design of OFDM radar signal with improved wideband ambiguity function. IEEE Trans. Sig. Process. **58**(2), 928–933 (2010)
10. Liu, Y., Liao, G., Yang, Z.: Ambiguity function analysis of integrated radar and communication waveform based on OFDM. Syst. Eng. Electron. **38**(09), 2008–2018 (2016)
11. Zhou, Y., Yang, R., Zuo, J.: Analysis on ambiguity function performance of multi-symbol OFDM radar communication shared signal. J. Air Force Early Warning Acad. **32**(05), 325–330 (2018)
12. Ding, L., Geng, F., Chen, J. Principles of radar. 5th edn. Publishing House of Electronics

Joint Power Allocation and Passive Beamforming Design for IRS-Assisted Cell-free Networks

Chen He[1(✉)], Xie Xie[1], Yangrui Dong[1], and Shun Zhang[2]

[1] School of Information Science and Technology, Northwest University, Xi'an, China
chenhe@nwu.edu.cn, {x_xie,dongyangrui}@stumail.nwu.edu.cn
[2] State Key Laboratory of Integrated Services Networks, Xidian University,
Xi'an, China
zhangshunsdu@xidian.edu.cn

Abstract. This paper investigates multiple intelligent reflecting surface (IRSs) assisted cell-free networks, where multiple single antenna access points (APs) and IRSs are connected to a network controller, to serve multiple user equipment (UEs) simultaneously. Our objective is to maximize the sum-rate of the cell-free network by jointly designing the power allocation of APs and the passive reflecting beamforming of IRSs, while the constraints on the maximum transmit power of each AP and the phase of each phase shifter (PS) of IRS are satisfied. However, the problem is non-convex and challenging to solve. To this end, we propose an efficient framework to jointly design the power allocation vectors and the passive reflecting beamforming matrices. Particularly, we first reformulate the problem as a more tractable form by employing the fractional programming methods and then decompose the transformed problem into two subproblems. Finally, we propose an alternating iteratively (AI) algorithm to solve the two subproblems, which is guaranteed to converge to locally optimal solutions. Simulation results indicate that the advantages of leveraging IRSs in improving the performance of the conventional cell-free networks.

Keywords: Intelligent reflecting surface · Power allocation · Fractional programming · Sequential Optimization

1 Introduction

Cell-free networks comprise a large number of randomly located single-antenna access points (APs) connected to a controller, and the controller optimizes the

This work was supported in part by the National Natural Science Foundation of China under Grant 61701401 and Grant 61872295, in part by the Key Research and Development Project of Shaanxi Province under Grant 2021KWZ-07, and in part by the Nova Program of Science and Technology of Shaanxi Province under Grant 2019KJXX-061.

H. Gao et al. (Eds.): ChinaCom 2021, LNICST 433, pp. 264–274, 2022.
https://doi.org/10.1007/978-3-030-99200-2_21

power allocation to improve the network performance [1]. Compared with the cellular system, cell-free networks are no cells or cell boundaries and do not assign the user equipment (UE) to the particular AP, and APs in cell-free networks serve multiple UEs simultaneously and cooperatively. Besides, the number of APs is larger than the number of UEs [2].

Recently, intelligent reflecting surfaces (IRSs) have emerged as a promising technology to improve the performance of the communication system via mitigating the detrimental propagation conditions and strengthening the desired signal powers [3]. Generally, IRSs are composed a large number of phase shifters (PSs), each of which can reconfigure the incident signals to desired directions with significant beamforming gains. The various IRS-assisted communication systems have been extensively investigated, and one of the key problem is jointly designing the power allocation (the active transmitting beamforming) and the passive reflecting beamforming [4–8]. The authors in [4] studied a joint power allocation and passive beamforming design problem to maximize the physical-layer security. The work [5] maximized the sum-rate of the IRS-assited non-orthogonal multiple access (NOMA) system by jointly optimizing the power allocation, passive and hybrid beamforming. The work [6] proposed an efficient algorithm based on vector forms of the fractional programming methods [9,10], i.e., Lagrangian dual transform (LDT) and quadratic transform (QT), and through joint optimizing the active transmit beamforming and the passive reflecting beamforming to maximize the sum-rate. The authors in [7] studied the sum-rate maximization problem in a multi-cell scenario, and they employed the weighted minimum mean-square error (WMMSE) [11] technique to transform the original problem into an equivalent form and proposed a block coordinate descent (BCD) and the majorization-minimization (MM) method based algorithm to solve it. The authors in [8] deployed the IRS to assist the joint processing coordinated multi-point (JP-CoMP) transmission. They studied two cases, i.e., the single UE case and the multiple UEs case. Particularly, the authors proposed a computational efficient MM-based algorithm to solve the problem in a single UE case and by capitalizing on the second-order cone programming (SOCP) and semi-definite relaxation (SDR) [12] techniques to solve the problem under multiple UEs case.

Motivated by the discussions as mentioned above, in this paper, we consider maximizing the sum-rate of IRSs-assisted cell-free communication system by jointly optimizing the power allocation of APs and the passive reflecting beamforming of IRSs. Note that the power allocation and the beamforming are intricately coupled, and the constant modulus constraints of IRS, the formulated problem is non-convex and challenging to solve. As a compromise approach, we provide an efficient framework to jointly design the power allocation and the beamforming. We first transform the problem to an equivalent form by employing the fractional programming methods, e.g., LDT and QT, and then decompose the reformulated problem into two subproblems, i.e., the power allocation optimization and the passive reflecting beamforming optimization. Then, we propose an alternating iteratively (AI) algorithm to solve the two subproblems. Particularly, for the former subproblem, we reformulate it as a convex quadratic programming (QP) problem and propose a Lagrangian dual sub-gradient based algorithm to obtain nearly closed-form solutions. While for the latter subprblem, we recast it as a constant modulus constrained quadratic programming (CMC-QP) problem, and then we propose a computational efficient sequential optimization (SO) algorithm to solve it with locally optimal solutions in closed-forms.

Notations: Low case letters denote vectors and upper case bold letters stand for matrices. Re $\{a\}$ is the real part of a. $\mathcal{CN}(0,1)$ is the distribution of a circularly symmetric complex Gaussian variable with zero mean and unit variance.

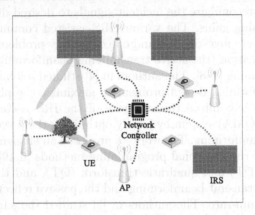

Fig. 1. System model of IRSs-assisted cell-free networks.

2   System Model and Problem Formulation

In this section, as shown in Fig. 1, we investigate an IRS-assisted cell-free networks, where $\mathcal{R}$ distributed IRSs are deployed to assisted $\mathcal{A}$ randomly distributed APs via connecting to a network controller for cooperatively serving $\mathcal{K}$ users. We assume that each AP and each UE are equipped with single antenna, and each IRS has $\mathcal{N}$ PSs.

The transmitted signal from the a-th AP is given as

$$x_a = \sum_{k=1}^{\mathcal{K}} \sqrt{p_{a,k}} t_{a,k}, \forall a \in \mathcal{A}, \tag{1}$$

where $t_{a,k} \sim \mathcal{CN}(0,1)$ represent the transmit signal from the a-th AP to the k-th UE and $p_{a,k}$ denotes the transmit power from the a-th AP to the k-th UE. Let $d_{a,k} \in \mathbb{C}$, $\mathbf{g}_{r,k} \in \mathbb{C}^{\mathcal{N} \times 1}$, and $\mathbf{s}_{a,r} \in \mathbb{C}^{\mathcal{N} \times 1}$ denote the channels of a-th AP to the k-th UE link, r-th IRS to the k-th UE link, and a-th AP to the r-th IRS link, respectively. Besides, the passive reflecting beamforming of IRSs are denoted as

$$\boldsymbol{\Theta}_r = \mathrm{diag}\left(e^{j\phi_{r,1}}, e^{j\phi_{r,1}}, \cdots, e^{j\phi_{r,\mathcal{N}}}\right), \forall r \in \mathcal{R}. \tag{2}$$

The received signal at the k-th UE can be expressed as

$$y_k = \sum_{a=1}^{\mathcal{A}} \left(d_{a,k} + \sum_{r=1}^{\mathcal{R}} \mathbf{g}_{r,k}^{\mathrm{H}} \mathbf{\Theta}_r \mathbf{s}_{a,r} \right) x_a + n_k, \forall k \in \mathcal{K}, \tag{3}$$

where n_k is the noise at the k-th UE with distribution $\mathcal{CN}\left(0, \sigma_k^2\right)$. Hence, the signal-to-interference-plus-noise ratio (SINR) at the k-th UE can be written as

$$\gamma_k = \frac{\sum_{a=1}^{\mathcal{A}} p_{a,k} \left| d_{a,k} + \sum_{r=1}^{\mathcal{R}} \mathbf{g}_{r,k}^{\mathrm{H}} \mathbf{\Theta}_r \mathbf{s}_{a,r} \right|^2}{\sum_{a=1}^{\mathcal{A}} \sum_{i \neq k}^{\mathcal{K}} p_{a,i} \left| d_{a,k} + \sum_{r=1}^{\mathcal{R}} \mathbf{g}_{r,k}^{\mathrm{H}} \mathbf{\Theta}_r \mathbf{s}_{a,r} \right|^2 + \sigma_k^2}. \tag{4}$$

To facilitate expression, we define $h_{a,k} = d_{a,k} + \sum_{r=1}^{\mathcal{R}} \mathbf{g}_{r,k}^{\mathrm{H}} \mathbf{\Theta}_r \mathbf{s}_{a,r}$, which denotes the equivalent channel spanning from the a-th AP to the k-th UE.

In this paper, our objective is to maximize the sum-rate of the cell-free network, i.e., $R\left(\mathbf{p}, \mathbf{\Theta}\right) = \sum_{k=1}^{\mathcal{K}} \log\left(1 + \gamma_k\right)$, by jointly designing the power allocation vector $\mathbf{p} = \left[p_{1,1}, \cdots, p_{1,\mathcal{K}}, p_{2,1}, \cdots, p_{\mathcal{A},\mathcal{K}}\right]^{\mathrm{T}}$, and the passive reflecting beamforming matrix $\mathbf{\Theta} = \mathrm{diag}\left\{\mathbf{\Theta}_1, \mathbf{\Theta}_2, \cdots, \mathbf{\Theta}_\mathcal{R}\right\}$. Therefore, the sum-rate maximization problem can be formulated as

$$\max_{p,\mathbf{\Theta}} \quad R\left(p, \mathbf{\Theta}\right) = \sum_{k=1}^{\mathcal{K}} \log\left(1 + \gamma_k\right) \tag{5}$$

$$\mathrm{s.t.} \quad \sum_{k=1}^{\mathcal{K}} p_{a,k} = P_{a,\max}, \forall a \in \mathcal{A}, \tag{5a}$$

$$\left|\mathbf{\Theta}_{r,n}\right| = 1, \forall r \in \mathcal{R}, \forall n \in \mathcal{N}, \tag{5b}$$

where the constraint (5a) limits the maximum transmitting power of a-th AP and the constraint (5b) represents the constant modulus constraint of each PS at IRSs. Since the variables $\mathbf{p}$ and $\mathbf{\Theta}$ are intricately coupled, the problem is non-convex and intractable. To this end, in the following section, we propose an efficient framework to solve the problem.

3 Jointly Design Framework

First, we employ the fractional programming methods, i.e., LDT [10, Theorem 4] and QT [10, Corollary 1], to equivalently transform the original problem to a more tractable form. By introducing auxiliary variable vectors $\alpha = \left[\alpha_1, \alpha_2, \cdots, \alpha_\mathcal{K}\right]^{\mathrm{T}}$ and $\beta = \left[\beta_1, \beta_2, \cdots, \beta_\mathcal{K}\right]^{\mathrm{T}}$, and according to the LDT and QT methods, the original problem can be reformulated as

$$\max_{\mathbf{p}, \mathbf{\Theta}, \alpha, \beta} \quad f\left(\mathbf{p}, \mathbf{\Theta}, \alpha, \beta\right)$$

$$\mathrm{s.t.} \quad (5a), (5b), \tag{6}$$

where $f(\mathbf{p}, \boldsymbol{\Theta}, \alpha, \beta)$ is given as $f(\mathbf{p}, \boldsymbol{\Theta}, \alpha, \beta) = \sum_{k=1}^{\mathcal{K}} \log(1 + \alpha_k) - \sum_{k=1}^{\mathcal{K}} \alpha_k$
$+ \sum_{k=1}^{\mathcal{K}} 2\beta_k \sqrt{(1 + \alpha_k) \sum_{a=1}^{\mathcal{A}} |h_{a,k}|^2 p_{a,k}} - \sum_{k=1}^{\mathcal{K}} |\beta_k|^2 \left(\sum_{a=1}^{\mathcal{A}} \sum_{i=1}^{\mathcal{K}} |h_{a,k}|^2 p_{a,i} \right.$
$\left. + \sigma_k^2 \right)$. For detailed transformation of Problem 6, the readers are referred to
[9]. Based on the fact that the objective function of the problem 6 is convex
with respect to any one of the four variables $\mathbf{p}$, $\boldsymbol{\Theta}$, α, and β, while fixing the
other three, the problem can be effectively solved in an alternating optimization
manner. The solutions after the u-th iteration are denoted by $(\cdot)^{(u+1)}$.

Note that α and β only appear in the objective function of Problem 6 and
do not exist in any constraints. Therefore, the optimal values of α and β can be
obtained by setting the partial derivatives of $f(\mathbf{p}, \boldsymbol{\Theta}, \alpha, \beta)$ with respect to $\alpha_k, \forall k$
and $\beta_k, \forall k$ to be zeros, respectively. First, with fixed $\mathbf{p}$, $\boldsymbol{\Theta}$, and β, we have

$$\alpha_k^{(u+1)} = \frac{\sum_{a=1}^{\mathcal{A}} |h_{a,k}|^2 p_{a,k}}{\sum_{a=1}^{\mathcal{A}} \sum_{i \neq k}^{\mathcal{K}} |h_{a,k}|^2 p_{a,i} + \sigma_k^2}, \forall k \in \mathcal{K}, \tag{7}$$

and then with fixed $\mathbf{p}$, $\boldsymbol{\Theta}$, and $\alpha_k^{(u+1)}$, we have

$$\beta_k^{(u+1)} = \frac{\sqrt{(1 + \alpha_k) \sum_{a=1}^{\mathcal{A}} |h_{a,k}|^2 p_{a,k}}}{\sum_{a=1}^{\mathcal{A}} \sum_{i \neq 1}^{\mathcal{K}} |h_{a,k}|^2 p_{a,i} + \sigma_k^2}, \forall k \in \mathcal{K}. \tag{8}$$

Meanwhile, with fixed $\boldsymbol{\Theta}$, and the so-obtained optimal $\alpha_k^{(u+1)}$ and $\beta_k^{(u+1)}$, the
objective function of Problem 6 can be reformulated as $f(\mathbf{p}, \boldsymbol{\Theta}, \alpha, \beta) = f_{\mathbf{p}}(\mathbf{p}) +$
const $(\boldsymbol{\Theta}, \alpha^{(u+1)}, \beta^{(u+1)})$, where const $(\boldsymbol{\Theta}, \alpha^{(u+1)}, \beta^{(u+1)}) = \sum_{k=1}^{\mathcal{K}} \log(1 + \alpha_k) -$
$\sum_{k=1}^{\mathcal{K}} \alpha_k - \sum_{k=1}^{\mathcal{K}} |\beta_k|^2 \sigma_k^2$, which is the irrelevant constant term with respect to
$\mathbf{p}$ and has no impact on the optimization of $\mathbf{p}$. Then, we have the following
optimization subproblem of $\mathbf{p}$, which is given as

$$\mathbf{p} = \arg\max_{\mathbf{p}} \quad f_{\mathbf{p}}(\mathbf{p})$$
$$\text{s.t.} \quad (5a), \tag{9}$$

where $f_{\mathbf{p}}(\mathbf{p}) = \sum_{k=1}^{\mathcal{K}} 2\beta_k \sqrt{(1 + \alpha_k) \sum_{a=1}^{\mathcal{A}} |h_{a,k}|^2 p_{a,k}} - \sum_{k=1}^{\mathcal{K}} |\beta_k|^2 \sum_{a=1}^{\mathcal{A}} \sum_{i=1}^{\mathcal{K}}$
$|h_{a,k}|^2 p_{a,i}$. It can be verified that the objective function and the constraints of
Problem 9 are both convex, which yields that Problem 9 is convex and can be
solved by employing convex solver tools, e.g. CVX. Note that the dual gap of
the above problem is guaranteed to be zero, therefore, instead of relying on the
generic solver with high computational complexity, we propose a Lagrangian
dual sub-gradient based algorithm to optimize $\mathbf{p}$.

The Lagrangian dual function of Problem 9 is defining as

$$\mathcal{L}(\mathbf{p}, \lambda) = f_{\mathbf{p}}(\mathbf{p}) - \sum_{a=1}^{\mathcal{A}} \lambda_a \left(\sum_{k=1}^{\mathcal{K}} p_{a,k} - P_{a,\max} \right), \tag{10}$$

where $\lambda = [\lambda_1, \lambda_2, \cdots, \lambda_{\mathcal{A}}]^\mathrm{T}$ is the dual vector with $\lambda_a \geq 0, \forall a \in \mathcal{A}$ is introduced for enforcing the maximal power in the a-th AP. The variables $\mathbf{p}$ and λ can be obtained in an alternating manner. First, with fixed λ^q, the optimal power allocation vector $\mathbf{p}^{\mathrm{q}+1}$ can be optimized by solving the following sub-problem

$$\mathbf{p}^{\mathrm{q}+1} = \arg\max_{\mathbf{p}} \mathcal{L}\left(\mathbf{p}, \lambda^\mathrm{q}\right). \tag{11}$$

By setting the first-order partial derivative of $\mathcal{L}\left(\mathbf{p}, \lambda^\mathrm{q}\right)$ with respect to $p_{a,k}, \forall \{\mathcal{A}, \mathcal{K}\}$ to be zeros, we have

$$p_{a,k}^{\mathrm{q}+1} = \frac{(1 + \alpha_k)\,\beta_k^2\,|h_{a,k}|^2}{\left(\sum_{i=1}^{\mathcal{K}} \beta_i^2\,|h_{a,i}|^2 + \lambda_a\right)^2}, \forall a \in \mathcal{A}, \forall k \in \mathcal{K}. \tag{12}$$

The dual variable λ can be determined by solving the following dual optimizing problem, which is given as

$$\lambda^{\mathrm{q}+1} = \arg\min_{\lambda}\max_{\mathbf{p}} \mathcal{L}\left(\mathbf{p}^{\mathrm{q}+1}, \lambda\right). \tag{13}$$

By defining $f_a\left(\lambda_a^\mathrm{q}\right) = \sum_{k=1}^{\mathcal{K}} p_{a,k} - P_{a,\max}$, which is a monotonically decreasing function for $\lambda_a \geq 0$, we propose a sub-gradient based method to update $\lambda_a^{\mathrm{q}+1}$. Particularly, with the fixed $p_{a,k}^{\mathrm{q}+1}$, the dual variable λ_a can be performed as follows

$$\lambda_a^{\mathrm{q}+1} = [\lambda_a^\mathrm{q} + \tau f_a\left(\lambda_a^\mathrm{q}\right)]^+, \forall a \in \mathcal{A}, \tag{14}$$

where τ_a denotes the positive step for updating $\lambda_a^{\mathrm{q}+1}$ and $[x]^+ = \max\{x, 0\}$. With the number of iterations increasing, the solutions finally converge to the locally optimal solution, and we have $\mathbf{p}^{(\mathrm{u}+1)} = \mathbf{p}^{\mathrm{q}^*}$.

Finally, we consider to optimize the passive reflecting beamforming matrix $\mathbf{\Theta}$. With the fixed $\mathbf{p}^{(\mathrm{u}+1)}$, $\alpha^{(\mathrm{u}+1)}$, and $\beta^{(\mathrm{u}+1)}$, the corresponding subproblem for optimizing $\mathbf{\Theta}_r, \forall r \in \mathcal{R}$ of IRSs are given as

$$\mathbf{\Theta} = \arg\max_{\mathbf{\Theta}} \quad f\left(\mathbf{p}^{(\mathrm{u}+1)}, \mathbf{\Theta}, \alpha^{(\mathrm{u}+1)}, \beta^{(\mathrm{u}+1)}\right)$$
$$\text{s.t.} \quad (5b). \tag{15}$$

By defining $\mathbf{d}_k = [d_{1,k}, d_{2,k}, \cdots, d_{\mathcal{A},k}]$, $\mathbf{g}_k = \left[\mathbf{g}_{1,k}^\mathrm{T}, \mathbf{g}_{2,k}^\mathrm{T}, \cdots, \mathbf{g}_{\mathcal{R},k}^\mathrm{T}\right]^\mathrm{T}$, $\mathbf{p}_k = [p_{1,k}, p_{2,k}, \cdots, p_{\mathcal{A},k}]^\mathrm{T}$,

$$\mathbf{S} = \begin{bmatrix} \mathbf{s}_{1,1}^\mathrm{T} & \mathbf{s}_{1,2}^\mathrm{T} & \cdots & \mathbf{s}_{1,\mathcal{R}}^\mathrm{T} \\ \mathbf{s}_{2,1}^\mathrm{T} & \mathbf{s}_{2,2}^\mathrm{T} & \cdots & \mathbf{s}_{2,\mathcal{R}}^\mathrm{T} \\ \vdots & \vdots & \vdots & \vdots \\ \mathbf{s}_{\mathcal{A},1}^\mathrm{T} & \mathbf{s}_{\mathcal{A},2}^\mathrm{T} & \cdots & \mathbf{s}_{\mathcal{A},\mathcal{R}}^\mathrm{T} \end{bmatrix}^\mathrm{T}, \tag{16}$$

we have $\sum_{a=1}^{\mathcal{A}}\left(d_{a,k} + \sum_{r=1}^{\mathcal{R}} \mathbf{g}_{r,k}^\mathrm{H} \mathbf{\Theta}_r \mathbf{s}_{a,r}\right)\sqrt{p_{a,k}} = \mathbf{d}_k\sqrt{\mathbf{p}_k} + \mathbf{g}_k^\mathrm{H}\mathbf{\Theta}\mathbf{S}\sqrt{\mathbf{p}_k}$. To make above expression more tractable, we further define $\theta =$

$[\theta_{1,1}, \theta_{1,2}, \cdots, \theta_{1,\mathcal{N}}, \cdots, \theta_{\mathcal{R},\mathcal{N}}]^{\mathrm{T}} \in \mathbb{C}^{\mathcal{RN} \times 1}$, and $\mathbf{H}_k = \operatorname{diag}(\mathbf{g}_k)\mathbf{S} \in \mathbb{C}^{\mathcal{RN} \times \mathcal{A}}$, and then $\mathbf{g}_k^{\mathrm{H}} \mathbf{\Theta} \mathbf{S} = \theta^{\mathrm{H}} \mathbf{H}_k$. Hence, with fixed $\mathbf{p}^{(\mathrm{u}+1)}$, $\alpha^{(\mathrm{u}+1)}$, and $\beta^{(\mathrm{u}+1)}$, we have $f\left(\mathbf{p}^{(\mathrm{u}+1)}, \theta, \alpha^{(\mathrm{u}+1)}, \beta^{(\mathrm{u}+1)}\right) = f_\theta(\theta) + \operatorname{const}\left(\mathbf{p}^{(\mathrm{u}+1)}, \alpha^{(\mathrm{u}+1)}, \beta^{(\mathrm{u}+1)}\right)$, where

$$
\begin{aligned}
f_\theta(\theta) = &\sum_{k=1}^{\mathcal{K}} 2\beta_k \sqrt{1+\alpha_k} \operatorname{Re}\left\{\theta^{\mathrm{H}} \mathbf{H}_k \sqrt{\mathbf{p}_k}\right\} - \sum_{k=1}^{\mathcal{K}} 2|\beta_k|^2 \left\{\theta^{\mathrm{H}} \mathbf{H}_k \sum_{i=1}^{\mathcal{K}} \sqrt{\mathbf{p}_i}\sqrt{\mathbf{p}_i}^{\mathrm{H}} \mathbf{d}_k^{\mathrm{H}}\right\} \\
&- \sum_{k=1}^{\mathcal{K}} |\beta_k|^2 \theta^{\mathrm{H}} \mathbf{H}_k \sum_{i=1}^{\mathcal{K}} \sqrt{\mathbf{p}_i}\sqrt{\mathbf{p}_i}^{\mathrm{H}} \mathbf{H}_k^{\mathrm{H}} \theta,
\end{aligned}
\tag{17}
$$

and $\operatorname{const}\left(\mathbf{p}^{(\mathrm{u}+1)}, \alpha^{(\mathrm{u}+1)}, \beta^{(\mathrm{u}+1)}\right)$ is the irreverent constant term about θ. After omitting the terms, the subproblem of optimizing θ can be expressed as

$$
\theta = \arg\max_\theta \; -\theta^{\mathrm{H}} \mathcal{Z}\theta + 2\operatorname{Re}\left\{\theta^H \omega\right\}
\tag{18}
$$

$$
\text{s.t. } |\theta_i| = 1, \forall i \in \hat{\mathcal{N}},
\tag{18a}
$$

where $\hat{\mathcal{N}} = \mathcal{RN}$, and

$$
\mathcal{Z} = \sum_{k=1}^{\mathcal{K}} |\beta_k|^2 \mathbf{H}_k \sum_{i=1}^{\mathcal{K}} \sqrt{\mathbf{p}_i}\sqrt{\mathbf{p}_i}^{\mathrm{H}} \mathbf{H}_k^{\mathrm{H}},
$$

$$
\omega = \sum_{k=1}^{\mathcal{K}} \beta_k \sqrt{1+\alpha_k} \mathbf{H}_k \sqrt{\mathbf{p}_k} - |\beta_k|^2 \mathbf{H}_k \sum_{i=1}^{\mathcal{K}} \sqrt{\mathbf{p}_i}\sqrt{\mathbf{p}_i}^{\mathrm{H}} \mathbf{d}_k^{\mathrm{H}}.
\tag{19}
$$

However, Problem 18 is a non-convex CMC-QP problem, which is still hard to obtain the optimal solution. Note that the objective function and the constant modulus constraints are separable with respect to $\theta_i, \forall i \in \hat{\mathcal{N}}$ [13], therefore, we can decompose Problem 18 into $\hat{\mathcal{N}}$ separate subproblems and solve them one-by-one. Particularly, we have

$$
\theta^{\mathrm{H}} \omega = \sum_{n=1}^{\hat{\mathcal{N}}} \theta_n^* \omega_n = \theta_i^* \omega_i + \sum_{n=1, n\neq i}^{\hat{\mathcal{N}}} \theta_n^* \omega_n.
\tag{20}
$$

Meanwhile, $\theta^{\mathrm{H}} \mathcal{Z}\theta$ can be expanded as

$$
\begin{aligned}
\theta^{\mathrm{H}} \mathcal{Z}\theta &= \sum_{\substack{n=1 \\ n\neq i}}^{\hat{\mathcal{N}}} \theta^{\mathrm{H}} \mathbf{z}_n \theta_n + \theta^{\mathrm{H}} \mathbf{z}_i \theta_i = \sum_{\substack{n=1 \\ n\neq i}}^{\hat{\mathcal{N}}} \theta_i^* z_{i,n} \theta_n + \theta^{\mathrm{H}} \mathbf{z}_i \theta_i + \sum_{\substack{m=1 \\ m\neq n}}^{\hat{\mathcal{N}}} \sum_{\substack{p=1 \\ p\neq i}}^{\hat{\mathcal{N}}} \theta_m^* z_{m,p} \theta_p \\
&= \theta_i^* z_{i,i} \theta_i + \sum_{\substack{n=1 \\ n\neq i}}^{\hat{\mathcal{N}}} \left(\theta_i^* z_{i,n} \theta_n + \theta_i z_{n,i} \theta_n^*\right) + \sum_{\substack{m=1 \\ m\neq n}}^{\hat{\mathcal{N}}} \sum_{\substack{p=1 \\ p\neq i}}^{\hat{\mathcal{N}}} \theta_m^* z_{m,p} \theta_p,
\end{aligned}
\tag{21}
$$

where $\mathcal{Z} = [\mathbf{z}_1, \mathbf{z}_2, \cdots, \mathbf{z}_{\hat{\mathcal{N}}}]$ and $\mathbf{z}_n = \left[z_{1,n}, z_{2,n}, \cdots, z_{\hat{\mathcal{N}},n}\right]^{\mathrm{T}} \in \mathbb{C}^{\mathcal{N} \times 1}$. By using the property $z_{i,n} = z_{n,i}^*$ and basing on the fact that $\mathcal{Z}$ is a positive semi-definite matrix, we have $-\theta^{\mathrm{H}}\mathcal{Z}\theta + \theta^{\mathrm{H}}\omega + \omega^{\mathrm{H}}\theta = 2\mathrm{Re}\left\{\theta_i^*\omega_i + \sum_{n=1,n\neq i}^{\hat{\mathcal{N}}}\theta_n^*\omega_n\right\} - \theta_i^* z_{i,i}\theta_i - 2\mathrm{Re}\left\{\sum_{\substack{n=1\\n\neq i}}^{\hat{\mathcal{N}}}\theta_i^* z_{i,n}\theta_n\right\} - \sum_{\substack{m=1\\m\neq n}}^{\hat{\mathcal{N}}}\sum_{\substack{p=1\\p\neq i}}^{\hat{\mathcal{N}}}\theta_m^* z_{m,p}\theta_p$. Then, $f_7(\theta)$ can be recast as

$$f_8(\theta) = \sum_{i=1}^{\hat{\mathcal{N}}} 2\mathrm{Re}\left\{\theta_i^*\mu_i\right\} + \xi, \tag{22}$$

where $\mu_i = \omega_i - \sum_{\substack{n=1\\n\neq i}}^{\hat{\mathcal{N}}} z_{i,n}\theta_n$ and $\xi = 2\mathrm{Re}\left\{\sum_{n=1,n\neq i}^{\hat{\mathcal{N}}}\theta_n^*\omega_n\right\} - \sum_{\substack{m=1\\m\neq n}}^{\hat{\mathcal{N}}}\sum_{\substack{p=1\\p\neq i}}^{\hat{\mathcal{N}}}\theta_m^* z_{m,p}\theta_p - \theta_i^* z_{i,i}\theta_i$, where ξ is the irreverent constant term with regard to θ_i (e.g., $\theta_i^* z_{i,i}\theta_i = z_{i,i}|\theta_i|^2 = z_{i,i}$), which do not affect the optimal value of θ_i. Therefore, we can only investigate $\mathrm{Re}\left\{\theta_i^*\mu_i\right\}$ for optimizing θ_i and sequentially optimize each element while fixing the remaining $\hat{\mathcal{N}} - 1$ elements. Problem 18 can be equivalently transformed as

$$\max_{\theta_i} \quad \mathrm{Re}\left\{\theta_i^*\mu_i\right\} \tag{23}$$

$$\text{s.t.} \quad |\theta_i| = 1, \tag{23a}$$

An equivalent expression for Problem 23 is given by

$$\max_{\phi_i} \quad \cos\left(-\phi_i + \eta_i\right) \tag{24}$$

$$\text{s.t.} \quad \phi_i \in [0, 2\pi], \tag{24a}$$

where η_i and $-\phi_i$ are the phases of μ_i and θ_i^*, respectively. Consequently, Problem 24 has a closed-form optimal solution whose phase is given by $\phi_i = \eta_i, \forall i \in \hat{\mathcal{N}}$. Consequently, we have

$$\theta_i = e^{j\eta_i}, \forall i \in \hat{\mathcal{N}}. \tag{25}$$

Based on the above discussions, the procedure of sequentially optimizing $\theta_1, \theta_2, \cdots, \theta_{\hat{\mathcal{N}}}$ and then repeatedly until convergence is attained. The complexity of the sequential optimization (SO) algorithm is $\mathcal{O}\left(\mathcal{I}_{\mathrm{SO}}\mathcal{R}^2\mathcal{N}^2\right)$, where $\mathcal{I}_{\mathrm{SO}}$ denotes the number of iterations when the SO algorithm converges.

The proposed AI algorithm is summarized in Algorithm 1, and we have the following lemma

Lemma 1. *The proposed Algorithm 1 is guaranteed to converge to locally optimal solutions.*

Proof. The proof is similar to that of [14], hence it is omitted for simplicity.

4   Numerical Simulation

As follows, simulation results are provided to evaluate the performance of the proposed alternating iteratively (AI) algorithm to evaluate its effectiveness and

---

Algorithm 1. Proposed AI algorithm

Input: $p_{a,k}^{(1)}, \forall \{a,k\}, \Theta^{(1)}$; threshold ε.

1: **Update** $\alpha_k^{(u+1)}, \forall k \in \mathcal{K}$ by using (7);
2: **Update** $\beta_k^{(u+1)}, \forall k \in \mathcal{K}$ by using (8);
3: **Update** $p_{a,k}^{(u+1)}, \forall a \in \mathcal{A}, \forall k \in \mathcal{K}$ by using (12);
4: **Update** $\Theta_i^{(u+1)}, \forall i \in \hat{\mathcal{N}}$ by using (25);
5: **IF** $\left| R^{(u+1)} - R^{(u)} \right| / R^{(u)} \leq v$, **terminate**;
  Otherwise, set $u = u + 1$ and go to Step 1.

---

show the performance gains achieved by the distributed IRSs. We consider six-APs and four-UEs are equipped with single antenna, and three-IRSs are composed of 60 PSs, and APs, UEs, and IRSs are randomly distributed in a circle with a radius of 100 m. Besides, we assumed that a uniform planar array (UPA) at the IRS. The large-scale path loss is given as $L\left(d\right) = L_0 + 10 \log \left(\frac{d}{D_0}\right)^{\rho}$, where L_0 is the path-loss at the reference distance $D_0 = 1\,\text{m}$, d denotes the distance, and ρ is the path loss exponent. We set the path loss exponent between APs and UEs, between APs and IRSs, and between IRSs and UEs are $\rho_{au} = 3.75$, $\rho_{ai} = 2.2$, $\rho_{iu} = 2.2$, respectively, which is based on the fact that the location of IRSs are appropriately chosen for ensuring a free-space. Meanwhile, the maximum transmit power of APs are $P_{a,\max} = 20\,\text{dBm}, \forall a \in \mathcal{A}$ and the noise power $\sigma^2 = \sigma_k^2 = -80\,\text{dBm}, \forall k \in \mathcal{K}$.

As shown in Fig. 2, we study the sum-rate achieved by the proposed AI algorithm under the different number of PSs, e.g. $\mathcal{N} = 30$ and $\mathcal{N} = 60$. It is observed that the sum-rate achieved by the AI algorithm converges to corresponding stationary point after a few iteration, which demonstrates that the proposed AI algorithm has good convergence behaviour. Besides, the convergence speed of "AI alg. $\mathcal{N} = 60$" is slower than "AI alg. $\mathcal{N} = 30$", which indicated that the convergence speed is sensitive to the size of IRSs.

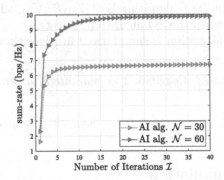

Fig. 2. Sum-rate versus the number of iterations.

Moreover, we introduce two baseline schemes to validate the performance of "AI alg.", i.e., "without IRS" and "random PSs" [4–8]. As illustrated in Fig. 3(a), we investigate the impact of the number of PSs of IRS on the performance. It is observed that the "AI alg." scheme achieves a significant performance gain compares with benchmark schemes. Besides, the "random PSs" scheme achieves a higher sum-rate than the "without IRS" scheme, which demonstrates that the IRS can improve the performance of the cell-free system even though the passive reflecting beamforming of IRSs are without optimizing. As shown in Fig. 3(b) illustrates the sum-rate achieved by the "AI alg." scheme and benchmark schemes over the maximum transmit power of APs. It is observed that the "AI alg." scheme outperforms both the "random PSs" and the "without IRS" scheme considerably, which also indicates the advantages of deploying IRSs into cell-free networks.

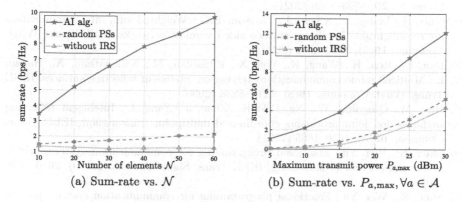

(a) Sum-rate vs. $\mathcal{N}$ (b) Sum-rate vs. $P_{a,max}, \forall a \in \mathcal{A}$

Fig. 3. Comparison of performance.

5   Conclusion

In this paper, we proposed an efficient jointly power allocation and passive reflecting beamforming design framework to maximize the sum-rate of the IRSs-assisted cell-free networks. Due to the problem was non-convex, we first transformed the original problem to an equivalent and tractable form, and then decomposed the reformulated problem into two subproblems. Finally, the subproblems were solved in an alternating optimization manner. The proposed AI algorithm was guaranteed to converge to locally optimal solutions. Simulation results demonstrated that IRSs can improve the performance of the conventional cell-free networks significantly and the proposed AI algorithm achieves considerably performance gains than benchmark schemes.

References

1. Nayebi, E., Ashikhmin, A., Marzetta, T.L., Yang, H., Rao, B.D.: Precoding and power optimization in cell-free massive mimo systems. IEEE Trans. Wirel. Commun. **16**(7), 4445–4459 (2017)
2. Ngo, H.Q., Ashikhmin, A., Yang, H., Larsson, E.G., Marzetta, T.L.: Cell-free massive mimo versus small cells. IEEE Trans. Wirel. Commun. **16**(3), 1834–1850 (2017)
3. Hu, X., Zhong, C., Zhu, Y., Chen, X., Zhang, Z.: Programmable metasurface-based multicast systems: design and analysis. IEEE J. Sel. Areas Commun. **38**(8), 1763–1776 (2020)
4. Ning, B., et al.: Joint power allocation and passive beamforming design for IRS-assisted physical-layer service integration. IEEE Trans. Wirel. Commun. **20**, 7286–7301 (2021)
5. Xiu, Y., et al.: Reconfigurable intelligent surfaces aided mmwave noma: joint power allocation, phase shifts, and hybrid beamforming optimization. IEEE Trans. Wirel. Commun. **20**, 8393–8409 (2021)
6. Guo, H., Liang, Y., Chen, J., Larsson, E.G.: Weighted sum-rate maximization for reconfigurable intelligent surface aided wireless networks. IEEE Trans. Wirel. Commun. **19**(5), 3064–3076 (2020)
7. Pan, C., Ren, H., Wang, K., Wei, X., Elkashlan, M., Nallanathan, A., Hanzo, L.: Multicell mimo communications relying on intelligent reflecting surfaces. IEEE Trans. Wirel. Commun. **19**(8), 5218–5233 (2020)
8. Hua, M., Qingqing, W., Ng, D.W.K., Zhao, J., Yang, L.: Intelligent reflecting surface-aided joint processing coordinated multipoint transmission. IEEE Trans. Commun. **69**(3), 1650–1665 (2021)
9. Shen, K., Wei, Yu.: Fractional programming for communication systems-part i: power control and beamforming. IEEE Trans. Signal Process. **66**(10), 2616–2630 (2018)
10. Shen, K., Wei, Yu.: Fractional programming for communication systems-part ii: uplink scheduling via matching. IEEE Trans. Signal Process. **66**(10), 2631–2644 (2018)
11. Shi, Q., Razaviyayn, M., Luo, Z., He, C.: An iteratively weighted mmse approach to distributed sum-utility maximization for a mimo interfering broadcast channel. IEEE Trans. Signal Process. **59**(9), 4331–4340 (2011)
12. Soltanalian, M., Stoica, P.: Designing unimodular codes via quadratic optimization. IEEE Trans. Signal Process. **62**(5), 1221–1234 (2014)
13. Cui, G., Xianxiang, Yu., Foglia, G., Huang, Y., Li, J.: Quadratic optimization with similarity constraint for unimodular sequence synthesis. IEEE Trans. Signal Process. **65**(18), 4756–4769 (2017)
14. He, J., Yu, K., Shi, Y.: Coordinated passive beamforming for distributed intelligent reflecting surfaces network. In: 2020 IEEE 91st Vehicular Technology Conference (VTC2020-Spring), pp. 1–5 (2020)

Deep Learning and Vehicular Communication

Text Error Correction Method in the Construction Industry Based on Transfer Learning

Zhenguo Hou[1]([✉]), Weitao Yang[1], Haiying He[1], Peicong Zhang[1], Ziyu Wang[2], and Xiaosheng Ji[3]

[1] China Construction Seventh Engineering Bureau Co., Ltd., Zhengzhou 450000, Henan, China
houzhenguo@cscec.com
[2] Hohai University Industrial Technology Research Institute, Changzhou 213022, Jiangsu, China
[3] College of IoT Engineering, Hohai University, Changzhou 213022, Jiangsu, China

Abstract. Text error correction is of great value in the review of texts in the construction industry. For construction industry texts, which are compound texts with multi-domain proper nouns, the lack of labeled data leads to poor error correction algorithms based on deep learning. For this reason, this paper proposes a text error correction method in the construction industry based on transfer learning. Based on the pre-trained BERT model, we transfer some parameters to the target error correction model after unsupervised training by unlabeled related field dataset, and then retrain the model through the training samples of the construction document corpus dataset to obtain better error correction effects. Meanwhile, we dynamically adjust the pre-training task in transfer learning to improve the performance of the word order correction task. Experimental results show that our proposed model has higher precision rate, recall rate and lower false positive rate in the error correction task than other models.

Keywords: Text error correction · Transfer learning · BERT model · Multi-domain text

1 Introduction

The construction industry [1] is a pillar industry of the national economy and has an irreplaceable position and role in social production and material life. In recent years, with the rapid development of our country's construction industry, the scale of the construction industry has continued to expand, and the demand of the construction market has also increased, such as industrial buildings, civil buildings and public buildings. The building construction scheme plays a pivotal role in the building construction process. It is necessary to consider multiple factors on the basis of a large number of theoretical analysis and field surveys to ensure the feasibility and rationality of the scheme. With the development of informatization in the construction industry, the building construction

H. Gao et al. (Eds.): ChinaCom 2021, LNICST 433, pp. 277–290, 2022.
https://doi.org/10.1007/978-3-030-99200-2_22

scheme is also transitioning from paper to electronic, but it inevitably brings about the problem of text errors in the scheme. These errors make the text information inaccurate and unauthoritative, and can even lead to grammatical and semantic errors in sentences, and reduce the readability of construction industry texts. In severe cases, ambiguity will occur and cause losses. For massive construction industry text data, it is unrealistic to rely solely on manual review. Therefore, it is of great significance to automatically detect and correct the content of construction industry texts, which is beneficial to improve the correctness and standardization of construction industry text language use and reduce the manual workload.

The BERT model is a deep learning language model with strong generalization ability and high accuracy among the current text error correction [2] models. Therefore, more and more scholars use the BERT model to perform text error correction, that is, first perform unsupervised training on a large amount of unlabeled data corpus which can learn a lot of a priori language, syntax, word meaning and other information, and then use the learned features for downstream tasks. However, for machine learning methods [3] to achieve good results, a large amount of data is required. In fact, compared with general documents, these composite professional documents such as building construction scheme documents, the amount of available data is relatively small, especially labeled data samples are more rare, which makes the training samples insufficient for complex machine learning algorithms to train and obtain a reliable error correction model. How to analyze and mine small-scale sample data is one of the most challenging frontiers of machine learning. In this context, transfer learning [4] came into being, and its advantage is that it avoids the restriction that there must be enough available training samples to be able to learn a good generative model. Therefore, transfer learning can mine the essential characteristics and internal structure of data between two different but related fields, which enables the transfer and reuse of supervised parameter information between fields.

This paper aims to solve the problem of the lack of data in the multi-domain compound document represented by the building construction scheme document, which causes the difficulty of the text error correction task. We base on the BERT model, and adopt the transfer learning method. After learning from the dataset of other related fields, we retrain the model. The model finally achieves better performance in the correction of construction industry texts.

2   Related Work

2.1   Text Error Correction

Natural Language Processing (NLP) is the study of theories and methods for information interaction between humans and machines using natural language. Text error correction is one of the most common tasks of NLP. In our work, we focus on Chinese text error correction.

Currently, there are many methods for correcting errors in Chinese texts. Traditional Chinese text error correction technology first uses rules, statistical models and other methods to check errors, and then perform error correction. Zhang et al. [5] proposed a text error correction method based on the combination of rules and statistical language models. Duan et al. [6] proposed to generate a corresponding confusion set for Chinese keywords, and then sorted the entries in the confusion set by features, and got the final error correction result. With the development of deep learning, various neural networks emerge in an endless stream, among which there are many models applied to Chinese text error correction. Yuan et al. [7] used a sequence-based decoding model to model the problem, and used the characteristics of long-term memory of the LSTM model to obtain the semantic information of the sentence, and then used the sequence decoding model for error correction. Fu et al. [8] used the Transformer model method based on the self-attention mechanism, and Li et al. [9] introduced the attention mechanism based on the Seq2Seq model to implement error correction tasks.

2.2 Transfer Learning

As a branch of machine learning, transfer learning applies the knowledge learned in the source domain to a different but related target domain to achieve cross-domain learning. Unlike traditional machine learning, transfer learning does not need to have the same data distribution in the source domain and target domain, and it can achieve good results even when the amount of data in the target domain is relatively small. Therefore, transfer learning has great advantages in solving the problem of insufficient data in the target domain. According to the transfer scenario, transfer learning can be divided into three categories: inductive transfer learning, transductive transfer learning and unsupervised transfer learning. According to the transfer content, transfer learning can be divided into four categories: sample transfer, feature transfer, parameter transfer and relationship transfer.

After the introduction of transfer learning, it has received widespread attention. Zhuang et al. [10] proposed a transfer learning method based on the supervised representation of a dual-coding layer autoencoder. Tan et al. [11] explored a new type of transfer learning problem oriented to remote domain transfer learning, and realized transfer learning when the target domain is completely different from the source domain. Ma et al. [12] proposed an algorithm which effectively performs the policy transfer using preference only and give the theoretical analysis on the convergence about the proposed algorithm. Fan et al. [13] proposed a transfer learning algorithm based on discriminative Fisher embedding and adaptive maximum mean discrepancy constraints.

3 BERT Model

3.1 Model Architecture

BERT [14] is a pre-trained language model proposed by Google for various NLP tasks, which has the ability of bidirectional encoding and feature extraction. Unlike traditional models, BERT uses bidirectional Transformers as the feature extractors for the language model. A large number of studies have shown that the two-way language model has a more accurate understanding of semantics during training than the one-way language model. With the emergence of Masked LM technology, the training of two-way language models was started. Since the goal of BERT is to generate language model, BERT uses the encoder part of Transformer. The BERT model architecture is divided into three layers: Embedding Layer, Transformer Layer and Prediction Layer, as shown in Fig. 1.

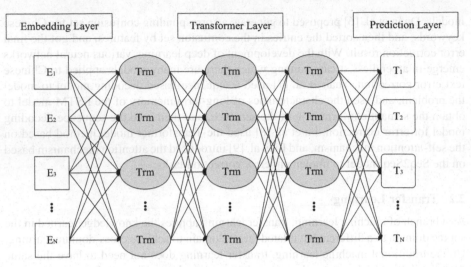

Fig. 1. BERT model architecture.

3.2 Pre-training Task

The highlight of the BERT model is focused on the task part of the pre-training process, including two unsupervised prediction tasks: Masked LM (MLM) and Next Sentence Prediction (NSP), as shown in Fig. 2. The MLM task is similar to cloze filling. For the input sentence, 15% of the tokens are randomly masked, and the masked tokens are predicted by an unsupervised learning method. Since there is no [MASK] token in the fine-tuning stage, there is a mismatch between pre-training and fine-tuning. In order to minimize this impact, the BERT model adopts the following strategy: For these 15% tokens, there is 80% probability to replace with [MASK], 10% probability to replace with random tokens and 10% probability to not change the current tokens. The essence of the NSP task is a binary classification task. In the pre-training process of the BERT model, the model receives paired sentences (such as A and B) as input, and then predict whether B is the next sentence of A. There is 50% probability that B is the next sentence of A, and 50% probability that B is a random sentence in the corpus.

When training the BERT model, in order to minimize the combined loss function of the two tasks, we jointly train MLM and NSP. The loss function is defined as

$$L(\theta, \theta_1, \theta_2) = L_1(\theta, \theta_1) + L_2(\theta, \theta_2) \tag{1}$$

Where θ is the parameter of the encoder part in the BERT model, θ is the parameter in the output layer connected to the encoders in the MLM task, and θ is the classifier parameter connected to the encoders in the NSP task. For the loss function of the first part, if the set of words to be masked is M, because it is a multi-classification problem with dictionary size |V|, the loss function used is a negative log-likelihood function.

And because it needs to be minimized, the loss function in the first part is equivalent to maximizing the log-likelihood function. The mathematical expression is

$$L_1(\theta, \theta_1) = -\sum_{i=1}^{M} \log p(m = m_i | \theta, \theta_1), m_i \in [1, 2, \ldots, |V|] \qquad (2)$$

For the loss function of the second part, since the NSP task is also a classification problem, the loss function expression is

$$L_2(\theta, \theta_2) = -\sum_{j=1}^{N} \log p(n = n_j | \theta, \theta_2), n_j \in [IsNext, NotNext] \qquad (3)$$

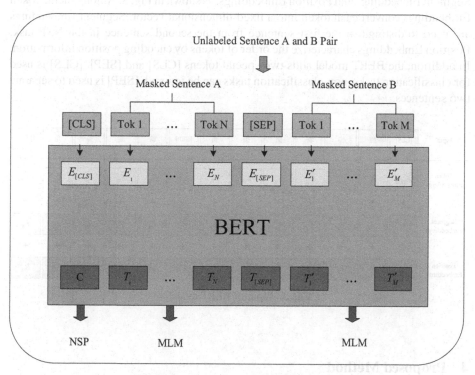

Fig. 2. Pre-training stage of BERT model.

However, in recent years, the necessity of the NSP task has been continuously questioned. As a model to question the NSP task earlier, RoBERTa [15] removes the NSP task. Research shows that if the input is FULL-SENTENCES or DOC-SENTENCES, the RoBERTa model can achieve the same or even higher accuracy on tasks such as SQuAD, MNLI-m and SST-2. Therefore, in the field of text error correction, we can achieve the same word order correction ability by removing the NSP task in the pre-training stage of the BERT model. This paper introduces the pre-training coefficient

λ, dynamically adjust pre-training tasks at different stages, and defines the following optional loss function

$$L'(\theta, \theta_1, \theta_2) = L_1(\theta, \theta_1) + \lambda L_2(\theta, \theta_2), \quad \lambda = 0 \ or \ 1 \tag{4}$$

After the pre-training is over, we get the BERT pre-training model. For different NLP tasks, fine-tuning is performed on the basis of the BERT pre-training model. Insert the input and output of a specific task into the BERT, and use Transformer's powerful attention mechanism to simulate downstream tasks.

3.3 Input of BERT Model

The input of the BERT model contains the sum of three vectors: Token Embeddings, Segment Embeddings and Position Embeddings, as shown in Fig. 3. Among them, Token Embeddings convert each token into a fixed-dimensional vector, Segment Embeddings are used to distinguish the first sentence from the second sentence in the NSP task, Position Embeddings characterize the order of tokens by encoding position information. In addition, the BERT model adds two special tokens [CLS] and [SEP]. [CLS] is used for classification tasks, non-classification tasks can be ignored. [SEP] is used to separate two sentences.

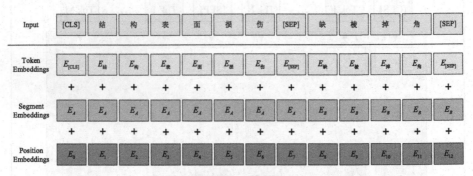

Fig. 3. Input of BERT model.

4   Proposed Method

4.1   Task Description

We focus on four types of text error in the text error correction task in construction industry texts: missing word completion, redundant word removal, word selection correction and word order correction. Examples of error types are shown in Table 1. There may be no errors in the input sentence or one or more errors. If no error is found, the program returns "correct", otherwise the program returns "wrong" and displays the location of the error and the sentence after the error correction.

Table 1. Error examples and types.

Types of text error	Original sentence	Corrected sentence
Missing word	基坑周围地面堆是否有超载情况	基坑周围地面堆载是否有超载情况
Redundant word	为保证桩的垂直度, 钢板桩桩要沿导向施打	为保证桩的垂直度, 钢板桩要沿导向施打
Word selection	复和单面自粘防水卷材	复合单面自粘防水卷材
Word order	大体积裂缝混凝土原因很复杂	大体积混凝土裂缝原因很复杂

4.2 Datasets

4.2.1 Construction Document Corpus Dataset

In our work, we use 41 building construction scheme documents from the construction industry, containing a total of 9,070 sentences. These sentences cover proper nouns in different fields, including geography, geology, chemical materials, construction equipments, building regulations, etc. We select 5,400 representative sentences and divide them into five groups, each with about 1,000 sentences. No modification is made to one of the groups, and the other four groups are artificially modified according to the above four types of errors. We name these corpora as the construction document corpus dataset (CDCD) and regard it as the target dataset, which is used to verify the model's ability in text error correction in the construction industry.

4.2.2 Unlabeled Related Field Dataset

Since the target task has fewer datasets and is rich in proper nouns from different fields, we try to use the unlabeled data of related fields by transfer learning, and fine-tune the parameters through training to make the model get the ability of vocabulary recognition closer to the target task. We select 1,538 professional documents in five fields, including geography, geology, chemical materials, construction equipments and building regulations, from Chinese literature databases such as CNKI and Wanfang, with a total of about 420,000 words, forming an unlabeled related field dataset (URFD).

4.3 Our Model

In this paper, we use the pre-trained BERT model based on Chinese corpus (including Renming daily 2014 edition data, CGED three-year competition data and some Chinese Wikipedia data), and adopt the progressive transfer learning method. The overall structure of our work is shown in Fig. 4.

In the first stage, we apply the pre-trained BERT model of massive Chinese corpus to the unlabeled related field dataset, so that the model can obtain vocabulary information from different subdivision fields. In the second stage, we use the training samples in the construction document corpus dataset for the retraining of the model after transfer learning, making it more targeted at construction industry texts involving multiple fields, and improving the ability to detect and correct text errors.

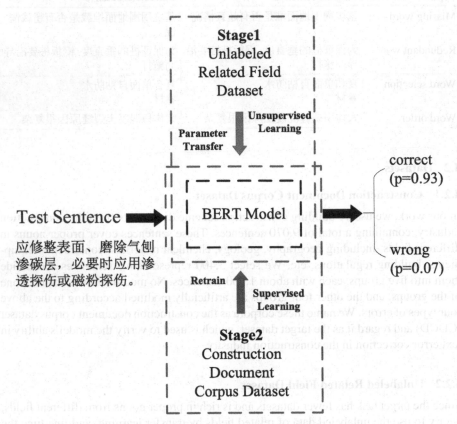

Fig. 4. The overall structure of our work.

As mentioned in Sect. 3.2, the BERT model has two pre-training tasks. The MLM task predicts the masked tokens, and the NSP task understands the relationship between sentences to predict the order between two sentences. After unsupervised training on a large number of unlabeled related field dataset, the ability of MLM has been significantly improved, and it can accurately predict proper nouns from different fields in the target dataset. However, since the unlabeled related field dataset is the entire article from each field, the sentence context in the same field has strong relevance, while building construction scheme documents of the target dataset are compound documents involving multiple fields, and the upper and lower sentences may come from different fields. In this case, the NSP task of the first stage will have a negative effect. Eventually, the correct word order will be recognized as word order error.

Inspired by the RoBERT model, in order to solve the problem of word order false positives in our proposed model, we make choices on the NSP task at different stages. In the first stage, when unsupervised training is carried out using unlabeled related field dataset, for formula (4), we set $\lambda = 0$ which removes the NSP task, and only train MLM to strengthen the model's learning of proper nouns in different fields. At the same time, it ensures that the model is not affected by strongly related sentences in the same field. In the second stage, for formula (4), we set $\lambda = 1$ to unlock the training of NSP. After the fine-tuning training of the target dataset, the word order false positive rate of the model can be effectively reduced. The transfer learning process of dynamically adjusting the pre-training tasks is shown in the Fig. 5.

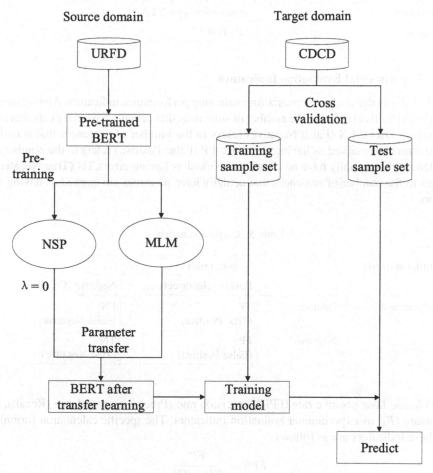

Fig. 5. Flow chart of transfer learning for dynamically adjusting pre-training tasks.

5 Experiments

5.1 Experimental Environment

The experimental environment of this paper is shown in Table 2.

Table 2. Experimental environment.

Experimental environment	Configuration
Operating system	Windows 10
CPU	AMD Ryzen 7 3700X
GPU	NVIDIA GeForce RTX 2070 SUPER
Random access memory	16G
Framework	tensorflow-gpu 2.2.0
Language	Python 3.7

5.2 Experimental Evaluation Indicators

Table 3 shows the confusion matrix for evaluating performance indicators. Among them, TP (True Positive) refers to the number of sentences that actually have errors are marked as having errors. FN (False Negative) refers to the number of sentences that actually have errors are marked as having no errors. FP (False Positive) refers to the number of sentences that actually have no errors are marked as having errors. TN (True Negative) refers to the number of sentences that actually have no errors are marked as having no errors.

Table 3. Confusion matrix.

Confusion matrix		System result	
		Positive (Incorrect)	Negative (Correct)
gold standard	Positive	TP (True Positive)	FN (False Negative)
	Negative	FP (False Positive)	TN (True Negative)

We use false positive rate (FPR), precision rate (Precision), recall rate (Recall), F-Measure (F_1) as experimental evaluation indicators. The specific calculation formulas of these indicators are as follows

$$FPR = \frac{FP}{FP + TN} \tag{5}$$

$$Precision = \frac{TP}{TP + FP} \tag{6}$$

$$Recall = \frac{TP}{TP + FN} \tag{7}$$

$$F_1 = \frac{2 * Precision * Recall}{Precision + Recall} \tag{8}$$

5.3 Experimental Results

For the division of the dataset, we set 70% of the construction document corpus dataset as the training sample set, and the remaining 30% as the test sample set.

In order to verify the accuracy and robustness of the method proposed in this paper, we first experimentally compare our model with commonly used deep learning text error correction models, such as LSTM, Transformer_self-attention, Seq2Seq-attention, etc. We conduct experiments on the following four models through the test samples in the construction document corpus dataset. From Table 4, it can be seen that the transfer learning method proposed in this paper has a higher precision rate than the commonly used deep learning text error correction model. The precision rate on the target dataset reaches 81.4%.

Table 4. Experimental results of different models.

Indicator	
Model	Precision
Our model	81.4%
LSTM	58.7%
Transformer_self-attention	60.2%
Seq2Seq-attention	63.4%

Then we perform ablation experiments with our model and the other three models. The three models are: (1) After freezing the parameters of the pre-trained BERT model, we apply it directly to the target text error correction task, set to M1; (2) After fine-tuning the pre-trained BERT model through the training samples in the construction document corpus dataset, we apply it to the target text error correction task, set to M2; (3) The pre-trained BERT model freezes the parameters after the transfer learning of the relevant field dataset, and we apply it directly to the target text error correction task, set to M3.

It can be seen from Fig. 6 that our proposed method significantly improves the ability of the text error correction task of the target dataset. It can be seen from Fig. 6(a) that because there are more proper nouns from different fields in the target dataset, if the pre-trained BERT model is used for text error correction without other operations, FPR will be as high as 19.8%. If the model is fine-tuned by the target dataset, FPR will drop slightly. This is because the target dataset has fewer training samples and the training effect is not ideal. After the transfer learning by unlabeled related domain dataset, the BERT model learns a large number of proper nouns from different fields. Then we fine-tune the model, the FPR of the model drops significantly, only 9.7%, which improves the robustness of the model. From Fig. 6(b) and Fig. 6(c), it can be seen that the model after transfer learning and fine-tuning improves the error location and error correction capabilities, which is reflected in that compared with the pre-trained BERT model, our model improves the precision rate by 13% and the recall rate by 12.8%. In addition, from the experimental comparison between M2 and M1 and the

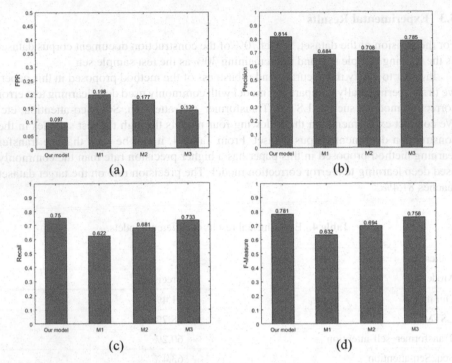

Fig. 6. Results of ablation experiments.

experimental comparison between our proposed model and M3, it can be inferred that fine-tuning for the target dataset is very important. The comparison of the experimental results of M3 and M1 shows that transfer learning of unlabeled related field dataset has a significant improvement in the effect of text error correction tasks, which shows that transfer learning greatly improves the situation of lack of target dataset.

Finally, in order to verify that our method makes significant progress in the word order correction task of the target dataset, we conduct an experimental comparison of the following two models: the model which removes the NSP task in the first stage after transfer learning and fine-tuning (the model proposed in this paper) and the model which reserves the NSP task in the first stage after transfer learning and fine-tuning (set as M4). And we use false positive rate as the evaluation indicator of the word order correction task. We classify the experimental results according to different types of text error, and get the false positive rate of the word order correction task. The final experimental results are shown in Table 5. Compared with M4, the false positive rate of our model in the word order correction task is reduced by 1.8%, which shows that removing the NSP task can effectively reduce the word order false positive.

Table 5. Comparison of the word order correction task in different models.

Model	FPR of word order correction task
Our model	3.72%
M4	5.52%

6   Conclusion

Due to the difficulty to locate the errors of construction industry texts by general methods, this paper proposes a text error correction method in the construction industry based on transfer learning. On the basis of the pre-trained BERT model, we perform transfer learning through unsupervised training of unlabeled related field dataset, and then use the training samples in the construction document corpus dataset to retrain the model after transfer learning to complete text error correction task. At the same time, in order to improve the performance of the word order correction task, we make choices on the NSP task at different stages. Finally, this paper analyzes the experimental results from multiple angles and verifies the accuracy and robustness of the method proposed in this paper.

References

1. You, Z., Feng, L.: Integration of industry 4.0 related technologies in construction industry: a framework of cyber-physical system. IEEE Access **8**, 122908–122922 (2020). https://doi.org/10.1109/ACCESS.2020.3007206
2. Singh, S., Singh, S.: Review of real-word error detection and correction methods in text documents. In: 2018 Second International Conference on Electronics, Communication and Aerospace Technology (ICECA), pp. 1076–1081 (2018). https://doi.org/10.1109/ICECA.2018.8474700
3. Mahima, Y., Ginige, T.N.D.S.: Graph and natural language processing based recommendation system for choosing machine learning algorithms. In: 2020 12th International Conference on Advanced Infocomm Technology (ICAIT), pp. 119–123 (2020). https://doi.org/10.1109/ICAIT51223.2020.9315570
4. Liang, H., Fu, W., Yi, F.: A survey of recent advances in transfer learning. In: 2019 IEEE 19th International Conference on Communication Technology (ICCT), pp. 1516–1523 (2019). https://doi.org/10.1109/ICCT46805.2019.8947072
5. Zhang, Y., Cao, Y., Yu, S.: Chinese text automatic error checking model and algorithm based on the combination of rules and statistics. J. Chin. Inf. Process. **4**, 1–7, 55 (2006)
6. Duan, J., Guan, X.: Research on query error correction method based on the combination of statistics and features. New Technol. Libr. Inf. Serv. **2**, 34–42 (2016)
7. Yuan, Z.: Research on Key Technologies of Text Proofreading on Government Websites. University of Electronic Science and Technology, Chengdu (2018)
8. Fu, K., Huang, J., Duan, Y.: A neural machine translation approach to chinese grammatical error correction. In: CCF International Conference on Natural Language Processing and Chinese Computing (2018)
9. Li, Q., Yang, H., Liu, H., Wu, X., Hu, H.: Research on intelligent error correction of civil aviation message based on Seq2Seq model. In: Yangtze River Information and Communication (2021)

10. Zhuang, F., Cheng, X., Luo, P., Pan, S.J., He, Q.: Supervised representation learning with double encoding-layer autoencoder for transfer learning. ACM Trans. Intell. Syst. Technol. **9**(2), 1–17 (2017)
11. Tan, B., Zhang, Y., Pan, S.J., Yang, Q.: Distant domain transfer learning. In: Thirty−First AAAI Conference on Artificial Intelligence, pp. 2604–2610. AAAI, San Francisco (2017)
12. Ma, X., Ming, X., Sun, F., Liu, P.: Adversarial Task Transfer from Preference. arXiv preprint arXiv:1805.04686 (2018)
13. Fan, Z., Shi, L., Liu, Q., Li, Z., Zhang, Z.: Discriminative fisher embedding dictionary transfer learning for object recognition. IEEE Trans. Neural Netw. Learn. Syst., 1–15 (2021). https://doi.org/10.1109/TNNLS.2021.3089566
14. Kaliyar, R.K.: A multi-layer bidirectional transformer encoder for pre-trained word embedding: a survey of BERT. In: 2020 10th International Conference on Cloud Computing, Data Science & Engineering (Confluence), pp. 336–340 (2020). https://doi.org/10.1109/Confluence47617.2020.9058044
15. Adoma, A.F., Henry, N.-M., Chen, W.: Comparative analyses of bert, roberta, distilbert, and xlnet for text-based emotion recognition. In: 2020 17th International Computer Conference on Wavelet Active Media Technology and Information Processing (ICCWAMTIP), pp. 117–121 (2020). https://doi.org/10.1109/ICCWAMTIP51612.2020.9317379

A Beam Tracking Scheme Based on Deep Reinforcement Learning for Multiple Vehicles

Binyao Cheng[✉], Long Zhao, Zibo He, and Ping Zhang

The Key Lab of Universal Wireless Communication, Ministry of Education,
Beijing University of Posts and Telecommunications (BUPT), Beijing, China
chengbinyao98@163.com

Abstract. In Internet of Vehicles (IoV), beam tracking for multiple
vehicles is a challenging topic due to the nonlinear mobility and inter-
vehicle interference (IVI). This paper considers the scenario that multiple
vehicles with high mobility are periodically served by the radiated beams
of millimeter wave (mmWave) massive multiple-input multiple-output
(MIMO) systems. The main objective is to maximize the probability
of successful information transmission in each beam tracking period,
where successful transmission is defined by signal-to-interference-plus-
noise ratio (SINR) exceeding a threshold. Based on deep reinforcement
learning, we propose a position prediction and joint selection (PPJS)
scheme for beam selection of multiple vehicles in consideration of both
the coverage and IVI. On one hand, long short-term memory (LSTM)
network is employed to predict the future trajectory in upcoming beam
tracking period for providing better beam coverage. On the other hand,
multi-layer perception (MLP) network is designed to select the served
beams by taking into account the IVI, where the vehicles are divided
into clusters and the objective of beam tracking in each cluster is decomposed to reduce the scheme complexity. Simulation results demonstrate
that the proposed PPJS scheme performs better than both the traditional position-based algorithm and deep Q-learning (DQN) algorithm.

Keywords: Internet of Vehicles · Beam tracking · Deep reinforcement
learning

1 Introduction

The Internet of Vehicles (IoV) is one solution to the Ultra-Reliable and Low
Latency Communications (URLLC) scenarios in the next generation mobile communication [1,2]. The main objectives of IoV are to improve the transportation
efficiency, traffic safety and quality of information services on vehicles. In order
to improve the communication quality of high-speed multi-vehicles, millimeter
wave (mmWave) [3] and massive multiple-input multiple-output (MIMO) are
adopted for IoV [4]. On one hand, mmWave frequency has a large bandwidth,

H. Gao et al. (Eds.): ChinaCom 2021, LNICST 433, pp. 291–305, 2022.
https://doi.org/10.1007/978-3-030-99200-2_23

which can provide high transmission rate according to Shannon formula; on the other hand, the extremely narrow beams generated by massive MIMO base station (BS) can provide high transmission gain. However, due to high-speed mobility of vehicles, it is difficult to guarantee the accurate coverage in time; meanwhile, dense vehicle distribution leads to the high inter vehicle interference (IVI) [5]. In order to balance the coverage and IVI for multiple vehicles, beam selection technology is significant for mmWave massive MIMO in IoV [6,7].

In the existing literature, beam selection for mmWave systems has been studied in the context of traditional communication scenario. A single beam selection scheme has been studied by tracking the users' path with analog beamforming architecture, which is realized by extended Kalman filter (EKF) [8]. But conventional filter scheme is mainly fit for stationary channels, which can not be directly applied for the terminals, such as vehicles or users with nonlinear trajectory. Therefore, the classical approaches, e.g., EKF and particle filter (PF), are modified for non-stationary scenarios, as well as reinforcement learning (RL)-based approaches are introduced for typical intersection scenario [9]. Taking into account mobility, a two-phase RL framework is studied to perform adaptive beam selection [10]. In this two-phase scheme, the inner agent adjusts the beam direction based on instantaneous signal-to-interference-plus-noise-ratio (SINR) reward, and the outer agent selects the number of utilized antennas based on long-term SINR reward. However, the IVI among vehicles has not been considered in both [9] and [10].

In this paper, the mmWave and massive MIMO are employed to serve multiple vehicles on the considered road segment. In order to improve the transmission quality by considering both the beam coverage and IVI, we propose a position prediction and joint selection (PPJS) scheme. In the proposed PPJS scheme, long short-term memory (LSTM) network is employed to predict the future vehicle trajectory, based on which we can improve the beam coverage; meanwhile the multi-layer perception (MLP) network is adopted to jointly select beams for minimizing the IVI among vehicles with low complexity, where multiple vehicles are divided into clusters and the objective in each cluster is decomposed. The simulation results indicate that the PPJS algorithm achieves better transmission quality than other compared schemes. In addition, the influence of beam width and the size of discrete beam direction set are discussed under different setups of moving condition, which provides a reference for practical systems.

The rest of the paper is organized as follows. Section 2 illustrates the IoV system model served by mmWave massive MIMO; Sect. 3 formulates the problem for beam selection based on reinforcement learning; The PPJS scheme is proposed in Sect. 4; Sect. 5 presents our simulations and corresponding analysis before summarizing our conclusion in Sect. 6.

2   System Model

As shown in Fig. 1(a), we consider a IoV scenario where the BS employing mmWave and N_{BS} uniform linear array (ULA) antennas simultaneously serves

K single-antenna vehicles. The BS is located on the top of a rectangle road with the coordinate of $[\frac{h_r}{2}, h_d]$, where $[0, 0]$ and $[h_r, 0]$ are the range of single lane road, and h_d is the distance between the BS and lane. The coordinates of the K moving vehicles are $[x_k, 0]$ ($k \in \{1, 2, \cdots, K\}$). We assume that the BS periodically changes the beams to provide coverage and reduce IVI for vehicles. As shown in Fig. 1(b), a beam selection period consists of I uplink intervals and J downlink intervals, and a time interval lasts for T_i seconds. The vehicles could periodically report their locations to the BS by uplink and the BS intelligently selects the beams for the downlink data transmission.

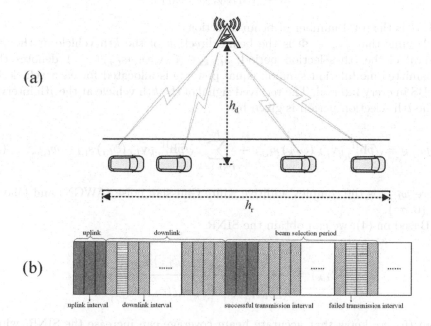

Fig. 1. Beam selection scenario and frame structure.

2.1 Channel Model

We assume that the channels from the BS to the vehicles satisfy the standard Line of Sight (LoS) path-loss model. Therefore, the complex channel vector between the BS and the kth vehicle at the jth downlink interval of the tth beam selection period can be written as

$$\mathbf{h}_{t,j,k} = \alpha_{t,j,k} \left[1, e^{-j\frac{2\pi}{\lambda} d \cos \delta_{t,j,k}}, \cdots, e^{-j\frac{2\pi}{\lambda} d(N_{BS}-1) \cos \delta_{t,j,k}} \right]^{T}, \quad (1)$$

where $\delta_{t,j,k}$ denotes the angle of departure (AoD) of the signal from the BS to the kth vehicle at the jth interval of the tth selection period; $\alpha_{t,j,k}$ is the channel fading coefficient; $d = \frac{\lambda}{2}$ denotes the antenna spacing of ULA antennas with the carrier wavelength λ [11].

2.2 Signal Model

In order to reduce the system complexity, the discrete beams are generally adopted in practical systems, i.e.,

$$\mathbf{v}(\phi) = \frac{1}{\sqrt{N_{\mathrm{BS}}}} \left[1, \mathrm{e}^{-j\frac{2\pi}{\lambda}d\cos\phi}, \cdots, \mathrm{e}^{-j\frac{2\pi}{\lambda}d(N_{\mathrm{BS}}-1)\cos\phi} \right]^{\mathrm{T}}, \qquad (2)$$

where ϕ is the beam direction, which belongs to

$$\boldsymbol{\Phi} = \{\phi_1, \phi_2, \cdots, \phi_M\}, \qquad (3)$$

and M is the total number of beam directions.

Assume that $\phi_{t,k} \in \boldsymbol{\Phi}$ is the beam direction of the kth vehicle at the jth interval of the tth selection period; $s_{t,j,k} \in \mathbb{C}$ with $|s_{t,j,k}| = 1$ denotes the transmitted modulation symbol; equal power p is allocated for each vehicle at the BS in every interval. The received signal of the kth vehicle at the jth interval of the tth selection period is given by

$$y_{t,j,k} = \sqrt{p}\mathbf{h}_{t,j,k}^{\mathrm{H}}\mathbf{v}_{t,k}\left(\phi_{t,k}\right)s_{t,j,k} + \sum_{i=1,i\neq k}^{K} \sqrt{p}\mathbf{h}_{t,j,k}^{\mathrm{H}}\mathbf{v}_{t,i}\left(\phi_{t,i}\right)s_{t,i} + w_{t,j,k}, \qquad (4)$$

where $w_{t,j,k}$ is the complex additive white Gaussian noise (AWGN) and follows $\mathcal{CN}\left(0, \sigma^2\right)$,

Based on (4), we can obtain the SINR

$$\zeta_{t,j,k}(\phi_{t,k}) = \frac{p\left|\mathbf{h}_{t,j,k}^{\mathrm{H}}\mathbf{v}_{t,k}\left(\phi_{t,k}\right)s_{t,j,k}\right|^2}{p\sum_{i=1,i\neq k}^{K}\left|\mathbf{h}_{t,j,k}^{\mathrm{H}}\mathbf{v}_{t,i}\left(\phi_{t,i}\right)s_{t,j,i}\right|^2 + \sigma^2}. \qquad (5)$$

From (5), we know that accurate beam coverage can increase the SINR, while IVI can decrease it.

Because the downlink interval is very short, the SINR can be regarded as a constant. If the SINR exceeds a given threshold ζ_{th}, the received signal can be considered to be decoded successfully, which is indicated by

$$a_{t,j,k}(\phi_{t,k}) = \begin{cases} 1, \ \zeta_{t,j,k}(\phi_{t,k}) \geq \zeta_{\mathrm{th}}, \\ 0, \qquad\qquad \text{else.} \end{cases} \qquad (6)$$

By improving the transmission quality for multiple vehicles, the objective of this paper is to maximize the successful transmission probability, i.e.,

$$\max_{\phi_{t,k}\in\Phi} \left\{ \frac{1}{KJ} \sum_{k=1}^{K} \sum_{j=1}^{J} a_{t,j,k}(\phi_{t,k}) \right\}. \qquad (7)$$

3   Problem Formulation

3.1   Deep Beam Selection Model

We use RL to formulate the beam selection problem. The basic elements of the beam selection problem are given as follows.

- State set: In order to efficiently cover the vehicles, the BS should predict the future vehicle positions according to their history trajectories. Therefore, we assume the state set to be

$$S = \{\mathbf{s}_t \mid \mathbf{s}_t = [\mathbf{x}_{t,1}, \mathbf{x}_{t,2}, \cdots, \mathbf{x}_{t,K}]\}, \tag{8}$$

where $\mathbf{x}_{t,k} = [x_{t-t_\mathrm{h},k}, x_{t-t_\mathrm{h}+1,k}, \cdots, x_{t,k}]^\mathrm{T}$, $x_{t_i,k}$ is the kth vehicle's location at the tth beam selection period, and t_h represents the number of history selection periods.

- Action set: The actions are defined to be a set of beam directions for K vehicles, i.e.,

$$A = \{\mathbf{a}_t \mid \mathbf{a}_t = [\phi_{t,1}, \phi_{t,2}, \cdots, \phi_{t,K}]^\mathrm{T}\}, \tag{9}$$

where $\phi_{t,k}$ denotes the beam direction of the kth vehicle at selection period t.

- Reward: The instant reward of K vehicles is defined as the average successful transmission probability in the tth beam selection period, i.e.,

$$r_t = \frac{1}{KJ} \sum_{k=1}^{K} \sum_{j=1}^{J} a_{t,j,k}(\phi_{t,k}). \tag{10}$$

3.2   Beam Selection Problem

Assume the beam selection strategy is $\pi : S \to A$, and based on the above model, the cumulative successful transmission probability in T beam selection periods can be defined as

$$R^\pi(\tau) = \sum_{t=0}^{T} \gamma^t r_t, \tag{11}$$

where $\tau = [\mathbf{s}_0, \mathbf{a}_0, r_0, \mathbf{s}_1, \mathbf{a}_1, r_1, \cdots, \mathbf{s}_T, \mathbf{a}_T, r_T]$ is a trail under beam selection strategy π, and γ represents the discount factor.

The objective of this paper is to maximize the average cumulative successful transmission probability $J(\pi)$ for any trail τ, i.e.,

$$J(\pi) = \mathbb{E}_{\tau \sim p(\tau)} [R^\pi(\tau)]$$

$$= \mathbb{E}_{\mathbf{s} \sim p(\mathbf{s}_0)} \left[\mathbb{E}_{\tau \sim p(\tau)} \left[\sum_{t=0}^{T} \gamma^t r_t \mid \tau_{\mathbf{s}_0} = \mathbf{s} \right] \right] \tag{12}$$

$$\triangleq \mathbb{E}_{\mathbf{s} \sim p(\mathbf{s}_0)} [V^\pi(\mathbf{s})],$$

where $p(\tau)$ is the probability of beam selection trail. Therefore the optimal beam selection strategy is

$$\pi^* = \arg\max_\pi J(\pi) = \arg\max_\pi V^\pi(\mathbf{s}). \tag{13}$$

4   PPJS Scheme Based on Deep Reinforcement Learning

As shown in Fig. 2(a) to solve the formulated problem in Sect. 3, we propose
a PPJS scheme by considering both the coverage and IVI, consists of three
modules. The state (8) is first fed into trajectory prediction module to obtain the
future positions; Based on the predicted and current positions, the K vehicles
can be divided into U clusters in vehicle division module; In beam selection
module, the joint objective of each cluster is decomposed to simplify the selection
complexity. The three modules will be discussed in details as follows.

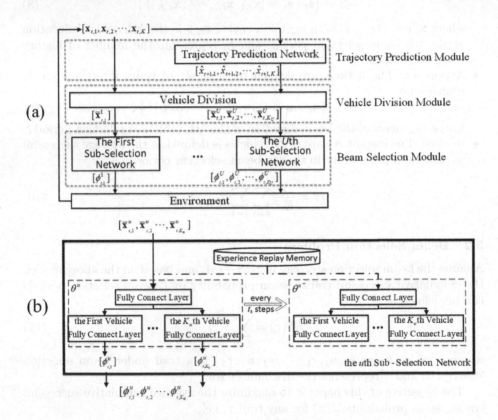

Fig. 2. The proposed PPJS scheme.

4.1   Trajectory Prediction Module

In order to ensure the accurate beam coverage, the positions of vehicles at the
beginning of current and next beam selection periods should be known. So we
use the LSTM network to predict the future positions according to the historical
trajectory in this module.

The input of this module is a series of historical positions, which is given in (8), and the output is the position at the beginning of the next beam selection period, i.e.,

$$[\hat{x}_{t+1,1}, \hat{x}_{t+1,2}, \cdots, \hat{x}_{t+1,K}]. \tag{14}$$

The distance loss function for LSTM training is adopted in our scheme, i.e.,

$$\text{Loss} = \frac{1}{K} \sum_{k=1}^{K} (x_{t+1,k} - \hat{x}_{t+1,k})^2, \tag{15}$$

where $x_{t+1,k}$ is the future position in practical.

In order to predict an accurate future position for optimal beams, the trajectory prediction module should be pretrained before being embedded into the PPJS scheme based on practical data set. Here we should note that the beams are not directly selected to point to the current position or the next position. Because the vehicles are moving during each selection period, each beam generally points to the position that between the beginning and ending positions. However, the accurate position should be selected by considering both the coverage gain and IVI, which is achieved by beam selection module.

4.2   Vehicle Division Module

According to (10), all the IVIs should been taken into account so the scheme needs to jointly select beams for all the served vehicles, which leads to overwhelming computational complexity. To simplify the PPJS scheme, the K vehicles are decomposed into U clusters according to vehicle spacing, where the IVI between any vehicles in different clusters is small and can be ignored. Specifically, the IVI is only considered in each cluster.

Therefore, vehicle division module first combines the kth vehicle's current position $x_{t,k}$ in state (8) and predicted position $\hat{x}_{t+1,k}$ in (14) into $\overline{\mathbf{x}}_{t,k} = [x_{t,k}, \hat{x}_{t+1,k}]^{\mathrm{T}}$, and then decompose the K vehicles into U clusters. The uth cluster's state can be written as

$$\mathbf{s}_t^u = \left[\overline{\mathbf{x}}_{t,1}^u, \overline{\mathbf{x}}_{t,2}^u, \cdots, \overline{\mathbf{x}}_{t,K^u}^u\right], \tag{16}$$

where K^u is the number of vehicles in the uth cluster. Hence, each cluster can respectively select their own joint beams

$$\mathbf{a}_t^u = \left[\phi_{t,1}^u, \phi_{t,2}^u, \cdots, \phi_{t,K^u}^u\right]^{\mathrm{T}}. \tag{17}$$

Moreover, the instance reward (10) can also be decomposed into each cluster's average successful transmission probability r_t^u in the tth selection period, i.e.,

$$
\begin{aligned}
r_t &= \frac{1}{KJ} \sum_{k=1}^{K} \sum_{j=1}^{J} a_{t,j,k}(\phi_{t,k}) \\
&= \frac{1}{KJ} \sum_{u=1}^{U} \sum_{k^u=1}^{K^u} \sum_{j=1}^{J} a_{t,j,k^u}^u(\phi_{t,k^u}^u) \\
&= \sum_{u=1}^{U} \frac{K^u}{K} \frac{1}{K^u J} \sum_{k^u=1}^{K^u} \sum_{j=1}^{J} a_{t,j,k^u}^u(\phi_{t,k^u}^u) \\
&= \sum_{u=1}^{U} p^u r_t^u,
\end{aligned}
\tag{18}
$$

where $p^u = K^u/K$ is the proportion the number of vehicles in the uth cluster to the total number of vehicles K.

Then the original objective in (12) can also be decomposed into multiple sub-objective $V^{\pi^u}(\mathbf{s}^u)$ for multiple clusters, i.e.,

$$
\begin{aligned}
V^{\pi}(\mathbf{s}) &= \mathbb{E}_{\tau \sim p(\tau)} \left[\sum_{t=0}^{T} \gamma^t r_t \mid \tau_{\mathbf{s}_0} = \mathbf{s} \right] \\
&= \mathbb{E}_{\tau \sim \prod_{u=1}^{U} p(\tau^u)} \left[\sum_{t=0}^{T} \gamma^t \sum_{u=1}^{U} p^u r_t^u \mid \tau_{\mathbf{s}_0} = \mathbf{s} \right] \\
&= \sum_{u=1}^{U} p^u \mathbb{E}_{\tau^u \sim p(\tau^u)} \left[\sum_{t=0}^{T} \gamma^t r_t^u \mid \tau_{\mathbf{s}_0^u}^u = \mathbf{s}^u \right] \\
&= \sum_{u=1}^{U} p^u V^{\pi^u}(\mathbf{s}^u),
\end{aligned}
\tag{19}
$$

where $\tau^u = [\mathbf{s}_0^u, \mathbf{a}_0^u, r_0^u, \mathbf{s}_1^u, \mathbf{a}_1^u, r_1^u \cdots, \mathbf{s}_T^u, \mathbf{a}_T^u, r_T^u]$ is the uth cluster's trail under beam selection strategy π^u.

Therefore, the optimal beam vectors for K vehicles can be alternatively obtained by selecting the optimal beams for the vehicles in each cluster, i.e.,

$$
\begin{aligned}
\pi^* &= [\pi^{1*}, \pi^{2*}, \cdots, \pi^{U*}] \\
&= [\arg\max_{\pi^1} V^{\pi^1}(\mathbf{s}^1), \arg\max_{\pi^2} V^{\pi^2}(\mathbf{s}^2), \cdots, \arg\max_{\pi^U} V^{\pi^U}(\mathbf{s}^U)].
\end{aligned}
\tag{20}
$$

After vehicle division module, the optimal beam selection for the vehicles only to be studied in each cluster.

4.3   Beam Selection Module

According to (17), the increasing action set of each cluster causes gradient explosion. Based on Bellman equation $V^{\pi}(\mathbf{s}) = \mathbb{E}_{\mathbf{a} \sim \pi(\mathbf{a}|\mathbf{s})} Q^{\pi}(\mathbf{s}, \mathbf{a})$

and value-decomposition networks (VDN) decomposing equation $Q^{\pi}(\mathbf{s}, \mathbf{a}) \approx \sum_{n=1}^{N} Q_{n}^{\pi}(\mathbf{s}, a_{n})$ under the condition that $r(\mathbf{s}, \mathbf{a}) = \sum_{n=1}^{N} r(\mathbf{s}, a_{n})$, we can reduce the complexity of the PPJS scheme by decomposing the objective of each cluster into multiple objectives of single vehicle, i.e.,

$$
V^{\pi^{u}}(\mathbf{s}^{u})
$$

$$
= \mathbb{E}_{\tau_{0:T}^{u} \sim p(\tau^{u})} \left[\sum_{t=0}^{T} \gamma^{t} r_{t}^{u} \mid \tau_{\mathbf{s}_{0}^{u}}^{u} = \mathbf{s}^{u} \right]
$$

$$
= \mathbb{E}_{\tau_{0:T}^{u} \sim p(\tau^{u})} \left[r(\mathbf{s}^{u}, \mathbf{a}^{u}, \mathbf{s}^{u'}) + \gamma \sum_{t=1}^{T} \gamma^{t-1} r_{t}^{u} \mid \tau_{\mathbf{s}_{1}^{u}}^{u} = \mathbf{s}^{u'} \right]
$$

$$
= \mathbb{E}_{\mathbf{a}^{u} \sim \pi^{u}(\mathbf{a}^{u}|\mathbf{s}^{u})} \mathbb{E}_{\mathbf{s}^{u'} \sim \pi^{u}(\mathbf{s}^{u'}|\mathbf{s}^{u}, \mathbf{a}^{u})}
$$

$$
\left[\frac{1}{K^{u} J} \sum_{k^{u}=1}^{K^{u}} \sum_{j=1}^{J} a_{t,j,k^{u}}^{u}(\phi_{t,k^{u}}^{u}) + \gamma \mathbb{E}_{\tau_{1:T}^{u} \sim p(\tau^{u})} \sum_{t=1}^{T} \gamma^{t-1} r_{t}^{u} \mid \tau_{\mathbf{s}_{0}^{u}}^{u} = \mathbf{s}^{u'} \right]
\tag{21}
$$

$$
= \frac{1}{K^{u}} \sum_{k^{u}=1}^{K^{u}} \mathbb{E}_{\mathbf{a}^{u} \sim \pi^{u}(\mathbf{a}^{u}|\mathbf{s}^{u})}
$$

$$
\mathbb{E}_{\mathbf{s}^{u'} \sim \pi^{u}(\mathbf{s}^{u'}|\mathbf{s}^{u}, \mathbf{a}^{u})} \left[r_{k_{u}}^{u} + \gamma \mathbb{E}_{\tau_{1:T}^{u} \sim p(\tau^{u})} \sum_{t=1}^{T} \gamma^{t-1} r_{t}^{u} \mid \tau_{\mathbf{s}_{0}^{u}}^{u} = \mathbf{s}^{u'} \right]
$$

$$
\triangleq \frac{1}{K^{u}} \sum_{k^{u}=1}^{K^{u}} Q_{k_{u}}^{\pi^{u}}(\mathbf{s}^{u}, \mathbf{a}^{u})
$$

$$
\approx \frac{1}{K^{u}} \sum_{k^{u}=1}^{K^{u}} Q_{k_{u}}^{\pi^{u}}(\mathbf{s}^{u}, \phi_{k_{u}}^{u}),
$$

where $r_{k_{u}}^{u}$ is the k_{u}th reward in the uth cluster, $\phi_{k_{u}}^{u}$ is the k_{u}th beam direction in the uth cluster. According to (21), we can know that the optimal beam selection strategy for each cluster is equivalent to choose the optimal beam for each vehicle, respectively, i.e.,

$$
\pi^{u*} = \arg\max_{\pi^{u}} V^{\pi^{u}}(\mathbf{s}^{u})
$$

$$
= \left[\arg\max_{\phi_{1}^{u}} Q_{1}^{\pi^{u}}(\mathbf{s}^{u}, \phi_{1}^{u}), \arg\max_{\phi_{2}^{u}} Q_{2}^{\pi^{u}}(\mathbf{s}^{u}, \phi_{2}^{u}), \cdots, \arg\max_{\phi_{K_{u}}^{u}} Q_{K_{u}}^{\pi^{u}}(\mathbf{s}^{u}, \phi_{K_{u}}^{u}) \right]
$$

$$
= [\phi_{1}^{u*}, \phi_{2}^{u*}, \cdots, \phi_{K_{u}}^{u*}].
\tag{22}
$$

Based on the above analysis, the uth cluster network is shown in Fig. 2(b). To realize the beam selection state of the proposed PPJS scheme, deep Q-learning network structure is employed in this paper. The beam selection network has two specialized structures for the purpose of stabilization. One is experience replay memory unit, which stores the set of beam selection transition $[\mathbf{s}_{t}^{u}, \mathbf{a}_{t}^{u}, r_{t}^{u}, \mathbf{s}_{t+1}^{u}]$ for the random-batch-training of clusters. The other is the quasi-static beam selection network, which contains two sub-networks with the same network structures, which is designed based on MLP by us, i.e., policy network and target network with different parameters θ and θ^-, respectively. The policy network is used to select beams in real time according to V^{π_θ}; the target network is used to provide a target y for the policy network's parameters update. And then we can obtain the loss function $(y - V^{\pi_\theta})^2$. Based on the loss function, we can use the gradient descent method to get the parameters θ, and the target network parameter θ^- is updated by θ every t_c steps. From the output layer of Fig. 2, we can see that beam selection is significantly simplified according to (19) and (21).

Algorithm 1: PPJS scheme

Input: $\mathbf{s}_t^u$ for the uth cluster ($u = \{1, 2, \cdots, U\}$).

1 **for** *cluster u=1 to U* **do**

2   Initialize buffer with capacity $C^u = C$, the policy parameter θ^u, the target network parameter θ^{u-}.

3 **end**

4 **for** *episode=1 to M* **do**

5   **for** *cluster u=1 to U* **do**

6   Initialize $\mathbf{s}_t^u = \left[\overline{\mathbf{x}}_{t,1}^u, \overline{\mathbf{x}}_{t,2}^u, \cdots, \overline{\mathbf{x}}_{t,K^u}^u\right]$.

7   **for** *k_u =1 to K_u* **do**

8   Generate a random number ξ in [0,1],

$$\phi_{t,k_u}^u = \begin{cases} \text{random beam direction } \phi_{t,k_u}^u \in \Phi & \xi \geq \epsilon, \\ \text{argmax } Q_{a_{t,k_u}}^u \left(\mathbf{s}_t^u, \phi_{t,k_u}^u; \theta_t^u\right) & \text{otherwise} . \end{cases}$$

9   **end**

10   Form the uth cluster action $\mathbf{a}_t^u = \left[\phi_{t,1}^u, \phi_{t,2}^u, \cdots, \phi_{t,K^u}^u\right]^{\mathrm{T}}$.

11   **end**

12   Execute action $\mathbf{a}_t = \left[\mathbf{a}_t^1, \mathbf{a}_t^2, \cdots, \mathbf{a}_t^U\right]$ to get the next selection period state $\mathbf{s}_{t+1}$ of K vehicles, and observe U clusters reward $\left[r_t^1, r_t^2, \cdots, r_t^U\right]$.

13   Feed $\mathbf{s}_{t+1}$ into the trained trajectory prediction module and obtain each cluster's own state $\mathbf{s}_{t+1}^u = \left[\overline{\mathbf{x}}_{t+1,1}^u, \overline{\mathbf{x}}_{t+1,2}^u, \cdots, \overline{\mathbf{x}}_{t,K^u}^u\right]$.

14   **for** *cluster u=1 to U* **do**

15   Store selection transition $\left(\mathbf{s}_t^u, \mathbf{a}_t^u, r_t^u, \mathbf{s}_{t+1}^u\right)$ in M_u.

16   Sample random minibatch of transitions $\left(\mathbf{s}_i^u, \mathbf{a}_i^u, r_i^u, \mathbf{s}_{i+1}^u\right)$ from M_u.

17   Set $y_i^u = \begin{cases} r_i^u & \text{if } i+1 = T, \\ r_i^u + \gamma \max V^{\pi^{\theta^{u-}}} \left(\mathbf{s}_{i+1}^u\right) & \text{otherwise} \end{cases}$

18   Perform a gradient descent on $\left(y_i^u - V^{\pi^{\theta^u}}\left(\mathbf{s}_i^u\right)\right)^2$ with respect to the network parameters θ^u.

19   Reset $\theta^{u-} = \theta^u$ every t_s steps .

20   **end**

21 **end**

5 Simulation Results and Performance Analysis

In this section, we first give the simulation parameters and then evaluate the performance of the proposed PPJS scheme under different scenarios.

Table 1. Environment parameters and network parameters.

Trajectory prediction network parameter	Value
Input size N_i	2
Output size N_o	2
Cluster size N_c	10
Learning rate l_t	0.06
Batch size L	500

Sub-selection network parameter	Value
Learning rate l_s	0.01
Discount factor γ	0.9
Threshold ϵ	0.9
Update time t_s	300
Batch size N_b	32
Memory size C	500

Environment parameter	Value
Number of antennas N_{BS}	32
Number of uplink intervals I	30
Number of downlink intervals J	70
Time interval T_i	10 ms
Road length h_r	200 m
Distance between BS and road h_d	100 m

5.1 Parameter Setup

The default parameters are listed in Table 1. Considering a practical traffic scenario [12,13], both the vehicle speed and distance distributions follow the truncated normal distribution. The relationship between them can be characterized by traffic flow theory [14], that is,

$$V = V_m \ln\left(\frac{K_m}{K}\right). \tag{23}$$

Moreover, the vehicles could speed up and down in order to avoid crash between vehicles.

5.2 Results and Analysis

Figure 3 depicts the successful transmission probability versus the number of iterations. We can see that the successful transmission probability gradually converges with the increasing number of iterations, which validates the convergence of the proposed PPJS.

The successful transmission probabilities versus different traffic parameters are shown in Fig. 4 under three beam tracking schemes. Apart from the proposed PPIA scheme, the DQN and position-based beam selection schemes are adopted for comparison, i.e., the beams for vehicles are directly selected based on DQN network and the current vehicle positions, respectively. Fig. 4 indicates that the proposed PPIA scheme performs better than the other two schemes,

302   B. Cheng et al.

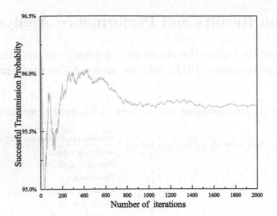

Fig. 3. Successful transmission probability versus iteration number

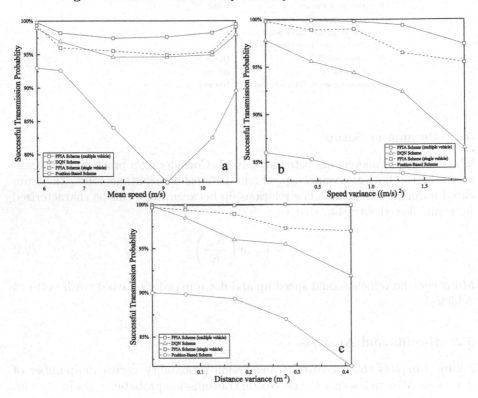

Fig. 4. The performance under different traffic parameters.

because both the beam coverage and IVI are taken into account. In Fig. 4(a), the successful transmission probability first increases and then decreases with the increasing vehicle mean speed. Because for a given number of antennas at the

BS, the beam width is fixed, then the average coverage gain will drop with the increasing speed mean of vehicles; meanwhile the mean distance between vehicles will increase based on traffic theory and therefore the IVI could be mitigated. In the low speed mean, the gain degradation dominates the performance, while the IVI mitigation dominates the performance in the high speed mean. Figure 4(b) indicates that the successful transmission probability becomes worse with the increasing speed variance. This is because some inter-vehicle distances decrease with the increasing speed variance, which results in severer IVI among vehicles. Although other inter-vehicle distances become large, the contribution of the decreasing IVI is tiny. Figure 4(c) also shows a decreasing tendency in successful transmission probability with the increasing vehicle distance variance, which could be explained by the same reason for Fig. 4(b).

Figure 5 illustrates the successful transmission probability versus the number of antennas at the BS under different moving speed setups. Based on the antenna theory, the beam width decreases with the increasing number of antennas. The wide beam could achieve better coverage with high interference, while the narrow beam has low IVI with limited coverage region. During a beam selection period, the coverage and beam gain should match the moving speed and inter vehicle distance. Therefore, the performance first increases and then decreases with the increasing number of antennas in Fig. 5.

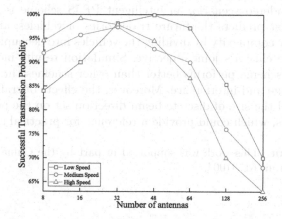

Fig. 5. Successful transmission probability versus the number of antennas.

Figure 6 indicates the relationship between the successful transmission probability and the size M of the beam direction set under different moving speed scenarios. With the increasing number of candidate beams in the beam direction set, the transmission performance gradually becomes better. Because the beam direction set with increasing number of discrete beams could gradually approximate to the continuous beams, then the optimal beam selection could be realized at cost of slight complexity increase.

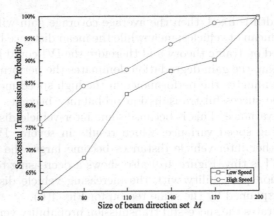

Fig. 6. Successful transmission probability versus the number of discrete beams.

6   Conclusion

This paper considers a scenario that an IoV system with mmWave and massive MIMO serves multiple mobile vehicles by periodical beam selection. In order to improve the transmission performance influenced by both beam coverage and IVI on our considered scenario, an intelligent PPJS scheme is proposed. The PPJS scheme first predicts the future trajectories of vehicles; and then selects beams with a low complexity by dividing the vehicles into decoupled clusters and decomposing the cluster's joint objective. Simulation results indicate that the proposed PPJS scheme performs better than other schemes through balancing the beam coverage and interference. Moreover, the effects of traffic parameters, beam width, and the size of discrete beam direction set on the performance are studied in details, which could provide a reference for practical systems.

Acknowledgment. This work was supported in part by the China Nature Science Funding under Grant 61731004.

References

1. Bujari, A., Gottardo, J., Palazzi, C.E., Ronzani, D.: Message dissemination in urban IoV. In: 2019 IEEE/ACM 23rd International Symposium on Distributed Simulation and Real Time Applications (DS-RT), pp. 1–4 (2019). https://doi.org/10.1109/DS-RT47707.2019.8958708
2. Hou, Z., She, C., Li, Y., Zhuo, L., Vucetic, B.: Prediction and communication co-design for ultra-reliable and low-latency communications. IEEE Trans. Wireless Commun. **19**(2), 1196–1209 (2020). https://doi.org/10.1109/TWC.2019.2951660
3. Bencivenni, C., Gustafsson, M., Haddadi, A., Zaman, A.U., Emanuelsson, T.: 5G mmWave beam steering antenna development and testing. In: 2019 13th European Conference on Antennas and Propagation (EuCAP), pp. 1–4 (2019)

4. Ali, M.Y., Hossain, T., Mowla, M.M.: A trade-off between energy and spectral efficiency in massive MIMO 5G system. In: 2019 3rd International Conference on Electrical, Computer Telecommunication Engineering (ICECTE), pp. 209–212 (2019). https://doi.org/10.1109/ICECTE48615.2019.9303551
5. Raza, A., Junaid Nawaz, S., Wyne, S., Ahmed, A., Javed, M.A., Patwary, M.N.: Spatial modeling of interference in inter-vehicular communications for 3-D volumetric wireless networks. IEEE Access 8, 108281–108299 (2020). https://doi.org/10.1109/ACCESS.2020.3001052
6. Zhao, L., Zhao, H., Zheng, K., Xiang, W.: Massive MIMO in 5G Networks: Selected Applications. Springer, Heidelberg (2018). https://doi.org/10.1007/978-3-319-68409-3
7. Zheng, K., Zhao, L., Mei, J., Shao, B., Xiang, W., Hanzo, L.: Survey of large-scale MIMO systems. IEEE Commun. Surv. Tutor. 17(3), 1738–1760 (2015). https://doi.org/10.1109/COMST.2015.2425294
8. Va, V., Vikalo, H., Heath, R.W.: Beam tracking for mobile millimeter wave communication systems. In: 2016 IEEE Global Conference on Signal and Information Processing (GlobalSIP), pp. 743–747 (2016). https://doi.org/10.1109/GlobalSIP.2016.7905941
9. Liu, Y., Jiang, Z., Zhang, S., Xu, S.: Deep reinforcement learning-based beam tracking for low-latency services in vehicular networks. In: ICC 2020–2020 IEEE International Conference on Communications (ICC), pp. 1–7 (2020). https://doi.org/10.1109/ICC40277.2020.9148759
10. Jeong, J., Lim, S.H., Song, Y., Jeon, S.W.: Online learning for joint beam tracking and pattern optimization in massive MIMO systems. In: IEEE INFOCOM 2020 - IEEE Conference on Computer Communications, pp. 764–773 (2020). https://doi.org/10.1109/INFOCOM41043.2020.9155475
11. Tran, X.V., Pham, V.H.: An analytical method for calculating the limitation of beam scanning in uniform linear array (ULA). In: 2009 International Conference on Advanced Technologies for Communications, pp. 257–260 (2009). https://doi.org/10.1109/ATC.2009.5349450
12. Cvetek, D., Muštra, M., Jelušić, N., Abramović, B.: Traffic flow forecasting at micro-locations in urban network using bluetooth detector. In: 2020 International Symposium ELMAR, pp. 57–60 (2020). https://doi.org/10.1109/ELMAR49956.2020.9219023
13. Dai, G., Ma, C., Xu, X.: Short-term traffic flow prediction method for urban road sections based on space-time analysis and GRU. IEEE Access 7, 143025–143035 (2019). https://doi.org/10.1109/ACCESS.2019.2941280
14. Abadi, A., Rajabioun, T., Ioannou, P.A.: Traffic flow prediction for road transportation networks with limited traffic data. IEEE Trans. Intell. Transp. Syst. 16(2), 653–662 (2015). https://doi.org/10.1109/TITS.2014.2337238

A Dynamic Transmission Design via Deep Multi-task Learning for Supporting Multiple Applications in Vehicular Networks

Zhixing He, Mengyu Ma, Chao Wang[✉], and Fuqiang Liu

Department of Information and Commnication Engineering, Tongji University,
Shanghai 201804, China
{1830275,mengyu_ma,chaowang,liufuqiang}@tongji.edu.cn

Abstract. We study a cross-layer transmission design problem for vehicular communication networks. Two source-destination links are considered to share the same spectrum resource. Each link intends to send two types of messages to support different delay-sensitive applications. The whole system operates in a dynamic environment in which the small-scale channel fading may change rapidly. Therefore, the sources need to vary their transmission strategies accordingly to efficiently use the available resources while keeping the performance requirements satisfactory. Conventional transmission design via mathematical tools in general demands an iterative computation process and results in high complexity unsuitable for rapid decision-making. In this paper, we propose tackling such a problem by first transforming the transmission design problem into a joint classification-regression problem, and then applying deep multi-task learning (MTL) to solve it. Through simulation results, we show that our method can achieve the similar performance as the transmission design found by mathematical optimizations, with a much faster inference process. The advantages would become even more notable when the network size increases and the environment becomes more complex.

Keywords: Cross-layer transmission design · Vehicular communication · Multi-task learning

1 Introduction

Improving road safety, alleviating traffic congestion, and reducing energy consumption have become the major concerns in the development of modern road transportation systems. Allowing high-quality communications among the elements on the road (including vehicles, roadside infrastructure, pedestrians, etc.) serves as one of the key technologies for establishing an intelligent transportation system (ITS) [1]. It can enable a rich amount of sensing, computing, storage and communication resources to be shared to enhance the capability of each

© ICST Institute for Computer Sciences, Social Informatics and Telecommunications Engineering 2022
Published by Springer Nature Switzerland AG 2022. All Rights Reserved
H. Gao et al. (Eds.): ChinaCom 2021, LNICST 433, pp. 306–321, 2022.
https://doi.org/10.1007/978-3-030-99200-2_24

individual [2]. However, transmission design in vehicular communication networks can be much more challenging than those in conventional cellular and WiFi networks [3]. First, vehicular communication often needs to support multiple safety-related applications, which have different quality of service (QoS) requirements. Second, the available spectrum resource is normally limited and may be shared by multiple links. Hence inter-user interference has to be taken into account, if efficient channel usage is desired. Finally, due to the rapid movement of vehicles, the channel fading changes dynamically. Continuous and rapid transmission decision-making should be conducted accordingly.

Efficient transmission design in wireless networks through power control to balance the desired performance of each link and mutual interference between links has been studied extensively in the past decade [4–6]. The optimization target can be very different, ranging from minimized energy consumption to maximized throughput, depending on the application demand. For instance, the amount of information that can be successfully delivered by one joule of energy (also termed energy efficiency, EE) has attracted recent attentions due to its fractional nature, practical meaning, and difficulties in solving the problem. For a multi-user interference channel, reference [5] proposes a quadratic transform technique for tackling the multiple-ratio concave-convex fractional problem for EE. Reference [6] also studies a successive pseudoconvex approximation framework to maximize the EE.

But these works may not be naturally applicable in vehicular communication networks because of a number of reasons. First, they consider only a single type of message between each source-destination link. In addition, transmission design is carried out based on information theory and considers only the channel state information (CSI) in the physical (PHY) layer. By this means, the queue state information (QSI) in the MAC layer is ignored such that message transmission delay cannot be properly taken into consideration in a dynamic environment [7]. To handle these issues, our earlier work [8] investigates a two-user vehicular communication network in which each link desires to deliver two types of different messages, representing periodic heart-beat status messages and random sensing messages respectively. Applying the Lyapunov optimization theory, sequential power control in a dynamic environment is transformed into a new optimization problem that can be solved greedily. It is shown that the EE can be notably improved by the proposed method compared with conventional approaches.

Nevertheless, the computation complexity of the above method is high since it applies an iterative process to identify the solution of the power control problem, and also exhaustively searches the best decision for determining which message should be transmitted. This hinders the method's operation in an online fashion. There is a need for an efficient and dynamic transmission design strategy for supporting multiple applications in vehicular communication networks.

In the past few years, using machine learning (ML) tools to solve complex optimization and decision-making problems in wireless systems has drawn a large amount of research interests. For example, reference [9] shows that the beamforming design in a multiple-input multiple-output (MIMO) system can be formulated as a multi-class classification problem, and thus be solved efficiently

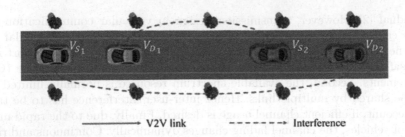

V_{S_1} V_{D_1} V_{S_2} V_{D_2}

→ V2V link   ----→ Interference

Fig. 1. System model

using ML algorithms without significant performance loss. References [10–12] prove that the power control problem and user scheduling problem can respectively be treated as regression and classification problems. Therefore, after being properly trained by datasets generated using time- and computation-demanding mathematical tools, ML models, especially deep learning models, can rapidly carry out the inference process and reach near-optimal solutions for unseen data. Advanced ML schemes, e.g. convolution neural networks [13] and ensemble learning [14], are applied to solve complex power control problems recently as well.

In this paper, we borrow the above ideas to solve the aforementioned dynamic transmission design problem. Specifically, we consider a vehicular communication system that consists of two source-destination links sharing the same spectrum resource. Each link intends to deliver two types of messages with different QoS requirements: The periodic heart-beat messages should be transmitted immediately with high reliability, while the sensor messages should be sent with limited delay. Targeting maximized EE, we first transform the non-convex fractional optimization problem with time average constraints into a series of solvable mixed integer optimization problems. Further, we apply a deep multi-task learning (MTL) network trained by data generated using mathematical optimization tools, and with loss function properly designed to balance individual learning tasks. Numerical results show that the proposed MTL-aided dynamic transmission design can achieve the similar performance as that found by mathematical optimizations, but with limited computational complexity.

The remainder of the paper is organized as follows. In Sect. 2, we present the system model, transmission design problem formulation, and the conventional solution. In Sect. 3, we develop the MTL-aided transmission design and elaborate the training process. Section 4 illustrates numerical performance assessments of the proposed method. Finally, Sect. 5 concludes the paper.

2 System Model and Transmission Design Problem

We consider a multi-user vehicular communication network, in which two pairs of information source and destination vehicles, denoted by V_{S_i} and V_{D_i} for $\forall i \in \{1, 2\}$, share the same spectrum resource, as shown in Fig. 1. Their operations can be coordinated by the serving roadside infrastructure. Each source V_{S_i} desires to send two types of delay-limited messages, which have different characteristics and

QoS requirements to support different road safety applications, to V_{D_i}. Messages generation and transmission are assumed to conduct in discretized unit time slots. Thus, V_{S_i} executes its transmission operation at the beginning of each time interval $[t, t+1)$, for $t \in \{1, 2, \dots\}$.

The first type of messages, termed *type-1 messages*, represent heartbeat messages and arrive in V_{S_i} periodically with a constant data rate r_i bits/slot. They normally carry real-time status of V_{S_i}. Hence they should be transmitted immediately. To guarantee V_{D_i} to attain a proper knowledge of the status of V_{S_i}, the successful transmission ratio should be sufficiently large. We use a binary indicator $\phi_i[t]$ to denote whether, at time slot t, V_{D_i} receives the message. The QoS requirement of the type-1 message can hence be expressed as

$$\bar{\phi}_i = \lim_{T \to \infty} \frac{1}{T} \sum_{t=1}^{T} \phi_i[t] \geq \phi_{i,\min}, \tag{1}$$

where $\phi_{i,\min}$ is the minimum threshold value for the ith link (e.g. 70%).

The second type of messages, termed *type-2 messages*, represent messages that randomly arrive in V_{S_i}. Such messages can be those that support environment sensing or on-board infotainment applications. Let $a_i[t]$ (in bits) denote the generated data volume (within time interval $[t, t+1)$), which, without loss of generality, is assumed to follow a stationary random process with expectation λ_i bits/slot. Different from type-1 messages, type-2 messages can normally tolerate certain delay. Hence, the data unable to delivery are temporarily stored in the source queue of V_{S_i} (denoted by $\mathbb{Q}_i$). Use $Q_i[t]$ and $b_i[t]$ to respectively denote the amounts of data stored in and leave $\mathbb{Q}_i$, at the beginning of time interval $[t, t+1)$. For some initial value $Q_i[1]$, the queue state $Q_i[t+1] = \max\{Q_i[t] - b_i[t], 0\} + a_i[t]$. To avoid the queuing delay to become unlimitedly large, we set the QoS requirement of the type-2 messages to be

$$\lim_{T \to \infty} \frac{1}{T} \sum_{t=1}^{T} b_i[t] \geq \lim_{T \to \infty} \frac{1}{T} \sum_{t=1}^{T} a_i[t] = \lambda_i. \tag{2}$$

This implies the queues being stable [15].

The message transmissions are conducted in a block-fading environment. The fading coefficient between V_{S_j} and V_{D_i} at time interval $[t, t+1)$ is denoted by $h_{ij}[t]$, which accounts for both the large-scale and small-scale fading phenomenon. We assume that the knowledge of $h_{ij}[t]$ is causally available at all terminals (through, e.g., channel training coordinated by the serving infrastructure). $|h_{ij}|^2$ denotes the channel power gains correspondingly. Use $p_i[t]$ (chosen from domain $[p_{i,\min}, p_{i,\max}]$) and $R_i[t]$ to denote the transmit power and data rate of V_{S_i} at time interval $[t, t+1)$. Encoding using unit-power capacity-achieving Gaussian random codes leads to

$$R_i[t] \leq B \log_2 \left(1 + \frac{|h_{ii}[t]|^2 p_i[t]}{|h_{ij}[t]|^2 p_j[t] + N_0} \right), \quad \forall i \in \{1, 2\}, \tag{3}$$

where B is the bandwidth and N_0 is the noise power.

We consider a dynamic vehicular communication environment such that the small-scale fading coefficients change independently across time slots. Therefore, a corresponding dynamic transmission design is expected to efficiently deliver the messages while keeping the QoS requirements satisfactory. In what follows, the transmission efficiency is evaluated by energy efficiency, i.e., the amount of information transmitted by one joule of energy consumption. Mathematically, it is defined as the ratio of the average of long-term aggregate data to the average of the corresponding long-term total energy consumption:

$$\bar{\eta} = \frac{\bar{R}}{\bar{P}} = \frac{\bar{R}_1 + \bar{R}_2}{\bar{p}_1 + \bar{p}_2}, \tag{4}$$

where the individual average transmission rate $\bar{R}_i = \lim_{T \to \infty} \frac{1}{T} \sum_{t=1}^{T} R_i[t]$, and the individual average power consumption $\bar{p}_i = \lim_{T \to \infty} \frac{1}{T} \sum_{t=1}^{T} P_i[t] = \lim_{T \to \infty} \frac{1}{T} \sum_{t=1}^{T} p_i[t] + p_{c,i}$, and $p_{c,i}$ is the constant circuit power of V_{S_i} [16].

In addition to controlling power $p_i[t]$, due to the requirement of supporting multiple applications, each source has an additional action to choose encoding which message, denoted mathematically by a binary indicator $\psi_i[t]$. Specifically, $\psi_i[t] = 0$ represents the case that V_{S_j} chooses excluding a type-1 message in its transmitted signal. As a result, all power consumption is used to satisfy the QoS requirement of the type-2 messages, i.e., emptying the queue $\mathbb{Q}_i$ with $b_i[t] = R_i[t]$. On the other hand, choosing $\psi_i[t] = 1$ means that V_{S_j} intends to send both types of messages. By this means, $b_i[t] = R_i[t] - r_i$ and the remaining r_i bits are allocated to the type-1 message. Note that in this case, V_{S_j} needs to use sufficient power to guarantee $R_i[t] \geq r_i$.

In summary, the dynamic transmission design in the considered vehicular communication network aims to, based on the knowledge of CSI in the PHY layer (i.e., channel fading coefficients) and QSI in the MAC layer (i.e., queue lengths), solve the following stochastic optimization problem:

$$\text{maximize} : \bar{\eta} = \frac{\bar{R}_1 + \bar{R}_2}{\bar{p}_1 + \bar{p}_2} \tag{5}$$

$$\text{s.t.} : \text{C1}: \bar{\phi}_i \geq \phi_{i,\min}, \quad \forall i \in \{1, 2\}, \tag{6}$$

$$\text{C2}: \mathbb{Q}_i \text{ is stable}, \quad \forall i \in \{1, 2\}, \tag{7}$$

$$\text{C3}: p_{i,\min} \leq p_i[t] \leq p_{i,\max}, \quad \forall i \in \{1, 2\}. \tag{8}$$

$$\text{C4}: \psi_i[t] \in \{0, 1\}, \quad \forall i \in \{1, 2\}. \tag{9}$$

The above problem is hard to solve exactly, since the objective function and constraints have time average operations. One can reach a suboptimal solution with the aid of Lyapunov optimization theory [8,15]. We define $\mathbf{p}^* = [p_1[1], p_2[1], p_1[2], p_2[2], \cdots]$ as the optimal power allocation vector. The optimal power allocation achieves the optimal EE $\bar{\eta}_{opt}$ if and only if $\max((\bar{R}_1 + \bar{R}_2) - \bar{\eta}_{opt}(\bar{p}_1 + \bar{p}_2)) = 0$ [17]. Thus, we can first transform the objective function at the beginning of time interval $[t, t + 1)$ into the form:

$$\text{maximize} : (R_1[t] + R_2[t]) - \bar{\eta}_{opt}(p_1[t] + p_{c,1} + p_2[t] + p_{c,2}) \tag{10}$$

$$\text{s.t.}: \text{C1, C2, C3, C4}.$$

Since $\bar{\eta}_{opt}$ is unknown, we define $\eta[t]$ as the objective function achieved so far:

$$\eta[t] = \frac{\sum_{\tau=1}^{t-1}(R_1[\tau] + R_2[\tau])}{\sum_{\tau=1}^{t-1}(p_1[\tau] + p_{c,1} + p_2[\tau] + p_{c,2})}. \tag{11}$$

We next replace $\bar{\eta}_{opt}$ in (10) by $\eta[t]$ and eliminate the constant part $p_{c,1}$ and $p_{c,2}$. Then problem (10) is cast as

$$\text{maximize} : (R_1[t] + R_2[t]) - \eta[t](p_1[t] + p_2[t]) \tag{12}$$
$$\text{s.t.}: \text{C1, C2, C3, C4}.$$

It is shown in [8,15] that the transformation is effective for solving the original problem (5), since the time average operation (over future transmission strategies) in the objective function is avoided.

Furthermore, we can transform the constraint C1 into an equivalent queue stability constraint [15], by defining a virtual queue for each source-destination pair $\mathbb{Y}_i$, $\forall i \in \{1, 2\}$. Use $Y_i[t]$ to represent the queue length of $\mathbb{Y}_i$ at the beginning of time interval $[t, t+1)$. For a certain initial value $Y_i[1]$, the virtual queue updates according to the following formula:

$$Y_i[t+1] = \max\{Y_i[t] - \phi_i[t], 0\} + \phi_{i,\min}, \quad \forall i \in \{1, 2\}. \tag{13}$$

Demanding the stability of the virtual queue eliminates the time average operation of the constraint C1 since it implies $\lim_{T\to\infty} \frac{1}{T}\sum_{t=1}^{T}\phi_i[t] = \bar{\phi}_i \geq \phi_{i,\min}$.

Based on the Lyapunov optimization theory, the constrains that the queues $\mathbb{Y}_1$, $\mathbb{Y}_2$, $\mathbb{Q}_1$, and $\mathbb{Q}_2$ are stable can be described by a Lyapunov drift-plus-penalty, and then transformed to a cost function in the objective function. In addition, finding a lower bound of the changing part of the drift-plus-penalty, the original transmission design problem (5) is transformed into a mixed integer optimization problem at each time interval $[t, t+1)$ as follows [8]:

$$\text{maximize} : V\sum_{i=1}^{2}(R_i[t] - \eta[t]p_i[t]) + 2\sum_{i=1}^{2}(u_iQ_i[t]b_i[t] + v_iY_i[t]\phi_i[t]) \tag{14}$$
$$\text{s.t.}: \text{C3 and C4},$$

where the parameter V can be tuned to trade off energy efficiency and queue lengths (both actual source queues $\mathbb{Q}_i$ and virtual queues $\mathbb{Y}_i$), and u_i and v_i are the weighting parameters of $Q_i[t]$ and $Y_i[t]$ controlling the power assignment between the two types of messages. All time average operations are eliminated.

Now, by individually considering each of the four encoding options in the constraint C4, the above optimization problem has only the constraint C3. Although it is not a convex optimization problem, it can be solved by the concave convex procedure algorithm [18]. By comparing the solutions corresponding to the four encoding options and using the one with the best performance, a sub-optimal solution to the original problem (5) is found. Through extensive simulations, it is shown in [8] that such a dynamic transmission design achieves better performance than a number of conventional methods. However, the concave convex procedure algorithm demands an iterative searching process to reach the solution. Taking enumerating the encoding actions also into account, at the beginning of each time interval, a complex computation is demanded to determine the dynamic transmission design strategy. This may prevent the strategy to be conducted in an online fashion. In what follows, we propose applying the deep MTL technique to tackle this problem.

3 Deep Multi-task Learning-Aided Transmission Design

To solve the dynamic transmission design problem for the considered vehicular communication network, we follow the idea presented in [12] and transform the solution of the optimization problem (14) at each time slot into a mapping problem, from the network CSI (taken as the form of channel power gains $|h_{ij}[t]|^2$), QSI (queue lengths $Q_i[t]$ and $Y_i[t]$), and achievable performance so far (energy efficiency $\eta[t]$), to the optimal encoding actions and transmission powers of both links. Specifically, let $\mathcal{F}$ denote the function that describes the mapping relationship. For fixed transmission requirement parameters $\phi_{1,\min}$, $\phi_{2,\min}$, $p_{1,\min}$, $p_{2,\min}$, $p_{1,\max}$, $p_{2,\max}$, and optimization weighting parameters u_1, u_2 v_1, v_2, V, the function $\mathcal{F}$ can be written as

$$\mathcal{F}: \left\{|h_{11}[t]|^2, |h_{12}[t]|^2, |h_{21}[t]|^2, |h_{22}[t]|^2, Q_1[t], Q_2[t], Y_1[t], Y_2[t], \eta[t]\right\}$$

$$\mapsto \left\{p_1^*[t], p_2^*[t], \psi_1^*[t], \psi_2^*[t]\right\}, \ \forall t \in \{1, 2, \cdots\}, \tag{15}$$

where the set $\{p_1^*[t], p_2^*[t], \psi_1^*[t], \psi_2^*[t]\}$ denotes the solution to problem (14).

If one knows the function $\mathcal{F}$, then making the dynamic transmission design decision is straightforward. However, the general form of function $\mathcal{F}$ is clearly unknown. On the other hand, for each possible realization of the input set $\{|h_{11}[t]|^2, |h_{12}[t]|^2, |h_{21}[t]|^2, |h_{22}[t]|^2, Q_1[t], Q_2[t], Y_1[t], Y_2[t], \eta[t]\}$, the associated output set $\{p_1^*[t], p_2^*[t], \psi_1^*[t], \psi_2^*[t]\}$ can be derived by the concave convex procedure algorithm, as explained above. But the computation process can be very time-consuming. Therefore, we considering integrating the advantages of the two approaches, by training a deep neural network (DNN), with data coming from mathematical derivations, to approximate $\mathcal{F}$. As long as the approximation is sufficiently accurate, inference using the DNN can guarantee a rapid and good decision-making process. Nevertheless, the mapping function presented in (15) is different from those considered in conventional works (e.g., [12]), since the problem (14) is a mixed integer optimization problem. The function output contains both quantitative (i.e., $p_1^*[t], p_2^*[t]$) and categorical (i.e., $\psi_1^*[t], \psi_2^*[t]$) values. The learning problem cannot be simply modeled as a regression or classification problem. To handle this issue, we propose applying the MTL technique to train the DNN and establish the approximate function.

MTL is a branch in ML that aims to exploit useful information contained in multiple related tasks to improve the generalization performance of all the tasks, [19]. DNN with hard parameter sharing is one of the frequently used structures in MTL networks [20]. As show in Fig. 2, there are two parts of hidden layers

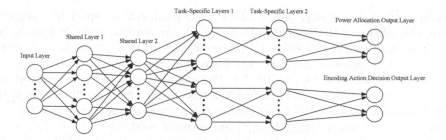

Fig. 2. Network structure with hard parameter sharing

between the input layer and the output layer, namely share layers and task-specific layers. The shared layers are used to learn the common information of different tasks. The task-specific layers are used to integrate information from the shared layers, learn the individual inherent characteristics, and output the specific form of each task. The network structure can be considered as a variant of a fully connected network. Having two different types of hidden layers simultaneously obtains input-output mapping of multiple tasks, and may also lead to a better generalization performance of each individual task [20].

For conventional regression and classification problems, one can use loss functions such as mean square errors (MSE) and cross entropy (CE) to respectively evaluate the difference between prediction and desired output results, to train DNNs. But in general it is difficult to find a single unified loss function for an MTL network. One apparent way is to add up all the losses with pre-defined weights, i.e.,

$$\mathcal{L}_{\text{total}} = \sum_i \omega_i \mathcal{L}_i, \tag{16}$$

where $\mathcal{L}_i$ and ω_i are the loss and weight parameters of the ith task, respectively.

However, the inference performance of such learning models are often strongly dependent on whether the weights are determined properly. Tuning these weight parameters manually, via e.g., grid search, is a difficult task. For practical applications, having a loss function that automatically finds the weights for different tasks are much more attractive. For this reason, reference [21] proposes a multi-task loss function based on maximizing the Gaussian likelihood with task-dependent uncertainty. Next we introduce how to transform the considered optimization problem into individual tasks and define the corresponding loss function.

In our vehicular communication network, the allocated power of each V2V link at each time slot is a real value between $p_{i,\min}$ and $p_{i,\max}$. We regard both links' powers at each time slot as a two-dimensional vector. The mapping from inputs to output with real values (i.e., from $|h_{11}[t]|^2$, $|h_{12}[t]|^2$, $|h_{21}[t]|^2$, $|h_{22}[t]|^2$, $Q_1[t], Q_2[t], Y_1[t], Y_2[t], \eta[t]$ to power assignments $p_1^*[t], p_2^*[t]$) is formulated as a regression problem. However, $p_1^*[t], p_2^*[t]$ can be chosen from a wide region (e.g., in our experiments in Sect. 4, the powers frequently change between 10^{-5} W to 10^{-1} W). Directly using them in the loss function may cause model training to mainly consider the impact of large power values, while small values, especially when the channel qualities of the desired links are good, are actually more important in achieving high energy efficiency. Therefore, we follow [10] and conduct a feature transformation process that takes the logarithm operation to transmit powers (values close to zero are set to 10^{-30}). The output vector of the power control regression problem in our MTL network is thus formed by $\log p_1^*[t]$ and $\log p_2^*[t]$. Due to similar reasons, the same feature transformation is carried out for the channel power gains, i.e. $\log\left(|h_{ij}[t]|^2\right)$ are actually used.

To define a loss function that can be used for training the MTL network, we follow [21] and consider the task-dependent uncertainty between the model output (denoted as $f_r^W(x)$, with x and W respectively being the input and network parameters of the MTL network) and desired output (denoted as y_r) of the regression task. Assume that $y_r - f_r^W(x)$ follows a zero-mean Gaussian distribution, i.e.,

$$y_r - f_r^W(x) \sim \mathrm{N}(0, \sigma_r^2), \tag{17}$$

where the standard deviation σ_r represents observation noise level. For the maximum likelihood estimation, we can obtain the log likelihood [21]:

$$\log \mathrm{Pr}\left\{y_r | f_r^W(x), \sigma_r\right\} = \log \mathrm{Pr}\left\{y_r - f_r^W(x) | \sigma_r\right\} \propto -\left(\frac{1}{2\sigma_r^2}\mathcal{L}_{\mathrm{MSE}}(W) + \log \sigma_r\right), \tag{18}$$

where $\mathcal{L}_{\mathrm{MSE}}(W) = \left\|y_r - f_r^W(x)\right\|^2$ is the regression output MSE.

In addition, the condition C4 in (14) represents the encoding actions in the considered network. This determines whether type-1 messages should be transmitted. Since $\psi_1^*[t]$ and $\psi_2^*[t]$ are binary indicators, seeking the optimal mapping from inputs $|h_{11}[t]|^2, |h_{12}[t]|^2, |h_{21}[t]|^2, |h_{22}[t]|^2, Q_1[t], Q_2[t], Y_1[t], Y_2[t], \eta[t]$ to them can be formulated as a classification problem. One simple way of the formulation is to treat it as a multi-class classification problem. For instance, in the considered 2-link network, the number of classes (all possible encoding actions at each time slot) is four. But as the network size becomes large, such a number increases exponentially. To avoid this issue, we deem the encoding action of each link as a single binary classification problem. This still fits in the MTL framework and leads to only a linear increase of decision space as the network size increases.

Let the MTL network have two classification output neurons, each applying the Sigmoid function to represent the probability that the input leads to the associated category. Again, following [21], assume that the likelihood of the output of the ith transmission link equals a scaled softmax function with scalar $\sigma_{c,i}$:

$$\mathrm{Pr}(y_{c,i} | f_{c,i}^W(x), \sigma_{c,i}) = \mathrm{Softmax}\left(\frac{1}{\sigma_{c,i}^2} f_{c,i}^W(x)\right), i \in \{1, 2\}, \tag{19}$$

where $f_{c,i}^W(x)$ and $y_{c,i}$ respectively denote the model output and actual value (encoding action derived using mathematical optimization) of the ith link. The log likelihood of multiple classification tasks can be simplified as

$$\log \Pr\left\{y_c | f_c^W(x), \sigma_c\right\} = \log \prod_{i=1}^{2} \Pr\{y_{c,i} | f_{c,i}^W(x), \sigma_{c,i}\}$$

$$\approx -\sum_{i=1}^{2} \frac{1}{\sigma_{c,i}^2} \mathcal{L}_{\mathrm{CE}}(y_{c,i}, f_{c,i}^W(x)) - \sum_{i=1}^{2} \log \sigma_{c,i}, \quad (20)$$

where $\mathcal{L}_{\mathrm{CE}}(y_{c,i}, f_{c,i}^W(x))$ denotes the ith classification output CE, and y_c and $f_c^W(x)$ denotes the vector representation of all desired outputs and all model outputs respectively. In our work, the two V2V links have the similar message transmission demands. To simplify the model training process, we set $\sigma_{c,1} = \sigma_{c,2} = \sigma_c$. Thus, the above log likelihood can be written as

$$\log \Pr\left\{y_c | f_c^W(x), \sigma_c\right\} \approx -2\left(\frac{1}{2\sigma_c^2} \sum_{i=1}^{2} \mathcal{L}_{\mathrm{CE}}\left(y_{c,i}, f_{c,i}^W(x)\right) + \log \sigma_c\right)$$

$$\propto -\left(\frac{1}{\sigma_c^2}\mathcal{L}_{\mathrm{BCE}}(W) + \log \sigma_c\right), \quad (21)$$

where $\mathcal{L}_{\mathrm{BCE}}(W) = \frac{1}{2}\sum_{i=1}^{2} \mathcal{L}_{\mathrm{CE}}(y_{c,i}, f_{c,i}^W(x))$.

To facilitate training one MTL network for making concurrent decisions of both encoding actions and power assignment, we formulate a single loss function as the negative sum of the above two log likelihood functions, i.e.,

$$\mathcal{L}_{\mathrm{total}}(W, \sigma_r, \sigma_c) = \frac{1}{2\sigma_r^2}\mathcal{L}_{\mathrm{MSE}}(W) + \frac{1}{\sigma_c^2}\mathcal{L}_{\mathrm{BCE}}(W) + \log \sigma_r + \log \sigma_c. \quad (22)$$

Minimizing $\mathcal{L}_{\mathrm{total}}(W, \sigma_r, \sigma_c)$ allows simultaneously increasing the performance of both the regression and classification tasks. The tradeoff between the impacts of the two tasks is determined by the uncertain weights $\frac{1}{2\sigma_r^2}$ and $\frac{1}{\sigma_c^2}$, as well as the term $\log \sigma_r + \log \sigma_c$, in which σ_r and σ_c are determined, together with W, in the training process as model parameters.

However, in (22), small values of σ_r and σ_c may cause the terms $\log \sigma_r$ and $\log \sigma_c$ to dominate the loss function and thus improperly influence the training performance. Following [22], we can slightly revise the loss function to mitigate the issue:

$$\mathcal{L}_{\mathrm{total}}(W, \sigma_r, \sigma_c) = \frac{1}{2\sigma_r^2}\mathcal{L}_{\mathrm{MSE}}(W) + \frac{1}{\sigma_c^2}\mathcal{L}_{\mathrm{BCE}}(W) + \log(1 + \sigma_r^2) + \log(1 + \sigma_c^2).$$

$$(23)$$

In what follows, we term the MTL model using loss function (22) as MTL with simple uncertainty weights, and that using (23) as MTL with revised uncertainty weights. The latter is mainly adopted for conducting the dynamic transmission design in the considered vehicular communication networks.

To establish an MTL-based transmission decision-making solution, we can design a deep MTL network structure as Fig. 2. The input layer contains 9 neurons, taking $\log |h_{11}[t]|^2$, $\log |h_{12}[t]|^2$, $\log |h_{21}[t]|^2$, $\log |h_{22}[t]|^2$, $Q_1[t]$, $Q_2[t]$, $Y_1[t]$, $Y_2[t]$, $\eta[t]$ at each time slot as inputs. Through shared layers, the network can learn the common representation from the inputs, and then is separated into two branches. The power allocation task-specific layers and output layer transform the common representation into two positive real values, denoting the log transformation of the allocated powers of the two links. The encoding action task-specific layers and output layer transform the common representation into two real values bounded between 0 and 1. They represent the probabilities whether the V2V links choose to send type-1 messages (when an output value is greater than 0.5) or not. All the training data can be generated using the mathematical optimization method described in Sect. 2. The network parameters together with $\log \sigma_r$ and $\log \sigma_c$ are trained using the loss function (23). Although the training procedure is time and computation consuming, the inference process is much faster than the mathematical optimization and hence suitable for rapid online decision making. In next section, we demonstrate the performance of the proposed dynamic transmission design method.

4   Numerical Results

Consider a V2V communication network with two pairs of vehicles. Each source intends to send type-1 messages with data rate $r_i = 7$ bits/Hz/slot and reliability requirement $\Phi_{i,\min} = 70\%$, for $\forall i \in \{1, 2\}$. The type-2 messages are generated from a Poisson process with average rate $\lambda_i = 3$ bits/Hz/slot $\forall i \in \{1, 2\}$. The source transmission powers are bounded below $p_{i,\max} = 0.1$ W and above $p_{i,\min} = 0$ W. At each time slot, the circuit power of each source is set to be $p_{c,i} = 10^{-3}$ W. The large-scale fading is considered to be path loss following model $PL_{ij} = 103.4 + 24.2 \log_{10}(d_{ij})$ dB with distance d_{ij} [23]. The small-scale fading is assumed to be Rayleigh. The data transmissions are operated in a spectral channel with bandwidth $B = 1$ MHz. The noise power spectral density is -174 dbm/Hz. The weighting parameters in the optimization objective are set to $u_i = 10$ and $v_i = 80$ for $i \in \{1, 2\}$.

In order to establish a sufficiently large dataset to train our MTL model, we apply the optimization algorithm described in Sect. 2 to generate 1000 blocks of transmission data. The distance between each source and its desired destination is 30 m. The distances for the V_{S_1}-V_{D_2} and the V_{S_2}-V_{D_1} interference links are 430 m and 460 m. For each block, the transmission between every source-destination pair contains a total of $T = 200$ time slots. We generate random small-scale fading coefficients and type-2 arrival data volume for each slot. The initial queue status $Q_i[1]$ and $Y_i[1]$ are sampled from an exponential distribution with parameter 80. The above process creates 200000 useful data samples, 180000 out of which are used as training set, and the remaining 20000 are used as test set.

The MTL network structure is shown in Fig. 2. There are 4 hidden layers, each containing 16 neurons. The first two are shared layers and the latter two are task-specific layers. ReLU is used as the activation function for hidden layers and the regression output layer, and Sigmoid is used for the classification output layer. We take the logarithmic transformation of the channel power gains

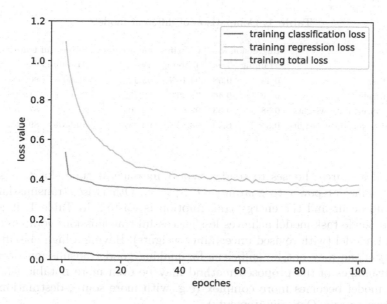

Fig. 3. Learning curves

(i.e., $|h_{ij}[t]|^2$) before sending them to the MTL network. The same is also conducted for the output power values (i.e., $p_1^*[t]$ and $p_2^*[t]$). By this means, failure of training can be notably reduced.

Figure 3 displays the learning curves of the MTL network with revised uncertainty weights for the case $V = 60$. In the experiments, the mini-batch size is chosen to be 10 (each epoch has 18000 mini-batches), and the learning rate is 0.001. From the figure we can see that loss values (for both the classification and regression problems, as well as the total loss) reduce as the training proceeds, and eventually converge to fixed values. The proposed MTL network thus can learn the patterns contained in the training data.

In addition to the revised uncertainty weights (23), we also conduct training of MTL networks (using the same network structure) with equal weights (i.e., setting as 1) and the simple uncertainty weight (22). Their training and test results (regarding classification and regression performance) are shown in Table 1. Clearly, MTL with the revised uncertainty weights performs better than the other two. Treating the encoding action selection and power control problems as independent classification and regression problems can also lead to two individual single-task models (one for encoding action selection and one for power control) that can be integrated to determine transmission decisions (termed Single-task in Table 1). This method has the similar classification accuracy as our MTL model, but with much worse performance in the regression problem (power control). Actually, the transmission power control and encoding action selection problems are strongly interrelated. Ignoring such a fact can cause serious problems in transmission decision-making. In Table 1, we also display the ratio of successful transmissions in the training and test experiments.

Table 1. Comparison of different models

Weights	Regression loss		Classification accuracy		Successful transmission	
	Training	Test	Training	Test	Training	Test
Single-task	0.45	0.58	94.16%	93.34%	82.60%	80.20%
Equal weights	0.40	0.56	78.39%	78.15%	72.68%	71.81%
Simple uncertainty weights	0.38	0.53	92.87%	92.50%	87.92%	87.73%
Revised uncertainty weights	0.36	0.49	93.84%	93.23%	89.53%	89.16%

Ideally, if a source chooses to send a type-1 message at time slot t, it should allocate sufficient power to guarantee $R_i[t] \geq r_i$. Otherwise, transmission failure would occur and the energy consumption is wasted. In Table 1, it is seen that the Single-task model achieves less successful transmissions compared with our MTL model (with revised uncertainty weights). Having a high classification accuracy alone may not be sufficient for making good transmission decisions. The advantages of the proposed method may be even more notable when the system model becomes more complex (e.g. with more source-destination links and heterogeneous QoS requirements).

Finally, we compare our MTL-aided transmission design scheme with three conventional solutions. The first considers only CSI to make transmission decisions. Since the QSI is not taken into consideration, both sources intend to transmit their messages with a fixed data rate of $R_i[t] = 10 + \theta$ bits/Hz/slot using the minimum power at each time slot. In such a fixed rate (FR) transmission, both messages are always transmitted and the parameter θ is chosen to counterbalance the impact of channel outage and keep queue length stable. Similar to our proposed method, the remaining two methods take into account both CSI and QSI. One of them orthogonalizes the message delivery processes of the two links and is termed orthogonal transmission (OT). In this case, the maximum transmission data rate between V_{S_i} and V_{D_i} within time interval $[t, t+1)$ is $R_i[t] = \frac{B}{2} \log \left(1 + \frac{2p_i[t]|h_{ii}[t]|^2}{N_0}\right)$ bits/slot. Inter-user interference is avoided with the cost of inefficient channel usage. The other method is the one that is described in Sect. 2. This method is termed Lyapunov optimization (LO) scheme and the concave convex procedure algorithm is applied to solve the optimization problem (5). In fact, this is also the method that we use to generate our training data.

For all the four schemes, we adjust the weighting factors so that the achievable QoS results of both types of messages are almost the same. Figure 4 illustrates the energy efficiency, the optimization objective in (5), for changing choices of V (for our MTL-aided design). It is seen that as V increases, the achievable energy efficiency of the OT, LO, and MTL schemes become larger, since more attentions have been placed to enhancing the energy efficiency, as expected. Because the spectrum resources are divided into 2 parts, the OT method needs larger powers to ensure the message QoS requirements, which leads to lower energy efficiency. The FR scheme intends to minimize the whole power consumption. However, a low power usage does not always mean a high energy efficiency. Thus,

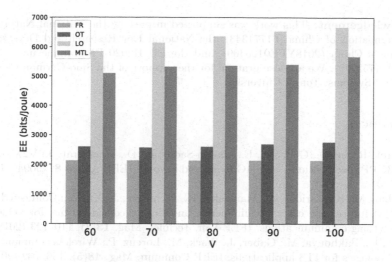

Fig. 4. Energy efficiency comparison

our MTL-aided design outperforms the FR and OT schemes. Finally, although our method does not perform as well as the LO scheme, the former demands much more computation time than the latter. Implementing the LO scheme on a PC with Inter Core i7-6800K CPU consumes around 10 hours for completing the transmission design of the 100 test transmission blocks (each consisting of 200 time slots). But the time consumption for the MTL-aided scheme (i.e., the inference process) is around only 30 s. The advantages of the proposed method can be clearly seen and are again expected to be even more significant in more complex systems.

5   Conclusion

We have investigated the potential of applying ML to facilitate dynamic cross-layer transmission design in vehicular communication networks. The system considered in this paper consists of two source-destination links. Each link desires to communicate information with the maximum energy efficiency to potentially support multiple applications, in a changing fading environment. Conventionally, to solve the formulated mixed integer optimization problem at each time slot, a time-consuming derivation procedure is needed. We have proposed transforming the optimization problem into a joint classification-regression problem, and then applying deep MTL to solve it. Simulation results have shown that our method can achieve similar performance as the mathematical optimization solution with a much faster decision-making process. The advantages are expected to be even more significant in more complex networks.

Acknowledgement. This work was supported in part by the National Natural Science Foundation of China (61771343), the National Key Research and Development Program of China (2018YFE0125400), and the EU H2020 Programme under Marie Curie IF (752979). We are also grateful for the support of the Sino-German Center of Intelligent Systems, Tongji University.

References

1. Garcia-Roger, D., González, E., Martín-Sacristán, D., Monserrat, J.: V2X support in 3GPP specifications: from 4G to 5G and beyond. IEEE Access **8**, 190946–190963 (2020)
2. Boban, M., Kousaridas, A., Manolakis, K., Eichinger, J., Xu, W.: Connected roads of the future: use cases, requirements, and design considerations for vehicle-to-everything communications. IEEE Veh. Technol. Mag. **13**(3), 110–123 (2018)
3. Dar, K., Bakhouya, M., Gaber, J., Wack, M., Lorenz, P.: Wireless communication technologies for ITS applications. IEEE Commun. Mag. **48**(5), 156–162 (2010)
4. Kha, H., Tuan, H., Nguyen, H.: Fast global optimal power allocation in wireless networks by local D.C. programming. IEEE Trans. Wirel. Commun. **11**(2), 510–515 (2011)
5. Shen, K., Yu, W.: Fractional programming for communication systems-part I: power control and beamforming. IEEE Trans. Signal Process. **66**(10), 2616–2630 (2018)
6. Yang, Y., Pesavento, M.: A unified successive pseudoconvex approximation framework. IEEE Trans. Signal Process. **65**(13), 3313–3328 (2017)
7. Cui, Y., Lau, V., Wang, R., Huang, H., Zhang, S.: A survey on delay-aware resource control for wireless systems-large deviation theory, stochastic Lyapunov drift, and distributed stochastic learning. IEEE Trans. Inf. Theory **58**(3), 1677–1701 (2012)
8. Lan, D., Wang, C., Wang, P., Liu, F., Min, G.: Transmission design for energy-efficient vehicular networks with multiple delay-limited applications. In: 2019 IEEE Global Communications Conference (GLOBECOM), pp. 1–6. IEEE (2019)
9. Joung, J.: Machine learning-based antenna selection in wireless communications. IEEE Commun. Lett. **20**(11), 2241–2244 (2016)
10. Matthiesen, B., Zappone, A., Besser, K.L., Jorswieck, E., Debbah, M.: A globally optimal energy-efficient power control framework and its efficient implementation in wireless interference networks. IEEE Trans. Signal Process. **68**, 3887–3902 (2020)
11. Zappone, A., Di Renzo, M., Debbah, M.: Wireless networks design in the era of deep learning: model-based, AI-based, or both? IEEE Trans. Commun. **67**(10), 7331–7376 (2019)
12. Zappone, A., Sanguinetti, L., Debbah, M.: User association and load balancing for massive MIMO through deep learning. In: 2018 Asilomar Conference on Signals, Systems, and Computers, pp. 1262–1266. IEEE (2018)
13. Lee, W., Kim, M., Cho, D.: Deep power control: transmit power control scheme based on convolutional neural network. IEEE Commun. Lett. **22**(6), 1276–1279 (2018)
14. Liang, F., Shen, C., Yu, W., Wu, F.: Towards optimal power control via ensembling deep neural networks. IEEE Trans. Commun. **68**(3), 1760–1776 (2019)
15. Neely, M.J.: Stochastic Network Optimization with Application to Communication and Queueing Systems. Morgan & Claypool Publishers (2010)

16. Li, Y., Sheng, M., Shi, Y., Ma, X., Jiao, W.: Energy efficiency and delay tradeoff for time-varying and interference-free wireless networks. IEEE Trans. Wirel. Commun. **13**(11), 5921–5931 (2014)
17. Dinkelbach, W.: On nonlinear fractional programming. Manage. Sci. **13**(7), 492–498 (1967)
18. Yuille, A., Rangarajan, A.: The concave-convex procedure. Neural Comput. **15**(4), 915–936 (2003)
19. Zhang, Y., Yang, Q.: A survey on multi-task learning. IEEE Trans. Knowl. Data Eng. 1 (2021, early access). https://ieeexplore.ieee.org/abstract/document/9392366
20. Ruder, S.: An overview of multi-task learning in deep neural networks. arXiv preprint arXiv:1706.05098 (2017)
21. Kendall, A., Gal, Y., Cipolla, R.: Multi-task learning using uncertainty to weigh losses for scene geometry and semantics. In: IEEE Conference on Computer Vision and Pattern Recognition, pp. 7482–7491 (2018)
22. Liebel, L., Körner, M.: Auxiliary tasks in multi-task learning. arXiv preprint arXiv:1805.06334 (2018)
23. Access, E.U.T.R.: Further advancements for E-UTRA physical layer aspects (release 9), 3GPP (2010)

Sub-carrier Spacing Detection Algorithm in 5G New Radio Systems

Tong Li$^{(\boxtimes)}$, Hang Long, and Li Huang

Wireless Signal Processing and Network Lab, Key Laboratory of Universal
Wireless Communication, Ministry of Education, Beijing University of Posts
and Telecommunications, Beijing, China
tttli@bupt.edu.cn

Abstract. In 5G New Radio (NR) systems, the configuration of frame is
flexible. There are several sub-carrier spacing (SCS) configurable which
is different from Long Term Evolution (LTE) systems. During the initial
access process, the user equipment cannot know the specific SCS directly
before obtaining time and frequency synchronization with the service cell
like in LTE systems. Therefore, it is necessary to investigate SCS detection. Based on the cyclostationary of Orthogonal Frequency Division
Multiplexing (OFDM) signals, we propose a SCS detection algorithm
with correlation operations on OFDM signals. The proposed algorithm
is theoretically analyzed and the theoretical expression for the detection
error rate is given by integrating theoretical models of the correlation
errors. The algorithm is evaluated and analyzed through simulation in
different scenarios. The analytical and simulation results show that the
proposed algorithm is effective and the performance is roughly consistent
with the theoretical analysis results.

Keywords: NR · SCS · Detection

1 Introduction

The 3rd Generation Partnership Project (3GPP) requires 5G New Radio (NR)
networks to provide user equipments (UE) access experience with ultra-high network capacity, ultra-high reliability and ultra-low network latency anytime and
anywhere [1]. Cell initial access is a process in which the UE can obtain time and
frequency synchronization with the servicing cell and detect the physical layer
cell ID of the servicing cell. In 5G NR systems, the concept of Synchronization
Signal Blocks (SSBs) is newly introduced [2]. In addition, in order to adapt to
different business scenarios, the design of frame structure is very flexible and
the sub-carrier spacing (SCS) of Orthogonal Frequency Division Multiplexing

Supported by the National Natural Science Foundation of China (NSFC) Project
61931005.

H. Gao et al. (Eds.): ChinaCom 2021, LNICST 433, pp. 322–334, 2022.
https://doi.org/10.1007/978-3-030-99200-2_25

(OFDM) symbols in different frequency bands is flexible and variable [3]. Therefore, the cell search process in 5G NR systems is somewhat different from that in Long Term Evolution (LTE) systems. UE cannot directly learn the SCS of the SSB. To takle this problem, it is necessary to carry out the research of SCS detection. After that, frequency domain synchronization can be obtained.

At present, few studies about SCS detection have been presented thus far in the literature. There are some literatures related to the estimation of OFDM parameters. Based on cyclostationary characteristics, a blind algorithm for estimating carrier frequency and symbol offset of OFDM system in Rayleigh multipath channels is proposed in [4]. In [5], the time symbol duration is determined by exploiting the cyclostationnarity of the symbol. In [6], a blind estimation method of OFDM system parameters is proposed. The method sequentially estimate the sampling frequency, the number of sub-carriers and the cyclic prefix (CP) length in OFDM systems. However, these literatures do not give a specific analysis of detection performance.

Against the above backdrop on SCS detection, we propose a new SCS detection algorithm based on the cyclostationarity of OFDM symbols. When the time delay is the number of effective symbol points corresponding to the actual SCS, the peak value of the correlation calculation can be obtained. Based on this feature, the SCS can be determined. The detection performance of the proposed algorithm is both theoretically analyzed and simulated under multiple scenarios. The simulation results show that the proposed detection algorithm is effective. Furethermore, the simulation results are consistent with their theoretical analysis results.

The remainder of this paper is organized as follows. Section 2 mainly describes the cyclostationarity of OFDM symbols. Section 3 designs and elaborates the proposed algorithm. And the performance is analyzed through theoretical derivation. In Sect. 4, the simulation results of the proposed algorithm are analized and compared with theoretical derivation. Finally, concluding remarks are drawn in Sect. 5.

2   System Model and the Cyclostationarity of OFDM Signals

The received OFDM signals after passing through the Additive White Gaussian Noise (AWGN) channel can be expressed as

$$r(t) = s(t) + n(t), \tag{1}$$

where $s(t)$ is the baseband OFDM signal, $n(t)$ is the zero-mean Gaussian white noise, $n(t) \sim CN(0, \sigma_n^2)$ and σ_n^2 represents the noise power. $s(t)$ is defined as follow

$$s(t) = \sum_k \sum_{l=0}^{G-1} c_{k,l} g(t - lT_c - kT_s), \tag{2}$$

where G represents the number of total sampling points of an OFDM symbol, $G = N + D$, N is the number of effective sampling points of an OFDM symbol, D is the length of the CP, T_c and T_s denote the sampling period and the OFDM symbol length, $T_s = GT_c$, $g(t)$ is the rectangular shaped pulse and $c_{k,l}$ represents the l-th sampling point of the k-th symbol in the time-domain. $c_{k,l}$ is defined as follow

$$c_{k,l} = \frac{1}{\sqrt{N}} \sum_{n=0}^{N-1} a_{k,n} e^{(j2\pi(l-D)n/N)}, l = 0, 1, \cdots, G-1, \tag{3}$$

where $a_{k,n}$ denotes the modulated data on the n-th sub-carrier of the k-th symbol in the frequency-domain. The mean and variance of $a_{k,n}$ are 0 and σ_a^2, respectively. $a_{k,n}$ is mutually independent and identically distributed. Moreover, $a_{k,n}$ has some distribution characteristics as follows

$$E[a_{k,n}] = 0, \tag{4}$$

$$E[a_{k,n} a_{m,l}] = 0, \tag{5}$$

$$E[a_{k,n} a_{m,l}^*] = \sigma_a^2 \delta(k-m) \delta(n-l), \tag{6}$$

where δ is the Dirac function.

OFDM signals have the characteristic of second-order cyclostationarity [7, 8,10]. Use $R_r(t,\tau)$ to denote the autocorrelation function of OFDM signals in time-domain. $R_r(t,\tau)$ can be expressed as follow

$$R_r(t,\tau) = \begin{cases} \sigma_a^2 \sum_k \sum_{l=0}^{G-1} g(t - lT_c - kT_s) \cdot g^*(t - lT_c - kT_s - \tau) + R_n(\tau), & |\tau| < T_c \\ \sigma_a^2 \sum_k \sum_{l=N}^{G-1} g(t - lT_c - kT_s) \cdot g^*(t - lT_c - kT_s - \tau_N) + R_n(\tau), & |\tau_N| < T_c \\ 0, & \tau = others \end{cases} \tag{7}$$

where $\tau_N = |\tau| - NT_c$. As can be seen from (7), the OFDM signal is second-order cyclostationary and the period of autocorrelation function is T_s. The autocorrelation function is formulated by

$$R_r(t,\tau) = \begin{cases} R_r(t + T_c, \tau), & |\tau| < T_c \\ R_r(t + T_s, \tau), & |\tau_N| < T_c \end{cases} \tag{8}$$

Hence, the autocorrelation function of the OFDM received signal can be expressed in the form of Fourier series. The absolute value of the autocorrelation function is as follow,

$$|R_s^\alpha(\tau)| = \begin{cases} \frac{\sigma_a^2}{T_c} \left| \frac{\sin[\pi\alpha(T_c - |\tau|)]}{\pi\alpha} \right|, & \alpha = m/T_c, |\tau| \le T_c, m \in Z \\ \frac{\sigma_a^2}{T_s} \left| \frac{\sin(\pi\alpha T_c D)}{\sin(\pi\alpha T_c)} \right| \cdot \left| \frac{\sin[\pi\alpha(T_c - |\tau_N|)]}{\pi\alpha} \right|, & \alpha = m/T_s, |\tau_N| \le T_c, m \in Z \end{cases} \tag{9}$$

where α represents a cycle frequency, $\alpha = m/T_s$, $m \in Z$. Therefore, the autocorrelation spectrum of the OFDM signal appears only when $\alpha = m/T_c$ or $\alpha = m/T_s$. If $\alpha = 0$, $R_s(\tau)$ has peaks only under the condition of $\tau = 0$ and $\tau = NT_c$, and the peaks are σ_a^2 and $\sigma_a^2 D/G$, respectively. The autocorrelation results with other time delays are all zero.

3 Description and Analysis of Sub-carrier Spacing Detection Algorithm

3.1 Algorithm Description

In 5G NR systems, SSB does not support the SCS configuration of 60 kHz [9]. In the frequency range 1 (FR1), the SCS that require detection are 15 kHz and 30 kHz. And in the frequency range 2 (FR2), the SCS that require detection are 120 kHz and 240 kHz. Therefore, there are two candidate SCS in each frequency range. Assuming that the number of effective sampling points of an OFDM symbol corresponding to the actual SCS is N_1, and the other is N_2. Use $r(m)$ to represent the sampling results of time-domain received sequence. The autocorrelation result $R_r(l)$ of $r(m)$ can be expressed as

$$R_r(l) = \frac{1}{M}\text{Re}\left[\sum_{m=0}^{M-1} r(m) \cdot r^*(m-l)\right] = \frac{1}{M}\text{Re}\left[\sum_{m=0}^{M-1} R(m,l)\right], \quad (10)$$

where M and l represent the number of the sampling points and time delay of the autocorrelation calculation, respectively. According to theoretical analysis, $R_r(l)$ obtains the peak value only when the time delay l equals N_1. If the time delay l equals N_2, the expectation of the autocorrelation calculation is zero.

The specific flow of the proposed algorithm is as follows.

- Carry out the autocorrelation calculation of the time-domain received sequence based on the time delay N_1 and N_2, respectively. The results are denoted as $R_r(N_1)$ and $R_r(N_2)$.
- Compare the values of $R_r(N_1)$ and $R_r(N_2)$ and decide the SCS corresponding to the larger one.

Apparently, if $R_r(N_2)$ were larger than $R_r(N_1)$, the detection result would definitely be wrong.

3.2 Analysis of Sub-carrier Spacing Detection Algorithm

Considering AWGN channels, there is additive white Gaussian noise $n(t)$ in the system, $n(t) \sim \mathcal{CN}(0, \sigma_n{}^2)$. The autocorrelation function of the received signal consists of three parts. It is expressed as

$$R_r(l) = R_s(l) + R_n(l) + R_{s,n}(l) + R_{n,s}(l), \quad (11)$$

where $R_s(l)$ and $R_n(l)$ represent the autocorrelation function of the useful signal $s(t)$ and the noise $n(t)$, $R_{s,n}(l)$ and $R_{n,s}(l)$ denote the cross-correlation function of useful signal and noise and $R_{s,n}(l) = R_{n,s}^*(l)$. Due to the limited sampling points at the receiving end for correlation calculations, $R_s(N_1)$ or $R_s(N_2)$ cannot get the theoretical peak values $\sigma_a^2 D/G$ and 0. In addition, the autocorrelation result of noise and the cross-correlation result between noise and useful signal cannot reach 0. These are all components of the error.

The autocorrelation function of the useful signal is as follow

$$R_s\left(l\right) = \frac{1}{M}\mathrm{Re}\left[\sum_{m=0}^{M-1} s\left(m\right)\cdot s^*\left(m-l\right)\right] = \frac{1}{M}\mathrm{Re}\left[\sum_{m=0}^{M-1} S\left(m,l\right)\right].\tag{12}$$

We suppose that the time-domain received signal sequence $s\left(m\right)$ follows a complex Gaussian distribution, i.e., $s\left(m\right) \sim \mathcal{CN}(0,\sigma_a{}^2)$, where $\sigma_a{}^2$ is the variance of the frequency-domain modulation symbols at the transmitter. It is assumed that the different sampling points are independent of each other. Then, the autocorrelation results $R_s\left(N_1\right)$ and $R_s\left(N_2\right)$ with time delay of N_1 and N_2 are assumed to obey Gaussian distributions. Correlation with delay N_1 involves the autocorrelation of the cyclic prefix (CP), so the mean value of $R_s\left(N_1\right)$ is positive. The distribution of $R_s\left(N_1\right)$ is expressed as

$$R_s\left(N_1\right) \sim \mathcal{N}\left(\frac{\sigma_a^2 D}{G}, \frac{2D\sigma_a{}^4 + N\sigma_a{}^4}{2MG}\right).\tag{13}$$

The mean value $\sigma_a^2 D/G$ is the expected theoretical peak of autocorrelation calculation as described in Sect. 3. While the mean value of the autocorrelation calculation results with N_2 as the time delay is 0. The distribution of $R_s\left(N_2\right)$ can be expressed as

$$R_s\left(N_2\right) \sim \mathcal{N}\left(0, \frac{\sigma_n{}^4}{2M}\right).\tag{14}$$

The autocorrelation function of noise is formulated by

$$R_n\left(l\right) = \frac{1}{M}\mathrm{Re}\left[\sum_{m=0}^{M-1} n\left(m\right)\cdot n^*\left(m-l\right)\right] = \frac{1}{M}\mathrm{Re}\left[\sum_{m=0}^{M-1} N\left(m,l\right)\right].\tag{15}$$

Since the noise at different sampling points is independent and identically distributed, the mean value and variance of $R_n\left(l\right)$ are respectively 0 and $\sigma_n^4/2M$. We suppose that $R_n\left(l\right)$ follows a Gaussian distribution, $R_n\left(l\right) \sim \mathcal{N}(0,\sigma_n{}^4/2M)$.

The cross-correlation function of useful signal and noise is as follow

$$R_{s,n}\left(l\right) = \frac{1}{M}\mathrm{Re}\left[\sum_{m=0}^{M-1} s\left(m\right)\cdot n^*\left(m-l\right)\right] = \frac{1}{M}\mathrm{Re}\left[\sum_{m=0}^{M-1} SN\left(m,l\right)\right].\tag{16}$$

The useful signal and noise are independent of each other. According to the approximate results of useful signal and noise, the cross-correlation function $R_{s,n}\left(l\right)$ is approximated as a Gaussian distribution, $R_{s,n}\left(\tau\right) \sim \mathcal{N}\left(0,\sigma_a{}^2\sigma_n{}^2/2M\right)$.

According to the above discription, the detection error rate of SCS under AWGN channel is calculated as

$$\begin{aligned}P_e &= P\left[R_r\left(N_2\right) > R_r\left(N_1\right)\right]\\ &= P\{R_n\left(N_2\right) + R_s\left(N_2\right) + 2\mathrm{Re}\left[R_{s,n}\left(N_2\right)\right]\\ &\quad -R_n\left(N_1\right) - R_s\left(N_1\right) - 2\mathrm{Re}\left[R_{s,n}\left(N_1\right)\right] > 0\}\end{aligned}.\tag{17}$$

Let

$$\xi = R_n(N_2) + R_s(N_2) + 2\text{Re}\left[R_{s,n}(N_2)\right] - R_n(N_1) - R_s(N_1) - 2\text{Re}\left[R_{s,n}(N_1)\right].$$
(18)

According to the above analysis, the distribution of ξ is expressed as

$$\xi \sim \text{N}\left(\frac{-\sigma_a^2 D}{G}, \frac{\sigma_n{}^4 + \sigma_a{}^4 + 2\sigma_a{}^2\sigma_n{}^2}{M} + \frac{D\sigma_a{}^4}{2MG}\right).$$
(19)

Therefore, the SCS detection error rate is calculated as

$$P_e = P(\xi > 0)$$
$$= Q\left(\frac{\sigma_a^2 D/G}{\sqrt{\frac{\sigma_n{}^4 + \sigma_a{}^4 + 2\sigma_a{}^2\sigma_n{}^2}{M} + \frac{D\sigma_a{}^4}{2MG}}}\right)$$
(20)
$$= Q\left(\frac{\sigma_a^2 D\sqrt{2M}}{\sqrt{2G^2(\sigma_n{}^4 + \sigma_a{}^4 + 2\sigma_a{}^2\sigma_n{}^2) + DG\sigma_a{}^4}}\right).$$

The detection error rate is related to signal-to-noise ratio (SNR) and the number of sampling points M. Moreover, the detection performance has nothing to do with the specific SCS. When other conditions are the same, the detection performance of different SCS is roughly consistent. In addition, the complexity of the proposed detection algorithm primarily depends on the number of conjugate multiplications M. Therefore, the larger the number of conjugate multiplications, the better the detection performance. However, the complexity also increases at the same time.

When SNR approaches infinity, the theoretical limit of detection error rate is formulated by

$$P_e \overset{\sigma_n{}^2 \to 0}{=} Q\left(\frac{\sigma_a^2 D\sqrt{2M}}{\sqrt{2G^2(\sigma_n{}^4 + \sigma_a{}^4 + 2\sigma_a{}^2\sigma_n{}^2) + DG\sigma_a{}^4}}\right) = Q\left(\frac{D\sqrt{2M}}{\sqrt{2G^2 + DG}}\right).$$
(21)

4 Algorithm Simulation and Result Analysis

This section presents the simulation and evaluation results of the SCS detection method. The impact of the number of sampling points and the frequency-domain resources occupied by the signal on performance is evaluated. The simulation results and theoretical analysis results are compared and analyzed. Simulation parameters are listed in Table 1.

Table 1. Simulation parameters.

Simulation parameter	Value
Frequency range	FR1
Modulation mode	QPSK
Bandwidth	10 MHz
Sampling frequency	15.36 MHz
Candidate SCS	15 KHz, 30 KHz
Channel modle	AWGN

4.1   Simulation Results

Simulation results and analysis are given below. The detection correct rate of different SCS with $M = 10^5$ is plotted in Fig. 1. As can be seen from Fig. 1, due to the huge number of sampling points, the proposed detection algorithm performs pretty well. When SNR is about -11 dB, the correct rate can reach 95%. When SNR is about -8 dB, the correct rate closes to 1. The detection performance of 15 kHz and 30 kHz is consistent. Furthermore, the simulation results are consistent with the theoretical analysis results.

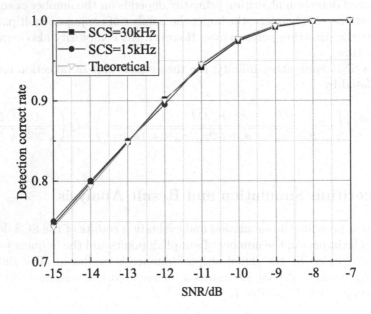

Fig. 1. The detection correct rate of different SCS with $M = 10^5$.

Figure 2 shows the detection correct rate of different SCS with $M = 1096$. As the number of sampling points used to calculate the aotocorrelation results is relatively small, the detection performance is degraded. When SNR is about 4dB, the correct rate can reach 95%. When the SNR is in small range (SNR < 0 dB) and the detection error is primarily caused by noise, the performance of 15 kHz and 30 kHz is almost consistent with the theoretical analysis results. However, there are some errors in the limit performance between the simulation results and theoretical analysis when SNR is large. In this case, the detection performance mainly depends on the error of the useful signal autocorrelation results $R_s(N_1)$ and $R_s(N_2)$. The sampling points of time-domain sequence for correlation calculations do not obey completely independent Gaussian distributions. However, such assumptions are made in theoretical analysis. Therefore, this will cause certain error between the simulation results and theoretical analysis.

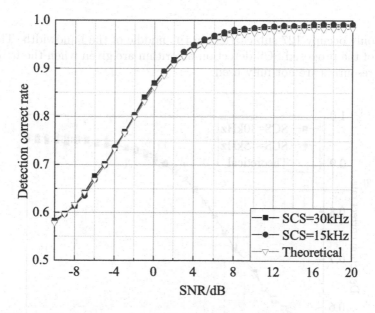

Fig. 2. The detection correct rate of different SCS with $M = 1096$.

It can be seen from Fig. 1 and Fig. 2 that the number of sampling points for autocorrelation calculation is a key factor impacting on the performance of SCS detection. Table 2 shows the detection correct rate with different M when SNR= −10 dB. Within a certain range, increasing M will significantly improve the detection performance and the simulation results are consistent with the conclusions of the theoretical analysis. On the other hand, the complexity of the detection algorithm $\mathcal{O}(M)$ will increase with the increase of M. Therefore, when selecting the number of sampling points for detection, both the detection performance and algorithmic complexity should be taken into account.

In the above simulation scenarios, the signal occupies all the frequency-domain resources. However, in the actual communication systems, the signal does not occupy all the sub-carriers in frequency-domain. In addition, the primary synchronization signals (PSS) in SSB block used in the synchronization

Table 2. Detection correct rate with different M when SNR = −10 dB.

Number of sampling points	Detection correct rate in simulation	Detection correct rate of theoretical analysis
1096	57.7%	57.8%
1096 × 5	67.8%	67.1%
1096 × 10	73.6%	73.4%
1096 × 50	91.7%	91.9%
1096 × 100	97.7%	97.6%
1096 × 120	98.5%	98.5%

process only occupy 127 sub-carriers in the middle of the bandwidth. The evaluations of the proposed SCS detection algorithm are given when the frequency-domain resources are not fully used.

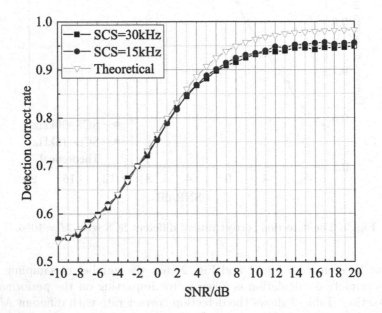

Fig. 3. The detection correct rate when the signals only occupy half of the frequency-domain resources.

Figure 3 and Fig. 4 plot the SCS detection correct rate when the signals only occupy half and one-quarter of the frequency-domain resources, respectively. The power of signal decreases as the frequency-domain resources reduces. Therefore, the theoretical analysis results are respectively 3 dB and 6 dB lower than the condition where all the frequency-domain resources are occupied. When SNR is high, the simulation performance deviates from the theoretical analysis.

When signal only occupies half of the frequency-domain resources, the theoretical highest detection correct rate is about 98.3%. And the simulation results of SCS with 15 kHz and 30 kHz are 95.5% and 94.8%, respectively. When the signal only occupies a quarter of the frequency-domain resources, the theoretical highest detection correct rate is still about 98.3%. However, the simulation results with SCS of 15 kHz and 30 kHz are 89.0% and 88.2%, respectively. As the frequency-domain resources decrease, the best performance of the proposed SCS detection algorithm gradually deteriorates. The sampling points of the signal sequence in time-domain are no longer independent when the number of sub-carriers drops significantly. Moreover, the distribution of the received signal does not obey Gaussian distribution as assumed in theoretical derivation process. As a result, the variance of the correlation results also increases relative to the theoretical analysis. It can be concluded that the reduction of frequency-domain resources undermines the performance of the SCS detection algorithm.

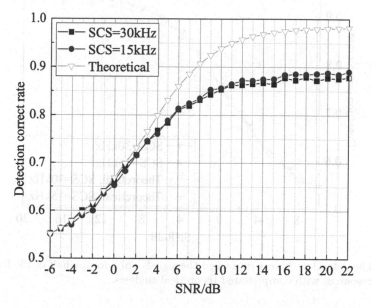

Fig. 4. The detection correct rate when the signals only occupy one-quarter of the frequency-domain resources.

Based on a large number of simulation results, the loss of performance can be analyzed quantitatively. When SNR is infinite, the noise can be ignored and the performance of SCS detection algorithm depends on the error of the correlation result $R_s(N_1)$ and $R_s(N_2)$ of the signal. According to the simulation statistics, if the useful signal occupies $\frac{1}{n}$ of all frequency-domain resources, $n \in Z$, the statistical variance of $R_s(N_1)$ increases to n times that of the theoretical analysis. For the case where the actual SCS is 30 kHz, the statistical variance of $R_s(N_2)$ increases to n times that of the theoretical analysis as well. While for the case where the actual SCS is 15 kHz, the statistical variance of $R_s(N_2)$ increases to $\frac{n}{2}$ times that of the theoretical analysis. When the actual SCS is 15 kHz, the variance of the simulation statistics result of $R_s(N_2)$ is smaller than that of 30 kHz. If the correlation calculation is based on OFDM symbols of 15 kHz SCS with the time delay of N_2, nearly half of the sampling points for conjugate multiplication are in the same symbol. Therefore, they are not completely independent. However, if the correlation calculation is based on OFDM symbols of 30 kHz SCS with the time delay of N_2, the sampling points for conjugate multiplication are in two different symbols and they are completely independent. This is the reason why the detection performance of 15 kHz is slightly better than that of 30 kHz.

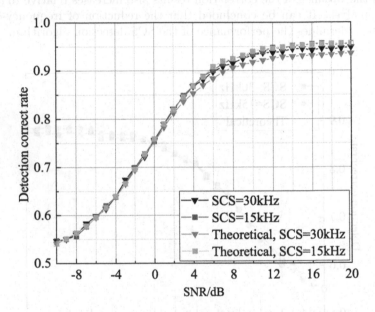

Fig. 5. The detection correct rate when the signals only occupy half of the frequency-domain resources with compensated theoretical analysis.

Based on the above simulation and analysis results, the variance of the theoretical analysis results with the SCS of 15 kHz and 30 kHz can be compensated for the situation that the signal does not occupy all the frequency-domain resources. Figure 5 and Fig. 6 show the detection performance after the compensation of the theoretical analysis results. It can be seen from the figures that the theoretical analysis result after compensation is closer to the actual simulation results, and the performance of 15 kHz is slightly better than that of 30 kHz.

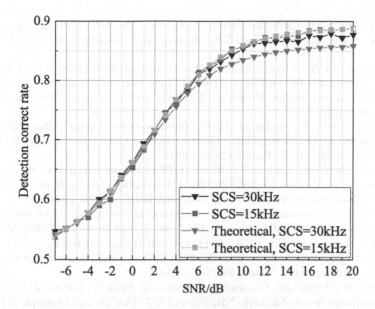

Fig. 6. The detection correct rate when the signals only occupy one-quarter of the frequency-domain resources with compensated theoretical analysis.

5 Conclusion

In 5G NR systems, UE can not get the actual SCS directly in the cell initial access process. Therefore, the SCS needs to be detected. A SCS detection algorithm based on the cyclostationarity of OFDM signals is proposed and the performance is evaluated with both simulations and theoretical analysis.

The simulation results are roughly consistent with the theoretical analysis and the detection performance of different SCS is almost symmetrical. The performance of the proposed algorithm improves as the number of sampling points and the SNR increase. According to the theoretical analysis, when the signal does not occupy all the frequency-domain resources, the performance deteriorates, which is only caused by the reduction of signal power. However, the errors excluded in theoretical analysis cannot be ignored during the actual detection process. Hence, the performance of the simulation is lower than the theoretical analysis. In this case, the theoretical analysis results can be compensated according to different SCS to make them roughly consistent with simulation results.

References

1. Liu, J., Au, K., Maaref, A.: Initial access, mobility, and user-centric multi-beam operation in 5G new radio. IEEE Commun. Mag. **56**(3), 35–41 (2018)
2. Lin, Z., Li, J., Zheng, Y.: SS/PBCH block design in 5G new radio (NR). In: 2018 IEEE Globecom Workshops (GC Wkshps), pp. 1–6 (2018)

3. 3GPP TS 38.213: 3rd Generation Partnership Project; Technical Specification Group Radio Access Network; NR; Physical layer procedures for control (Release 16)
4. Zhang, J., Li, Y.: Cyclostationarity-based symbol timing and carrier frequency offset estimation for OFDM system. In: International Conference on Computer Application and System Modeling (ICCASM 2010), Taiyuan, China, pp. V5-546–V5-550 (2010)
5. Walter, A., Eric, K., Andre, Q.: OFDM parameters estimation a time approach. In: Conference Record of the Thirty-Fourth Asilomar Conference on Signals, Systems and Computers. IEEE, Pacific Grove (2000)
6. Shi, M., Bar-Ness, Y., Su, W.: Blind OFDM systems parameters estimation for software defined radio. In: 2007 2nd IEEE International Symposium on New Frontiers in Dynamic Spectrum Access Networks. IEEE, Dublin (2007)
7. Heath, R.W., Giannakis, G.B.: Exploiting input cyclostationarity for blind channel identification in OFDM systems. IEEE Trans. Signal Process. 47, 848–856 (1999)
8. Al-Habashna, A., Dobre, O. A.: Cyclostationarity-based detection of LTE OFDM signals for cognitive radio systems. In: 2010 IEEE Global Telecommunications Conference GLOBECOM. Miami, FL, USA (2010)
9. 3GPP TS 38.300: 3rd Generation Partnership Project; Technical Specification Group Radio Access Network; NR; NR and NG-RAN Overall Description (Release 16)
10. Li, Z.: The research about spectral correlation of subcarrier of OFDM and blind recoginition about modulation type. In: 2011 International Conference on Consumer Electronics, Communications and Networks. IEEE, Xianning (2011)

Pre-handover Mechanism in the Internet of Vehicles Based on Named Data Networking

Gaixin Wang$^{(\boxtimes)}$, Zhanjun Liu, and Qianbin Chen

School of Communication and Information Engineering, Chongqing University of Posts and Telecommunications, Chongqing, China
wgx1513073331@163.com

Abstract. Named Data Networking (NDN) is considered to be one of the most promising designs for the next generation of network architecture. Vehicles are typically interested in the content itself than the location of the host. Owing to the content-centric attributes, NDN can support highly dynamic topologies more than the traditional host-centric communication mode. So it will be widely used in vehicluar networks. The moving vehicles are constantly handover between different roadside units (RSU), which affects the backhaul of data packets and increases transmission delay. Therefore handover in Internet of Vehicles (IoV)-based NDN is a challenge to be solved. In order to improve communication quality, this paper proposes a pre-handover forwarding mechanism based on RSU level and content popularity by establishing a probability model. Routing data packets in advance by predicting whether the content requested by the vehicle is cached in the RSU after handover. Thus the mechanism not only reduces delay, but also solves the problem of packet loss caused by network topology changes. In addition, in order to further improve content retrieval efficiency and reduce delay, this paper also proposes a method based on the tabu search (TS) algorithm to quickly search for the target vehicle within the coverage of the RSU. Simulation results show that the mechanism proposed in this paper outperforms existing related mechanisms in terms of average delay and delivery rate.

Keywords: Named data networking · Internet of Vehicles · Predication · Forwarding

1 Introduction

The Internet of Vehicles (IoV), a technical basis for intelligent transportation systems (ITS), plays an important role in the construction of smart cities. With the increasing demand for vehicle service, the IoV is attracting extensive attention in the field of communications. Each vehicle is equipped with an on-board unit (OBU) for vehicle-to-vehicle (V2V) and vehicle-to-infrastructure (V2I) communications in the IoV [1]. When the IoV is applied to a host-centric communication mode, the data packets are routed using IP addresses [2]. Due

© ICST Institute for Computer Sciences, Social Informatics and Telecommunications Engineering 2022
Published by Springer Nature Switzerland AG 2022. All Rights Reserved
H. Gao et al. (Eds.): ChinaCom 2021, LNICST 433, pp. 335–347, 2022.
https://doi.org/10.1007/978-3-030-99200-2_26

to vehicles with specific addresses constantly change their locations, IP-based communication mode can no longer achieve stable and fast data access in such a dynamic network. To solve this problem, Named Data Networking (NDN) is applied to the IoV [3]. NDN changes the traditional host-centric communication mode to data-centric communication mode. The IoV-based NDN can not only improve the efficiency of content retrieval [4], but also improve the scalability of the network. Therefore the application of NDN in the IoV has become a research hotspot. In the IoV-based NDN, some areas concerning data packet forwarding [5,6], cache strategy [7,8], and naming method [9], etc., have attracted the wildly attention of researchers.

In the IoV-based NDN, consumer (vehicle) sends interest packets to the road-side unit (RSU) connected to it [10]. RSU is a communication infrastructure that mainly provides information such as road conditions and forwards data. In the IoV-based NDN, RSU is considered as a node in the NDN network and has the function of in-network caching. Consumer sends interest packets to the network, then the router in the network retrieves the corresponding content according to the interest packet. The content publisher (RSU or base station (BS)), encapsulates the content data into a data packet, then the data packet will return to the consumer along the forwarding path of the interest packet [11]. Due to the mobility of vehicles, the symmetric routing mechanism of NDN is destroyed, which may make the data packets loss in the forwarding process. Therefore the quality of experience (QoE) and quality of service (QoS) will be affected. For this reason, the research focus of the IoV-based NDN is to improve the success rate of backhaul data packet and communication efficiency in a mobile environment. To reduce the transmission delay, several mechanisms have been proposed. Handover of V-NDN based on RSU-assisted is proposed in [12], this mechanism is only considered V2R (vehicle-to-RSU) communication, without considering BS. In order to not interrupt the communication service, R. Chen et al. [13] proposed a fast handover mechanism in named data networking-railway, but this mechanism is only applicable to high-speed rail scenario. The schemes that reduce the authentication latency are proposed in [14,15], so that mobile node can be authenticated in advance. An efficient handover [16] reduces the number of redundant handovers based on the received signal strength and wireless transmission loss in hierarchical cell networks.

Although the handover mechanisms for NDN and IP can reduce delay and improve communication quality. However, when the continuous changes in network topology occurs, these mechanisms cannot cope with the backhaul of data packets when consumers change frequently. They also cannot achieve seamless handover. Therefore, in this paper we propose a pre-handover mechanism based on RSU level and content popularity. The main contributions in this paper can be summarized as follows:

1. In the IoV-based NDN scenario, consumers are constantly moving, which destroys the symmetric routing mechanism of NDN and affects the backhaul of data packets. To solve the problem, this paper proposes a probabilistic model that considers both RSU level and content popularity, which used for predicting whether the content requested by the vehicle is cached in the RSU after handover.

2. When the RSU is going to return a data packet to the requesting vehicle, the idea of the tabu search (TS) algorithm [17] can be applied to search for the target vehicle within the coverage of the RSU to improve the backhaul efficiency.
3. Simulation is performed to show that the proposed mechanism in this paper can not only effectively reduce the delay of the vehicle handover between different RSUs, but also improve the success rate of the backhaul data packets.

The remainder of this paper is organized as follows. In Sect. 2, we illustrate the flow of the pre-handover mechanism and explain the probabilistic model for prediction in detail. In Sect. 3, a method of searching for the target vehicle quickly within the coverage of the RSU is proposed. Simulation results and analyses are presented is Sect. 4. Conclusions are drew in Sect. 5.

2 Proposed Pre-handover Mechanism

2.1 Pre-handover Application Scenario

At the road intersection, Due to the movement of the vehicle, the pre-handover process between different RSUs is shown in Fig. 1, which includes three parts: vehicle, RSU and BS. RSU is considered as a node in the NDN network. The data structure in each NDN node consists of three parts: Content Store (CS), Pending Interest Table (PIT) and Forwarding Information Base (FIB), where CS is used to store data, PIT is used to record the data request interface information that the node has forwarded interest packet but not yet satisfied, FIB stores forwarding the interface information and routes the interest packet according to the routing table. The process of node processing data packet is shown in Fig. 2.

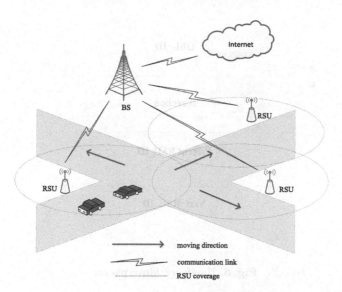

Fig. 1. Application scenario

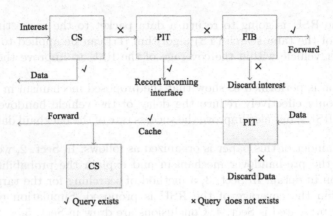

Fig. 2. The routing process of NDN node

2.2 The Routing Process of Data Packets

When the consumer sends an interest packet within the RSU coverage, the RSU will forward the interest packet to the producer (the content provider). According to the characteristic that the data packet returns along the forwarding path of the interest packet in NDN, the data packet will be forwarded back to the vehicle. Due to the mobility of the vehicle, vehicles in the RSU coverage change dynamically. At this time, the vehicle is going to leave the current RSU coverage, but has not received the backhaul data packet, the vehicle will send a Handover_Move packet to the RSU that is going to leave. The specific data packet format is shown in Fig. 3.

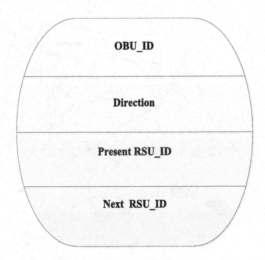

Fig. 3. Handover_Move packet

The Handover_Move data packet contains the vehicle number (OBU_ID), the direction of the vehicle (Direction), the present RSU_ID and the next RSU_ID. Among them, OBU_ID is the identity of the consumer, it facilitates the RSU to search for the target vehicle within its coverage. Due to the vehicle may be in a straight lane or at intersection, the direction is the direction of motion of the consumer after the handover. The present RSU_ID and the next RSU_ID are to prevent ping-pong handover [18]. For the ease of presentation, we denote RSU_A as the RSU before the handover, RSU_B as the RSU after the handover. After prediction, if the RSU_B is cached the content requested by the vehicle, the RSU_A will directly forward the interest packet to the RSU_B. The RSU_B will return a data packet to the requesting vehicle. If the RSU_B is not cached, the RSU_A will forward the interest packet to the BS while the vehicle is being switched. The BS returns the data packet to the RSU_B, then the data packet will be forwarded to the target vehicle through the RSU_B. The pre-handover routing processing is shown in Fig. 4.

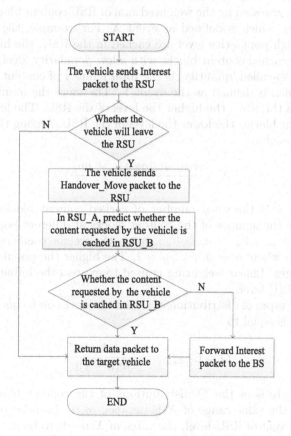

Fig. 4. Pre-handover routing process

2.3 Probabilistic Prediction Model

Because content caching is at the block level, the set of information items are cached in each RSU is denoted by $I = \{1, ..., i\}$, where i is the number of cached information items in RSU, there are j content blocks in each information item. Therefore there is a total of $n = j * i$ content blocks in each RSU. The popularity of content is characterized by the Zipf distribution. The content blocks cached in the RSU are divided into K categories according to their popularity, the set of the content category about popularity is denoted by $K = \{1, ...k\}$, where k is the popularity of the content block, where the popularity of 1 is greater than 2, etc.

Firstly, this paper classifies the RSU according to the two parameters of the 'quality' and 'quantity' of the cached content in the RSU. The so-called 'quality' refers to the popularity of cached content in the RSU. The quality of content in RSU increases as the popularity level of cached content increases. Since the content blocks in RSU are divided into K categories according to popularity, the 'quality' can be expressed by the weighted sum of RSU content blocks in different popularity levels, which is defined as event X. For example, the more content blocks with a high popularity level are cached in the RSU, the higher the RSU level. The more cached content blocks with a low popularity level, the lower the RSU level. The so-called 'quantity' is the total number of content blocks cached in the RSU, which is defined as the event Y. The more the number of content blocks cached in the RSU, the higher the level of the RSU. The less the number of cached content blocks, the lower the level of the RSU. Among them, the event X can be expressed as:

$$X = \frac{\alpha n_1 + \beta n_2 + ... + \eta n_k}{n} \tag{1}$$

In formula (1), n is the total number of cached content blocks in the RSU. $n_1, n_2, ..., n_k$ are the number of the content blocks at different popularity levels, where $n_1 + n_2 + ... + n_k = n$. $\alpha, \beta, ..., \eta$ are the weight coefficient of different popularity level, where $\alpha + \beta + ... \eta = 1$. The higher the popularity level, the greater the weight. Linear weighting is used to express the influence of content popularity on RSU level.

There are Q types of distribution of n content blocks on k content popularity levels, where Q is equal to

$$Q = C_{n+k-1}^{k-1} = \frac{(n + k - 1)!}{n!(k - 1)!} \tag{2}$$

through the analysis of the Q distributions of the content blocks. It can be concluded that the value range of X is between $[\eta, \alpha]$. In order to facilitate the analysis of the event of RSU level, the value of X needs to be normalized to the interval of 0 to 1, which can be shown as follows:

$$X_{nor} = X' = a + k(X - X_{\min}) \tag{3}$$

where $k = \frac{(b-a)}{X_{max}-X_{min}}$, $a = 0$, $b = 1$, X_{min} and X_{max} are the minimum and maximum values of X respectively.

Event Y can be expressed as:

$$Y = m * n + l \tag{4}$$

where m and l are the coefficient of the linear function. The relationship between the number of cached content blocks in the RSU n and the RSU level G is linear positive correlation. Consequently, the RSU level increases with the increase of the number of cached content blocks in the RSU. In order to comprehensively evaluate the event of RSU level, the results of event Y need to be standardized, which can be expressed as follows:

$$Y' = \frac{\log_{10}(m * n + 1)}{\log_{10}(m * n_{\max} + l)} \tag{5}$$

where n_{max} is the maximum value of the cached content in the RSU. Performing logarithmic conversion standardization processing for Y, to make the range of Y' between 0–1.

The RSU level is jointly determined by the event X and the event Y. Consequently the RSU level G can be expressed as:

$$G = X' * Y' = \left(\frac{\alpha n_1 + \beta n_2 + ... + \eta n_k}{n}\right)_{nor} * \frac{\log_{10}(m * n + 1)}{\log_{10}(m * n_{\max} + l)} \tag{6}$$

where $\left(\frac{\alpha n_1 + \beta n_2 + ... + \eta n_k}{n}\right)_{nor}$ represents the influence of the content blocks at different popularity levels on the RSU level. The value range of X' is 0–1. In order to make the two events X and Y to determine RSU-level together, $\frac{\log_{10}(m*n+l)}{\log_{10}(m*n_{\max}+l)}$ represents the influence of the number of content blocks in the RSU. It needs to be standardized by logarithmic transformation so that its value is also between 0-1. Only in this way can a comprehensive evaluation of the RSU level be carried out. The quantitative processing of the RSU level is shown as follows:

$$G' = \begin{cases} 1 & if \ 0 < G' \leq 0.2 \\ 2 & if \ 0.2 < G' \leq 0.4 \\ 3 & if \ 0.4 < G' \leq 0.6 \\ 4 & if \ 0.6 < G' \leq 0.8 \\ 5 & if \ 0.8 < G' \leq 1 \end{cases} \tag{7}$$

P is the probability of the content requested by the vehicle is cached in the RSU_B. It can be calculated as follows:

$$P = v * G + \mu * p \tag{8}$$

where G is the level of RSU, p is the popularity of the requested content. The probability of the content is cached in RSU is determined by the level of RSU G and the popularity of the requested content p. ν and μ are the weight of event

X and event Y, and $\mu > \nu$, $\nu + \mu = 1$. The distribution characteristic of the content request conforms to the Zipf-Mandelbrot distribution. Most users access a little part of the most popular content in the network, while most part of the content is rarely accessed. Hence, the higher the popularity of the content cached in the RSU, the greater the probability of being requested. The above probability model obtains the probability of the content requested by the vehicle is cached in the RSU_B. We can judge whether the RSU has cached the requested content through a preset threshold θ_p. If $P \geq \theta_p$, the requested content is cached in the RSU, if $P < \theta_p$, otherwise.

3   Routing of Backhaul Data Packets

The TS algorithm is an extension of the local search algorithm. In order to improve the shortcomings that the local search is easy to fall into the local optimal point. The TS algorithm introduces a tabu table to record the local optimal points that have been searched. In the next search, the information in the tabu table is no longer searched or selectively search, to jump out of the local best, and finally achieve global optimization. The so-called tabu is to prevent the repetition of the previous work. The algorithm applied to this paper is to quickly search and find the random mobile consumer within the coverage of the RSU. Since each vehicle is equipped with an OBU, it has a unique OBU_ID. Under the coverage of a RSU, the TS algorithm is used to search and match the OBU_ID of the vehicle, so that the backhaul data packet can be returned to the requesting vehicle at the fastest speed, which can ultimately reduce the communication delay. Moreover, the loss of data packet caused by vehicle movement will be mitigated. The specific search process is shown in Algorithm 1.

Algorithm 1. Routing of backhaul data packets

1: Collecting global nodes(vehicles) under the coverage of the RSU
2: Establishing dynamic candidate list SET_N(x) and an empty tabu list SET_T(x)
3: Once the vehicle leaves the coverage of the RSU, the corresponding OBU_ID of the vehicle needs to be deleted from the SET_N(x) in real time
4: Randomly select a node from the dynamic candidate list SET_N(x) for OBU_ID matching
5: **for** there are unsearched nodes in SET_N(x) and the target node is not found **do**
6:    **if** matching succeeded **then**
7:        the data packet is transmitted to the corresponding vehicle
8:    **else**
9:        the OBU_ID is added to the tabu list SET_T(x) prevent repeated searches
10:       In SET_N(x), randomly select a node from the neighborhood of the failed node to match
11:   **end if**
12: **end for**

---

4 Simulation Results and Discussions

In the simulation, this paper assumes that the position of the vehicle is randomly placed. The RSU is set to have a certain buffer to store data, while the BS is set to an infinite buffer to store data. In addition, there is no transmission conflict between them. The simulation is mainly aimed at analyzing the performance of traditional handover, V-NDN handover and pre-handover. The traditional handover based on IP, the vehicle needs to re-establish the connection with the RSU after the handover. The V-NDN handover based on NDN, the vehicle need to resend the request interest packet to the RSU_B after the handover. If the requested content is not cached in RSU_B, the interest packet must be forwarded to the content producer, this will increase the delay. The pre-handover based on NDN is to predict whether the requested content is cached in the RSU_B before the handover, the RSU_A routes data packets in advance to reduce transmission delay. This part mainly analyzes the average delay consumed during data packet transmission and the successful delivery rate of data packets. The simulation parameters are shown in Table 1.

Table 1. Simulation parameters

Parameter	Value
Number of consumers	5, 10, 15, 20, 25, 30, 35, 40, 45, 50
Number of RSUs	4
Number of BSs	2
RSU Range	400 m
Cache replacement	Least Recently Used (LRU)
Simulation time	100 s
Communication delay between RSUs	10 ms
Packet lifetime	2 s

The average delay can reflect the real-time nature of data transmission. Smaller average delay means better communication service quality. The successful delivery rate of data packets can reflect the stability of the data communication link. The higher the packet delivery rate, the better the stability of routing data transmission.

In order to evaluate the proposed pre-handover mechanism, we compare the average delay of different handover mechanisms. As shown in Fig. 5, as the popularity of the requested content increases, the handover performance of pre-handover is significantly better than the other two handover mechanisms. In the pre-handover mechanism, the RSU_A predicts whether the content requested by the vehicle is cached in RSU_B. If the requested content is not cached in RSU_B, the average delay is 40 ms. While if the RSU_B is cached, the average delay is 20 ms. The reason is that the pre-handover mechanism can handover the data packet in advance in RSU_A, which reduces the delay and improves content retrieval efficiency.

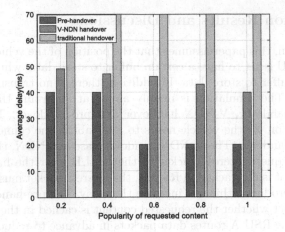

Fig. 5. The effect of requested content popularity on average delay

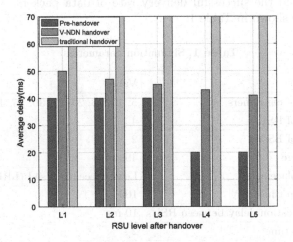

Fig. 6. The effect of RSU level after handover on average delay

Figure 6 shows the changes of the average delay with the level of the RSU_B increases. Compared with traditional handover, if the content is not cached, the pre-handover can reduce about 43% of the average delay. If the content is cached, the average delay can reduce about 71%. Compared with V-NDN handover, the pre-handover can reduce about 53% of the average delay in the highest RSU level. The reason is that traditional handover is based on IP, the change of RSU level has no effect on it. While the V-NDN handover and the pre-handover are both based NDN, the pre-handover can route data packets in advance through prediction, which can reduce transmission delay.

Figure 7 compares the delivery rate of different handover mechanisms as the number of requested vehicles increases. As shown in Fig. 7, the delivery rate of the pre-handover is significantly higher than that of the V-NDN handover.

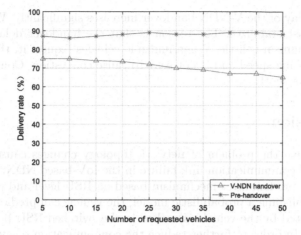

Fig. 7. The effect of the number of requested vehicles on delivery rate

As the number of requested vehicles increases, the pre-handover delivery rate hardly changes. While the delivery rate of the V-NDN handover has a certain downtrend. The reason is that the pre-handover process is completed through the cooperation of two RSUs. the handover delay is short. When there are many requesting vehicles, there is less pressure on the link, therefore the success rate of data packet delivery is high.

Figure 8 compares the change of the average delay of the pre-handover and the V-NDN handover, as the number of requested vehicles increases. In terms of delay, the pre-handover performance is better than V-NDN handover. When there are less requesting vehicles, the average delay of the two handover mechanisms changes the same. However, as the number of requesting vehicles increases,

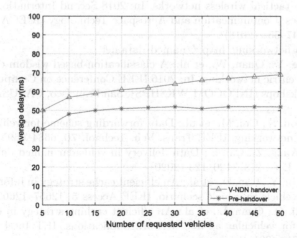

Fig. 8. The effect of the number of requested vehicles on average delay

the average delay of the V-NDN handover increases significantly. While the pre-handover almost stabilizes. The reason is that the handover delay of the pre-handover mechanism is little. When multiple vehicles request it, the mechanism has fast processing speed and is not easy to cause congestion. Consequently the average delay is short.

5   Conclusion

In order to solve the problem of network topology changes caused by vehicle movement and communication link failure in the IoV-based NDN. In this paper we proposed a pre-handover mechanism based on RSU level and content popularity. By establishing a probabilistic model, the vehicle can predict whether the content requested by the vehicle is cached in the switched RSU before the handover process. In order to further reduce the communication delay and increase the delivery rate of data packets, this paper also proposed a search method based on the TS algorithm, which is used to effectively search for the target vehicle within the coverage of the RSU. Finally, we analyzed the performance of the pre-handover mechanism from two aspects: average delay and delivery rate. In future research, we will further study the data transmission rules when both the requester and the publisher are on the move, and reduce the network consumption as much as possible in the combined V2R and V2V communication mode to improve the network transmission performance.

References

1. Wang, A., Chen, T., Chen, H.: NDNVIC: named data networking for vehicle infrastructure cooperation. IEEE Access **7**, 62231–62239 (2019)
2. Selvakumar, G., Ramesh, S., Ungati, S., et al.: IP packet forwarding mechanism in multi-hop tactical wireless networks. In: 2018 Second International Conference on Electronics, Communication and Aerospace Technology (ICECA), Coimbatore, India, pp. 547–551 (2018)
3. Named data networking. http://named-data.net
4. Guan, J., He, Y., Quan, W., et al.: A classification-based wisdom caching scheme for content centric networking. In: 2016 IEEE Conference on Computer Communications Workshops (INFOCOM WKSHPS), San Francisco, CA, USA, pp. 822–827 (2016)
5. Hou, R., Zhou, S., Cui, M., et al.: Data forwarding scheme for vehicle tracking in named data networking. IEEE Trans. Veh. Technol. **70**, 6684–6695 (2021)
6. Wang, X., Wang, Z., Cai, S.: Data delivery in vehicular named data networking. IEEE Netw. Lett. **2**(3), 120–123 (2020)
7. Zhao, W., Qin, Y., Gao, D., et al.: An efficient cache strategy in information centric networking vehicle-to-vehicle scenario. IEEE Access **5**, 12657–12667 (2017)
8. Doan, D., Ai, Q., Quan, L., et al.: An efficient caching strategy in content centric networking for vehicular ad-hoc network applications. IET Intel. Transp. Syst. **12**(7), 703–711 (2018)

9. Wang, Q., Wu, Q., Zhang, M., et al.: Learned bloom-filter for an efficient name lookup in information-centric networking. In: 2019 IEEE Wireless Communications and Networking Conference (WCNC), Morocco, pp. 1–6 (2019)
10. Khelifi, H., Luo, S., Nour, B.: Named data networking in vehicular ad hoc networks: state-of-the-art and challenges. IEEE Commun. Surv. Tutor. **22**(1), 320–351 (2020)
11. Selvan, E., Manisha, G., Ramkumar, M., et al.: Interest forwarding strategies in vehicular named data networks. In: International Conference on Computation of Power, Energy, Information and Communication, pp. 53–57 (2019)
12. Rui, H., Shou, Z., Yu, J., et al.: Research on RSU-assisted data forwarding method in V-NDN. J. SouthCent. Univ. Natl. (Nat. Sci. Edn.) **40**(1), 74–79 (2021)
13. Chen, R., Wang, Y., Fan, W., et al.: Fast handover for high-speed railway via NDN. In: 2018 1st IEEE International Conference on Hot Information-Centric Networking (HotICN), Shenzhen, pp. 167–172 (2018)
14. Li, C., Nguyen, T., Nguyen, L., et al.: Efficient authentication for fast handover in wireless mesh networks. Comput. Secur. **37**(1), 124–142 (2013)
15. Choi, J., Jung, S.: A handover authentication using credentials based on chameleon hashing. IEEE Commun. Lett. **14**(1), 54–56 (2010)
16. Xu, P., Fang, X., He, R., et al.: An efficient handoff algorithm based on received signal strength and wireless transmission loss in hierarchical cell networks. Telecommun. Syst. **52**(1), 317–325 (2013)
17. Tao, W., Hongzhen, W.: The TS&SS algorithm for vehicle routing problem. In: 2013 25th Chinese Control and Decision Conference (CCDC), Guiyang, pp. 396–400 (2013)
18. Cao, S., Tao, X., Hou, Y.: Joint handover decision to relieve ping-pong effect in V2X communication. In: 2015 International Conference on Connected Vehicles and Expo (ICCVE), pp. 56–59 (2016)
19. Nour, B., Ibn-Khedher, H., Moungla, H., et al.: Internet of things mobility over information-centric/named-data networking. IEEE Internet Comput. **24**(1), 14–24 (2020)
20. Wang, J., Luo, J., Zhou, J., et al.: A mobility-predict-based forwarding strategy in vehicular named data networks. In: IEEE Global Communications Conference, pp. 1–6 (2020)

Edge Computing and Deep Learning

Selective Modulation and Cooperative Jamming for Secure Communication in Untrusted Relay Systems

Li Huang[✉], Xiaoxu Wu, and Hang Long

Beijing University of Posts and Telecommunications, Beijing, China
huangliqiaoyi@163.com

Abstract. In this paper, an innovative secure communication scheme for untrusted relay systems is proposed. The source and jamming signals are jointly designed based on the selective modulation. The source signal adopts selective modulation, composed of one selection bit and one or more signal bits. The selection bit is employed to carry confidential information, while the signal bits are generated randomly and are modulated the random signal. The selection bit determines the transmit slot/frequency of signal bits, namely, the random signal. The design of the cooperative jamming signal includes two parts, adjusting the signal transmit power and rotating the signal phase. The simulation results demonstrate that our proposed scheme can make the bit error ratio of the selection bit at the untrusted relay to be 0.5. That is, there is no confidential information leakage at the relay node. Besides, the proposed scheme is superior to the Gaussian jamming signal in terms of the secrecy capacity.

Keywords: Physical layer security · Cooperative systems · Selective modulation

1 Introduction

In wireless communication systems, propagation paths are severely affected by the environment. As a result, the signal fading is more obvious and the performance of direct link is relatively terrible. With the collaboration of relays, the communication performance can be improved and the communication range can be expanded. Meanwhile, the transmissions in wireless networks have broadcast nature, which makes transmitted information more vulnerable to be eavesdropped.

In cooperative relay systems, the relay nodes are commonly assumed to be trusted. During the transmission, the relay is responsible for amplifying and forwarding the signal [1,2] or serving as a cooperative jamming node to interference

Supported by National Nature Science Foundation of China (NSFC) Project 61931005.

H. Gao et al. (Eds.): ChinaCom 2021, LNICST 433, pp. 351–365, 2022.
https://doi.org/10.1007/978-3-030-99200-2_27

the eavesdropper [3–5]. Nevertheless, the relay may be a potential eavesdropper, intercepting confidential information. Thus, studying the security of untrusted relay systems is extremely significant.

Ref. [6–13] have researched untrusted relay systems, where the source and the destination nodes exchange information through untrusted relays. In 2007, Oohama took the lead in studying relay channel coding to prevent wiretapping [6]. The destination node is employed to send cooperative jamming signals, proving that the positive secrecy rate is achievable in [7]. A novel beamforming design is proposed in [8], directing the cooperative jamming signal to an untrusted relay. The authors of [9] propose a joint design for the precoding of useful and jamming signals to maximize the secrecy capacity. Besides the precoding at source and destination nodes, the precoding at relay is also designed to maximize the secrecy rate in [10].

The existing works mostly adopt the secrecy capacity or the secrecy rate or the secrecy outage probability to estimate the security of untrusted relay systems [6–10]. In addition, Ref. [11–13] adopt the bit error rate (BER) and the symbol error rate (SER) as evaluation criteria. In [11], the BER at the untrusted relay can approximate 0.5 by employing a greedy algorithm. The authors of [12] consider designing the amplitude and phase of the jamming signal. The simulation results demonstrate that the SER at the trusted relay will approach 15/16 while the signal-to-noise ratio (SNR) is greater than 25 dB. Ref. [13] proposes a constellation rotation aided scheme to prevent eavesdropping. However, there are some deficiencies in these schemes. The greedy algorithm complexity is high [11], the required SNR is comparatively high [12] and a certain amount of confidential information may be intercepted at untrusted relay [13].

In this paper, selective modulation is proposed to improve the security combined with cooperative jamming. Selective modulation refers to extending the source signal by one dimension. The source signal is composed of one selection bit and signal bits, where the selection bit carries confidential information and determine the transmit time/frequency of signal bits; signal bits are used to bear the random signal and do not carry useful information. The cooperative jamming signal is designed according to channel state information (CSI). The main contributions of this paper can be summarized as follows:

(1) Different from existing signal designs in untrusted relay systems, the source signal adopts selective modulation by extending the signal dimension to transmit confidential information.
(2) The selective modulated source signal and cooperative jamming are jointly designed to realize zero confidential information leakage at the untrusted relay. The algorithm complexity is decreased compared with [11]; the required SNR is reduced compared with [12] and in contrast to [13], confidential information leakage at untrusted relay is completely suppressed.

The remainder of this paper is organized as follows. In Sect. 2, the untrusted relay system model with destination-aided cooperative jamming is presented. Section 3 presents the proposed selective modulation and cooperative jamming.

The security scheme design without leakage is proposed in Sect. 4. Section 5 provides simulation results and conclusions are drawn in Sect. 6.

Notation: $\mathcal{E}(\cdot)$ denotes mathematical expectation. P(A) is the probability that the event A occurs. $\min(\cdot)$ is the minimum function.

2   System Model

As described in Fig. 1, there is a two-hop untrusted relay system based on cooperative jamming. The system consists of a source node S, a destination node D and an untrusted relay node R. S and D solely communicate under the collaboration of R. R amplifies and forwards the received information without altering it. Besides, R acts as a potential eavesdropper, intercepting confidential information. Define the *S-R*, *R-S*, *R-D* and *D-R* channels as h_{SR}, h_{RS}, h_{RD} and h_{DR}, respectively.

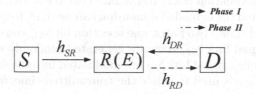

Fig. 1. Untrusted relay system with destination-aided cooperative jamming.

Assuming that a typical two-phase protocol is adopted. In the first phase, S sends the signal x_S to R with the transmit power P_S, while D cooperates in sending jamming signals with the transmit power P_J. Thus, the received signal at untrusted relay R is obtained as

$$y_R = \sqrt{P_S}h_{SR}x_S + \sqrt{P_J}h_{DR}J_D + n_R, \tag{1}$$

where x_S is the signal transmitted by S, J_D is the jamming signal sent by D, and $n_R \sim \mathcal{CN}(0,\sigma_R^2)$ is the noise at R.

In the second phase, R amplifies and forwards the received signal

$$x_R = \beta y_R, \tag{2}$$

where β is the power constraint coefficient at R, and the power of R is constrained by $\mathcal{E}(x_R^H x_R) = P_R$.

The signal received at D is

$$y_D = h_{RD}x_R + n_D, \tag{3}$$

where $n_D \sim \mathcal{CN}(0,\sigma_D^2)$ is the additive noise at D.

After receiving the signal, D can cancel the jamming signal using self-interference cancellation. Thus, the performance of the legitimate node is almost unaffected by jamming signals.

3 Selective Modulation and Cooperative Jamming

In this section, we propose the selective modulation for the source signal and analyze the security of untrusted relay systems based on selective modulation and cooperative jamming.

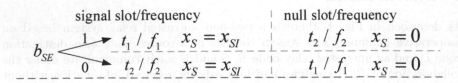

Fig. 2. Selective modulation.

The selective modulation refers to the fact that the source signal is extended by one dimension and the extended dimension can be time, frequency. The source signal bits are divided into two parts, one selection bit b_{SE} and one or more signal bits b_{SI}. As for signal bits b_{SI}, they are generated randomly without protected information. Assuming signal bits b_{SI} are modulated into the random signal x_{SI}. The selection bit b_{SE} is used to select the transmitted time/frequency of signal bits b_{SI}, namely, the random signal x_{SI}.

It is worth noting that only the selection bit b_{SE} bears confidential information in our proposed scheme. The signal bits b_{SI} are merely applied to carry the random signal x_{SI}, preventing the untrusted relay from eavesdropping on the selection bit b_{SE}.

As depicted in Fig. 2, S generates one selection bit b_{SE} to determine the transmitted slot/frequency of the random signal x_{SI}. S only selects one slot/frequency to send the random signal in every two slots/frequencies, and a null signal is transmitted at the other slot/frequency, that is, to send nothing. Assuming that if $b_S = 1$, the modulated signals x_S sent by S are x_{SI} and 0 at two slots t_1 and t_2 or at two frequencies f_1 and f_2, respectively. If $b_S = 0$, x_S are 0 and x_{SI} at t_1 and t_2 or f_1 and f_2.

According to whether S sends the random signal or not, the transmitted slots/frequencies are divided into signal slot/frequency and null slot/frequency. That is, if $b_S = 1$, t_1/f_1 is signal slot/frequency and t_2/f_2 is null slot/frequency; if $b_S = 0$, t_2/f_2 is signal slot/frequency and t_1/f_1 is null slot/frequency.

Figure 3 describes the system model with selective modulation and jamming signal, including signal slot/frequency in Fig. 3(a) and null slot/frequency in Fig. 3(b). The source signal adopts selective modulation, while the jamming signal is sent by the destination node D at each slot/frequency.

As shown in Fig. 3(a), at signal slot/frequency, $x_S = x_{SI}$. With (1), in the first phase, the received signal at R is

$$y_{RS} = \sqrt{P_S} h_{SR} x_{SI} + \sqrt{P_J} h_{DR} J_{DS} + n_{RS}, \qquad (4)$$

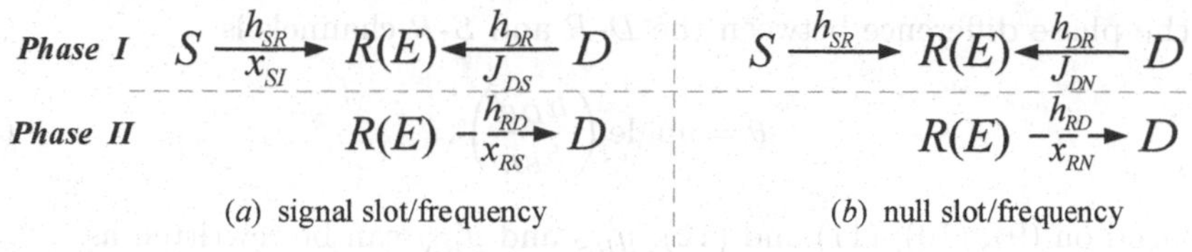

Phase I	$S \xrightarrow[x_{SI}]{h_{SR}} R(E) \xleftarrow[J_{DS}]{h_{DR}} D$	$S \xrightarrow{h_{SR}} R(E) \xleftarrow[J_{DN}]{h_{DR}} D$
Phase II	$R(E) -\xrightarrow[x_{RS}]{h_{RD}} D$	$R(E) -\xrightarrow[x_{RN}]{h_{RD}} D$
	(a) signal slot/frequency	(b) null slot/frequency

Fig. 3. System with selective modulation and cooperative jamming.

where x_{SI} is the signal transmitted by S with $\mathcal{E}(x_{SI}^H x_{SI}) = 1$, J_{DS} is the jamming signal sent by D with $\mathcal{E}(J_{DS}^H J_{DS}) = 1$, and $n_{RS} \sim \mathcal{CN}(0, \sigma_R^2)$ is the noise at R.

With (2), at signal slot/frequency, R amplifies and forwards the received signal

$$x_{RS} = \beta y_{RS}. \tag{5}$$

Similarly, at null slot/frequency, $x_S = 0$, as shown in Fig. 3(b). The received signal at R is

$$y_{RN} = \sqrt{P_J} h_{DR} J_{DN} + n_{RN}, \tag{6}$$

where J_{DN} is the jamming signal with $\mathcal{E}(J_{DN}^H J_{DN}) = 1$, and $n_{RN} \sim \mathcal{CN}(0, \sigma_R^2)$.

With (2), at null slot/frequency, R amplifies and forwards the received signal

$$x_{RN} = \beta y_{RN}. \tag{7}$$

According to (2), β can be expressed as

$$\beta = \sqrt{\frac{2P_R}{E\left(y_{RS}^H y_{RS}\right) + E\left(y_{RN}^H y_{RN}\right)}}$$

$$= \sqrt{\frac{2P_R}{E\left(h_{SR}^H h_{SR}\right) P_S + 2E\left(h_{DR}^H h_{DR}\right) P_J + 2\sigma_R^2}}. \tag{8}$$

Assuming that the channel fading is relatively slow and CSI remains identical during two slots/frequencies. According to (4), at signal slot/frequency, the equivalent received signal at R is

$$y_{RS}' = \frac{y_{RS}}{\sqrt{P_S} h_{SR}} = x_{SI} + \frac{\sqrt{P_J} h_{DR}}{\sqrt{P_S} h_{SR}} J_{DS} + \frac{n_{RS}}{\sqrt{P_S} h_{SR}}. \tag{9}$$

With (6), at null slot/frequency, the equivalent received signal at R is

$$y_{RN}' = \frac{y_{RN}}{\sqrt{P_S} h_{SR}} = \frac{\sqrt{P_J} h_{DR}}{\sqrt{P_S} h_{SR}} J_{DN} + \frac{n_{RN}}{\sqrt{P_S} h_{SR}}. \tag{10}$$

Define the ratio of the amplitude of the cooperative jamming signal to that of the source signal at the untrusted relay as

$$a = \left| \frac{\sqrt{P_J} h_{DR}}{\sqrt{P_S} h_{SR}} \right|, \tag{11}$$

and the phase difference between the D-R and S-R channels is

$$\theta = \text{angle}\left(\frac{h_{DR}}{h_{SR}}\right). \tag{12}$$

Based on (9), (10), (11) and (12), y'_{RS} and y'_{RN} can be rewritten as

$$y'_{RS} = x_{SI} + ae^{j\theta}J_{DS} + n'_{RS}, \tag{13}$$

$$y_{RN}' = ae^{j\theta}J_{DN} + n'_{RN}, \tag{14}$$

where n'_{RS}, n'_{RN} are the equivalent noise at R, $n_{RS}, n_{RN} \sim \mathcal{CN}\left(0, \sigma^2\right)$, $\sigma^2 = \sigma_R^2 / \left(P_S|h_{SR}|^2\right)$.

In general, the receiver judges the slot/frequency with higher power as the slot/frequency where the signal is sent. This signal demodulation method is simple. The proposed selective modulation aims to decrease the complexity of communication systems. Thus, the untrusted relay and the destination node adopt this signal demodulation method in this paper.

If the power magnitude of the received signal at R is uncertain at signal and null slots/frequencies, so that the untrusted relay cannot distinguish signal slot/frequency. Therefore, the probability of R guessing the signal slot/frequency correctly is 0.5, that is, BER of the selection bit at R is 0.5.

According to (13) and (14), the power of equivalent signals at the untrusted relay R are $\left|x_{SI} + ae^{j\theta}J_{DS} + n_{RS}'\right|^2$ and $\left|ae^{j\theta}J_{DN} + n_{RN}'\right|^2$ at signal and null slots/frequencies, respectively.

Compared with null slot/frequency, if the probability of the equivalent signal at R with higher and lower power at signal slot/frequency is equal, the error probability of R distinguishing signal slot/frequency will be 0.5. That is, the BER of the selection bit at R is 0.5. Under this condition, zero secret information leakage can be achieved at untrusted relay R.

On this basis, the joint design of the source signal and the jamming signal is proposed in next sections. It aims to make the power of signals received by R at signal and null slots/frequencies satisfy

$$\begin{aligned}
&\text{P}\left(\left|x_{SI} + ae^{j\theta}J_{DS} + n_{RS}'\right|^2 > \left|ae^{j\theta}J_{DN} + n_{RN}'\right|^2\right) \\
&= \text{P}\left(\left|x_{SI} + ae^{j\theta}J_{DS} + n_{RS}'\right|^2 < \left|ae^{j\theta}J_{DN} + n_{RN}'\right|^2\right).
\end{aligned} \tag{15}$$

4   Security Scheme Design

The previous section presents the system security analysis with selective modulation and cooperative jamming. To achieve absolute secure communication, the BER of the selection bit at R is required to be 0.5. The magnitude of received signals power at R must be uncertain between signal and null slots/frequencies. The signal power at signal slot/frequency is related to the random signal, the jamming signal and noise, while the signal power at null slot/frequency is solely related to jamming signal and noise.

In this section, the random signal x_{SI} and the jamming signal J_D are designed jointly on the basis of the system security analysis. Furthermore, the security scheme design without leakage is proposed.

4.1   Signal Design Based on MPSK

As for selective modulation, we find that MPSK is optimal to improve the security compared with amplitude shift keying (ASK) and quadrature amplitude modulation (QAM). Assuming that the random and jamming signals are MPSK modulated. The random signal x_{SI} is M_1PSK modulated,

$$x_{SI} = e^{j\theta_S},$$

where $M_1 = 2^{n_1}$, $n_1 = 1, 2, 3, 4...$, θ_S is the phase of x_{SI}, $\theta_S = (2l_1 - 1)\,\pi/M_1$, $l_1 \in \{1, 2, 3...M_1\}$.

The jamming signal J_D is M_2PSK modulated,

$$J_D = e^{j\theta_D},$$

where $M_2 = 2^{n_2}$, $n_2 = 1, 2, 3, 4...$, θ_D is the phase of J_D, $\theta_D = (2l_2 - 1)\,\pi/M_2$, $l_2 \in \{1, 2, 3...M_2\}$.

For brevity of exposition, we define

$$M_0 = \max\left(M_1, M_2\right). \tag{16}$$

The complex Gaussian noise may be complicated to analyze. Without loss of generality, theoretical analysis ignores the noise received at R but simulation results consider the effect of noise. Thus, the signals power received by R are simplified as $\left|x_{SI} + ae^{j\theta}J_{DS}\right|^2$ at signal slot/frequency and $\left|ae^{j\theta}J_{DN}\right|^2$ or a^2 at null slot/frequency. Based on (15), if the complex Gaussian noise is not considered, the condition for zero confidential information leakage at R is

$$\mathrm{P}\left(\left|x_{SI} + ae^{j\theta}J_{DS}\right|^2 > a^2\right) = \mathrm{P}\left(\left|x_S + ae^{j\theta}J_{DS}\right|^2 < a^2\right), \tag{17}$$

where J_{DS} is the cooperative jamming signal sent by D and have the same distribution as J_D. Thus, we have $J_{DS} = J_D$. Without the noise being considered, at signal slot/frequency, the signal power received by R can be expressed as

$$\begin{aligned}
\left|x_{SI} + ae^{j\theta}J_{DS}\right|^2 &= \left(x_{SI} + ae^{j\theta}J_D\right)\left(x_{SI} + ae^{j\theta}J_D\right)^{\mathrm{H}} \\
&= x_{SI}x_{SI}^H + \left(ae^{j\theta}J_D\right)\left(ae^{j\theta}J_D\right)^{\mathrm{H}} + x_{SI}\left(ae^{j\theta}J_D\right)^{\mathrm{H}} + ae^{j\theta}J_Dx_{SI}^H \\
&= 1 + a^2 + ae^{j\theta_S - \theta - \theta_D} + ae^{j\theta + \theta_D - \theta_S} \\
&= 1 + a^2 + 2a\cos\left(\theta_S - \theta_D - \theta\right).
\end{aligned} \tag{18}$$

Since the random signal x_{SI} is $M_1\text{PSK}$, the period of θ_S is $2\pi/M_1$. Similarily, the period of θ_D is $2\pi/M_2$ as for jamming signal J_D. Thus, x_{SI} and J_D can be expressed as followed:

$$x_{SI} = e^{j(\theta_S + 2\pi/M_1)}, \tag{19}$$

$$J_D = e^{j(\theta_D - 2\pi/M_2)}. \tag{20}$$

Substituting (19) or (20) for (18), the signal power at signal slot/frequency can be rewritten as

$$\begin{aligned}
\left| x_{SI} + ae^{j\theta} J_{DS} \right|^2 &= 1 + a^2 + 2a\cos\left(\theta_S - \theta_D - \theta + 2\pi/M_1\right) \\
&= 1 + a^2 + 2a\cos\left(\theta_S - \theta_D - \theta + 2\pi/M_2\right).
\end{aligned} \tag{21}$$

Comparing (18) with (21), it can be shown that the signal power at signal slot/frequency is a periodic function of the channel phase difference θ. If $M_1 \neq M_2$, the minimum positive period of θ is

$$\min\left(\frac{2\pi}{M_1}, \frac{2\pi}{M_2}, \left| \frac{2\pi}{M_1} - \frac{2\pi}{M_2} \right| \right) = \frac{2\pi}{M_0}; \tag{22}$$

if $M_1 = M_2 = M_0$, the minimum positive period is

$$\min\left(\frac{2\pi}{M_1}, \frac{2\pi}{M_2} \right) = \frac{2\pi}{M_0}. \tag{23}$$

According to the characteristics of MPSK modulation, if the signal phase is reversed, it is still MPSK. Thus, x_{SI} and J_D can be expressed as $x_{SI} = e^{-j\theta_S}$, $J_D = e^{-j\theta_D}$. Integrating them into (18), we can obtain the expression

$$\begin{aligned}
\left| x_{SI} + ae^{j\theta} J_{DS} \right|^2 &= 1 + a^2 + 2a\cos\left(-\theta_S + \theta_D - \theta\right) \\
&= 1 + a^2 + 2a\cos\left(\theta_S - \theta_D + \theta\right).
\end{aligned} \tag{24}$$

Compared with (18), the signal power at signal slot/frequency is also an even function of θ.

In summary, the signal power received by R at signal slot/frequency is function of θ and is also an even function of θ. It can be concluded that the period of θ is $2\pi/M_0$ and the symmetry axis is π/M_0. Thus, the effective range of θ is $[0, \pi/M_0]$.

4.2   Security Analysis Based on MPSK

According to (18), as the values of the amplitude ratio a and the channel phase difference θ are certain, the signal power at signal slot/frequency depends on the difference $\theta_S - \theta_D$. Since $\theta_S, \theta_D \in [0, 2\pi]$, $\theta_S - \theta_D \in [-2\pi, 2\pi]$.

For brevity of analysis, we restrict the difference $\theta_S - \theta_D$ to the range of $0 \sim 2\pi$. If $\theta_S - \theta_D \notin [0, 2\pi]$, it can be mapped into the corresponding interval since the period of phase is 2π.

<1> $M1 = M2 = M_0$

Assuming that both the random signal x_{SI} and the jamming signal J_D are M_0PSK. The phase difference of x_{SI} and J_D can be expressed as

$$\theta_S - \theta_D = 2l_3\pi/M_0, \tag{25}$$

where, $l_3 = (l_1 - l_2 + M_0) \bmod M_0$.

Thus, $l_3 = 0, 1, 2, 3...M_0 - 1$ and l_3 is a uniform distribution.

1) $M_0 = 2$

Assuming that x_{SI} and J_D are BPSK, so the effective range of θ is $[0, \pi/2]$, $l_3 = 0, 1$. If $l_3 = 0$, $\theta_S - \theta_D - \theta = -\theta$; if $l_3 = 1$, $\theta_S - \theta_D - \theta = \pi - \theta$.

If $l_3 = 0$, we have $\cos(\theta_S - \theta_D - \theta) \geq 0$, $1 + a^2 + 2a\cos(\theta_S - \theta_D - \theta) > a^2$. That is, the signal power at signal slot/frequency is greater than null slot/frequency as $l_3 = 0$ and the probability is 0.5. In order to realize zero confidential information leakage at R, $1 + a^2 + 2a\cos(\theta_S - \theta_D - \theta) < a^2$ is required if $l_3 = 1$ according to (17). That is, $1 + a^2 - 2a\cos\theta < a^2$ is required to satisfy. Thus, we can obtain $\cos\theta > 1/2a$, namely, $\theta < \arccos(1/2a)$. Besides, $\cos\theta \leq 1$, so $a > 1/2$ is required.

To summarize, if x_{SI} and J_D are BPSK, the BER of the selection bit b_{SE} at R is 0.5 under the condition that $a > 1/2$ and $0 \leq \theta < \arccos(1/2a)$.

2) $M_0 > 2$

Both x_{SI} and J_D are M_0PSK, the effective range of θ is $[0, \pi/M_0]$ and $l_3 = 0, 1, 2, 3...M_0 - 1$.

If $l_3 \in [0, M_0/4]$, $-\pi/2 < \theta_S - \theta_D - \theta \leq \pi/2$; if $l_3 \in [3M_0/4 + 1, M_0 - 1]$, $3\pi/2 \leq \theta_S - \theta_D - \theta < 2\pi$. Thus, if $l_3 \in [0, M_0/4] \cup [3M_0/4 + 1, M_0 - 1]$, we have $1 + a^2 + 2a\cos(\theta_S - \theta_D - \theta) > a^2$.

Since the length of $[0, M_0/4] \cup [3M_0/4 + 1, M_0 - 1]$ is half the length $[0, M_0 - 1]$, we can obtain that $P\left(\left|x_{SI} + ae^{j\theta}J_{DS}\right|^2 > a^2\right) = 0.5$.

Similarly, if $l_3 \in [M_0/4 + 1, 3M_0/4]$, we have $\pi/2 + 2\pi/M_0 - \theta \leq \theta_S - \theta_D - \theta \leq 3\pi/2 - \theta$. Meanwhile, $1 + a^2 + 2a\cos(\theta_S - \theta_D - \theta) < a^2$ is required. If $\theta \in [0, \pi/M_0]$, the maximum value of $1 + a^2 + 2a\cos(\theta_S - \theta_D - \theta)$ will be $1 + a^2 + 2a\cos(3\pi/2 - \theta)$. Thus, $1 + a^2 + 2a\cos(3\pi/2 - \theta) < a^2$, namely, $1 - 2a\cos(\pi/2 - \theta) < 0$ is required. On the basis of this, we can obtain the condition for zero confidential information leakage at R is that $a > 1/2\cos(\pi/2 - \theta)$ and $\theta > \pi/2 - \arccos(1/2a)$.

To summarize, if x_{SI} and J_D are M_0PSK, the BER of the selection bit b_{SE} at R is 0.5 under the condition that $a > 1/2\cos(\pi/2 - \pi/M_0)$ and $\pi/2 - \arccos(1/2a) < \theta \leq \pi/M_0$,

<2> $M_1 \neq M_2, M_0 = \max(M_1, M_2)$

Suppose $M_1 > M_2$, we can have $M_1 = M_0$ and $M_2 = M_0/k$, where $k = 2, 4, 8, ..M_0/2$. The phase of x_{SI} and J_D can be expressed as $\theta_S = (2l_4 - 1)\pi/M_0$ and $\theta_D = (2l_5 - 1)\pi/M_2 = (2l_5 - 1)k\pi/M_0$, where $l_4 \in \{1, 2, 3...M_0\}$, $l_5 \in \{1, 2, 3...M_0/k\}$.

The phase difference of x_{SI} and J_D can be expressed as

$$\theta_S - \theta_D = (2l_6 + 1)\,\pi/M_0, \tag{26}$$

where $l_6 = [l_4 - 1 - (2l_5 - 1)\,k/2 + M_0]\bmod M_0$. Thus, $l_6 = 0, 1, 2, 3 \ldots M_0 - 1$ and l_6 is a uniform distribution.

Compared with (18), there only exists a phase shift of π/M_0. Thus, the required range of a is identical and the required range of θ has the phase shift of π/M_0. On this basis, we can obtain that the required range of θ is $\pi/2 + \pi/M_0 - \arccos(1/2a) < \theta \le 2\pi/M_0$. Since the symmetry axis of θ is π/M_0, it can be mapped into the interval $[0, \pi/M_0]$ and the range is $0 \le \theta < \arccos(1/2a) + \pi/M_0 - \pi/2$.

To summarize, if x_{SI} and J_D are M_1PSK and M_2PSK, respectively, $M1 \ne M2$, the BER of the selection bit b_{SE} at R is 0.5 under the condition that $a > 1/2\cos(\pi/2 - \pi/M_0)$ and $0 \le \theta < \arccos(1/2a) + \pi/M_0 - \pi/2$.

In conclusion, if $M_1 = M_2 = M_0 > 2$, the BER of the selection bit b_{SE} at R is 0.5 under the condition that $a > 1/2\cos(\pi/2 - \pi/M_0)$ and $\pi/2 - \arccos(1/2a) < \theta \le \pi/M_0$; if $M_1 = M_2 = 2$ or $M_1 \ne M_2$ and $M_0 = \max(M_1, M_2)$, the BER of the selection bit b_{SE} at R is 0.5 under the condition that $a > 1/2\cos(\pi/2 - \pi/M_0)$ and $0 \le \theta < \arccos(1/2a) + \pi/M_0 - \pi/2$.

Through the above analysis, the signal power at null slot/frequency depends on the amplitude ratio a, while the signal power at signal slot/frequency is related to the values of a, θ and $\theta_S - \theta_D$. It can be observed from (11) and (12) that the values of a and θ are independent of the signal modulation.

Besides, the scheme with M_1PSK-modulated x_{SI} and M_2PSK-modulated J_D has the same distribution of $\theta_S - \theta_D$ as the scheme with M_2PSK-modulated x_{SI} and M_1PSK-modulated J_D. Thus, the two schemes are equivalent in system security with selective modulation and cooperative jamming. Compared to the scheme with $M_1 = M_2 = M_0$, there only exists the phase shift of π/M_0 in the scheme with $M_1 \ne M_2$, $M_0 = \max(M_1, M_2)$.

4.3   Security Scheme Design Without Leakage

In this section, we propose the security scheme without leakage. Through the security analysis based on MPSK, it can be concluded that if the amplitude ratio a is greater than the corresponding threshold and the channel phase difference θ is limited in a certain range, zero confidential information leakage at R can be achieved. However, the values of a and θ are arbitrary in practical communication system.

In the presented system model, the jamming signal J_D is sent by D. D can cancel it by self-interference cancellation so that the design of J_D has little effect on the received performance of D. Thus, the jamming signal is designed to realize zero confidential information leakage.

According to (11), the transmit power P_J of the jamming signal can be adjusted to make $a > 1/2\cos(\pi/2 - \pi/M_0)$. Based on (12), θ depends on the channels in practical system and cannot be changed directly. Thus, the phase of the jamming signal is designed as

$$J'_D = e^{j\theta'_D} = J_D e^{j\theta_{\Delta D}}, \tag{27}$$

where $\theta_{\Delta D}$ is a variable related to θ.

The phase difference can be rewritten as

$$\theta_S - \theta'_D - \theta = \theta_S - \theta_D - (\theta_{\Delta D} + \theta). \tag{28}$$

If $M_1 = M_2 = M_0 > 2$, $\theta_{\Delta D}$ is designed to make

$$\pi/2 - \arccos(1/2a) - \theta < \theta_{\Delta D} \leq \pi/M - \theta; \tag{29}$$

if $M_1 = M_2 = 2$ or $M_1 \neq M_2$ and $M_0 = \max(M_1, M_2)$, $\theta_{\Delta D}$ is designed to make

$$-\theta \leq \theta_{\Delta D} < \arccos(1/2a) + \pi/M - \pi/2 - \theta. \tag{30}$$

In this paper, a security scheme without leakage is proposed. The source signal adopts selective modulation, including one selection bit and signal bits. The selection bit bears confidential information and the signal bits carry modulated signals x_{SI}. The source node S generates one selection bit b_{SE} to determine the transmitted slot/frequency of MPSK modulated x_{SI}. Meanwhile, the jamming signal J_D is designed according to the related CSI and the modulation order. The proposed scheme can achieve zero confidential information leakage at the untrusted relay.

5   Simulation Results

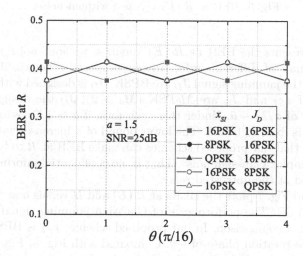

Fig. 4. BER at R (E) versus θ with $a = 1.5$ and SNR $= 25$ dB.

The demodulation BER and secrecy capacity are adopted as the evaluating metrics in this section. Figure 4 depicts the BER at R (E) versus θ with $a = 1.5$

and SNR = 25 dB. Assuming that x_{SI} is M_1PSK and J_D is M_2PSK, where $M_0 = \max(M_1, M_2) = 16$. As shown in Fig. 4, the BER at R (E) is a periodic function of the channel phase difference θ. The period of θ is $2\pi/16$ and the symmetry axis is $\pi/16$. The BER at R (E) with 16PSK-modulated x_{SI} and J_D is plotted in the red line. Compared with the other schemes, there only exists the difference of the phase shift $\pi/16$. If the modulation orders of x_{SI} and J_D are different, the BER at R depends on the signal with larger modulation order. If the larger modulation order is identical, these schemes with $M_1 \neq M_2$ are equivalent in system security with selective modulation and cooperative jamming. The simulation results are in accordance with the theoretical analysis.

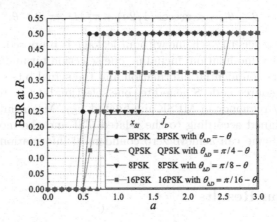

Fig. 5. BER at R (E) versus a without noise.

Figure 5 presents the BER at R (E) versus a without noise under MPSK modulated signals of different order. In our proposed schemes, if the random signal x_{SI} and the jamming signal J_D are BPSK, J_D is designed with the rotation phase of $-\theta$; if x_{SI} and J_D are M_0PSK ($M_0 > 2$), J_D the designed with the rotation phase of $\pi/M - \theta$. Under these schemes, if $a > 1/2\cos(\pi/2 - \pi/M_0)$, the BER at R is fixed to be 0.5. The lower bound of a increases with M_0. It can be clearly seen that our proposed schemes can make BER at R to be 0.5 without considering complex Gaussian noise. That is, no confidential information leakage can be achieved at the untrusted relay.

Figure 6 and Fig. 7 plots the BERs at R (E) and D versus a at SNR = 20 dB and SNR = 15 dB. The performance of Gaussian jamming signal is also given in the figure as a comparison. In our proposed scheme, x_{SI} is BPSK and J_D is BPSK with the rotation phase of $-\theta$. Compared with Fig. 5, Fig. 6 and Fig. 7 considers the complex Gaussian noise. It can be shown that at SNR = 20 dB, the BER at R (E) is 0.5 if $a \geq 0.7$; at SNR = 15 dB, the BER at R (E) is 0.5 if $a \geq 0.6$. That is, the untrusted relay cannot eavesdrop any useful information. In contrast to Fig. 5, the lower bound of a being 0.5 increases in Fig. 6 and Fig. 7. With the increase of SNR, the lower bound of a will approach 0.5. However, the BER at R (E) is less than 0.5 with Gaussian jamming signal. Besides, the BER at D are identical and is not affected by the jamming signal. In brief, the proposed scheme performs better than Gaussian jamming signal.

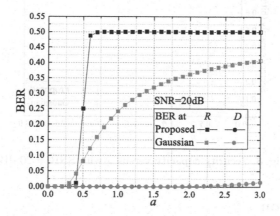

Fig. 6. BERs at R (E) and D versus a with SNR = 20 dB.

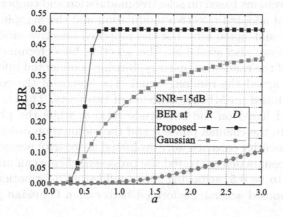

Fig. 7. BERs at R (E) and D versus a with SNR = 15 dB.

Figure 8 compares the security capacity of the proposed scheme and the Gaussian jamming signal at SNR = 20 dB. In our proposed scheme, x_{SI} is BPSK and J_D is BPSK with the rotation phase of $-\theta$. It can be observed from the figure that the overall security capacity of the proposed scheme is better than that of Gaussian jamming. If $0.6 \leq a \leq 1.6$, the secret capacity with the proposed scheme tends to approach 1. As $a > 1.6$, the security capacity of both schemes tend to decrease. This is because with the increase of a, the BER at D tends to increase as $a > 1.6$. Compared with Fig. 6, as SNR = 20 dB and $a \geq 0.6$, the BER at R is 0.5. That is, even if the security capacity is less than 1 as $a > 1.6$, zero confidential information leakage can still be achieved.

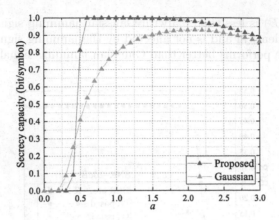

Fig. 8. Security capacity versus a with SNR = 20 dB.

6   Conclusion

In this paper, we proposed an innovative secure communication scheme for untrusted relay systems based on selective modulation and cooperative jamming. The source signal adopted selective modulation and the mapping bits included one selection bit and signal bits. For the proposed selective modulation, multi-bit selective bits are also available. The selection bit determined the transmit slot/frequency of the random signal x_{SI}, bearing confidential information, while the signal bits carried the random signal. The design of the cooperative jamming signal J_D based on MPSK consisted of two parts. The transmit power of J_D was adjusted based on the amplitude ratio a, while the phase of J_D was rotated related to the channel phase θ and the modulation order. The demodulated BER and security capacity were evaluated as security performance metrics. The simulation results demonstrated the proposed scheme can make BER at the untrusted relay to be 0.5, achieving zero confidential information leakage at R. Besides, the proposed scheme performed better than Gaussian jamming signal in the secrecy.

References

1. Long, H., Xiang, W., Li, Y.: Precoding and cooperative jamming in multi-antenna two-way relaying wiretap systems without eavesdroppers channel state information. IEEE Trans. Inf. Forensics Secur. **12**(6), 309–1318 (2017)
2. Xie, W., Liao, J., Yu, C., Zhu, P., Liu, X.: Physical layer security performance analysis of the FD-based NOMA-VC System. IEEE Access **7**, 115568–115573 (2019)
3. Lin, Z., Wang, L., Cai, Y., Yang, W., Yang, W.: Robust secure switching transmission in MISOSE relaying networks with channel uncertainty. In: 2015 International Conference on Wireless Communications Signal Processing, pp. 1–6. IEEE, Nanjing (2015)

4. Sinha, R., Jindal, P.: Performance analysis of cooperative schemes under total transmit power constraint in single hop wireless relaying system. In: 2016 2nd International Conference on Communication Control and Intelligent Systems, pp. 28–31. IEEE, Mathura (2016)
5. Divya, T., Gurrala, K. K., Das, S.: Performance analysis of hybrid decode amplify-forward (HDAF) relaying for improving security in cooperative wireless network. In: 2015 Global Conference on Communication Technologies, pp. 682–687. IEEE, Thuckalay (2015)
6. Oohama, Y.: Capacity theorems for relay channels with confidential messages. In: 2007 IEEE International Symposium on Information Theory, pp. 926–930. IEEE, Nice (2007)
7. He, X., Yener, A.: Two-hop secure communication using an untrusted relay: a case for cooperative jamming. In: IEEE GLOBECOM 2008–2008 IEEE Global Telecommunications Conference, pp. 1–5. IEEE, New Orleans (2008)
8. Mekkawy, T., Yao, R., Tsiftsis, T.A., Xu, F., Lu, Y.: Joint beamforming alignment with suboptimal power allocation for a two-way untrusted relay network. IEEE Trans. Inf. Forensics Secur. 13(10), 2464–2474 (2018)
9. Zhang, L., Long, H., Huang, L.: Precoding and destination-aided cooperative jamming in MIMO untrusted relay systems. In: 2020 IEEE/CIC International Conference on Communications in China, pp. 605–610. IEEE, Chongqing (2020)
10. Li, Q., Yang, L.: Artificial noise aided secure precoding for MIMO untrusted two-way relay systems with perfect and imperfect channel state information. IEEE Trans. Inf. Forensics Secur. 13(10), 2628–2638 (2018)
11. Zhang, Q., Gao, Y., Zang, G., Zhang, Y., Sha, N.: Physical layer security for cooperative communication system with untrusted relay based on jamming signals. In: 2015 International Conference on Wireless Communications Signal Processing (WCSP), pp. 1–4. IEEE, Nanjing (2015)
12. Liu, Y., Li, L., Pesavento, M.: Enhancing physical layer security in untrusted relay networks with artificial noise: a symbol error rate based approach. In: 2014 IEEE 8th Sensor Array and Multichannel Signal Processing Workshop (SAM), pp. 261–264. IEEE, La Coruna (2014)
13. Xu, H., Sun, L., Ren, P., Du, Q.: Securing two-way cooperative systems with an untrusted relay: a constellation-rotation aided approach. IEEE Commun. Lett. 19(12), 2270–2273 (2015)

Deep CSI Feedback for FDD MIMO Systems

Zibo He[✉], Long Zhao, Xiangchen Luo, and Binyao Cheng

The Key Lab of Universal Wireless Communication, Ministry of Education,
Beijing University of Posts and Telecommunications (BUPT), Beijing, China
zibohe@bupt.edu.cn

Abstract. With the increasing number of antennas at the base station
(BS), the feedback overhead of traditional codebook in frequency division duplexing (FDD) mode becomes overwhelming, since the number of
codewords in codebook increases quickly. Alternatively, we can directly
feedback the channel state information (CSI) to the BS for precoding. To
reduce the overhead of CSI feedback, this paper proposes three CSI compression models based on autoencoder network. The first two of them,
adopting deep learning (DL) structure, are named FCNet and CNet,
respectively. FCNet employs full-connected network architecture, while
CNet is designed based on convolutional neural network with lightweight
convolution kernels and multi-channel architecture. By applying principal component analysis (PCA) on CSI feedback, the third one, i.e.,
PCANet, is also studied and analyzed in details. Experiments show that
CNet has best accuracy performance at the cost of high computational
complexity, while FCNet shows medium accuracy and complexity among
the three models. Besides, the accuracy of PCANet is nearly the same as
CNet in some specific channel conditions. Compared with the state-of-
the-art of CsiNet, the proposed models have their own advantages and
limitations in different scenarios.

Keywords: MIMO · CSI feedback · Deep learning · PCA

1 Introduction

In frequency division duplexing (FDD) multiple-input multiple-output (MIMO)
systems, the user equipment (UE) selects suitable codeword in the codebook and
transmits its index to the base station (BS) with few bits since the number of
codewords is small [1,2]. However, with the increasing number of antennas at the
BS, the number of codebooks increases and becomes overwhelming for MIMO
systems [3,4]. Alternatively, we can directly feedback the downlink channel state
information (CSI) to the BS through the feedback link for precoding [5].

In recent years, deep learning (DL) is widely used in image compression and
has achieved great success, such as convolutional neural network (CNN) [6–8].

© ICST Institute for Computer Sciences, Social Informatics and Telecommunications Engineering 2022
Published by Springer Nature Switzerland AG 2022. All Rights Reserved
H. Gao et al. (Eds.): ChinaCom 2021, LNICST 433, pp. 366–376, 2022.
https://doi.org/10.1007/978-3-030-99200-2_28

By taking into account that CSI compression is similar to image compression, DL has been applied for CSI feedback. Based on CNN, CsiNet has been proposed and shows remarkable performance [5]. With larger size of convolution kernels, CsiNetPlus is further introduced and shows better performance than CsiNet [9]. By leveraging multi-resolution architecture, another network called CRNet is also studied to improve the performance of CsiNet [10]. Combined with superimposed coding (SC), the CSI feedback performance of DL can also be improved [11].

Although the nonlinear representation ability of DL has been extensively studied, the feedback accuracy needs to be further improved for practical systems. On the other hand, it is true that DL for CSI feedback has outstanding performance, however the linear compression methods have advantages in reducing computational complexity, which is critical for UE with limited computing resources.

This paper adopts the antoencoder architecture for CSI feedback, where the encoder compresses the downlink CSI into bit information and feedbacks it to the BS; then the bit information is recovered to the CSI by the decoder for precoding. To achieve better feedback accuracy or low complexity, three different models, i.e., CNet, FCNet and PCANet, are proposed for the design of both encoder and decoder. By optimizing the number of neurons and network layers as well as designing a new activation function, FCNet based on full-connected network (FCN) can achieve low computational complexity at the cost of little accuracy degradation. By leveraging lightweight convolution kernels and multi-channel architecture, CNet based on CNN can achieve better accuracy than CsiNet. One linear compression scheme, i.e., principal component analysis (PCA), is also studied for designing the encoder and decoder, denoted as PCANet, which shows medium accuracy with extremely low complexity. The number of quantization bits are further analyzed for different schemes, by which we can determine the optimal number of quantization bits for CSI feedback.

The rest of this paper is organized as follows. Section 2 describes the system model and the overall structure of CSI feedback. The design of FCNet, CNet and PCANet as well as the quantization scheme are detailedly studied in Sect. 3. The numerical results and analysis are given in Sect. 4, while the conclusion is drawn in Sect. 5.

2   System Model

For simplification, we consider a narrow band FDD MIMO system, where the BS and UE are equipped with N_t and N_r antennas, respectively. Assuming that the transmitted pilot and data symbols are $\mathbf{x}_p \in \mathbb{C}^{N_t \times 1}$ and $x_d \in \mathbb{C}$, respectively, then the received pilot signal $\mathbf{y}_p$ and data signal $\mathbf{y}_d$ can be respectively expressed as:

$$\begin{aligned} \mathbf{y}_p &= \mathbf{H}\mathbf{x}_p + \mathbf{z}_p \\ \mathbf{y}_d &= \mathbf{H}\mathbf{w}x_d + \mathbf{z}_d \end{aligned} \tag{1}$$

where $\mathbf{H} \in \mathbb{C}^{N_r \times N_t}$, $\mathbf{w} \in \mathbb{C}^{N_t \times 1}$ and $\mathbf{z}_p, \mathbf{z}_d \in \mathbb{C}^{N_r \times 1}$ denote the channel matrix, precoding vector and additive white Gaussian noise (AWGN), respectively.

As shown in Fig. 1, when the UE received the pilot signal $\mathbf{y}_p$, it can estimate channel $\mathbf{H}$, then compresses and feedbacks it to the BS via the uplink, so that the BS can design the precoder $\mathbf{w}$ for downlink transmission. Downlink channel estimation [12] is beyond the scope of this paper, so we assume that the perfect channel can be acquired by UE and only focus on the feedback scheme.

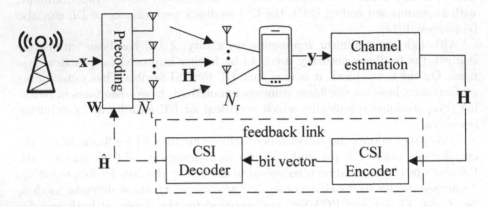

Fig. 1. Channel feedback model

In order to achieve better performance of CSI feedback, we borrow the autoencoder network, which generally consists of an encoder and a decoder. The objective of the encoder is to compress the channel matrix $\mathbf{H}$ to a (N/B)-dimensional compressed vector $\mathbf{s}$, i.e.,

$$\mathbf{s} = \left[s_1, s_2, \cdots, s_{N/B} \right]^{\mathrm{T}} = \text{Encoder} \left(\mathbf{H} \right) \tag{2}$$

where N and B denote the numbers of total feedback bits and quantization bits per element in $\mathbf{s}$, respectively. Then, $\mathbf{s}$ can be quantized into a bit vector $\mathbf{c}$ and transmitted to the BS:

$$\mathbf{c} = \left[c_1, c_2, \cdots, c_N \right]^{\mathrm{T}} = \text{Quantization} \left(\mathbf{s} \right) \tag{3}$$

Once the BS received $\mathbf{c}$, it is dequantized into the estimated $\mathbf{s}$, i.e.,

$$\hat{\mathbf{s}} = \left[\hat{s}_1, \hat{s}_2, \cdots, \hat{s}_{N/B} \right]^{\mathrm{T}} = \text{Dequantization} \left(\mathbf{c} \right) \tag{4}$$

which is further recovered to the reconstructed channel matrix $\hat{\mathbf{H}}$ by the decoder:

$$\hat{\mathbf{H}} = \text{Decoder} \left(\hat{\mathbf{s}} \right) \tag{5}$$

The objective of the entire network is to minimize the normalized distance between the original $\mathbf{H}$ and the reconstructed $\hat{\mathbf{H}}$, i.e., normalized mean square error (NMSE), given by:

$$\text{NMSE} = \mathrm{E} \left\{ \frac{\left\| \mathbf{H} - \hat{\mathbf{H}} \right\|_2^2}{\left\| \mathbf{H} \right\|_2^2} \right\} \tag{6}$$

3   Design of Deep CSI Feedback

By leveraging different methods or DL network structures, three schemes for CSI feedback are proposed in this section and detailedly given as follows.

3.1   FCNet

FCN is first considered for CSI feedback due to its low computational complexity. Based on the FCN, Fig. 2 illustrates the proposed FCNet architecture, the encoder and decoder of which consist of a series of full-connected layers.

The encoder consists of one input layer, three hidden layers and one quantization layer, which have L, P and N/B neurons, respectively. Firstly, we reshape the original CSI matrix $\mathbf{H}$ into a vector $\mathbf{h} \in \mathbb{C}^{M \times 1}$, where $M = 2N_t N_r$ and 2 represents the real and imaginary parts of the CSI matrix. The CSI vector $\mathbf{h}$ serves as the input of encoder and the output is a bit vector $\mathbf{c}$. Once the BS received $\mathbf{c}$ from UE through feedback link, it utilizes the decoder to reconstruct the CSI matrix $\hat{\mathbf{H}}$, where the decoder contains three hidden layers with P neurons and one output layer with M neurons.

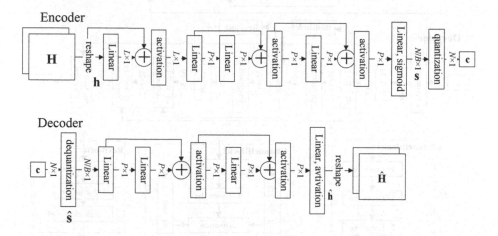

Fig. 2. Architecture of FCNet for CSI feedback

In order to further extract the channel feature and avoid gradients vanishing or exploding caused by deep FCNet, we adopts the shortcut connection [13] in FCNet. What's more, we create a new avtivation function in hidden layers to improve the nonlinear representation ability of the model, and it can be written as:

$$\text{activation}\,(x) = x \cdot \tanh\,(\text{softplus}\,(x)) \tag{7}$$

Moreover, uniform quantization is developed as the quantizer of FCNet. Uniform quantization is a rounding operation, where each value is rounded to the nearest value in a finite set of predefined quantization levels [14], which can be written as:

$$\hat{s}_k = \frac{\text{round}\left(s_k \cdot 2^B - 0.5\right) + 0.5}{2^B}, k \in [1, 2, \cdots, N/B] \tag{8}$$

where B represents the number of quantization bits and the function round() can convert its input to its nearest integer value[1].

3.2 CNet

Although FCNet can achieve low computational complexity, it shows low feedback accuracy. In order to improve the accuracy of CSI feedback, CNet is designed based on CNN, the detailed architecture of which is shown in Fig. 3.

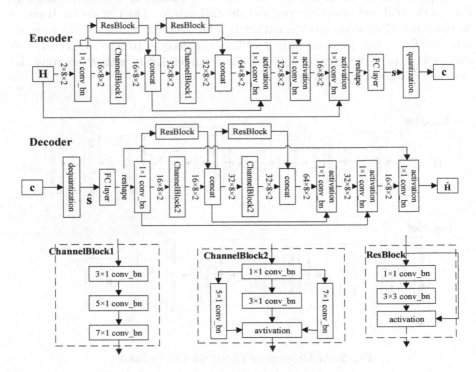

Fig. 3. Architecture of CNet for CSI feedback

Compared with CsiNet relying on CNN, we introduce lightweight convolution kernels into our CNet. It can efficiently extract the channel features for the dense CSI matrix with low computational complexity. On the other hand, we

---

[1] Since each value in **s** belongs to (0, 1) (the sigmoid function), we minus 0.5 before round function, quantize the rounded value to bits and then transmit them to the decoder. Once the decoder received these bits, it dequantizes them, plus 0.5 and then divide them by 2^B to get the quantization value.

apply multi-channel architecture on CNet, i.e., ChannelBlock1, ChannelBlock2 and ResBlock in Fig. 3. The multi-channel architecture convolutes the channel matrix along different paths to extract the channel features, which are further concatenated into a three-dimension channel feature matrix.

Similar to FCNet, the shortcut connection and the proposed activation function (7) are also adopted in CNet. The quantization process in CNet is also the same to that in FCNet.

3.3 PCANet

PCA is a linear method for dimension reduction, and it can map the original data onto a new feature space, the orthogonal basis of which is called principal component.

In the proposed PCANet, the encoder first calculates the covariance matrix $\mathbf{D}$ of the original CSI vector $\mathbf{h}$, which is reshaped from $\mathbf{H}$. Then, the eigenvalues and eigenvectors of $\mathbf{D}$ are calculated and sorted. The first N/B largest eigenvectors are selected to constitute the new orthogonal basis $\mathbf{P}$. Finally the CSI vector $\mathbf{h}$ is mapped on the new orthogonal basis $\mathbf{P}$ to generate the weight vector $\mathbf{s}$. After quantization of $\mathbf{s}$, the obtained bit vector $\mathbf{c}$ will be transmitted to the decoder.

Once the decoder received $\mathbf{c}$, it can be dequantized and then mapped into the weight vector $\hat{\mathbf{s}}$. Further, the decoder projects $\hat{\mathbf{s}}$ on the transposed orthogonal basis, i.e., $\mathbf{P}^{\mathrm{T}}$, to acquire the reconstructed CSI matrix $\hat{\mathbf{H}}$.

The detailed PCANet for CSI feedback is given in Algothrim 1.

Algorithm 1 : PCANet algorithm

Encoder

1: Get the original CSI matrix $\mathbf{H}$ and convert it to a vector $\mathbf{h}$;
2: Calculate the covariance matrix $\mathbf{D} = \mathbf{h}\mathbf{h}^{\mathrm{T}}$;
3: Compute the eigenvalues $\boldsymbol{\lambda}$ and eigenvectors $\{\mathbf{v}_1, \mathbf{v}_2, \cdots, \mathbf{v}_M\}$ of $\mathbf{D}$;
4: Sort $\boldsymbol{\lambda}$ in descending order and choose corresponding eigenvectors to the first N/B largest eigenvalues as the new orthogonal basis $\mathbf{P}$;
5: Project $\mathbf{h}$ on the new orthogonal basis $\mathbf{P}$ and obtain the mapping vector $\mathbf{s} = \mathbf{h}\mathbf{P} = [s_1, s_2, \cdots, s_{N/B}]$;
6: Each value in $\mathbf{s}$ is quantized into B bits and form $\mathbf{C} = [\mathbf{c}_1, \mathbf{c}_2, \cdots, \mathbf{c}_{N/B}]^{\mathrm{T}}$, where $\mathbf{c}_i$ denotes the quantization bit vector of s_i $(i = 1, 2, \cdots, N/B)$;
7: Reshape $\mathbf{C}$ into the bit vector $\mathbf{c}$ and is transmitted to the decoder;

Decoder

8: Dequantize $\mathbf{c}$ and get the estimated weight vector $\hat{\mathbf{s}} = [\hat{s}_1, \hat{s}_2, \cdots, \hat{s}_{N/B}]$;
9: Calculate the reconstructed CSI vector $\hat{\mathbf{h}} = \hat{\mathbf{s}}\mathbf{P}^{\mathrm{T}}$ and is then converted to the CSI matrix $\hat{\mathbf{H}}$.

Since each value in the weight vector $\mathbf{s}$ does not belong to $(0, 1)$, quantization formula in (8) cannot be directly applied on PCANet. In order to map $\mathbf{s}$ into $(-1, 1)$, we divide each element of $\mathbf{s}$ by $\max\{s_i, i = 1, \cdots, N/B\}$. After quantization and dequantization, these values will be multiplied by $\max\{s_i, i = 1, \cdots, N/B\}$ in order to obtain the original values.

4   Simulation Results and Analysis

4.1   Experiment Setup

We consider two types of typical channels: CDL-C and CDL-D. The carrier frequency is 2.6 GHz and the delay spreads of CDL-C and CDL-D channels are 300 ns and 100 ns, respectively. The number of antennas at the UE and BS are $N_r = 2$ and $N_t = 8$, respectively. We generate 70000 independent channel samples, which is divided into 50000 training samples, 10000 validation samples and 10000test samples.

The whole simulation is implemented in Pytorch. Both FCNet and CNet adopt random initialization and Adam optimizer. To improve the feedback accuracy of FCNet and CNet, the NMSE is directly adopted as the loss function instead of mean square error (MSE). The initial learning rate is 0.001 and the batch size is 32. Moreover, FCNet and CNet are trained for 50 and 100 epochs, respectively.

4.2   Simulation Results and Analysis

The NMSEs of both FCNet and CNet versus the number of epochs are shown in Fig. 4a and Fig. 4b. From the two figures, we can know that the NMSEs on both the training set and test set gradually decrease and finally become convergent with the increasing of training epochs. Figure 4c shows that the largest eigenvalue of the covariance matrix gradually decreases and finally becomes convergent with the increasing of training samples, which shows the feasibility of PCANet.

Figure 5 shows the NMSEs of different models versus the number of total feedback bits under $B = 2$. It can be seen that the NMSE of each model decreases with the increasing number of total feedback bits, which means the accuracy of CSI feedback could be improved at the cost of high feedback overhead.

The number of quantization bits B determines the accuracy of uniform quantization and the number of elements in $\mathbf{s}$, i.e., N/B, and further affects the feedback accuracy of the proposed models, especially for PCANet. In order to find the optimal number of quantization bits for different models, Fig. 6 illustrates the NMSE versus the number of quantization bits under $N = 12$. It can be seen from Fig. 6 that the NMSE of each model first decreases and then increases with the increasing number of quantization bits, therefore there should exist an optimal number of quantization bits B for each model. Moreover, different models have different optimal numbers B; even for the same model, the optimal numbers B may be different under different channels. Taking CDL-D channel as an example, the optimal numbers of quantization bits for FCNet, CNet and PCANet are 4, 6 and 2, respectively. As for FCNet, the optimal numbers under CDL-C and CDL-D channels are 6 and 4, respectively.

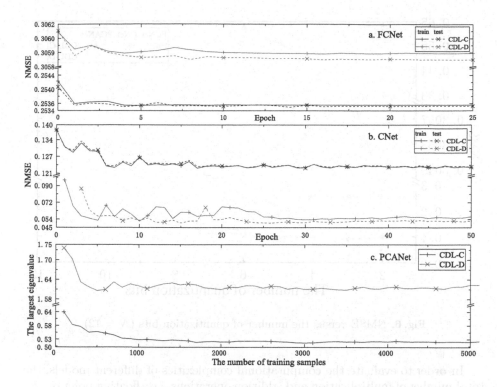

Fig. 4. Convergence of the proposed models

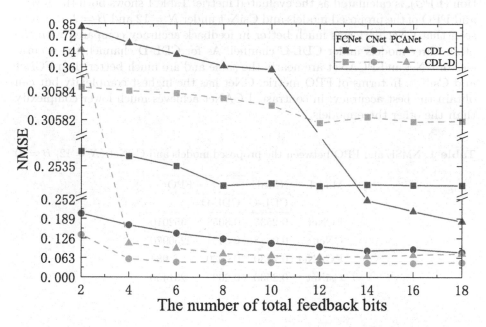

Fig. 5. NMSE versus the number of feedback bits ($B = 2$)

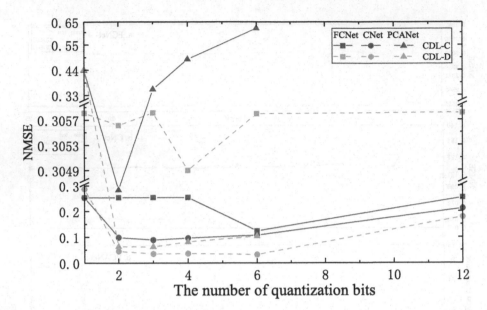

Fig. 6. NMSE versus the number of quantization bits ($N = 12$)

In order to evaluate the computational complexities of different models, the total number of multiplication and addition operations, i.e., floating point operation (FPO), is calculated as the evaluated metric. Table 1 shows both the NMSE and FPO of the proposed models and CsiNet under $N = 12$ and $B = 2$. It can be seen that CNet performs much better in feedback accuracy compared with the other three models under CDL-C channel. As for CDL-D channel, the accuracies of CNet and PCANet are nearly the same and are much better than FCNet and CsiNet. In terms of FPO metric, CNet has the highest complexity but can obtain the best accuracy; in contrast, PCANet achieves much lower complexity than the other three models.

Table 1. NMSE and FPO between the proposed models and CsiNet ($N = 12$, $B = 2$)

Model	NMSE		FPO
	CDL-C	CDL-D	
FCNet	0.2535	0.3058	352010
CNet	0.0983	0.0454	2250970
PCANet	0.2830	0.0630	984
CsiNet	0.2530	0.2633	231450

5   Conclusion

In order to efficiently feedback the CSI from the UE to BS, this paper proposes three models, named FCNet, CNet and PCANet, respectively, based on the autoencoder network in FDD MIMO systems. By optimizing the numbers of neurons and network layers as well as designing a new activation function, we first propose FCNet based on FCN architecture. By leveraging lightweight convolution kernels and multi-channel architecture for extracting channel feature, CNet is designed to further improve the accuracy of CSI feedback. Moreover, based on the linear method of PCA for dimension reduction, PCANet is studied and analyzed for CSI feedback with low complexity. Simulation results indicate that CNet performs best in the feedback accuracy at the cost of high complexity, while FCNet shows medium accuracy and complexity among the three models. Besides, the accuracy of PCANet is nearly the same as CNet in some specific channel conditions. Compared with the state-of-the-art of CsiNet, the feedback accuracy of CNet (or FCNet) is much better (or slightly worse). With much lower complexity, PCANet performs better in different scenarios.

Acknowledgment. This work was supported in part by the China Natural Science Funding under Grant 61731004.

References

1. 3GPP Technical Specification (TS) 38.211: NR; Physical channels and modulation (Release 15) (2017)
2. 3GPP Technical Specification (TS) 38.214: NR; Physical layer procedure for data (Release 15) (2017)
3. Zhao, L., Zhao, H., Zheng, K., Xiang, W.: Massive MIMO in 5G Networks: Selected Applications, 1st edn. Springer, Heidelberg (2018). https://doi.org/10.1007/978-3-319-68409-3
4. Zheng, K., Zhao, L., Mei, J., Shao, B., Xiang, W., Hanzo, L.: Survey of large-scale MIMO systems. IEEE Commun. Surv. Tutor. **17**(3), 1738–1760 (2015)
5. Wen, C., Shih, W., Jin, S.: Deep learning for massive MIMO CSI feedback. IEEE Wirel. Commun. Lett. **7**(5), 748–751 (2018)
6. Akbari, M., Liang, J., Han, J.: DSSLIC: deep semantic segmentation-based layered image compression. In: IEEE International Conference on Acoustics, Speech and Signal Processing (ICASSP), Brighton, pp. 2042–2046 (2019)
7. Huang, C., Nguyen, T., Lai, C.: Multi-channel multi-loss deep learning based compression model for color images. In: IEEE International Conference on Image Processing (ICIP), Taipei, pp. 4524–4528 (2019)
8. Cai, C., Chen, L., Zhang, X., Gao, Z.: Efficient variable rate image compression with multi-scale decomposition network. IEEE Trans. Circuits Syst. Video Technol. **29**(12), 3687–3700 (2019)
9. Guo, J., Wen, C., Jin, S., Li, G.Y.: Convolutional neural network-based multiple-rate compressive sensing for massive MIMO CSI feedback: design, simulation, and analysis. IEEE Trans. Wirel. Commun. **19**(4), 2827–2840 (2020)

10. Lu, Z., Wang, J., Song, J.: Multi-resolution CSI feedback with deep learning in massive MIMO system. In: IEEE International Conference on Communications (ICC), Dublin, pp. 1–6 (2020)
11. Qing, C., Cai, B., Yang, Q., Wang, J., Huang, C.: Deep learning for CSI feedback based on superimposed coding. IEEE Access **7**, 93723–93733 (2019)
12. Liu, Y., You, S.D., Chung, C.: Channel estimation method for LTE downlink MIMO transmission. In: IEEE 8th Global Conference on Consumer Electronics (GCCE), Osaka, pp. 719–722 (2019)
13. He, K., Zhang, X., Ren, S., Sun, J.: Deep residual learning for image recognition. In: IEEE Conference on Computer Vision and Pattern Recognition (CVPR), Las Vegas, pp. 770–778 (2016)
14. Chen, T., Guo, J., Jin, S., Wen, C., Li, G.Y.: A novel quantization method for deep learning-based massive MIMO CSI feedback. In: IEEE Global Conference on Signal and Information Processing (GlobalSIP), Ottawa, pp. 1–5 (2019)

Joint Computation Offloading and Wireless Resource Allocation in Vehicular Edge Computing Networks

Jiao Zhang[1(✉)], Zhanjun Liu[1], Bowen Gu[2], Chengchao Liang[1], and Qianbin Chen[1]

[1] School of Communication and Information Engineering,
Chongqing University of Posts and Telecommunications, Chongqing, China
zjj.0502@foxmail.com

[2] Faculty of Information Technology, Macau University of Science and Technology,
Avenida Wai Long, Taipa, Macau, China

Abstract. In vehicular edge computing (VEC) networks, a promising strategy for in-vehicle user equipments (UEs) with limited battery capacity is to offload data-intensive or/and latency-sensitive services to VEC servers via vehicle-to-infrastructure (V2I) links to reduce their energy consumption (EC). However, limited power supply, inadequate computation capability, and dynamic task latency make it extremely challenging to implement. In this work, we focus on designing a system-centric EC scheme in a VEC network. In particular, a joint computation offloading and wireless resource allocation problem is formulated to minimize the system EC by optimizing power control, offloading decision as well as subcarrier allocation while taking into account the dynamic and fickle time delay of each UE's task. In light of the intractability of the problem, we propose an effective block coordinate descent (BCD)-based algorithm with greedy search to find a high-quality sub-optimal solution. Simulation results illustrate that the proposed algorithm has better energy savings than the benchmarks.

Keywords: Vehicular network · Resource allocation · Energy consumption · Edge computing

1 Introduction

The recent advancement of vehicular networks has spurred various new applications in the domains of autonomous driving, video-aided real-time navigation, video streaming, and augmented reality [1,2]. However, it also means higher requirements for vehicular networks to handle such latency-sensitive and data-intensive services [3]. Although the processing capacity of the vehicle's central processing unit (CPU) becomes stronger, it is still unable to handle enormous applications in a short time, and the quality of service (QoS) is hard to be effectively guaranteed, which has become one of the bottlenecks for the further implementation of the vehicular networks [4].

To tackle this challenge, vehicular edge computing (VEC), as a promising technology, has been proposed. Specifically, the computation tasks generated by applications can be allowed to offload to the edges of the networks, so that long-distance transmission and excessive network hops can be eliminated [5]. In addition, the computational response time is reduced to better ensure the QoS of UEs. As a result, VEC-based computation offloading has attracted a large amount of attentions and has been extensively investigated from different perspectives, e.g., task computation delay minimization [6], vehicular terminals and servers cost minimization [7] and system utility maximization [8].

On the other hand, VEC also provides an opportunistic solution for energy-limited in-vehicle UEs to reduce their energy consumption (EC) and extend their lifetime [9]. Specifically, traditional solutions are tending to process all computing tasks locally by UEs, which leads to tremendous EC and ephemeral lifetime for UE. Thanks to the flourish of VEC, users' tasks can be offloaded to VEC servers for computation with the help of vehicle-to-infrastructure (V2I) links, which not only improves computational efficiency but also compensates for the inherent weaknesses in the traditional approach.

Resource allocation, as an essential technology in VEC networks, can achieve spectrum management, power allocation, and resource scheduling, etc. Such as the authors in [10] maximized the sum computation rate of all UEs via the binary offloading mode in wireless-powered mobile-edge computing networks. However, binary offloading may no longer be suitable for data-ultra-intensive computing tasks in practical. To this end, the authors in [11] minimized the weighted sum computation delay by invoking partial offloading for D2D-enabled mobile edge computing networks. We notice that [10] and [11] mainly focus on computation rate and latency, ignoring system energy consumption, which may be contrary to the green communication initiative. To deal with it, EC has been studied in [12,13]. In [12], the authors studied the resource-management policy to minimize the EC of all tasks for a mobile-edge computation offloading system under time constraints for mobile UEs. In [13], the authors jointly optimizing the transmit beamforming, the processing unit frequencies, and the offloading decision for energy minimization in wireless-powered multi-user mobile-edge computing systems. Although the aforementioned works [10–13] investigated resource allocation for UEs based on static cellular networks, the reached conclusions are not applicable to vehicular networks, due to the highly dynamic and unstable connectivity. To solve this issue, the authors in [14] minimized the EC of in-vehicle UEs and proposed an effective workload offloading scheme in VEC networks. However, the power allocation has not been involved, which may lead to biased resource scheduling and mismatched EC.

Motivated by the mentioned facts, in this paper, we investigate a joint computation offloading and wireless resource allocation problem for VEC networks, where energy savings of UEs is evaluated. And the contributions of this paper are summarized as follows.

- Our goal is to minimize the total EC of UEs, by jointly optimizing offloading decision, power allocation, and subcarrier selection, where transmit power threshold, the latency constraint, and computing ability of UEs are taking into account. The resulting problem is non-convex and challenging to solve.
- To tackle its non-convexity, we decompose the original problem into three subproblems based on block coordinate descent (BCD) method and solved them respectively. In specific, the interior-point method is used to address the subproblem of offloading decision. And for the subproblem of transmit power, we transform it into an equivalent convex optimization problem via Dinkelbach's method and successive convex approximation (SCA) method and then solve it by using Lagrangian dual decomposition theory. Besides, the greedy search method is employed to solve the subproblem of subcarrier selection.
- Simulation results demonstrate that the proposed algorithm not only has a close-to-optimal solution but also is better than the benchmarks in energy saving.

The rest of this paper is organized as follows. In Sect. 2, we introduce the system model. In Sect. 3, we formulate the EC minimization problem and propose a resource allocation algorithm to solve this problem. After that, simulation results are provided in Sect. 4. Finally, the paper is concluded in Sect. 5.

2   System Model

2.1   The Overall System Model

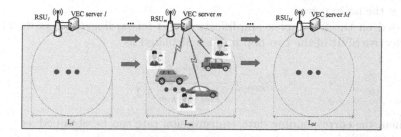

Fig. 1. System model

The task offloading framework for vehicular networks is shown in Fig. 1. There is a unidirectional road, where M RSUs are located along the road. Each RSU is equipped with a VEC server for remote edge computing. The road is divided into M segments based on the coverage areas of the M RSUs, represented as $\{L_1, L_2, ..., L_M\}$. A set of vehicles $\mathcal{K} = \{1, 2, ..., K\}$ in the m-th segment traveling towards the same direction. Define $U_{k,m}$ and $V_{k,m}$ as the UE k and the vehicle k within the RSU m coverage, respectively. We assume that there is only one task denoted by $\phi_{k,m} = \{D_{k,m}, C_{k,m}, \tau_{k,m}\}$ needs to be executed within a period of time for UE $U_{k,m}$, where $D_{k,m}$ represents the data size of the task, $C_{k,m}$ is the required computing ability for processing the task, and $\tau_{k,m}$ is the maximum tolerable latency of the task. In this paper, similar to [11], we assume that the task-input bits are bit-wise independent and can be arbitrarily divided into different groups. Thus the system support data offloading and local computing simultaneously.

2.2 The Transmission Model

The OFDMA is introduced to our system thus the co-channel interference among different vehicles can be ignored. The total bandwidth is divided into N sub-carriers, denoted as $n \in N = \{1, 2, ... N\}$, where $N \geq K$. To ensure the quality of service (QoS) and reduce the EC of UE $U_{k,m}$, the task of UE $U_{k,m}$ can be offloaded to the VEC Server m through vehicle $V_{k,m}$. Specifically, the data transmission is from UE $U_{k,m}$ to RSU m in a two-hop. (i.e., the data is firstly sent from UE $U_{k,m}$ to the vehicle $V_{k,m}$, then is forwarded from vehicle $V_{k,m}$ to RSU m). For the UE $U_{k,m}$, the signal to noise ratio (SNR) of the first-hop link and the second-hop link are calculated as

$$\gamma_{k,m}^{n,1} = \frac{p_{k,m}^n h_{k,m}^n}{\sigma^2} \tag{1}$$

$$\gamma_{k,m}^{n,2} = \frac{P_{k,m}^n g_{k,m}^n}{\sigma^2} \tag{2}$$

where $p_{k,m}^n$ and $P_{k,m}^n$ are the transmit power over the subcarrier n of UE $U_{k,m}$ and vehicle $V_{k,m}$, respectively. $h_{k,m}^n$ and $g_{k,m}^n$ are the channel gain between UE $U_{k,m}$ and vehicle $V_{k,m}$ and between vehicle $V_{k,m}$ and RSU m, respectively. σ^2 denotes the noise power.

Based on the full-duplex amplification and forwarding (AF) protocol [15], the effective SNR of the two-hop link is calculated as

$$\gamma_{k,m}^n = \frac{\gamma_{k,m}^{n,1} \gamma_{k,m}^{n,2}}{\gamma_{k,m}^{n,1} + \gamma_{k,m}^{n,2} + 1} \tag{3}$$

Then, the corresponding data transmission rate can be obtained as

$$R_{k,m}^n = \alpha_{k,m}^n B \log_2(1 + \gamma_{k,m}^n) \tag{4}$$

where B denotes the subcarrier bandwidth, and $\alpha_{k,m}^n \in \{0, 1\}$ is the subcarrier selection indicator with $\alpha_{k,m}^n = 1$ implying the data of UE $U_{k,m}$ is transmitted over the subcarrier n and $\alpha_{k,m}^n = 0$ otherwise.

Defining $\lambda_{k,m}$ as the offloading ratio of task $\phi_{k,m}$, the corresponding offloading time is expressed as

$$T_{k,m}^{\text{trans}} = \frac{\lambda_{k,m} D_{k,m}}{\sum_{n=1}^N R_{k,m}^n} \tag{5}$$

The EC of UE $U_{k,m}$ transmitting the data to the vehicle m is expressed as

$$E_{k,m}^{\text{trans}} = \sum_{n=1}^{N} \alpha_{k,m}^n p_{k,m}^n T_{k,m}^{\text{trans}} = \frac{\sum_{n=1}^{N} \alpha_{k,m}^n p_{k,m}^n \lambda_{k,m} D_{k,m}}{\sum_{n=1}^{N} R_{k,m}^n} \tag{6}$$

2.3 The Computation Model

In what follows, the specific process of task computation with partial offloading mode is introduced, which includes local computing and VEC.

1) Local Computing Model

The UE $U_{k,m}$ executes $(1 - \lambda_{k,m})C_{k,m}$ parts of its computation task $\phi_{k,m}$ locally, and the local computing time $T_{k,m}^{\text{loc}}$ can be obtained as

$$T_{k,m}^{\text{loc}} = \frac{(1 - \lambda_{k,m})C_{k,m}}{f_{k,m}} \tag{7}$$

where $f_{k,m}$ is the CPU cycle frequency of UE $U_{k,m}$.

Motivated by the observation on [16], the power consumption for local computing can be modeled as $P_{k,m}^{\text{loc}} = \delta(f_{k,m})^3$, where δ is the coefficient depending on the chip architecture. Based on (7), the EC of local computation is expressed as

$$E_{k,m}^{\text{loc}} = P_{k,m}^{\text{loc}} T_{k,m}^{\text{loc}} = \delta (1 - \lambda_{k,m}) C_{k,m} f_{k,m}^2 \tag{8}$$

The overall EC of UE $U_{k,m}$, which contains the energy consumed by local computing and data offloading, is expressed as

$$\begin{aligned} E_{k,m} &= E_{k,m}^{\text{trans}} + E_{k,m}^{\text{loc}} \\ &= \sum_{n=1}^{N} \frac{\alpha_{k,m}^n p_{k,m}^n \lambda_{k,m} D_{k,m}}{R_{k,m}^n} + \delta (1 - \lambda_{k,m}) C_{k,m} f_{k,m}^2 \end{aligned} \tag{9}$$

2) The VEC Model

The UE $U_{k,m}$ offloads the rest of computation task $\lambda_{k,m} C_{k,m}$ to the VEC server m, thus, the processing time of VEC server m can be calculated as

$$T_{k,m}^{\text{comp}} = \frac{\lambda_{k,m} C_{k,m}}{f_{k,m}^{\text{VEC}}} \tag{10}$$

where $f_{k,m}^{\text{VEC}}$ is the CPU-cycle frequency of VEC Server m when executing the edge computing task $\phi_{k,m}$ [17].

According to (5) and (10), the total time for computation offloading is $T_{k,m}^{\text{vec}} = T_{k,m}^{\text{trans}} + T_{k,m}^{\text{comp}}$. To ensure the QoS of UE $U_{k,m}$, we have $T_{k,m}^{\text{vec}} \leq \tau_{k,m}$. However, in vehicular networks, the effective execution time for one task may be different under different vehicle velocities. Specifically, the practical execution time $T_{k,m}^{\mathrm{pra}}$ relies on the dwell time $\tau_{k,m}^{\mathrm{o}}$ and the maximum tolerance delay $\tau_{k,m}$, such as

$$T_{k,m}^{\mathrm{pra}} = \min\left\{\tau_{k,m}, \tau_{k,m}^{\mathrm{o}}\right\} \tag{11}$$

where $\tau_{k,m}^{\mathrm{o}} = \frac{d_{k,m}}{v_{k,m}}$, $d_{k,m}$ is the distance between vehicle $V_{k,m}$ and the edge of RSU m in the vehicle heading direction, $v_{k,m}$ is the velocity of vehicle $V_{k,m}$.

3   Problem Formulation and Solution

Our goal is to minimize the total EC of UEs by jointly optimizing the transmit power, subcarrier selection, offloading ratio, and computing ability. Mathematically, the optimization problem is formulated as

$$\min_{p_{k,m}^n, \alpha_{k,m}^n, \lambda_{k,m}, f_{k,m}} \sum_{k=1}^{K} E_{k,m}$$

$$s.t.\ C_1: \sum_{k=1}^{K} \alpha_{k,m}^n \leq 1,\ \alpha_{k,m}^n \in \{0,1\},\ \forall n \in \mathcal{N}$$

$$C_2: \sum_{n=1}^{N} \alpha_{k,m}^n p_{k,m}^n \leq P_{\max},\ \forall k \in \mathcal{K} \tag{12}$$

$$C_3: f_{k,m} \leq f_{k,m}^{\max},\ \forall k,m$$

$$C_4: T_{k,m}^{\mathrm{loc}} \leq \tau_{k,m},\ \forall k,m$$

$$C_5: T_{k,m}^{\mathrm{vec}} \leq T_{k,m}^{\mathrm{pra}},\ \forall k,m$$

$$C_6: \lambda_{k,m} \in [0,1],\ \forall k \in \mathcal{K}$$

where C_1 denotes that each subcarrier is only allocated to one UE. C_2 represents the maximum transmit power constraint. C_3 is computing ability constraint. C_4 and C_5 ensure the latency requirements of local computing and remote executions. And C_6 enforces the offloaded task ratio of UE cannot exceed 1.

3.1   The Local Rescource Allocation

According to (7) and C3, we have $f_{k,m} \in \left[\frac{(1-\lambda_{k,m})C_{k,m}}{\tau_{k,m}}, f_{k,m}^{\max}\right]$. From (8), it is evident that the EC for local computing is monotonically increasing with $f_{k,m}$. To reduce the EC of local computing as much as possible, $f_{k,m}$ should take the smallest value within the feasible range. As a result, the optimal computing ability of UE $U_{k,m}$ is

$$f_{k,m}^* = \frac{(1 - \lambda_{k,m})\,C_{k,m}}{\tau_{k,m}} \tag{13}$$

Correspondingly, the optimal local computing time is expressed as

$$T_{k,m}^{\mathrm{loc},*} = \tau_{k,m} \tag{14}$$

Substituting $T_{k,m}^{\mathrm{loc},*}$ and $f_{k,m}^*$ into (12), as a result, problem (12) can be rewritten as

$$\min_{p_{k,m}^n, \alpha_{k,m}^n, \lambda_{k,m}} \overline{E}_{k,m} = \sum_{k=1}^K \left(\frac{\sum\limits_{n=1}^N \alpha_{k,m}^n p_{k,m}^n \lambda_{k,m} D_{k,m}}{\sum\limits_{n=1}^N R_{k,m}^n} + \frac{\delta(C_{k,m})^3}{\tau_{k,m}^2}(1-\lambda_{k,m})^3 \right)$$

$$s.t.\ C1, C2, C5, C6$$

(15)

However, problem (15) is non-convex due to the coupled variables and the binary variable. There are no standard methods to address such problems optimally. Therefore, Motivated by the idea of the BCD method [18], we decompose problem (15) into three subproblems, namely, offloading ratio, transmit power, and subcarrier selection.

3.2 The Subproblem of Offloading Ratio

With the given $\alpha_{k,m}^n, p_{k,m}^n$, the subproblem of offloading ratio is expressed as

$$\min_{\lambda_{k,m}} \sum_{k=1}^K \left(\frac{\sum\limits_{n=1}^N \alpha_{k,m}^n p_{k,m}^n \lambda_{k,m} D_{k,m}}{\sum\limits_{n=1}^N R_{k,m}^n} + \frac{\delta(C_{k,m})^3}{\tau_{k,m}^2}(1-\lambda_{k,m})^3 \right)$$

$$s.t.\ C6, C5: \lambda_{k,m} \left(\frac{D_{k,m}}{\sum\limits_{n=1}^N R_{k,m}^n} + \frac{C_{k,m}}{f_{k,m}^{\mathrm{VEC}}} \right) \le T_{k,m}^{\mathrm{pra}}$$

(16)

Observing problem (16), we find that the objective function and constraints are linear about $\lambda_{k,m}$, it is a convex problem. Thus, the interior-point method can be employed to solve it.

3.3 The Subproblem of Transmit Power

With the given $\lambda_{k,m}$ and $\alpha_{k,m}^n$, the subproblem of transmit power is

$$\min_{p_{k,m}^n} \sum_{k=1}^K \sum_{n=1}^N \left(\frac{\alpha_{k,m}^n p_{k,m}^n \lambda_{k,m} D_{k,m}}{R_{k,m}^n} \right)$$

$$s.t.\ C2, \bar{C}5: \sum_{n=1}^N R_{k,m}^n \ge \frac{D_{k,m}\lambda_{k,m} f_{k,m}^{\mathrm{VEC}}}{T_{k,m}^{\mathrm{pra}} f_{k,m}^{\mathrm{VEC}} - C_{k,m}\lambda_{k,m}}$$

(17)

However, problem (17) is a non-convex problem since the objective function is non-smooth, which is challenging to solve. To this end, Dinkelbach's method

[19] is used to transform the objective function with fractional form into one with non-fractional form, e.g.,

$$\min_{p_{k,m}^n} \sum_{k=1}^K \sum_{n=1}^N \left(\alpha_{k,m}^n p_{k,m}^n \lambda_{k,m} D_{k,m} - A R_{k,m}^n \right) \tag{18}$$
$$s.t. \ C2, \bar{C}5$$

where $A \geq 0$. However, there are still some coupled variables in $R_{k,m}^n$, which makes problem (18) difficult to solve. Then, the SCA method [20] is employed. We have the following inequality

$$\log_2(1 + \gamma_{k,m}^n) \geq a_{k,m}^n \log_2(\gamma_{k,m}^n) + b_{k,m}^n \tag{19}$$

where $a_{k,m}^n$ and $b_{k,m}^n$ are defined as

$$\begin{cases} a_{k,m}^n = \frac{\bar{\gamma}_{k,m}^n}{1 + \bar{\gamma}_{k,m}^n} \\ b_{k,m}^n = \log_2(1 + \bar{\gamma}_{k,m}^n) - a_{k,m}^n \log_2(\bar{\gamma}_{k,m}^n) \end{cases} \tag{20}$$

The lower bound is tight when $\gamma_{k,m}^n = \bar{\gamma}_{k,m}^n$, we define $\bar{\gamma}_{k,m}^n$ as the SINR for UE $U_{k,m}$ from the last iteration. We use the initialized value to calculate $\bar{\gamma}_{k,m}^n$ for the first iteration. As a result, the data transmission rate with the lower bound is

$$\bar{R}_{k,m}^n = B \alpha_{k,m}^n \left(a_{k,m}^n \log_2(\bar{\gamma}_{k,m}^n) + b_{k,m}^n \right) \tag{21}$$

To make the problem tractable, we define $\mathbf{p} = 2^{\mathbf{q}}$, where $\mathbf{q} = \left\{ q_{k,m}^n \right\}$. Substituting (3) into (21), the lower bound of $\bar{R}_{k,m}^n$ can be expressed as

$$\bar{R}_{k,m}^n = B \alpha_{k,m}^n \left[a_{k,m}^n \left(q_{k,m}^n + C_{k,m}^n - \log_2(2^{q_{k,m}^n} \frac{h_{k,m}^n}{\sigma^2} + \frac{P_{k,m}^n g_{k,m}^n}{\sigma^2} + 1) \right) + b_{k,m}^n \right] \tag{22}$$

where $C_{k,m}^n = \log_2(\frac{h_{k,m}^n}{\sigma^2}) + \log_2(\frac{P_{k,m}^n g_{k,m}^n}{\sigma^2})$.

Based on (21) and (22), problem (18) can be transformed as

$$\min_{q_{k,m}^n} \sum_{k=1}^K \sum_{n=1}^N \left(\alpha_{k,m}^n 2^{q_{k,m}^n} \lambda_{k,m} D_{k,m} - A \bar{R}_{k,m}^n \right) \tag{23}$$
$$s.t. \ C2, \bar{C}6 : \sum_{n=1}^N \bar{R}_{k,m}^n \geq \frac{D_{k,m} \lambda_{k,m} f_{k,m}^{\mathrm{VEC}}}{T_{k,m}^{\mathrm{pra}} f_{k,m}^{\mathrm{VEC}} - C_{k,m} \lambda_{k,m}}$$

Problem (23) is a convex optimization problem, which can be solved by using the Lagrangian dual decomposition theory. The Lagrangian function of problem (23) is

$$L_m(\mathrm{q}_{k,m}^n, \mu_{k,m}^n, \varphi_{k,m}^n) = \sum_{k=1}^{K} \sum_{n=1}^{N} (\alpha_{k,m}^n 2^{q_{k,m}^n} \lambda_{k,m} D_{k,m} - \mathrm{A}\bar{R}_{k,m}^n)$$
$$+ \sum_{k=1}^{K} \varphi_{k,m}^n \left(\frac{D_{k,m} \lambda_{k,m} f_{k,m}^{\mathrm{VEC}}}{T_{k,m}^{\mathrm{pra}} f_{k,m}^{\mathrm{VEC}} - C_{k,m} \lambda_{k,m}} - \sum_{n=1}^{N} \bar{R}_{k,m}^n \right) \quad (24)$$
$$+ \sum_{k=1}^{K} \mu_{k,m}^n \left(\sum_{n=1}^{N} \alpha_{k,m}^n 2^{q_{k,m}^n} - P_{\max} \right)$$

where $\varphi_{k,m}^n$ and $\mu_{k,m}^n$ are the non-negative Lagrange multipliers. For ease of handling, (24) can be rewritten as

$$L_m \left(q_{k,m}^n, \mu_{k,m}^n, \varphi_{k,m}^n \right) = L_{k,m}^n \left(q_{k,m}^n, \mu_{k,m}^n, \varphi_{k,m}^n \right)$$
$$- \sum_{k=1}^{K} \mu_{k,m}^n P_{\max} + \sum_{k=1}^{K} \varphi_{k,m}^n \frac{D_{k,m} \lambda_{k,m} f_{k,m}^{\mathrm{VEC}}}{T_{k,m}^{\mathrm{pra}} f_{k,m}^{\mathrm{VEC}} - C_{k,m} \lambda_{k,m}} \quad (25)$$

where

$$L_{k,m}^n \left(q_{k,m}^n, \mu_{k,m}^n, \varphi_{k,m}^n \right) = \alpha_{k,m}^n 2^{q_{k,m}^n} \lambda_{k,m} D_{k,m} - q\bar{R}_{k,m}^n$$
$$+ \alpha_{k,m}^n 2^{q_{k,m}^n} \mu_{k,m}^n - \bar{R}_{k,m}^n \varphi_{k,m}^n \quad (26)$$

The dual problem of (23) is

$$\max_{\mu_{k,m}^n, \varphi_{k,m}^n} \quad D_m \left(\mu_{k,m}^n, \varphi_{k,m}^n \right)$$
$$s.t. \ \mu_{k,m}^n \geq 0, \varphi_{k,m}^n \geq 0 \quad (27)$$

where

$$D_m \left(\mu_{k,m}^n, \varphi_{k,m}^n \right) = \min_{q_{k,m}^n, \mu_{k,m}^n, \varphi_{k,m}^n} L_m \left(q_{k,m}^n, \mu_{k,m}^n, \varphi_{k,m}^n \right) \quad (28)$$

According to KKT condition [21], we can derive the optimal transmit power of UE $U_{k,m}^n$, i.e.,

$$p_{k,m}^{n,*} = \left[\frac{-\psi_{k,m}^n + \sqrt{\left(\psi_{k,m}^n \right)^2 - 4h_{k,m}^n \xi_{k,m}^n}}{2h_{k,m}^n} \right]^+ \quad (29)$$

where $\psi_{k,m}^n = P_{k,m}^n g_{k,m}^n + \sigma^2$, $[\cdot]^+ = \max(0, \cdot)$, and $\xi_{k,m}^n = \frac{a_{k,m}^n (A + \varphi_{k,m}^n)(P_{k,m}^n g_{k,m}^n + \sigma^2)}{\ln 2 (\alpha_{k,m}^n \lambda_{k,m} D_{k,m} + \alpha_{k,m}^n \mu_{k,m}^n)}$.

By applying the subgradient method [22], the Lagrange multipliers can be updated as

$$\mu_{k,m}^n (l+1) = \left[\mu_{k,m}^n (l) - \Delta_{\mu_{k,m}^n} \left(P_{\max} - \sum_{n=1}^{N} \alpha_{k,m}^n 2^{q_{k,m}^n} \right) \right]^+ \quad (30)$$

$$\varphi_{k,m}^n(l+1) = \left[\varphi_{k,m}^n(l) - \Delta_{\varphi_{k,m}^n}\left(\sum_{n=1}^{N}\bar{R}_{k,m}^n - \frac{D_{k,m}\lambda_{k,m}f_{k,m}^{VEC}}{T_{k,m}^{pra}f_{k,m}^{VEC}-C_{k,m}\lambda_{k,m}}\right)\right]^+$$

(31)

where l is the iteration number, $\Delta_{\mu_{k,m}^n}$ and $\Delta_{\varphi_{k,m}^n}$ are the positive gradient steps.

Algorithm 1. A BCD-based resource allocation algorithm

Initialize: K, N, $D_{k,m}$, $C_{k,m}$, $\tau_{k,m}$, $P_{k,m}$, N_0, B, $h_{k,m}^n$, $g_{k,m}^n$, δ, $v_{k,m}$, $d_{k,m}$, $f_{k,m}^{VEC}$;

Output:

  The task offloading ratio $\lambda_{k,m}^*$, the transmission power $p_{k,m}^{n,*}$
  and the subcarrier selection decision $\alpha_{k,m}^{n,*}$

1. **for** $k=1$ to $k=K$ **do**
2.   Initialize $E_G = 0$.
3.   **for** $n=1$ to $n=N$ **do**
4.   Set $\alpha_{k,m}^n = 1$
5.   **for** $l=1$ to $l=L$ **do**
6.   Set $E_c(l) = 0$
7.   With the fixed $p_{k,m}^n(l)$, solve the problem (16) via the
          interior-point method and obtain the solution $\lambda_{k,m}(l+1)$.
8.   With the fixed $\lambda_{k,m}(l+1)$, solve optimization problem (23)
          via CVX tools to obtain $p_{k,m}^n(l+1)$.
9.   Calculate the total EC of UEs $E_c(l+1)$.
10.   **if** $|E_c(l+1) - E_c(l)| > \zeta_{k,m}$, **then**
11.   update $l=l+1$ and return step 7;
12.   **else**
13.   update $p_{k,m}^{n,*} = p_{k,m}^n(l)$, $\lambda_{k,m}^* = \lambda_{k,m}(l)$, $E_c^* = E_c(l)$;
14.   **end if**
15.   **end for**
16.   **if** $E_c^* \leq E_G$ **then**
17.   set $\alpha_{k,m}^n = 0$
18.   **else**
19.   update $E_G = E_c^*$, $p_{k,m}^n = p_{k,m}^{n,*}$, $\lambda_{k,m} = \lambda_{k,m}^*$;
20.   **end if**
21. **end for**
22. **end for**

Similarly, based on the given $\lambda_{k,m}$ and $p_{k,m}$, the subcarrier selection decision can be obtained. Specifically, owing to the $\alpha_{k,m}^n$ is a binary variable, the optimization problem is difficult to solve. Motivated by [23], the greedy algorithm is employed to tackle this problem. The detail of the proposed algorithm is shown in Algorithm 1.

4 Simulation Results

In this section, simulation results are provided to verify the effectiveness of the proposed algorithm. We assume that there is a unidirectional road, where four RSUs are randomly located along a 2000-m road. There are 20 vehicles driving on the road at a speed of 50 km/h–90 km/h within the coverage of the same RSU. The path-loss model is $PL = 128.1 + 37.6\log_{10}d$, where $d \in (0.05, 0.1)$ km is the distance between one receiver and one transmitter. The other parameters used in our simulations are summarized in Table 1. To better show the effectiveness of the proposed algorithm, we compare two other algorithms, namely, the local computing algorithm and the full offloading algorithm.

Table 1. Parameters

Parameter	Value
Number of UEs	15
Task data size/Mb	6–10
Task tolerance latency/s	0.8–1.3
Velocity of vehicles/km/h	50–90
Transmission power of vehicles /w	0.02–0.27
Noise power/dbm	−70
$f_{k,m}^{\text{VEC}}$ for VEC server/GHz	6
Maximum transmit power of UEs/w	0.9
The bandwidth of each UE/MHz	0.9
Diameter of RSU coverage/m	500
The bandwidth of each UE/MHz	0.9

Figure 2 shows the convergence performance of the proposed algorithm. It can be seen that the proposed algorithm can quickly converge to the stationary point within 3 iterations. Besides, the obtained solution of the proposed algorithm is closed to that of an exhaustive search algorithm, which shows the effectiveness of the proposed algorithm.

Figure 3 displays the EC of UEs versus the amount of input data bits. As expected, with the increase of $D_{k,m}$, the EC of UEs under three resource allocation algorithms increases. The reason is that as the input data bits increase, the required computing resources also increase, which leads to a higher EC for each UE via (9). Besides, it is also observed that the proposed algorithm consumes the least energy because the proposed algorithm adopts partial offloading mode, in which a balance between local computing and data offloading can be achieved.

Figure 4 shows the relationship between the EC of UEs and the delay of tasks. In this figure, the EC decreases with the increasing task tolerance latency

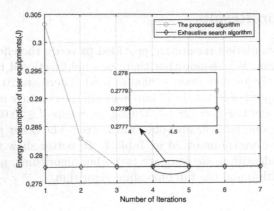

Fig. 2. The convergence performance of the proposed algorithm

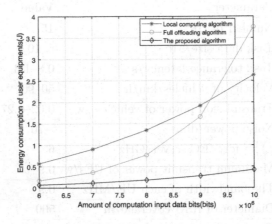

Fig. 3. EC vs input data bits

under three resource allocation algorithms. The reason is that a larger latency means that the computation burden per unit time decreases. Furthermore, the proposed algorithm has the lowest EC. This is because both the local computing algorithm and the full offloading algorithm only consider a single computation strategy, while leads to a non-flexible resource scheduling. However, this case can be overcome by the proposed algorithm with partial offloading.

Figure 5 shows vehicle velocity versus the EC of UEs. It can be seen from the figure that with the increase of vehicle velocity, the EC under all resource allocation algorithms keeps unchanged and begins to increase when vehicle velocity is bigger than 70 km/h. The reason is that the latency decreases with the increasing vehicle velocity, which leads to a higher EC for each UE accordingly. Moreover, the EC under the proposed algorithm is lower than other resource allocation algorithms. This is because the proposed algorithm can reduce the EC of UEs by dynamically adjusting the offloading portion.

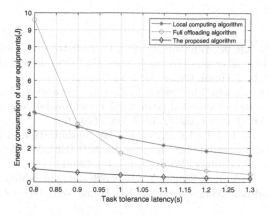

Fig. 4. EC vs delay

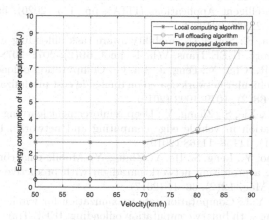

Fig. 5. EC vs velocity

5 Conclusion

In this paper, we have studied the joint computation offloading and wireless resource allocation problem for minimizing the total EC of UEs in vehicular networks with the constraints of the impact of vehicle mobility, the limited power supplies, computational capabilities of UEs, and the maximum latency of tasks. To solve this problem, a BCD-based iterative algorithm with greedy search was developed by jointly optimizing the computing ability of UEs, offloading decisions, transmit power, and subcarrier selection. Simulation results demonstrated that the proposed algorithm not only obtained a high-quality sub-optimal solution but also outperformed the benchmarks in terms of EC.

References

1. Lin, C., Han, G., Qi, X., Guizani, M., Shu, L.: A distributed mobile fog computing scheme for mobile delay-sensitive applications in SDN-enabled vehicular networks. IEEE Trans. Veh. Technol. **69**(5), 5481–5493 (2020)
2. Chen, Y., Wang, Y., Liu, M., Zhang, J., Jiao, L.: Network slicing enabled resource management for service-oriented ultra-reliable and low-latency vehicular networks. IEEE Trans. Veh. Technol. **69**(7), 7847–7862 (2020)
3. Wang, H., Lin, Z., Lv, T.: Energy and delay minimization of partial computing offloading for D2D-assisted MEC systems. In: 2021 IEEE Wireless Communications and Networking Conference (WCNC), pp. 1–6 (2021)
4. Zhang, J., Guo, H., Liu, J., Zhang, Y.: Task offloading in vehicular edge computing networks: a load-balancing solution. IEEE Trans. Veh. Technol. **69**(2), 2092–2104 (2020)
5. Peng, E., Li, Z.: Optimal control-based computing task scheduling in software-defined vehicular edge networks. In: 2020 International Conference on Internet of Things and Intelligent Applications (ITIA), pp. 1–5 (2020). https://doi.org/10.1109/ITIA50152.2020.9312356
6. Misra, S., Bera, S.: Soft-VAN: mobility-aware task offloading in software-defined vehicular network. IEEE Trans. Veh. Technol. **69**(2), 2071–2078 (2020)
7. Du, J., Yu, F.R., Chu, X., Feng, J., Lu, G.: Computation offloading and resource allocation in vehicular networks based on dual-side cost minimization. IEEE Trans. Veh. Technol. **68**(2), 1079–1092 (2019)
8. Liu, Y., Yu, H., Xie, S., Zhang, Y.: Deep reinforcement learning for offloading and resource allocation in vehicle edge computing and networks. IEEE Trans. Veh. Technol. **68**(11), 11158–11168 (2019)
9. Zhang, K., Mao, Y., Leng, S., He, Y., Zhang, Y.: Mobile-edge computing for vehicular networks: a promising network paradigm with predictive off-loading. IEEE Veh. Technol. Mag. **12**(2), 36–44 (2017)
10. Bi, S., Zhang, Y.J.: Computation rate maximization for wireless powered mobile-edge computing with binary computation offloading. IEEE Trans. Wirel. Commun. **17**(6), 4177–4190 (2018)
11. Saleem, U., Liu, Y., Jangsher, S., Tao, X., Li, Y.: Latency minimization for D2D-enabled partial computation offloading in mobile edge computing. IEEE Trans. Veh. Technol. **69**(4), 4472–4486 (2020)
12. You, C., Zeng, Y., Zhang, R., Huang, K.: Asynchronous mobile-edge computation offloading: energy-efficient resource management. IEEE Trans. Wirel. Commun. **17**(11), 7590–7605 (2018)
13. Wang, F., Xu, J., Wang, X., Cui, S.: Joint offloading and computing optimization in wireless powered mobile-edge computing systems. IEEE Trans. Wirel. Commun. **17**(3), 1784–1797 (2018)
14. Zhou, Z., Feng, J., Chang, Z., Shen, X.: Energy-efficient edge computing service provisioning for vehicular networks: a consensus ADMM approach. IEEE Trans. Veh. Technol. **68**(5), 5087–5099 (2019)
15. Gupta, A., Singh, K., Sellathurai, M.: Time-switching eh-based joint relay selection and resource allocation algorithms for multi-user multi-carrier AF relay networks. IEEE Trans. Green Commun. Netw. **3**(2), 505–522 (2019)
16. Xu, Y., Gu, B., Hu, R.Q., Li, D., Zhang, H.: Joint computation offloading and radio resource allocation in MEC-based wireless-powered backscatter communication networks. IEEE Trans. Veh. Technol. **70**(6), 6200–6205 (2021)

17. Song, Z., Liu, Y., Sun, X.: Joint radio and computational resource allocation for NOMA-based mobile edge computing in heterogeneous networks. IEEE Commun. Lett. **22**(12), 2559–2562 (2018)
18. Dai, Y., Xu, D., Maharjan, S., Zhang, Y.: Joint load balancing and offloading in vehicular edge computing and networks. IEEE Internet Things J. **6**(3), 4377–4387 (2019)
19. Dinkelbach, W.: On nonlinear fractional programming. Manage. Sci. **13**, 492–498 (1967)
20. Papandriopoulos, J., Evans, J.S.: SCALE: a low-complexity distributed protocol for spectrum balancing in multiuser dsl networks. IEEE Trans. Inf. Theory **55**(8), 3711–3724 (2009)
21. Boyd, S., Vandenberghe, L.: Convex Optimization. Cambridge University Press, U.K. (2004)
22. Fang, F., Cheng, J., Ding, Z.: Joint energy efficient subchannel and power optimization for a downlink NOMA heterogeneous network. IEEE Trans. Veh. Technol. **68**(2), 1351–1364 (2019)
23. Lan, Y., Wang, X., Wang, D., Zhang, Y., Wang, W.: Mobile-edge computation offloading and resource allocation in heterogeneous wireless networks. In: 2019 IEEE Wireless Communications and Networking Conference (WCNC), pp. 1–6 (2019)

Non-coherent Receiver Enhancement Based on Sequence Combination

Xiaoxu Wu[✉], Tong Li, and Hang Long

Wireless Signal Processing and Network Lab, Key Laboratory of Universal Wireless Communications, Ministry of Education, Beijing University of Posts and Telecommunications, Beijing, China
wuxiaoxu@bupt.edu.cn

Abstract. Physical Uplink Control Channel (PUCCH) transmits Uplink Control Information (UCI) from the User Equipment to the next generation NodeB. Coverage enhancement is an item studied in the procedure of Rel-17. PUCCH is one of bottleneck channels according to the simulation results in several scenarios. There are several proposals on PUCCH coverage enhancement including Demodulation Reference Signal-less transmission scheme. In this paper, based on PUCCH sequence combination design at the transmitter, an enhanced non-coherent receiver is proposed. Firstly, the specific sequence combination design is described. Then, the procedure description and theoretical analysis are given for this enhanced receiver. Finally, simulation results and complexity analysis on different schemes verify that the proposed receiver reduces computation complexity significantly with slight performance loss. Therefore, the proposed receiver has high practical value for PUCCH coverage enhancement.

Keywords: PUCCH · Sequence combination · Non-coherent receiver

1 Introduction

Physical Uplink Control Channel (PUCCH) is one of uplink channels in the physical layer of 5G wireless communication. Uplink Control Information (UCI) including Hybrid -Automatic-Repeat-reQuest Acknowledgement (HARQ ACK /NACK), Scheduling Requests (SR), and Channel State Information (CSI), is carried from the User Equipment (UE) to the next generation NodeB (gNB) [1]. Though 5G develops at a high speed and millimeter-wave frequency band is used, the space propagation loss is high which brings challenge to the coverage of 5G and the investment cost is expected to be a large amount [2]. Therefore, it's necessary to accelerate the promotion of 5G coverage enhancement scheme to reduce 5G construction cost in the future. Ref. [3] proposes that potential techniques for PUCCH coverage enhancements need to be studied. PUCCH format 3 with 11bits UCI needs to be enhanced according to the observations from Ref. [4] and [5] which provide the baseline performance in FR1 and FR2.

Supported by National Natural Science Foundation of China (NSFC) Project 61931005.

There are two methods for UCI to transmit in PUCCH. One is coherent transmission based on Demodulation Reference Signal (DMRS), and the other is non-coherent transmission without DMRS. DMRS-based coherent transmission is used in PUCCH format 1/2/3/4 in Rel-16, and DMRS-less non-coherent transmission is used in PUCCH format 0. It is well-known that the performance comparison between coherent and non-coherent transmission depends on operating Signal-to-Noise Ratio (SNR) region. In high SNR region, coherent transmission is better. However, in low SNR region, non-coherent transmission should have better link level performance than the other [6]. Therefore, considering 5G coverage scenarios, DMRS-less non-coherent transmission can be adopted as one of potential techniques for PUCCH coverage enhancement.

For specific solution of DMRS-less coherent transmission, there are already some studies. Ref. [7] summarizes some PUCCH enhancements based on sequence design and gives the direction of future study. It includes reusing Rel-15/16 Zadoff-Chu (ZC) sequence by current specification or specifying new sequences based on Rel 15/16 existing sequence, and reducing the complexity of receiver. Ref. [6] proposes two methods to generate sequences based on orthogonal sequences and non-orthogonal sequences. However, the complexity of receiver is high for the latter method. Ref. [8] proposes a new method called sequence combination to generate sequences and comprise sequence pool which represents UCI. This method also has the problem of high computation complexity.

In this paper, non-coherent receiver enhancement based on PUCCH sequence combination is proposed. The conventional non-coherent receiver based on sequence has high reliability but with high complexity, which limits the utilization of non-coherent receiver. The sequence combination design at the transmitter adopts the design derived from Ref. [8]. The main idea of enhanced non-coherent receiver in this study is dividing the received signal into two parts and choosing several candidate sequences of each part to determine the target sequence combination. The computation complexity can be reduced significantly with smaller solution domain at the receiver. The simulation results show that the enhanced receiver has slight performance loss compared with baseline schemes. Comprehensively considering complexity and performance, the enhanced receiver in this study can be used as a potential technique in PUCCH coverage enhancement.

The rest of this paper is organized as follows. In Sect. 2, the system model of transmission based on PUCCH sequence is described. In Sect. 3, the enhanced non-coherent receiver is introduced and analyzed. In Sect. 4, simulation results and analysis are given to show the performance of proposed scheme. Finally, the conclusion of the paper is drawn in Sect. 5.

2   System Model

2.1   Transmitter

A sequence without DMRS based on PUCCH to enhance the coverage of PUCCH was proposed during Study Item (SI) [9]. One approach to generate sequence for DMRS-less PUCCH is reusing Rel-15/16 ZC sequence [7].

In Rel-15/16, the low-Peak-to-Average Power Ratio (PAPR) sequence $r_{(u,v)}^{(\alpha,\delta)}(n)$ is defined by a cyclic shift α and a base sequence $\bar{r}_{(u,v)}(n)$ [10]. The sequence is generated by

$$r_{(u,v)}^{(\alpha,\delta)}(n) = e^{j\alpha n}\bar{r}_{(u,v)}(n), 0 \leq n < M_{ZC} \tag{1}$$

where M_{ZC} is the length of the sequence. Multiple sequences are defined from a single base sequence through different values of α and δ. When the length of the base sequences is larger than 36, the base sequence $\bar{r}_{(u,v)}(n)$ is given in Ref. [10].

This method of a long sequence needs to generate too much sequences and consume much resource. For the improved scheme called sequence combination in Ref. [8], a sequence with length M that is equal to the number of Resource Element (RE) is transmitted in the PUCCH resource. Two short sequences are combined to be a long sequence carrying UCI payloads. The whole sequence $r(n)$ can be defined by

$$r(n) = \begin{cases} e^{j\alpha_1 n}\bar{r}_{u_1,v}(n), & 0 < n \leq \frac{M}{2} \\ e^{j\alpha_2 n}\bar{r}_{u_2,v}(n), & \frac{M}{2} < n \leq M \end{cases} \tag{2}$$

where $M/2$ represents the length of short sequence, $\alpha_1\&\alpha_2$ represent different cyclic shifts, and $u_1\&u_2$ represent different base sequences. Different sequence combinations carry different sets of UCI. Supposing there are 14 Orthogonal Frequency Division Multiplexing (OFDM) symbols, two parts each with 7 symbols carry two sequences to transmit UCI. The specific procedure is shown in Fig. 1.

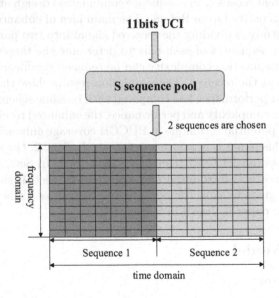

Fig. 1. Procedure of sequence combination in PUCCH at the transmitter

When two sequences are selected from the sequence pool, the number of combinations is $C_S^2 = S!/2!/(S-1)!$. If UCI payloads are 11 bits, there are 2080 kinds of combinations according to formula C_S^2 and the former 2048 kinds are chosen to represent UCI. Then, 65 short sequences are generated by 13 cyclic shifts and 5 base sequences. The sequence combination scheme uses less sequences to represent UCI payloads. What's more, it provides possibility of enhancement on receiver.

Alternatively, if the order of two short sequences also represents information, choosing two sequences from S sequence pool will have more choices. Thus, the integer value of S is 46 which is enough for composing the sequence pool.

2.2   Receiver

At the receiver side, a sequence detection is performed and UCI payloads could be determined. For a long sequence, the non-coherent receiver adopts the correlator to detect the received signal, which is also called blind detection. The procedure of detection is shown in Fig. 2. The received signal can be expressed in frequency domain as

$$Y(m) = H(m)X(m) + N(m) \tag{3}$$

where H is the channel coefficient, and X is the transmitted signal. N is a complex Gaussian noise vector with mean 0 and variance σ^2. m represents every RE that is occupied by signal.

Firstly, the received sequence is correlated with all possible sequences X_s in the sequence pool. The correlation value $R(s)$ can be given by

$$R(s) = \sum_{m=1}^{M} X_s(m)^* Y(m), \ m = 1, 2, ..., M \tag{4}$$

where M represents the length of sequence, and $*$ represents conjugate of a complex symbol. The module $U(s)$ is taken by

$$U(s) = |R(s)|^2 \tag{5}$$

Then, the candidate sequence with the largest power value U is determined as the transmitted sequence s'.

$$U = \max_s (U(s)) \tag{6}$$

$$s' = \arg\max_s (U(s)) \tag{7}$$

Finally, the index of the sequence in the form of decimal is converted to the form of binary. For sequence combination scheme, the non-coherent receiver can also adopt the blind detection to determine the transmitted sequence. For each correlation, the received sequence is correlated with a long sequence comprised by two short sequences in the sequence pool. This detection method has high reliability but high complexity at the receiver.

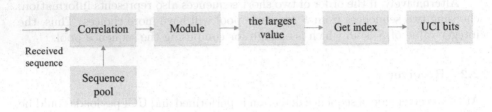

Fig. 2. Procedure of non-coherent receiver

For coverage enhancement study on PUCCH with more than 2bits UCI, in addition to 1% Block Error Rate (BLER) performance metric agreed in RAN1#101e, the Discontinuous Transmission (DTX) detection is necessary and the following performance metric can be considered [11]:

– 1% DTX to ACK error rate which is also called false alarm rate;
– 1% ACK miss detection (including ACK to NACK and ACK to DTX) error rate;
– 0.1% NACK to ACK error rate.

3   Enhanced Non-coherent Receiver

3.1   Design of Enhanced Non-coherent Receiver

As it is mentioned in the previous section, the existing non-coherent receiver has high complexity and the sequence combination scheme provides possibility of non-coherent receiver enhancement. In this part, the non-coherent receiver enhancement based on PUCCH sequence combination is proposed. The specific design of the enhanced receiver is shown in Fig. 3.

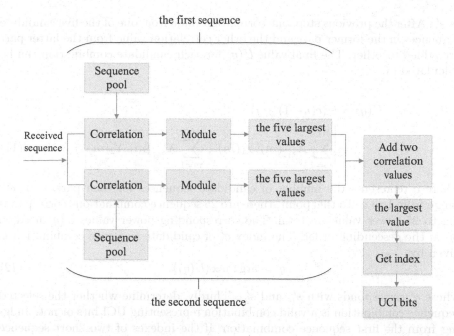

the first sequence

Sequence pool

Received sequence

Correlation → Module → the five largest values

Correlation → Module → the five largest values

Add two correlation values

the largest value

Sequence pool

Get index

the second sequence

UCI bits

Fig. 3. Procedure of enhanced non-coherent receiver

The transmitted signal is comprised of two short sequences. Thus, the received signal can be divided into two parts and correlated with sequences within the sequence pool respectively at the receiver without being correlated with all possible sequence combinations. In this way, the computation complexity at the receiver can be reduced. For example, when 11bits UCI are transmitted, there are 2048 alternative sequence combinations and 65 candidate short sequences. Firstly, the receiver divides the received signal into two parts $Y_1 \& Y_2$ of the same length. For each part, the half of the received sequence is correlated with 65 short sequences, and the module values $U_1(s) \& U_2(s)$ are taken. The specific calculation is given by

$$U_1(s) = \left| \sum_{m=1}^{M/2} X_s(m)^* Y_1(m) \right|^2 , \quad m = 1, 2, ..., \frac{M}{2} \tag{8}$$

where $U_2(s)$ is the same as $U_1(s)$. Secondly, the five largest power values which represent five candidate sequences $s'_{11} - s'_{15}$ can be retained.

$$s'_{11} = \arg\max_{s} (U_1(s)) \tag{9}$$

$$s'_{12} = \arg\max_{s-1} (U_1(s)), s \neq s'_{11} \tag{10}$$

where s'_{13}, s'_{14} and s'_{15} are almost the same as s'_{12}. Every time finding the largest value, there needs to exclude the sequence s' that has been found. s'_2 is the same as s'_1. After the previous step, one correlation value from one of the five candidate sequences in the former part and the other correlation value from the latter part are added together. The final value $\hat{U}(q)$ for each candidate combination can be calculated by

$$\hat{U}(q) = \hat{U}(c(i-1)+j)$$

$$= \left| \sum_{m=1}^{M/2} X_{s'_{1i}}(m)^* Y_1(m) + \sum_{m=1}^{M/2} X_{s'_{2j}}(m)^* Y_2(m) \right|^2 \qquad (11)$$

where c represents the number of candidate sequences for each part, and $i\&j$ range from 1 to c. To this point, there are 25 sequence combinations corresponding to 25 power values in total. The corresponding power values $\hat{U}(q)$ need to be in the descending order. The index q' of candidate sequence combination is given by

$$q' = \arg\max_q (\hat{U}(q)) \qquad (12)$$

where q' corresponds with s'_{1i} and s'_{2j}. Thirdly, determine whether the selected sequence combination is a valid combination representing UCI bits or not. Judging from the first sequence combination. If the indexes of two short sequence are valid indexes arranged in the sequence pool, the corresponding sequence combination will be judged as a valid sequence combination; otherwise, the next sequence combination will be judged until the valid combination is found. Finally, the index of the sequence combination in the form of decimal is converted to the form of binary.

The size of the solution domain on conventional non-coherent receiver and enhanced receiver can be evaluated and analyzed. The definition of the solution domain is that there are several kinds of alternative sequences at the receiver. The solution domain size of the former one is 2048, which is larger than that of the latter one with 25. When the size of the solution domain is greatly reduced, the complexity of the non-coherent receiver is also reduced which can be expected. Supposing that 10 candidate sequences are reserved for each part, the combinations of two parts will be 100 kinds of choices. In this assumption, the solution domain size of enhanced receiver is 100/2048 of that of original receiver. The enhanced receiver reduces the computation complexity but at the cost of reducing the reliability and accuracy, which can be predicted. The more candidate sequences are reserved in each part, the more combinations can be selected. Due to the enhanced receiver containing the sort of power values, the latency of computation increases gradually as the solution domain is enlarged. It's necessary to determine appropriate number of candidate sequences for each part to make balance between latency and reliability.

3.2 Complexity

In this paper, three schemes are chosen as the baseline, including transmission scheme on PUCCH format 3 in Rel-16, a long sequence scheme based on PUCCH with conventional non-coherent receiver and sequence combination scheme with conventional receiver mentioned above.

The receiver complexity between existing receiver and proposed receiver is compared in this section. For PUCCH format 3 in Rel-16, the coherent receiver comprises channel estimation, equalization, demodulation and decoding. The complexity of PUCCH format 3 will not be listed here. The reason is that the schemes based on sequence are mainly compared and the complexity of PUCCH format 3 is difficult to evaluate. For the schemes based on sequence, the non-coherent receiver comprises three steps as follows:

- Correlation between received signal and candidate sequences in the sequence pool
- Non-coherent combination across receiving antennas
- Select the final sequence that yields the largest power after combination.

For two existing schemes based on sequence and proposed scheme, the second and the third steps are almost the same. Thus, the computation complexity of the first step on different schemes needs to be compared. In addition, the conventional non-coherent receiver of the long sequence scheme is the same as the sequence combination scheme. Supposing 11 bits UCI are received in one antenna, the number of multiplications, additions and modular squares between conventional non-coherent receiver and enhanced non-coherent receiver are listed in Table 1. The following notations are used:

- c: number of candidate sequences for each part at the receiver,
- M: number of REs in the PUCCH.

Table 1. Complexity of different configuration

Type of non-coherent receiver	Convention	Enhancement		
		$c = 1$	$c = 5$	$c = 10$
Multiplications	$2048 * M$	$65 * M$	$65 * M$	$65 * M$
Additions	$2048 * (M - 1)$	$65 * M - 130 + 1$	$65 * M - 130 + 5 * 5$	$65 * M - 130 + 10 * 10$
Modular squares	$2048 * M$	$65 * M + 1$	$65 * M + 5 * 5$	$65 * M + 10 * 10$

As shown in Table 1, the computation of the enhanced receiver is about 65/2048 of that of conventional scheme. The statistics of complexity verifies that the enhanced receiver reduces complexity of non-coherent receiver dramatically. The more candidate sequences are reserved in each part, the more combinations can be selected. In addition, when the solution domain is enlarged, the reliability of the receiver can be increased with slightly higher complexity. From the point of the view of computation complexity, the non-coherent receiver enhancement based on PUCCH sequence combination has great advantages. It's too complicated to evaluate the reliability of the scheme in theoretical calculation. Thus, the reliability of the scheme needs to be evaluated by computer simulation.

4   Simulation Results and Analysis

In this section, simulations are performed to evaluate the detection performance of non-coherent receiver enhancement based on PUCCH sequence combination. The detection performance of enhanced receiver is compared with the baseline, including PUCCH format 3 in Rel-16 with coherent receiver, a long sequence scheme with non-coherent receiver and sequence combination scheme with non-coherent receiver. The simulation parameters are shown in Table 2.

Table 2. Parameters configuration

Simulation parameters	Values
Length of UCI	11 bits
Bandwidth	20 MHz
Carrier frequency	4 GHz
Number of RB	1
Number of OFDM symbol	4–14
SCS	30 kHz
Channel model	TDL-C 300 ns-11 Hz
Antenna configuration	1Tx * 4Rx
Channel estimation for PUCCH format 3	LMMSE with ideal PDP
Sequence generation for the long sequence scheme	2048 sequences based on 16 base sequences and 128 cyclic shifts
Sequence generation for sequence combination scheme	65 sequences based on 5 base sequences and 13 cyclic shifts

In this paper, metrics are determined by whether DTX detection is considered or not. For transmission without DTX detection, 1% BLER is taken as the metric to evaluate the performance of the scheme. 1% DTX to ACK error rate, 1% ACK miss detection error rate and 0.1% NACK to ACK error rate are taken as metrics for different schemes with DTX detection, which is mentioned in Sect. 2. Simulation results simulated by MATLAB are shown below.

Firstly, the simulation results of enhanced non-coherent receiver based on PUCCH sequence combination with different candidate sequences are shown in Fig. 4. The integer value c represents the number of candidate sequences for each part which is mentioned in Sect. 3. The horizontal axis represents the number of OFDM symbols and the vertical axis represents SNR of 1% BLER metric. As can be seen in Fig. 5, the performance of enhanced non-coherent receiver with 10 candidate sequences is better than that with 5 and 1 candidate sequence. Compared with conventional non-coherent receiver, the enhanced receiver has 0.2–0.4 dB performance loss. The enhanced receiver reduces computation complexity with slightly performance loss. Comprehensively considering latency of choosing best candidate sequence combination and reliability of enhanced receiver, 10 candidate sequences is the best choice.

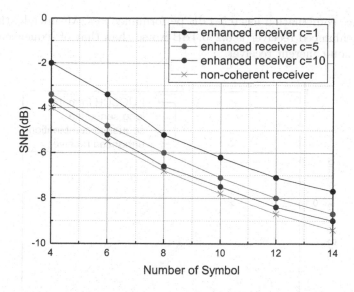

Fig. 4. Enhanced non-coherent receiver with different c, c represents the number of candidate sequences for each part.

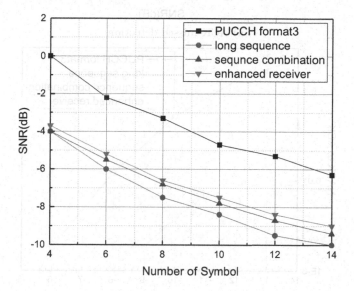

Fig. 5. Simulation results of different schemes without DTX detection

Then, the enhanced receiver needs to be compared with the baseline. As can be seen in Fig. 5, the performance of three schemes based on sequence is much better than PUCCH format 3 with coherent receiver. Compared with the long sequence scheme with non-coherent receiver, the enhanced receiver based on sequence combination has 0.3–1 dB performance loss. Meanwhile, the performance of enhanced receiver is 0.2–0.4 dB worse than that of sequence combination with non-coherent receiver.

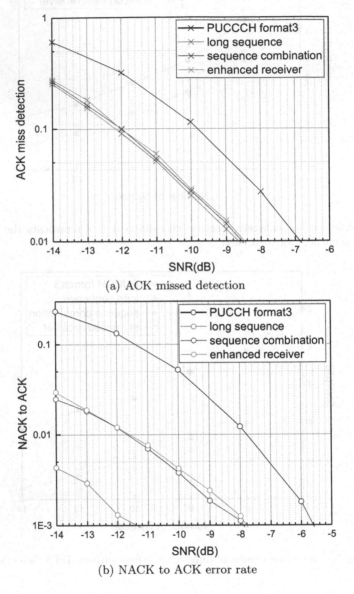

(a) ACK missed detection

(b) NACK to ACK error rate

Fig. 6. Simulation results of different schemes with DTX detection

Finally, as mentioned in Sect. 2, DTX detection needs to be considered in PUCCH UCI transmission. Two metrics including ACK miss detection and

NACK to ACK error rate are considered and analyzed together in this part. The case of 14 OFDM symbols with DTX detection is simulated and the results are shown in Fig. 6. As can be seen in Fig. 6, the performance of three schemes except for the long sequence scheme is determined by the metric of NACK to ACK error rate. The performance of three schemes based on sequence design is much better than PUCCH format 3 with coherent receiver, which is the same as the results without DTX detection in Fig. 5. The performance of enhanced receiver is about 0.8dB worse than that of long sequence with non-coherent receiver, and 0.1dB worse than that of sequence combination with non-coherent receiver, which is almost the same as results above. To sum up, simulation results with DTX detection is the same as that without DTX detection.

Though the performance loss can not be ignored, the design of sequence combination reduces resource consumption which is mentioned in Sect. 2. What's more, the enhanced receiver can reduce the computation complexity significantly.

5   Conclusion

In this paper, non-coherent receiver enhancement based on PUCCH sequence combination is proposed for PUCCH coverage enhancement item in Rel-17. Compared with transmission scheme based on a long sequence with non-coherent receiver, the enhanced receiver based on sequence combination has less resource consumption and lower computation complexity. The enhanced non-coherent receiver reduces the size of solution domain compared with conventional receiver. The simulation results show that whether DTX detection is considered or not, the enhanced non-coherent receiver has slight performance loss compared with conventional coherent receiver. Comprehensively considering complexity, latency and reliability of performance, the number of candidate sequences of each part can be determined due to specific scenarios. Therefore, according to simulation results and analysis, non-coherent receiver enhancement based on PUCCH sequence combination can be considered to be used in the scenarios of PUCCH coverage enhancement. For the future work, in addition to the design of sequence combination, there may exists better method to generate sequence. For the receiver, the features of sequence can be used to reduce complexity such as orthogonality.

References

1. 3GPP TS 38.300, 3rd Generation Partnership Project, Technical Specification Group Radio Access Network, NR, NR and NG-RAN Overall Description, Release 16 (2020)
2. Yuan, Y., Tang, W., Sun, C.: 5G NR control channel coverage enhancement base on segmented sequence detection and narrow beam. In: 2020 IEEE 20th International Conference on Communication Technology (ICCT), pp. 493–498 (2020). https://doi.org/10.1109/ICCT50939.2020.9295929
3. RP-200861, Revised SID on study on NR overage enhancements, 3GPP TSG RAN WG1, Meeting#88e, 29th June–3rd July 2020

4. R1-2006224, Discussion on the baseline performance in FR1, 3GPP TSG RAN WG1, Meeting#102e, 17th–28th August 2020
5. R1-2006225, Discussion on the baseline performance in FR2, 3GPP TSG RAN WG1, Meeting#102e, 17th–28th August 2020
6. R1-2990315, Potential coverage enhancement techniques for PUCCH, 3GPP TSG RAN WG1, Meeting#103e, 26th October–13th November 2020
7. 3GPP TR 38.830, 3rd Generation Partnership Project, Technical Specification Group Radio Access Network, Study on NR coverage enhancements, Release 17 (2020)
8. R1-2008027, Discussion on the PUCCH coverage enhancement, 3GPP TSG WAN WG1, Meeting#102e, 17th–28th August 2020
9. R1-2004499, Potential techniques for coverage enhancement, 3GPP TSG RAN WG1, Meetng#101e, 25th May–5th June 2020
10. 3GPP TS 38.211, 3rd Generation Partnership Project, Technical Specification Group Radio Access Network, NR, Physical channels and modulation, Release 16 (2020)
11. RAN1#103e Chairman's Notes, 3GPP TSG RAN WG1, Meeting#103e, 26th October–13th November 2020

MCAD-Net: Multi-scale Coordinate Attention Dense Network for Single Image Deraining

Pengpeng Li[iD], Jiyu Jin[✉][iD], Guiyue Jin[iD], Jiaqi Shi[iD], and Lei Fan[iD]

Dalian Polytechnic University, Dalian 116034, China
{jiyu.jin,guiyue.jin,fanlei}@dlpu.edu.cn

Abstract. Single image rain removal is an urgent and challenging task. Images acquired under natural conditions are often affected by rain, which leads to a serious decline in the visual quality of images and hinders some practical applications. Therefore, the research of image rain removal has attracted much attention. However, both the model based method and the deep learning based method can not adapt to the spatial and channel changes of rain feature information. In order to solve these problems, this paper proposes an end-to-end Multi-scale Coordinate Attention Dense Network (MCAD-Net) for single image deraining. MCAD-Net can accurately identify and characterize rain streaks and remove them, while preserving image details. To better solve the problem, the Multi-scale Coordinate Attention Block (MCAB) is introduced into the MCAD-Net to improve the ability of feature extraction and representation of rain streaks. MCAB first uses different convolution kernels to extract and fuse multi-scale rain streaks features, and then uses the coordination attention module with adaptation module to recognize rain streaks in different spaces and channels. A large number of experiments have been carried out on several commonly used synthetic datasets and real datasets. The quantitative and qualitative results show that the proposed method is superior to the recent state-of-the-art methods in improving the performance of image rain removal and preserving image details.

Keywords: Deraining · Coordinate attention · Multi-scale

1 Introduction

The images captured from the outdoor vision system are often affected by rain, resulting in a serious degradation of the visual quality of the captured images. Generally, raindrops and rain streaks in the vicinity tend to block or distort the content of the background scene, while raindrops and rain streaks in the distance tend to produce atmospheric texture effects [1–4]. Especially in heavy rain, rain fog, rain streaks and rain particles in the air are superimposed on the background to form a veil-like visual degradation, which greatly reduces the contrast and visibility of the scene. Therefore, rain removal

Fund Project: Scientific Research Project of the Education Department of Liaoning Province (LJKZ0518, LJKZ0519, LJKZ0515).

H. Gao et al. (Eds.): ChinaCom 2021, LNICST 433, pp. 405–421, 2022.
https://doi.org/10.1007/978-3-030-99200-2_31

has become a necessary preprocessing step for subsequent tasks, such as object tracking [5], scene analysis [6], personnel reidentification [7], and road condition detection of automatic driving to further improve its performance. As an important research topic, image rain removal has attracted widespread attention in the field of computer vision and pattern recognition in recent years [3, 8–11]. In many practical application scenarios, there is an urgent need to restore rain-free images in rainy days.

In recent years, with the much application of artificial intelligence in the field of computer vision, especially deep learning has achieved extraordinary results in image processing. The use of deep learning and neural networks to solve the problem of image enhancement and restoration in harsh environments has become research hotspots. The single-image rain removal method has gradually transitioned from model-driven to data-driven [12]. Traditional methods based on model-driven include filtering-based methods and prior-based methods. The method based on filtering is to use physical filtering to restore the rainless image [13–16]. The prior-based method considers single image rain removal as an optimization problem, which includes sparse prior [17] Gaussian Mixture Model (GMM) [10] and low-rank representation [18]. Compared with the model-based method, the data-driven method regards the removal of a single image as a process of learning a nonlinear function [19]. Motivated by the success of deep learning in recent years, researchers have begun to use convolutional neural networks (CNN) or Generative Adversarial Networks (GAN) to model mapping functions [20]. Both the CNN-based design method [21–23] or the GANs-based design method have achieved better results than traditional methods.

Although the methods mentioned above have achieved good results in many application scenarios, there are still many limitations. Because image rain removal is a complex process, most rain removal models based on filtering and prior methods are not enough to cover some important factors in real rainfall images, such as rain streaks. As for the method based on deep learning, it ignores the internal mechanism of rain, making it easy to fall into the process of over-adapting training. Many existing algorithms are limited to some image blocks or limited receptive fields, and can not obtain image feature connection between large regions, resulting in usually unable to restore structure and details. But this kind of information that is ignored due to the limited receptive field has been proved to be very helpful to the image to remove the rain.

To address the above-mentioned issues, the paper present a Multi-scale Coordinate Attention Dense Network called MCAD-Net. The proposed MCAD-Net is based on DenseNet. The DenseNet has several compelling advantages: they alleviate the vanishing-gradient problem, strengthen feature propagation, encourage feature reuse, and substantially reduce the number of parameters. Multi-scale Coordinate Attention Block (MCAB) is introduced to better utilize multi-scale information and feature attention for improving the rain feature representation capability. Combing the features of different scales and layers, multi-scale manner is an efficient way to capture various rain streak components especially in the heavy rainy conditions. Recent studies on single image deraining network design have demonstrated the remarkable effectiveness of channel attention (e.g., the Squeeze-and-Excitation Attention) for lifting model performance, but they generally neglect the positional information, which is important for generating spatially selective attention maps. Therefore, coordinate attention module is involved in the MCAB. In this way, long-range dependencies can be captured along one spatial direction and meanwhile precise positional information can be preserved along the other spatial direction. The coordinate attention module helps the proposed network to adjust the three-color channels respectively and identify the rainy region properly. We evaluate the proposed network on the public competitive benchmark synthetic and real-world datasets and the results significantly outperform the current outstanding methods on most of the deraining tasks.

In summary, the contributions of this work may be summarized as follows:

1) We propose a MCAD-Net to address the single image deraining problem, which can effectively remove the rain streaks while well preserve the image details. The modified DenseNet is applied to boost the model performance via multi-level features reuse and maximum information flow between layers. It can alleviate the vanishing-gradient problem, strengthen feature propagation, encourage feature reuse, while fully utilizing the features of different layers to restore the details.
2) To our knowledge, the Multi-scale Coordinate Attention Block (MCAB) is first constructed to improve the representation of rain streaks. This hierarchical structure uses different convolution kernels to generate features of different scales that contain rich hierarchical information. In addition, the inherent correlation of multi-scale features can be used to gain a deeper understanding of the image layout and improve the performance of feature extraction to a large extent. Then, by introducing the coordinate attention module added to the adaptation module, the color channel and spatial location information are used to extract features better.
3) We perform experiments on both synthetic and real-world rain datasets (4 synthetic and 2 real-world datasets). Our proposed network outperforms the state-of-the-art methods in visually and quantitatively comparisons. Furthermore, ablation research is provided to verify the rationality and necessity of the important modules involved in our network.

2   Related Work

In this section, some image deraining methods are reviewed. Compared with the rain removal in video, single image rain removal is more challenging because of less available information. Therefore, more and more researchers pay attention to the algorithm design of single image rain removal in recent years. At present, the existing single image rain removal methods can be divided into three categories: filtering-based methods, prior-based methods and deep learning-based methods. The method based on filtering and prior is also called model-based method.

Model Based Methods. Xu et al. [13] proposed a single image rain removal algorithm with guided filtering kernel [14]. In short, it first uses the chromaticity characteristics of rain streaks to obtain a rain-free image with lower accuracy, and then filters the rain image to obtain a rain-free image with higher accuracy. Ding et al. [24] designed a guided L0 smoothing filter to obtain a rain-free image to improve the performance of a single image to remove rain.

In recent years, the Maximum A Posteriori (MAP) has been widely used in the method of removing rain from a single image [25, 26], which can be mathematically described as:

$$\max_{B,R\in\Omega} p(B, R|O) \propto p(O|B, R) \cdot p(B) \cdot (R) \tag{1}$$

where $O \in \mathbb{R}^{h\times w}$, $B \in \mathbb{R}^{h\times w}$, and $R \in \mathbb{R}^{h\times w}$ denote the observed rainy image, rain free image, and rain streaks, respectively. $p(B, R|O)$ is the posterior probability and $p(O|B, R)$ is the likelihood function. $\Omega: = \{B, R|0 \leq B_i, R_i \leq O_i, \forall i \in [1, M \times N]\}$ is the solution space.

Various methods have been proposed for designing the forms of all terms involved in (1). Fu et al. [27] used morphological component analysis (MCA) to describe image removal as an image decomposition problem. First, bilateral filtering is used to divide the rainy image into two parts: low-frequency component and high-frequency component, and then the low-frequency component and the non-rain component are combined to obtain the rain removal result. Recently, Gu et al. [11] proposed a joint convolution analysis and synthesis (JCAS) sparse representation model, using analytical sparse representation (ASR) to approximate the large-scale structure of the image, and using synthetic sparse representation (SSR) to describe the image Fine texture. The complementarity of ASR and SSR enables JCAS to effectively extract the image texture layer without excessively smoothing the background layer.

Deep Learning Based Methods. Single image rain removal based on deep learning started in 2017. Yang et al. [3] constructed a joint rain detection and removal network (JORDER) to focus on the removal of overlapping rain streaks under heavy rain. The network can better monitor the rain and locate the rain through prediction. This method has achieved impressive results under heavy rain conditions, but it may delete some texture details by mistake. Qian et al. [28] designed an Attention Generation Network whose basic idea is to inject visual information. Since the deep residual network (ResNet) [29] has achieved the greatest success in the field of deep learning, Fu et al. [2] further proposed a deep detail network (DDN) to achieve better Rain effect. Existing deep learning methods usually treat the network as an end-to-end mapping module, instead of studying the rationality of removing rain streaks [30, 31]. Li et al. [32] proposed a non-local enhancement encoder-decoder network, which can effectively learn more abstract features, so as to achieve more accurate image removal while retaining image details.

In order to alleviate the problem that the deep network structure is difficult to reproduce, Ren et al. [33] presented a simple and effective progressive recurrent deraining network (PReNet). Chen et al. [34] present a Multi-scale Hourglass Hierarchical Fusion Network (MH2F-Net) in end-to-end manner, this network accurately obtain rain trace features through multi-scale extraction, hierarchical extraction and information fusion. However, most of the existing single-image rain-removing networks have not well noticed the internal connection of rain streaks at different scales. And although they have tried to introduce the attention module in the rain-removing network, whether it is channel attention or spatial attention, they have ignored the position information, which is important for generating spatially selective attention map.

3 Proposed Method

In this section, the MCAD-Net is described in Sect. 3.1 and the Multi-scale Coordinate Attention Block (MCAB) is introduced in Sect. 3.2. The proposed loss function is described in Sect. 3.3.

3.1 Proposed MCAD-Net

This paper proposes an MCAD network based on the DenseNet, which is an end-to-end network that can input any rain image for training. The overall architecture of the MCAD network is shown in Fig. 1. The network effectively achieves the maximum reuse of the rain image features, thereby reducing the number of parameters and improving the efficiency of training and testing. In order to speed up the training process, global skip connections are introduced between different MCAB modules. These skip connections can not only help the back-propagation gradient to update the parameters, but also directly spread the lossless information through the entire network, so it is very helpful for the estimation of the final rain removal image. In addition, because different rain traces have different directions and shapes, MCAB is introduced into the network, which uses multi-scale and the latest feature attention module to obtain various rain trace characteristics and structure information. The details will be described in the following section.

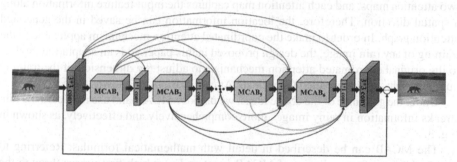

Fig. 1. The overall architecture of proposed MCAD-Net for image deraining. MCAB is shown in Fig. 3. The goal of the MCAD-Net is to recover the corresponding rainless image from the rainy image.

3.2 Design of Multi-scale Coordinate Attention Block (MCAB)

Multi-scale Feature. Multi-scale feature acquisition methods effectively combine image features at different scales, and are currently widely used to acquire useful information about objects and their vicinity. In order to further improve the ability of network representation, inter-layer multi-scale information fusion is applied in MCAB, which realizes the information fusion between features of different scales. This structure also ensures that the input information can be propagated through all parameter layers, so that the characteristic information of the original image can be better learned.

Coordinate Attention. It is well known that a color image is composed of RGB three channels, and the rain density of a rainy image is different on each channel in actual conditions. Most traditional rain removal methods often ignored this situation, until the channel attention mechanism was proposed, which can effectively obtain the important characteristics of rain on different channels. For example, in the attention mechanism network SENet, channel characteristic information is obtained through global average pooling (GAP). In addition, the study also found that the distribution of rain streaks in space is also uneven, so the spatial attention mechanism proposed later is also very important for removing rain from images.

However, whether it is channel attention or spatial attention mechanism, they usually ignore the specific position information, and the position information is very important for generating spatial selective attention feature maps. Therefore, the coordinated attention mechanism proposed by Qi et al. [35] is introduced in MACB to solve the problem of rain removal better, enhance the ability of the network to extract characteristic information, and improve network performance and accuracy. The coordinated attention mechanism is different from the previous channel attention mechanism. It does not convert multiple feature vectors into a single feature vector through 2D global pooling, but decomposes the channel attention into two 1D feature encoding processes, thus achieve the effect of gathering feature information along the X and Y spatial directions respectively. Then, the two feature maps with specific direction information are coded into two attention maps, and each attention map captures the input feature information along a spatial direction. Therefore, the location information can be saved in the generated attention graph. In order to make the coordinated attention mechanism applicable to the training of any rain image, the design proposed in this paper adds an adaptation module to the original coordinated attention mechanism to adjust the dimension of the training output weight, which is called Coordinate Attention-Block as shown in the Fig. 2.

Under the guidance of the above ideas, we propose MCAB and use it to learn the rain streaks information in rainy images more comprehensively and effectively, as shown in Fig. 3.

The MCAB can be described in detail with mathematical formulas. Referring to Fig. 3, the input feature image of MCAB is set as F_{in}, which first passes through the convolutional layers with the convolution kernel sizes of 1×1, 3×3, and 5×5, and the output is expressed as follows:

$$F_a^{1\times1} = Conv_{1\times1}\left(F_{in}; \theta_a^{1\times1}\right) \tag{2}$$

$$F_a^{3\times3} = Conv_{3\times3}\left(F_{in}; \theta_a^{3\times3}\right) \tag{3}$$

$$F_a^{5\times5} = Conv_{5\times5}\left(F_{in}; \theta_a^{5\times5}\right) \tag{4}$$

where $F_a^{n\times n}$ presents the first layer output of multi-scale convolution with the convolution size of $n \times n$, $Conv_{n\times n}(\cdot)$ presents convolution operation, and $\theta_a^{n\times n}$ means the hyperparameter formed by the first multi-scale convolutional layer with the convolution

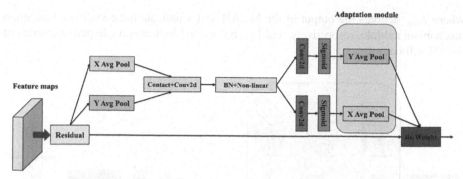

Fig. 2. The overall architecture of our proposed Coordinate Attention-Block for image deraining. Compared with the original coordinated attention, an adaptation module is added. The goal of Coordinate Attention-Block is to capture cross-channel information, while also capturing direction perception and position-sensitive information, which helps the overall network model to more accurately locate and identify more important objects.

kernel size of $n \times n$. The image features can be further extracted by using the convolution kernel size to be $1 \times 1, 3 \times 3$, and 5×5

$$F_b^{1\times1} = Conv_{1\times1}\left(\left(F_a^{1\times1} + F_a^{3\times3} + F_a^{5\times5}\right); \theta_b^{1\times1}\right) \tag{5}$$

$$F_b^{3\times3} = Conv_{3\times3}\left(\left(F_a^{1\times1} + F_a^{3\times3} + F_a^{5\times5}\right); \theta_b^{3\times3}\right) \tag{6}$$

$$F_b^{5\times5} = Conv_{5\times5}\left(\left(F_a^{1\times1} + F_a^{3\times3} + F_a^{5\times5}\right); \theta_b^{5\times5}\right) \tag{7}$$

where $F_b^{n\times n}$ presents the output of the second layer of multi-scale convolution with size $n \times n$, $Conv_{n\times n}(\cdot)$ presents a convolution of size $n \times n$, and $\theta_b^{n\times n}$ means the hyperparameter formed by the second multi-scale convolutional layer with a size of $n \times n$. Similarly, we can express the output of the multi-scale third layer as follows:

$$F_c^{1\times1} = ((Conv_{1\times1}\left(F_b^{1\times1} + F_b^{3\times3} + F_b^{5\times5} + F_a^{1\times1}\right) + F_a^{1\times1}); \theta_c^{1\times1}) \tag{8}$$

$$F_c^{3\times3} = ((Conv_{3\times3}\left(F_b^{1\times1} + F_b^{3\times3} + F_b^{5\times5} + F_a^{1\times1}\right) + F_a^{3\times3}); \theta_c^{3\times3}) \tag{9}$$

$$F_c^{5\times5} = ((Conv_{5\times5}\left(F_b^{1\times1} + F_b^{3\times3} + F_b^{5\times5} + F_a^{5\times5}\right) + F_a^{5\times5}); \theta_c^{5\times5}) \tag{10}$$

As shown in Fig. 3, MCAB realizes multi-scale information fusion through convolutional layers with the convolution kernel sizes of 1×1 and 3×3, and finally introduces a coordinated attention mechanism module to improve feature fusion. We can express the final output of MCAB as follows:

$$F_{out} = ca\left(\left(Conv_{3\times3}\left(Conv_{1\times1}\left(Cat\left(F_c^{1\times1}, F_c^{3\times3}, F_c^{5\times5}\right); \eta_1\right); \eta_2\right); \eta_3\right); \eta_4\right) + F_{in} \tag{11}$$

where F_{out} denotes the output of the MCAB, $ca(\cdot)$ indicate the coordinated attention mechanism module, respectively, and $\{\eta_1; \eta_2; \eta_3; \eta_4\}$ indicates the hyperparameters of the MCAB output.

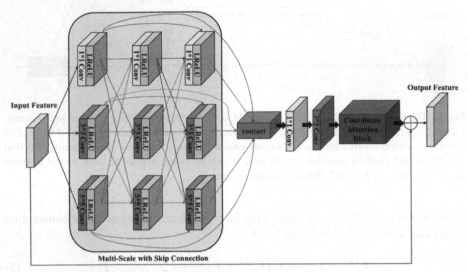

Fig. 3. The overall architecture of our proposed Multi-scale Coordinate Attention Block (MCAB). It is mainly composed of a multi-scale feature acquisition module and a coordinated attention module. The coordinated attention mechanism has been shown in Fig. 2. This architecture enables the network to better explore and reorganize features in different scales.

3.3 Loss Function

The commonly used loss function in image processing research is MSE loss function, because it has better results in most application scenarios. However, the disadvantages of the MSE loss function are also obvious. When the application task involves image quality evaluation, it is assumed that the influence of noise is independent of the local features of the image. This is contrary to the human visual system (HVS), so the correlation between MSE as a loss function and image quality is poor. In order to solve the above shortcomings, we combine MSE and SSIM to propose a loss function, so as to achieve a balance between the effect of image rain removal and image quality evaluation.

In this paper, the rain model we refer to is as follows:

$$I = B + R \tag{12}$$

we can obtain the rain-free background by subtracting rain streaks R from the rainy image I.

MSE Loss. A pixel-wise MSE loss can be defined as follows

$$L_{MSE} = \frac{1}{HWC} \sum_{i=1}^{H} \sum_{j=1}^{W} \sum_{k=1}^{C} \left\| \hat{B}_{i,j,k} - B_{i,j,k}^2 \right\|^2 \tag{13}$$

where H, W and C represent height, width, and number of channels respectively. $\hat{B}$ and B denote the restored rain-free image and the groundtruth, respectively.

SSIM Loss. SSIM is an important indicator to measure the structural similarity between two images [36], with the equation as follows:

$$SSIM\left(\hat{B}, B\right) = \frac{2\mu_{\hat{B}}\mu_B + C_1}{\mu_{\hat{B}}^2 + \mu_B^2 + C_1} \cdot \frac{2\sigma_{\hat{B}}\sigma_B + C_2}{\sigma_{\hat{B}}^2 + \sigma_B^2 + C_2} \quad (14)$$

where μ_x, σ_x^2 are the mean and the variance value of the image: x. The covariance of two images is σ_{xy}, C_1 and C_2 are constants value used to maintain equation stability. The value range of SSIM is from 0 to 1. In the image rain removal problem, the larger the value obtained by SSIM in the interval means that the recovered rain-free image is closer to the real image. Therefore, the loss function based on SSIM can be defined as:

$$L_{SSIM} = 1 - SSIM\left(\hat{B}, B\right) \quad (15)$$

Total Loss. The total loss is defined by combing the MSE loss and the SSIM loss as follows:

$$L = L_{MSE} + \lambda L_{SSIM} \quad (16)$$

where λ is a hyperparameter that balances the weight between MSE loss and SSIM loss. By properly setting λ, the similarity of each pixel can be ensured while maintaining the global structure. This helps to get a better rain image.

4 Experiments

In this section, the dataset used in the experiment is introduced. Then some details of the experimental environment and settings are described. In order to prove that the proposed MCAD-Net has a good effect on the image rain removal problem, we have carried out a quantitative and qualitative evaluation of the proposed method on the synthetic dataset and the real dataset, and compared the results with the recent state-of-the-art methods. At the same time, complete ablation studies are conducted to prove the importance of each component in the proposed network.

4.1 Experiment Details

As shown in Fig. 1, in order to achieve the best image removal effect, the number of MCAB in MCAD-Net is set to 6. The reason will be explained in Sect. 4.6 Ablation Experiment. For all the datasets, we randomly crop 64 × 64 patch from each input image. The training workstation is configured as Ubuntu 18.04, memory 16G, NVIDIA P102 GPU (10G). During training, the batch size is set to 32 and use the Adam optimizer to train a total of 100 epochs. The initial learning rate is set to 1×10^{-3}, and the learning rate is reduced by half every 25 epochs.

4.2 Datasets and Evaluation Metrics

Four synthetic datasets and two real-world datasets will be used to evaluate the performance of the proposed method. The composition of the datasets is shown in Table 1.

Table 1. Synthetic and real-world datasets.

Datasets	Training set	Testing set	Type
Rain100L	200	100	Synthetic
Rain100H	1800	100	Synthetic
Rain800	700	100	Synthetic
Rain1400	12600	1400	Synthetic
MPID	–	185	Real-world
Li et al.	–	34	Real-world

As shown in Table 1, experiments are conducted on the proposed MCAD-Net on the four synthetic datasets Rain100L [3], Rain100H [3], Rain800 [37] and Rain1400 [2]. The four datasets include rain streaks of various sizes, shapes, and directions. Among them, Rain100L is a light rain dataset which contains only one kind of rain streaks, which is composed of 200 training image pairs and 100 test image pairs. The Rain100H dataset contains 5 rain streaks in different directions, consisting of 1800 training image pairs and 100 test image pairs. Rain800 consists of 700 training image pairs and 100 test image pairs. Rain1400 contains 14 rain streaks of different directions and sizes, from which 12600 image pairs with rain are selected as training data, and the other 1400 image pairs are used for testing. The real-world dataset is very important for evaluating the performance of image rain removal, so we further conducted experiments on two real-world datasets: One of them is the MPID dataset proposed by Li et al., and the other is also proposed by Li et al. in 2019. They are composed of 185 and 34 real pictures of rain [38].

The performance of image rain removal methods is usually evaluated according peak signal-to-noise ratio (PSNR) and structural similarity (SSIM). The higher the PSNR value, the better the performance of recovering the rainless image from the rainy image. The SSIM value means the similarity of two different images to each other, and the value range is 0 to 1. When the SSIM is closer to 1, the rain removal performance is good. Since it is a basic fact that there is no completely clean image in the real world, which makes it impossible to quantitatively analyze the rain removal effect, we will intuitively evaluate the performance on the real world datasets from the visual effect.

4.3 Results on Synthetic Datasets

In this section, a large number of experiments are conducted on the synthetic datasets Rain100L, Rain100H, Rain800 and Rain1400, which are commonly used in the problem of image rain removal. The results of the MCAD-Net proposed in this paper on the datasets compared with some recent mainstream advanced methods: GCANet [39], LPNet [40], RESCAN [22], JORDER [3], JORDER-R [3], JORDER-E [3], DID-MDN [41], SPANet [42], ReHEN [43], PReNet [34], RCDNet [44], MPRNet [45].

Table 2 shows the quantitative results of the proposed method on the four synthetic datasets. It can be seen that the method proposed in this paper has improved PSNR and SSIM value compared with the advanced method of reference. It shows that MCAD-Net has better robustness and versatility.

Table 2. The quantitative results in the table are evaluated based on the PSNR and SSIM average results of the synthetic benchmark datasets (Rain100L, Rain100H, Rain800 and Rain1400), and the best results are shown in bold.

Methods	Datasets			
	Rain100L (PSNR/SSIM)	Rain100H (PSNR/SSIM)	Rain800 (PSNR/SSIM)	Rain1400 (PSNR/SSIM)
Rainy	26.91/0.838	13.35/0.388	21.16/0.652	25.24/0.810
GCANet	31.70/0.932	24.80/0.814	–	27.84/0.841
LPNet	33.39/ 0.958	24.39/0.820	25.26/0.781	22.03/0.800
RESCAN	36.12/ 0.970	27.88/0.816	24.09/0.841	29.88/0.905
DDN	–	24.95/0.781	22.16/0.732	27.61/0.901
JORDER	36.55/0.974	22.79/0.697	26.24/0.850	27.55/0.853
JORDER-R	36.71/0.980	23.45/0.749	26.73/0.859	–
JORDER-E	37.20/0.979	24.54/0.802	27.10/0.863	–
DID-MDN	28.27/0.857	17.39/0.612	21.89/0.795	27.99/0.869
SPANet	35.33/0.970	25.11/0.833	24.37/0.861	28.57/0.891
ReHEN	–	27.97/0.864	26.96/0.854	30.63/0.918
PReNet	37.11/0.971	28.06/0.888	22.83/0.790	30.73/0.920
RCDNet	35.28/0.971	26.18/0.835	24.59/0.821	–
MPRNet	37.10/0.975	26.31/0.846	–	–
Ours	**37.33/0.982**	**28.51/0.901**	**27.54/0.865**	**30.92/0.927**

In addition to the quantitative evaluation of the rain removal effect of the images, some images are provided for intuitive comparison. As shown in Fig. 4 and Fig. 5, Rain100H and Rain100L images are provided for visual comparison. We selected a detail in the image and enlarged it. By observing the enlarged local area, although JORDER, LPNet, PReNet and RESCAN have removed a lot of rain patterns, they will all cause different degrees of background blur and there are certain shortcomings in the preservation of the image background details. Compared with Ground Truth, the results obtained by the method in this paper have achieved good results. Therefore, by comparing the method proposed in this paper, the rain pattern can be effectively removed while preserving the background details on the synthetic datasets.

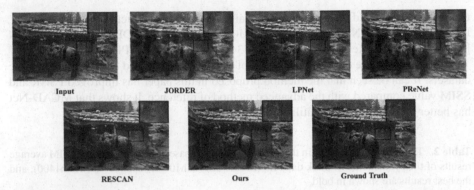

Fig. 4. Visual quality comparisons of all competing methods on synthetic dataset (Rain100H)

Fig. 5. Visual quality comparisons of all competing methods on synthetic dataset (Rain100L)

4.4 Results on Real-World Datasets

In order to evaluate the effectiveness of the proposed method in practical application, the proposed method is compared with the reference method on two real-world rainy datasets mentioned in Sect. 4.2 for further experimental evaluation. In order to compare the fairness, all methods use the weight of the pretraining model obtained from Rain100H dataset to remove the rain streaks from the real rain dataset. As shown in Fig. 6 and Fig. 7, compared with the three most advanced methods, the proposed method produces a more natural and pleasant rain removal image. Specifically, from the enlarged local details, it can be seen that DDN can not completely remove the rain streaks in most cases, while JORDER and LPNet blur the details of the rain removal results more. The method in this paper can remove the rain streaks in the real world rain image more effectively and retain more texture details.

4.5 Ablation Study

In order to prove the validity and rationality of the structure configuration and parameter setting in MACD-Net proposed in this paper, ablation experimental studies are

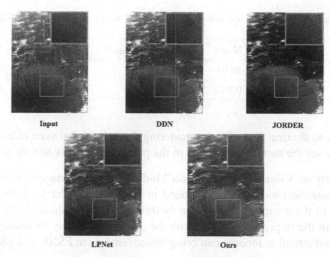

Fig. 6. Visual quality comparisons of all competing methods on Real-world dataset

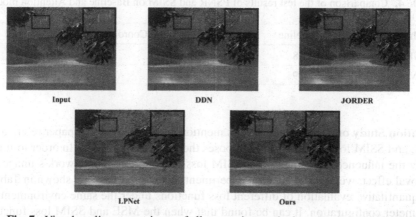

Fig. 7. Visual quality comparisons of all competing methods on Real-world dataset

conducted. All the studies involved use the Rain100L dataset and are guaranteed to be carried out in the same environment.

Ablation study on Multi-scale Coordinate Attention Block Numbers. In order to study the influence of different numbers of MCAB in MCAD-Net on network image rain removal. Experiments are conducted with Rain100L as the reference dataset, where the number of MCAB ranges from 2 to 10, and the experimental results are shown in Table 3.

It can be seen from Table 3 that in the ablation experiment of Multi-Scale Coordinate Attention Block Numbers, the PSNR and SSIM values can be improved as the number of MCAB is increased. When n = 6, the evaluation value reaches the highest, and

Table 3. Comparison of the test results of PSNR and SSIM on different MCAB numbers

MCAB no.	N = 2	N = 4	N = 6 (default)	N = 8	N = 10
PSNR	35.76	36.91	**37.33**	37.10	36.81
SSIM	0.977	0.980	**0.982**	0.981	0.979

then it begins to decline. Therefore, considering the balance between effectiveness and efficiency, we set the number of MCAB in the proposed network to 6 by default.

Ablation study on Coordinate Attention Modules. As introduced in Sect. 3.2, the coordinated attention module is introduced in MCAB. In order to further verify the effectiveness of the coordinated attention module, the basic multi-scale fusion network compared with the improved network with the attention module. As shown in Table 4, the coordinated attention module can bring improvements in PSNR and SSIM.

Table 4. Comparison of the test results of PSNR and SSIM on Baseline and Attention module

Methods	Baseline	Baseline + Coordinate Attention
PSNR	36.48	**37.33**
SSIM	0.979	**0.982**

Ablation Study on Loss Function. As mentioned in Sect. 3.3, this paper refers to the MSE and SSIM loss functions and proposes the Total loss function. In order to further study the influence of the MSE and SSIM loss functions on the network's image rain removal effect, we also did ablation experiments for comparison. As shown in Table 5, the quantitative evaluation of different loss functions under the same environment and parameter configuration. It can be found that when the MSE and SSIM loss functions are used alone, the MSE loss function makes the PSNR value higher, while the SSIM loss function focuses on the attack and structural similarity of the picture, which leads to a higher SSIM value. Therefore, comprehensively considering the quantitative combination of the MSE and SSIM loss functions, the Total loss function has achieved better results.

Table 5. Comparison of test results between PSNR and SSIM on different Loss Functions

Loss	MSE loss	SSIM loss	MSE+SSIM (Total loss)
PSNR	36.20	35.13	**37.33**
SSIM	0.968	0.9758	**0.982**

5 Conclusion

In this paper, a Multi-scale Coordinate Attention Dense Network (MCAD-Net) is proposed to remove rain from a single image. MCAD-Net is based on DenseNet and uses multiple skip connections to achieve feature reuse and full dissemination of feature information. In order to better identify and characterize the feature information of rain streaks, multi scale feature acquisition and coordinated attention module are introduced to form a Multi-scale Coordinated Attention Block (MCAB). First, get rain streaks features of different scales through convolution kernels of different sizes and fuse them. Then the attention module is used to further explore the rain streaks features in different channels and spaces. The rationality of module introduction and parameter setting is proved by ablation experiments. Finally, the proposed method is tested under common synthetic datasets and real datasets, and compared with the latest advanced methods, both showed better performance. In addition, the method proposed in this paper also better restores the details of the rainless image. In the future, we plan to extend our research work to video deraining under autonomous driving. For example, adding an additional module to our network to capture time information.

References

1. Yang, W., Tan, R.T., Feng, J., Guo, Z., Yan, S., Liu, J.: Joint rain detection and removal from a single image with contextualized deep networks. IEEE Trans. Pattern Anal. Mach. Intell. **42**, 1377–1393 (2020)
2. Fu, X., Huang, J., Zeng, D., Huang, Y., Ding, X., Paisley, J.: Removing rain from single images via a deep detail network. In: 2017 IEEE Conference on ComputerVision and Pattern Recognition (CVPR), pp. 1715–1723 (2017)
3. Yang, W., Tan, R.T., Feng, J., Liu, J., Guo, Z., Yan, S.: Deep joint rain detectionand removal from a single image. In: 2017 IEEE Conference on Computer Vision and Pattern Recognition (CVPR), pp. 1685–1694 (2017)
4. Simonyan, K., Zisserman, A.: Very deep convolutional networks for large-scaleimage recognition. In: ICLR 2015: International Conference on Learning Representations 2015 (2015)
5. Comaniciu, D., Ramesh, V., Meer, P.: Kernel-based object tracking. IEEE Trans. Pattern Anal. Mach. Intell. **5**, 564–575 (2003)
6. Itti, L., Koch, C., Niebur, E.: A model of saliency-based visual attention for rapid scene analysis. IEEE Trans. Pattern Anal. Mach. Intell. **11**, 1254–1259 (1998)
7. Farenzena, M., Bazzani, L., Perina, A., Murino, V., Cristani, M.: Person re-identification by symmetry-driven accumulation of localfeatures. In: IEEE Computer Society Conference on Computer Vision and Pattern Recognition, pp. 2360–2367 (2010)
8. Zhang, X., Li, H., Qi, Y., Leow, W., Ng, T.: Rain removal in video by combiningtemporal and chromatic properties. In: 2006 IEEE International Conference on Multimedia and Expo, pp. 461–464 (2006)
9. Luo, Y., Xu, Y., Ji, H.: Removing rain from a single image via discriminative sparsecoding. In: 2015 IEEE International Conference on Computer Vision (ICCV), pp. 3397–3405 (2015)
10. Li, Y., Tan, R.T., Guo, X., Lu, J., Brown, M.S.: Rain streak removal using layerpriors. In: 2016 IEEE Conference on Computer Vision and Pattern Recognition (CVPR), pp. 2736–2744 (2016)

11. Gu, S., Meng, D., Zuo, W., Zhang, L.: Joint convolutional analysis and synthesissparse representation for single image layer separation. In: 2017 IEEE International Conference on Computer Vision (ICCV), pp. 1717–1725 (2017)
12. Li, S., et al.: Single image deraining: a comprehensivebenchmark analysis. In: 2019 IEEE/CVF Conference on Computer Vision and Pattern Recognition (CVPR), pp. 3838–3847 (2019)
13. Xu, J., Zhao, W., Liu, P., Tang, X.: Removing rain and snow in asingle image using guided filter. In: IEEE International Conference on Computer Science and Automation Engineering (CSAE), vol. 2, pp. 304–307 (2012)
14. He, K., Sun, J., Tang, X.: Guided image filtering. IEEE Trans. Pattern Anal. Mach. Intell. **35**(6), 1397–1409 (2012)
15. Xu, J., Zhao, W., Liu, P., Tang, X.: An improved guidance image based method to remove rain and snow in a single image. Comput. Inf. Sci. **5**(3), 49 (2012)
16. Zheng, X., Liao, Y., Guo, W., Fu, X., Ding, X.: Single-image-based rain and snow removal using multi-guided filter. In: Lee, M., Hirose, A., Hou, Z.-G., Kil, R.M. (eds.) ICONIP 2013. LNCS, vol. 8228, pp. 258–265. Springer, Heidelberg (2013). https://doi.org/10.1007/978-3-642-42051-1_33
17. Zhang, H., Patel, V.M.: Convolutional sparse and low-rank coding-based rainstreak removal. In: 2017 IEEE Winter Conference on Applications of Computer Vision (WACV), pp. 1259–1267 (2017)
18. Chen, Y.L., Hsu, C.T.: A generalized low-rank appearance model for spatio temporally corelated rain streaks. In: 2013 IEEE International Conference onComputer Vision, pp. 1968–1975 (2013)
19. Yang, W., Tan, R.T., Wang, S., Fang, Y., Liu, J.: Single image deraining: from model-based to data-driven and beyond. IEEE Trans. Pattern Anal. Mach. Intell. **43**, 4059–4077 (2020)
20. Goodfellow, I., et al.: Generative adversarial nets. In: Advances in NeuralInformation Processing Systems, vol. 27, pp. 2672–2680 (2014)
21. Ren, D., Zuo, W., Hu, Q., Zhu, P., Meng, D.: Progressive image deraining networks:a better and simpler baseline. In: 2019 IEEE/CVF Conference on Computer Vision and Pattern Recognition (CVPR), pp. 3937–3946 (2019)
22. Li, X., Wu, J., Lin, Z., Liu, H., Zha, H.: Recurrent squeeze-and-excitation context aggregation net for single image deraining. In: Proceedings of the European Conference on Computer Vision (ECCV), pp. 262–277 (2018)
23. Yang, W., Liu, J., Yang, S., Guo, Z.: Scale-free single image deraining via visibility enhanced recurrent wavelet learning. IEEE Trans. Image Process. **28**, 2948–2961 (2019)
24. Mu, P., Chen, J., Liu, R., Fan, X., Luo, Z.: Learning bilevel layerpriors for single image rain streaks removal. IEEE Signal Process. Lett. **26**(2), 307–311 (2019)
25. Mu, P., Chen, J., Liu, R., Fan, X., Luo, Z.: Learning bilevel layer priors for single image rain streaks removal. IEEE Signal Process. Lett. **26**(2), 307–311 (2019)
26. Gauvain, J.-L., Lee, C.-H.: Maximum a posteriori estimation for multivariate Gaussian mixture observations of Markov chains. IEEE Trans. Speech Audio Process. **2**(2), 291–298 (1994)
27. Fu, Y.-H., Kang, L.-W., Lin, C.-W., Hsu, C.-T.: Single-frame-basedrain removal via image decomposition. In: 2011 IEEE International Conference on Acoustics, Speech and Signal Processing (ICASSP) (2011)
28. Qian, R., Tan, R.T., Yang, W., Su, J., Liu, J.: Attentive generativeadversarial network for raindrop removal from a single image. In: Proceedings of the IEEE Conference on Computer Vision and Pattern Recognition, pp. 2482–2491 (2018)
29. He, K., Zhang, X., Ren, S., Sun, J.: Deep residual learning for imagerecognition. In: Proceedings of the IEEE Conference on Computer Vision and Pattern Recognition, pp. 770–778, pp. 1453–1456 (2016)

30. Pan, J., et al.: Learning dual convolutional neural networks for low-level vision. In: Proceedings of the IEEE Conference on Computer Vision and Pattern Recognition, pp. 3070–3079 (2018)
31. Wang, Y.-T., Zhao, X.-L., Jiang, T.-X., Deng, L.-J., Chang, Y., Huang, T.-Z.: Rain streak removal for single image via kernel guided CNN. arXiv preprint arXiv:1808.08545 (2018)
32. Li, G., He, X., Zhang, W., Chang, H., Dong, L., Lin, L.: Non-locally enhanced encoder-decoder network for single image de-raining. arXiv preprint arXiv:1808.01491 (2018)
33. Ren, D., Zuo, W., Hu, Q., Zhu, P., Meng, D.: Progressive image deraining networks: a better and simpler baseline. In: 2019 IEEE/CVF Conference on Computer Vision and Pattern Recognition (CVPR), pp. 3937–3946 (2019)
34. Chen, X., Huang, Y., Xu, L.: Multi-scale hourglass hierarchical fusion network for single image deraining. In: IEEE/CVF Conference on Computer Vision and Pattern Recognition Workshops (CVPRW) (2021)
35. Hou, Q., Zhou, D., Feng, J.: Coordinate attention for efficient mobile network design (2021)
36. Wang, Z., Bovik, A., Sheikh, H., Simoncelli, E.: Image quality assessment: from error visibility to structural similarity. IEEE Trans. Image Process. 13, 600–612 (2004)
37. Zhang, H., Sindagi, V., Patel, V.M.: Image de-raining using a conditional generative adversarial network. IEEE Trans. Circuits Syst. Video Technol. 30, 3943–3956 (2017)
38. Wang, H., Li, M., Wu, Y., Zhao, Q., Meng, D.: A survey on rain removal from video and single image. arXiv preprint arXiv:1909.08326 (2019)
39. Chen, D., et al.: Gatedcontext aggregation network for image dehazing and deraining. In: 2019 IEEE Winter Conference on Applications of Computer Vision (WACV), pp. 1375–1383 (2019)
40. Fu, X., Liang, B., Huang, Y., Ding, X., Paisley, J.: Lightweight pyramid networks for image deraining. IEEE Trans. Neural Netw. 31, 1794–1807 (2020)
41. Zhang, H., Patel, V.M.: Density-aware single image de-raining using a multi-stream dense network. In: Proceedings of the IEEE Conference on Computer Vision and Pattern Recognition, pp. 695–704 (2018)
42. Wang, T., Yang, X., Xu, K., Chen, S., Zhang, Q., Lau, R.W.: Spatial attentivesingle-image deraining with a high quality real rain dataset. In: 2019 IEEE/CVF Conference on Computer Vision and Pattern Recognition (CVPR), pp. 12270–12279 (2019)
43. Yang, Y., Lu, H.: Single image deraining via recurrent hierarchy enhancement network. In: Proceedings of the 27th ACM International Conference on Multimedia, pp. 1814–1822. ACM (2019)
44. Wang, H., Xie, Q., Zhao, Q., Meng, D.: A model driven deep neural network for single image rain removal. In: Proceedings of the IEEE/CVF Conference on Computer Vision and Pattern Recognition, pp. 3103–3112 (2020)
45. Mehri, A., Ardakani, P., Sappa, A.: MPRNet: multi-path residual network for lightweight image super resolution (2020)

Finite Blocklength and Distributed Machine Learning

6G mURLLC over Cell-Free Massive MIMO Systems in the Finite Blocklength Regime

Zhong Li[1]($\boxtimes$), Jie Zeng[2], Wen Zhang[1], Shidong Zhou[2], and Ren Ping Liu[3]

[1] School of Communication and Information Engineering, Chongqing University of
Posts and Telecommunications, Chongqing 400065, China
{s190131188,s190131003}@stu.cqupt.edu.cn
[2] Department of Electronic Engineering, Tsinghua University, Beijing 100084, China
{zengjie,zhousd}@tsinghua.edu.cn
[3] School of Electrical and Data Engineering, University of Technology Sydney,
Sydney 2007, Australia
RenPing.Liu@uts.edu.au

Abstract. In order to support the new service requirements-massive
ultra-reliable low-latency communications (mURLLC) in the six-
generation (6G) mobile communication system, finite blocklength (FBL)
information theory has been introduced. Furthermore, cell-free massive
multiple input multiple output (MIMO) has emerged as one of the 6G
essential promising technologies. A great quantity of distributed access
points (APs) jointly serve massive user equipment (UE) at the same time-
frequency resources, which can significantly improve various quality-of-
service (QoS) metrics for supporting mURLLC. However, as the number of UE grows, the orthogonal pilot resources in the coherent time are
insufficient. This leads to serious non-orthogonal pilot contamination and
pilot allocation imbalance. Therefore, we propose an analytical cell-free
massive MIMO system model and precisely characterize the error probability metric. In particular, we propose a FBL based system model,
formulate and resolve the error probability minimization problem, given
the latency requirement. Simulation results verify the effectiveness of the
proposed scheme and show that the error probability can be improved
by up to 15.9%, compared with the classic pilot allocation scheme.

Keywords: Cell-free massive multiple input multiple output
(MIMO) · Finite blocklength (FBL) · mURLLC · Pilot allocation

1 Introduction

With the rapid expansion of wireless communication systems, the amount of
mobile users has shown an explosive trend. The growth of mobile users has

Supported by the National Key R&D Program (No. 2018YFB1801102), the Natural
Science Foundation of Beijing (No. L192025), the National Natural Science Foundation of China (No. 62001264), and the China Postdoctoral Science Foundation (No.
2020M680559 and No. 2021T140390).

H. Gao et al. (Eds.): ChinaCom 2021, LNICST 433, pp. 425–437, 2022.
https://doi.org/10.1007/978-3-030-99200-2_32

benefited from the popularization of smart terminals and the emergence of various new mobile services. However, due to multiple constraints such as energy, spectral efficiency, and cost, the current fifth-generation (5G) mobile communication system cannot meet the requirements for ultra-large traffic, ultra-low latency, ultra-large connections, and ultra-high reliability. Therefore, the sixth-generation (6G) mobile communication system is regarded as the future research focus.

Although it is far from reaching the unified stage of 6G definition, based on the research progress [1–4] of various countries, it can be predicted that 6G will adopt transformative technologies from the cell-free network architecture [5,6].

As we all know, the cellular network architecture is an epoch-making concept proposed by Bell Labs in the 1970s. It adopts frequency reuse and cell splitting technologies to improve the exploitation of spectrum resources and support the rapid development of mobile communications. In order to satisfy the ever-increasing requirements for services, the entire evolution of mobile communication systems from the first-generation (1G) mobile communication system to 5G is based on cellular networks, that is, using macro cell splitting and vertical micro cell network layering. However, as the cell area continues to shrink, problems such as inter-cell interference and frequent handovers have become more and more serious, resulting in a bottleneck in system performance improvement. In order to overcome these challenges, a cell-free massive multiple input multiple output (MIMO) network architecture [7] which completely reforms the cellular network architecture has become one of the feasible solutions.

The cell-free massive MIMO system distributes a great quantity of access points (APs) with one or more antennas in a wide area, transmits data to the central processing unit (CPU) through the backhaul links, and serves massive user equipment (UE) at the same time-frequency resources. Cell-free massive MIMO combines the advantages of distributed antenna systems and centralized massive MIMO, that is, introducing the idea of "user-centric" [7,8]. It can reduce the distances between APs and UE, obtain spatial macro diversity gain, greatly cut down the path loss, and use the favorable propagation brought by a great quantity of APs to reduce the interference among massive users, so that the entire area is covered evenly and the user experience is significantly improved [8]. Due to these advantages, cell-free massive MIMO is very suitable for major hospitals, stadiums, high-speed rail stations, office buildings, shopping malls and other hot spot scenes, and is considered to be one of the important research directions in future mobile communication systems.

Since the amount of users rapidly increases, the orthogonal pilot resources in the coherent time are insufficient, which leads to serious non-orthogonal pilot contamination and pilot allocation imbalance. The authors of [9] designed random and structured non-orthogonal pilot allocation schemes to maximize the user pilot reuse distance. In [10], the authors used a dynamic pilot multiplexing method to make two users share the same pilot resource to maximize the uplink sum rate. Literature [11] divided users into groups based on available pilot resources, and assigned a pilot sequence to each user group, but there was still pilot contamination among user groups. Moreover, the authors of [12] proposed random and orthogonal pilot allocation schemes, but there was a large waste of resources.

In addition, since the wireless fading channels has the stochastic nature, it is difficult to satisfy the requirements for ultra-reliable low-latency communications (URLLC). Moreover, as one promising technique for supporting latency-sensitive services, massive ultra-reliable and low-latency communications (mURLLC) combines URLLC with massive access. Meanwhile, massive short-packet data transmissions are required in mURLLC to support latency-sensitive applications [13]. This indicates that the classical Shannon theory with infinite blocklength coding is not applicable to the new scenario any more. Therefore, finite blocklength (FBL) information theory [14] has been developed to satisfy both ultra-reliable and low-latency requirements by short-packet data transmissions. Therefore, we analyze the error probability performance over 6G cell-free massive MIMO systems and optimize the pilot length to minimize the error probability.

The remainder of this paper is organized as follows. Section 2 describes the cell-free massive MIMO system model. Section 3 formulates the downlink error probability minimization and presents the golden section search algorithm for the pilot optimization problem. Section 4 presents the simulation and numerical results. Finally, Sect. 5 concludes this paper.

2   System Model

Take into consideration a cell-free massive MIMO system which consists of one CPU, K APs and M UE, as shown in Fig. 1. M UE are served by K APs at the same time-frequency resources. It is assumed that the APs and UE are located randomly in a wide area. Assume that each AP has N antennas, while each UE has a single antenna. Moreover, all APs are linked to a CPU through infinite perfect backhaul links. Furthermore, we assume that the system is operated under time-division duplexing (TDD) mode, which can permit channel reciprocity to require the downlink channel state information (CSI) by uplink pilot training. In addition, we let n_p represent the number of channel uses for uplink pilot training, n_d for downlink data transmission, and n for the whole transmission, i.e., $n = n_p + n_d$. The channel coefficient between the kth AP (AP-k) and the mth UE (UE-m), denoted by $g_{k,m} \in \mathbb{C}^{N \times 1}$, can be characterized as

$$g_{k,m} = h_{k,m} \sqrt{\beta_{k,m}} \tag{1}$$

where $\beta_{k,m}$ denotes the large-scale fading coefficient and $h_{k,m}$ denotes the small-scale vector.

2.1   Uplink Pilot Training

The pilot training sequence for UE-m is defined as $\varphi_m^{n_p} = \left[\varphi_m^{(1)}, \ldots, \varphi_m^{(n_p)} \right] \in \mathbb{C}^{1 \times n_p}$ and $\left\| \varphi_m^{n_p} \right\|^2 = 1$ where $\| \cdot \|$ represents the Euclidean norm. Therefore, the received signal at AP-k, denoted by $\mathbf{Y}_k^{n_p} \in \mathbb{C}^{N \times n_p}$, is derived as

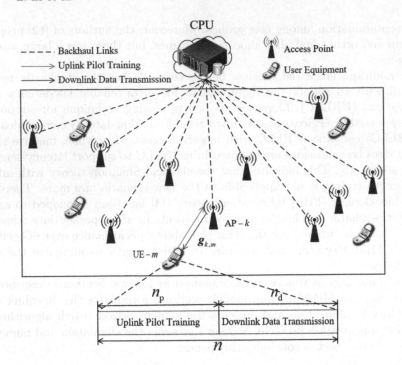

Fig. 1. The cell-free massive MIMO system model.

$$\mathbf{Y}_k^{n_\mathrm{p}} = \sqrt{n_\mathrm{p}\mathcal{P}_\mathrm{p}} \sum_{m=1}^{M} \boldsymbol{g}_{k,m} \boldsymbol{\varphi}_m^{n_\mathrm{p}} + \mathbf{N}_k \tag{2}$$

where $\mathcal{P}_\mathrm{p}$ denotes the pilot transmit power at the UE and $\mathbf{N}_k \in \mathbb{C}^{N \times n_\mathrm{p}}$ denotes the additive white Gaussian noise (AWGN) matrix with zero mean and covariance $\mathbf{I}_N$, where $\mathbf{I}_N$ denotes the identity matrix with size N.

The projection of $\mathbf{Y}_k^{n_\mathrm{p}}$ onto $\boldsymbol{\varphi}_m^{n_\mathrm{p}}$, denoted by $\boldsymbol{y}_{k,m}^{n_\mathrm{p}} \in \mathbb{C}^{N \times 1}$, can be derived as

$$\boldsymbol{y}_{k,m}^{n_\mathrm{p}} = \mathbf{Y}_k^{n_\mathrm{p}} \left(\boldsymbol{\varphi}_m^{n_\mathrm{p}}\right)^H = \sqrt{n_\mathrm{p}\mathcal{P}_\mathrm{p}} \boldsymbol{g}_{k,m} + \sum_{\substack{m'=1 \\ m' \neq m}}^{M} \sqrt{n_\mathrm{p}\mathcal{P}_\mathrm{p}} \boldsymbol{g}_{k,m'} + \boldsymbol{n}_k, \tag{3}$$

where $(\cdot)^H$ denotes the conjugate transpose and $\boldsymbol{n}_k \triangleq \mathbf{N}_k \left(\boldsymbol{\varphi}_m^{n_\mathrm{p}}\right)^H \in \mathbb{C}^{N \times 1}$ is an independent and identically distributed (i.i.d.) Gaussian vector with zero mean and covariance $\mathbf{I}_N$.

Denote $\mathbf{G}_k \triangleq \left[\boldsymbol{g}_{k,1}, \ldots, \boldsymbol{g}_{k,M}\right]$ as the channel coefficient matrix from the AP-k to all UEs. Furthermore, we denote $\mathbf{R}_{\mathbf{G}_k} \triangleq \mathbb{E}\left[\mathbf{G}_k \left(\mathbf{G}_k\right)^H\right] = \mathrm{diag}\left(\beta_{k,1}, \ldots, \beta_{k,M}\right)$ as the covariance matrix of $\mathbf{G}_k$, where $\mathbb{E}[\cdot]$ represents the expectation operation and diag($\cdot$) denotes the diagonal matrix. Then, the minimum mean-squared error (MMSE) estimator which is represented by $\widehat{\mathbf{G}}_k$, is given by [15]

$$\widehat{\mathbf{G}}_k = \sqrt{n_{\mathrm{p}} \mathcal{P}_{\mathrm{p}}} \mathbf{R}_{\mathbf{G}_k} \left(n_{\mathrm{p}} \mathcal{P}_{\mathrm{p}} \mathbf{R}_{\mathbf{G}_k} + \mathbf{I}_N \right)^{-1} \boldsymbol{y}_{k,m}^{n_{\mathrm{p}}}. \tag{4}$$

Furthermore, the channel estimator $\hat{\boldsymbol{g}}_{k,m}$ is derived as

$$\hat{\boldsymbol{g}}_{k,m} = \frac{\sqrt{n_{\mathrm{p}} \mathcal{P}_{\mathrm{p}}} \beta_{k,m}}{n_{\mathrm{p}} \mathcal{P}_{\mathrm{p}} \sum_{m'=1}^{M} \beta_{k,m'} \left| \boldsymbol{\varphi}_m \left(\boldsymbol{\varphi}_{m'} \right)^H \right|^2 + 1} \boldsymbol{y}_{k,m}^{n_{\mathrm{p}}}. \tag{5}$$

It is noted that the channel estimator $\hat{\boldsymbol{g}}_{k,m}$ consists of N independent elements. The variance of each element of $\hat{\boldsymbol{g}}_{k,m}$ is given by

$$\sigma_{k,m} = \frac{n_{\mathrm{p}} \mathcal{P}_{\mathrm{p}} \beta_{k,m}^2}{n_{\mathrm{p}} \mathcal{P}_{\mathrm{p}} \sum_{m'=1}^{M} \beta_{k,m'} \left| \boldsymbol{\varphi}_m \left(\boldsymbol{\varphi}_{m'} \right)^H \right|^2 + 1}. \tag{6}$$

The MMSE estimation error is $\tilde{\boldsymbol{g}}_{k,m} = \boldsymbol{g}_{k,m} - \hat{\boldsymbol{g}}_{k,m}$, which is independent of the true channel. Moreover, each element of the channel estimate error follows $\mathcal{CN} \left(0, \beta_{k,m} - \gamma_{k,m} \right)$.

2.2 Downlink FBL Data Transmission

We let Q_m denote the multiplexing order of UE-m. Furthermore, we define a Q_m-dimensional beamformer $\mathbf{B}_m \triangleq \mathbf{I}_{Q_m} \otimes \mathbf{1}_{N/Q_m}$ for UE-m, where $\otimes$ represents the Kronecker product, $\mathbf{I}_{Q_m}$ represents the identity matrix with size Q_m, and $\mathbf{1}_{N/Q_m}$ represents the all one vector with size N/Q_m. Denote $\mathbf{X}_k^{n_{\mathrm{d}}} \triangleq \left[\boldsymbol{x}_k^{(1)}, \ldots, \boldsymbol{x}_k^{(n_{\mathrm{d}})} \right]$ as the transmitted signal matrix from AP-k and $\boldsymbol{y}_m^{n_{\mathrm{d}}} \triangleq \left[y_m^{(1)}, \ldots, y_m^{(n_{\mathrm{d}})} \right]$ as the received signal vector at UE-m. Based on the MMSE estimator matrix which is represented by $\widehat{\mathbf{G}}_k = \left[\hat{\boldsymbol{g}}_{k,1}, \ldots, \hat{\boldsymbol{g}}_{k,M} \right]$, the transmitted signal which is represented by $\boldsymbol{x}_k^{(l)}$, for transmitting lth data block by using conjugate beamforming [7], is derived as

$$\boldsymbol{x}_k^{(l)} = \mathbf{W}_k \left(\boldsymbol{\Sigma}_k \right)^{\frac{1}{2}} s_m^{(l)}, \quad l = 1, \ldots, n_{\mathrm{d}} \tag{7}$$

where $s_m^{(l)}$ is the lth data block to UE-m. $\boldsymbol{\Sigma}_k \triangleq \mathrm{diag} \left(\eta_{k,1}, \ldots, \eta_{k,M} \right)$ represents the power coefficient matrix where $\eta_{k,m} (m = 1, \ldots, M)$ represents the power coefficient for transmitting lth data block from AP-k to UE-m. $\mathbf{W}_k$ represents the downlink precoder, which is given by

$$\mathbf{W}_k \triangleq \widehat{\mathbf{G}}_k \left[\left(\widehat{\mathbf{G}}_k \right)^H \widehat{\mathbf{G}}_k \right]^{-1} \mathbf{B}_m \left(\boldsymbol{\Xi}_k \right)^{\frac{1}{2}} \tag{8}$$

where $\boldsymbol{\Xi}_k = \mathrm{diag} \left(\chi_1, \ldots, \chi_M \right)$ represents the normalization matrix. Therefore, the columns of $\mathbf{W}_k$ have unit norm and the normalization variable $\chi_k (k = 1, \ldots, M)$ which are following the central chi-square distribution with

(2ℓ) degrees of freedom, where $\ell = K - M + 1$. The probability density function (PDF) of χ_k is given by [16]

$$f_\ell(\chi_k) = \frac{1}{\Gamma(\ell)}\chi_k^{\ell-1}e^{-\chi_k} \tag{9}$$

where $\Gamma(z) = \int_0^{+\infty} t^{z-1}e^{-t}dt$ represents the Gamma function. Define $\mathbf{W}_k = [\boldsymbol{w}_{k,1}, \ldots, \boldsymbol{w}_{k,M}]$ where $\boldsymbol{w}_{k,m}$ represents the downlink precoder. Furthermore, the power constraint at each AP on power coefficients is given by

$$\frac{1}{n_\text{d}}\sum_{l=1}^{n_\text{d}}\mathbb{E}\left[\left\|\boldsymbol{x}_k^{(l)}\right\|^2\right] \leq \overline{\mathcal{P}}_\text{d} \tag{10}$$

where $\overline{\mathcal{P}}_\text{d}$ denotes the average transmit power for each AP and $\boldsymbol{x}_k^{(l)}$ is derived by (7). In addition, with the number of APs K growing, the system will experience only small variations (with respect to the average) in the achievable data rate, which is known as the channel hardening [17]. Then, although the instantaneous CSI is not available at the UE, $\mathbb{E}\left[\left(\boldsymbol{g}_{k,m}\right)^T \boldsymbol{w}_{k,m}\right]$ can be used to calculate the channel coefficient, where $(\cdot)^T$ represents the transpose of a vector.

The received signal, denoted by $y_m^{(l)}$, for transmitting the lth finite-blocklength data block from AP-k to UE-m, is derived as

$$y_m^{(l)} = \underbrace{\sum_{k=1}^{K}\sqrt{\overline{\mathcal{P}}_\text{d}\eta_{k,m}}\mathbb{E}\left[\left(\boldsymbol{g}_{k,m}\right)^T \boldsymbol{w}_{k,m}\right]s_m^{(l)}}_{\text{DS}_m}$$

$$+ \underbrace{\sqrt{\overline{\mathcal{P}}_\text{d}}\left\{\sum_{k=1}^{K}\sqrt{\eta_{k,m}}\left(\boldsymbol{g}_{k,m}\right)^T \boldsymbol{w}_{k,m} - \mathbb{E}\left[\sum_{k=1}^{K}\sqrt{\eta_{k,m}}\left(\boldsymbol{g}_{k,m}\right)^T \boldsymbol{w}_{k,m}\right]\right\}s_m^{(l)}}_{\text{BU}_m}$$

$$+ \underbrace{\sum_{\substack{m'=1 \\ m'\neq m}}^{M}\sum_{k=1}^{K}\sqrt{\overline{\mathcal{P}}_\text{d}\eta_{k,m'}}\left(\boldsymbol{g}_{k,m}\right)^T \boldsymbol{w}_{k,m'}s_{m'}^{(l)}}_{\text{UI}_{m'}} + n_m^{(l)} \tag{11}$$

where $s_m^{(l)}$ and $s_{m'}^{(l)}$ represent the signals sent to UE-m and UE-m', respectively; $\eta_{k,m}$ and $\eta_{k,m'}$ represent the power coefficients for UE-m and UE-m', respectively; $\boldsymbol{g}_{k,m} \in \mathbb{C}^{1\times N}$ represents the channel coefficient vector from AP-k to UE-m; $n_m^{(l)}$ represents the AWGN; and DS_m, BU_m, and $\text{UI}_{m'}$ are the strength of the desired signal, the beamforming gain uncertainty, and the interference caused by UE-m', respectively.

Correspondingly, the signal to noise plus interference ratio (SINR) which is represented by γ_m, from the APs to UE-m, is derived by

$$\gamma_m = \frac{\|\text{DS}_m\|^2}{\mathbb{E}\left[\|\text{BU}_m\|^2\right] + \sum_{\substack{m'=1 \\ m'\neq m}}^{M}\mathbb{E}\left[\|\text{UI}_{m'}\|^2\right] + 1} \tag{12}$$

3   Minimize the Error Probability in the FBL Regime

In the FBL regime, the accurate approximation of the achievable data rate for UE-m, denoted by R_m (bits per channel use), with error probability, denoted by $\epsilon_m (0 \le \epsilon_m < 1)$, and coding blocklength, denoted by n_d, is given by [14]

$$R_m (n_d, \epsilon_m) \approx C(\gamma_m) - \sqrt{\frac{V(\gamma_m)}{n_d}} Q^{-1}(\epsilon_m) \tag{13}$$

where $Q(x) = \int_x^{+\infty} \frac{1}{\sqrt{2\pi}} e^{-\frac{1}{2}t^2} dt$ denotes the Q-function and $Q^{-1}(\cdot)$ denotes the inverse of Q-function. $C(\gamma_m)$ and $V(\gamma_m)$ represent the channel capacity and channel dispersion, respectively, which are given by [14]

$$\begin{cases} C(\gamma_m) = \log_2 (1 + \gamma_m) \\ V(\gamma_m) = 1 - \frac{1}{(1+\gamma_m)^2} \end{cases} \tag{14}$$

Since $n = Bt_D = n_p + n_d$ where B denotes the bandwidth and t_D denotes the latency, the achievable data rate R_m for UE-m, given the pilot length n_p, is given by

$$R_m (t_D, \epsilon_m) \approx C(\gamma_m) - \sqrt{\frac{V(\gamma_m)}{Bt_D - n_p}} Q^{-1}(\epsilon_m). \tag{15}$$

In the case where the achievable data rates for all UE are given, i.e., $R_m = \frac{D}{Bt_D - n_p}$ $(m = 1, \ldots, M)$, where D is the size of downlink data packet (measured in bits). The error probability for UE-m can be derived as

$$\epsilon_m (t_D, n_p) \approx Q \left(\sqrt{\frac{Bt_D - n_p}{V(\gamma_m)}} \left[C(\gamma_m) - \frac{D}{Bt_D - n_p} \right] \right) \tag{16}$$

Given the latency which can satisfy mURLLC ($t_D \le 0.5$ ms), the error probability is a function of the pilot length n_p. Therefore, the optimization problem to minimize the error probability can be modeled as

$$n_p^* = \underset{M \le n_p \le Bt_D - 1}{\arg \min} \epsilon_m (n_p). \tag{17}$$

In order to find the optimal pilot length n_p^*, the exhaustive method is often used. However, the low-complexity golden section search algorithm can be used as an effective solution to quickly converge to the optimal pilot length to reduce computational complexity. The detailed steps of the golden section search algorithm are listed in Algorithm 1.

The complexity of the golden section search algorithm is $\mathcal{O}(log(Bt_D - K))$, which is much lower than the complexity of the exhaustive method $\mathcal{O}(Bt_D - K)$.

Algorithm 1. Golden section search algorithm

1: **Input:** Number of UE M, bandwidth B, latency t_D
2: **Initialization:** Search interval $[n_\mathrm{L}, n_\mathrm{R}]$, $n_\mathrm{L} = M$, $n_\mathrm{R} = Bt_\mathrm{D} - 1$, tolerance $t_\mathrm{tol} = 0.5$, golden ratio $\rho = 0.618$
3: **do**
4:     $n_1 = n_\mathrm{L} + (1 + \rho)(n_\mathrm{R} - n_\mathrm{L})$ and $n_2 = n_\mathrm{L} + \rho(n_\mathrm{R} - n_\mathrm{L})$
5:     Calculate the error probability $\epsilon_k(n_1)$ and $\epsilon_k(n_2)$
6:     **if** $\epsilon_k(n_1) < \epsilon_k(n_2)$ **then**
7:         $n_\mathrm{R} = n_2$, $n_2 = n_1$, and $n_1 = n_\mathrm{D} + (1 - \rho)(n_\mathrm{R} - n_\mathrm{L})$
8:     **else**
9:         $n_\mathrm{L} = n_1$, $n_1 = n_2$, and $n_2 = n_\mathrm{L} + \rho(n_\mathrm{R} - n_\mathrm{L})$
10:     **end if**
11: **while** $|n_2 - n_1| > t_\mathrm{tol}$
12: **Output:** Optimal pilot length $n_\mathrm{p}^* = \arg\min_{n_\mathrm{p}} \epsilon_m(n_\mathrm{p})$ where $n_\mathrm{p} \in \{[(n_\mathrm{L} + n_\mathrm{R})/2], [(n_\mathrm{L} + n_\mathrm{R})/2] + 1\}$

Table 1. Simulation parameters

Parameters	Values
Amount of transmit antennas N	$[2, 10]$
Amount of APs K	$[100, 800]$
Amount of UE M	$[50, 400]$
Uplink pilot transmit power $\overline{\mathcal{P}}_\mathrm{p}$ for each UE	$[1, 10]$ W
Average downlink transmit power $\overline{\mathcal{P}}_\mathrm{d}$	$[1, 40]$ W

4   Numerical Results

MATLAB-based simulations are carried out to validate and evaluate our proposed cell-free massive MIMO based schemes for minimizing the error probability in the finite blocklength regime. The simulation parameters are set as Table 1.

In Fig. 2, we set the amount of transmit antennas $N = 4$, the multiplexing order $Q_m = 2$, the uplink pilot transmit power $\overline{\mathcal{P}}_\mathrm{p} = 1$ W, and the average downlink transmit power $\overline{\mathcal{P}}_\mathrm{d} = 10$ W. The error probability ϵ_m varies within $[10^{-7}, 10^{-5}]$, and the latency t_D varies within $[0.1\,\mathrm{ms}, 0.5\,\mathrm{ms}]$. Fig. 2 shows that the achievable data rate $R_m(t_\mathrm{D}, m)$ varies with latency t_D and error probability ϵ_m. It can be seen from Fig. 2 that when the error probability is given, the achievable data rate is obviously a monotonically increasing function of latency. As the amount of APs increases, the favorable propagation brought by a great quantity of APs can reduce the interference among massive users. Specifically, when the error probability given by 10^5, by increasing the amount of APs form 100 to 200, the achievable data rate can be improved by 9.8%.

In Fig. 3, we set the amount of APs $K = 200$, the amount of UE $M = 100$, the amount of transmit antennas $N = 10$, the multiplexing order $Q_m = 2$, the uplink pilot transmit power $\overline{\mathcal{P}}_\mathrm{p} = 1$ W, and the average downlink transmit power

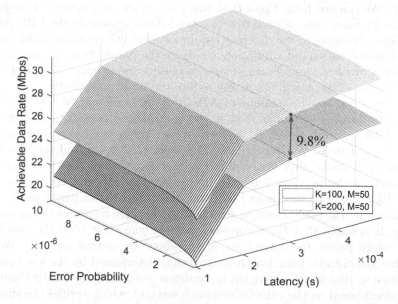

Fig. 2. The achievable data rate $R_m(t_D, \epsilon_m)$ vs. latency t_D and error probability ϵ_m in the FBL regime.

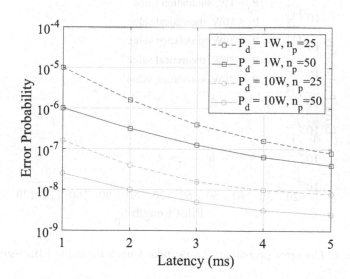

Fig. 3. The error probability ϵ_m vs. latency t_D for the proposed cell-free massive MIMO system in the FBL regime.

$\overline{\mathcal{P}}_d = 10$ W. Figure 3 demonstrates the trade-off between error probability and latency. We can see from Fig. 3 that there is a trade-off between error probability and latency over 6G cell-free massive MIMO system in the FBL regime. Therefore, according to the different requirements of 6G applications for latency and reliability, we should reasonably configure the parameters to meet different application requirements. Furthermore, increasing the transmit power of downlink data transmission can effectively reduce the error probability.

In Fig. 4, we set the amount of APs $K = 200$, amount of UE $M = 100$, amount of transmit antennas $N = 8$, the multiplexing order $Q_m = 4$, and the uplink pilot transmit power $\overline{\mathcal{P}}_p = 1$ W. We set the average downlink transmit power $\overline{\mathcal{P}}_d = 1$ or 10 W, respectively. The theoretical value and the simulation value of error probability under different pilot lengths are shown in Fig. 4. As mentioned in the previous analysis, proper pilot length can reduce the error probability. It can be seen from Fig. 4 that the optimal pilot length is 50, which is obtained by the golden section search algorithm. By means of the simulation results, it is found that the theoretical value is consistent with the simulation value, which proves the correctness of the above theoretical analysis. We also find that the optimal pilot length (pentagram) determined by the low complexity golden section search algorithm is consistent with the pilot length (diamond) determined based on the exhaustive search method, which verifies the effectiveness of the golden section search algorithm.

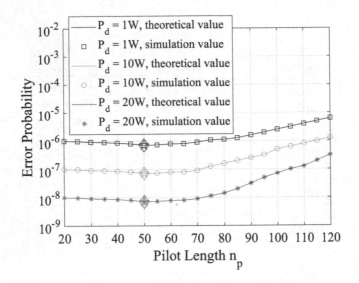

Fig. 4. The error probability ϵ_m vs. pilot length n_p in the FBL regime.

In Fig. 5, we set the average downlink transmit power $\overline{\mathcal{P}}_d \in \{1\,\text{W}, 10\,\text{W}\}$, and the multiplexing order $Q_m \in \{2, 4\}$. Figure 5 plots the achievable data rate with varying amount of APs for the proposed 6G cell-free massive MIMO networks.

From Fig. 5, we can find that the achievable data rate increases with the amount of APs. Figure 5 reveals that a larger multiplexing order Q_m can increase the achievable data rate. Figure 5 also demonstrates that the gap between different multiplexing orders increases with the amount of APs, which is the result from the channel hardening effect. We can also find that increasing the multiplexing order Q_m can improve the achievable data rate. Specifically, by increasing the multiplexing order Q_m from 2 to 4, the achievable data rate is improved by up to 6% and 9.9%, respectively.

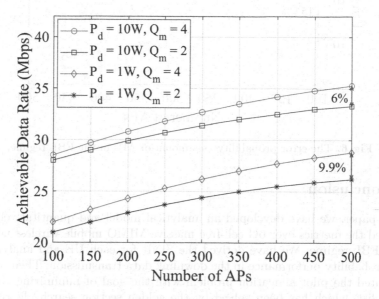

Fig. 5. The achievable data rate vs. amount of APs in the FBL regime.

In Fig. 6, we let the amount of UE $M = 100$, the amount of transmit antennas $N = 10$, the multiplexing order $Q_m = 2$, and the uplink pilot transmit power $\overline{\mathcal{P}}_p = 1$ W. Given different average downlink transmit power and pilot length, Fig. 6 depicts the error probability with varying amounts of APs. Obviously, we can find that the error probability decreases with the amounts of AP increasing. We can see from Fig. 6 that the pilot length n_p can make an impact on the error probability, and the proper pilot length can significantly reduce the error probability, which makes the pilot optimization problem more necessary. Specifically, compared with the case where the pilot length is equals to the amount of UE ($n_p = M = 100$), the optimal pilot length can improve the error probability by up to 7.2% and 15.9% when $K = 100$, respectively.

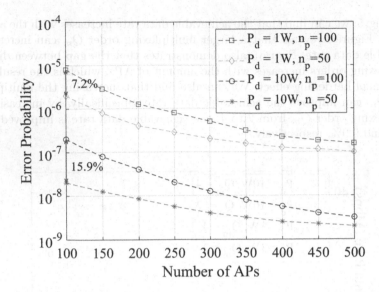

Fig. 6. The error probability vs. amount of APs in the FBL regime.

5   Conclusion

In this paper, we have developed an analytical model and quantitatively characterized the metrics over 6G cell-free massive MIMO mobile wireless networks in the FBL regime. We have derived the SINR for each UE and analyzed the error probability performance in the downlink data transmission. Then, we have formulated the pilot allocation problem with the goal of minimizing the error probability, which has been solved by the golden section search algorithm. A number of simulations have been conducted to verify and evaluate the proposed cell-free massive MIMO scheme in the FBL regime. Our research can effectively meet the requirements for mURLLC in 6G networks, and is of great significance to the theoretical research and actual deployment of cell-free massive MIMO systems.

References

1. Tariq, F., et al.: A speculative study on 6G. Wirel. Commun. **27**(4), 118–125 (2020)
2. Saad, W., Bennis, M., Chen, M.: A vision of 6G wireless systems: applications, trends, technologies, and open research problems. IEEE Netw. **34**(3), 134–142 (2020)
3. Yang, H., et al.: Artificial-intelligence-enabled intelligent 6G networks. IEEE Netw. **34**(6), 272–280 (2020)
4. Matthaiou, M., et al.: The road to 6G: ten physical layer challenges for communications engineers. IEEE Commun. Mag. **59**(1), 64–69 (2021)
5. Bjrnson, E., Sanguinetti, L.: Making cell-free massive MIMO competitive with MMSE processing and centralized implementation. IEEE Trans. Wirel. Commun. **19**(1), 77–90 (2020)

6. Buzzi, S., Andrea, C.D., Zappone, A., Elia, C.D.: User-centric 5G cellular networks: resource allocation and comparison with the cell-free massive MIMO approach. IEEE Trans. Wirel. Commun. **19**(2), 1250–1264 (2020)

7. Ngo, H.Q., Ashikhmin, A., Yang, H., Larsson, E.G., Marzetta, T.L.: Cell-free massive MIMO versus small cells. IEEE Trans. Wirel. Commun. **16**(3), 1834–1850 (2017)

8. Wang, D., et al.: Implementation of a cloud-based cell-free distributed massive MIMO system. IEEE Commun. Mag. **58**(8), 61–67 (2020)

9. Attarifar, M., Abbasfar, A., Lozano, A.: Random vs structured pilot assignment in cell-free massive MIMO wireless networks. In: 2018 IEEE International Conference on Communications Workshops (ICC Workshops), Kansas City, MO, USA, pp. 1–6 (2018)

10. Sabbagh, R., Pan, C., Wang, J.: Pilot allocation and sum-rate analysis in cell-free massive MIMO systems. In: 2018 IEEE International Conference on Communications (ICC), Kansas City, MO, USA, pp. 1–6 (2018)

11. Li, Y., Baduge, G.A.A.: NOMA-aided cell-free massive MIMO systems. IEEE Wirel. Commun. Lett. **7**(6), 950–953 (2018)

12. Bashar, M., Cumanan, K., Burr, A. G., Debbah, M., Ngo, H. Q.: Enhanced max-min SINR for uplink cell-free massive MIMO systems. In: 2018 IEEE International Conference on Communications (ICC), Kansas City, MO, USA, pp. 1–6 (2018)

13. Chen, X., et al.: Massive access for 5G and beyond. IEEE J. Sel. Areas Commun. **39**(3), 615–637 (2021)

14. Polyanskiy, Y., Poor, H.V., Verdu, S.: Channel coding rate in the finite blocklength regime. IEEE Trans. Inf. Theory **56**(5), 2307–2359 (2010)

15. Zhang, X., Wang, J., Poor, H.V.: Statistical delay and error-rate bounded QoS provisioning for mURLLC over 6G CF M-MIMO mobile networks in the finite blocklength regime. IEEE J. Sel. Areas Commun. **39**(3), 652–667 (2021)

16. Caire, G., Shamai, S.: On the achievable throughput of a multiantenna Gaussian broadcast channel. IEEE Trans. Inf. Theory **49**(7), 1691–1706 (2003)

17. Hochwald, B.M., Marzetta, T.L., Tarokh, V.: Multiple-antenna channel hardening and its implications for rate feedback and scheduling. IEEE Trans. Inf. Theory **50**(9), 1893–1909 (2004)

Cloud-Edge Collaboration Based Data Mining for Power Distribution Networks

Li An and Xin Su[✉]

College of IOT Engineering, Hohai University, Changzhou 213022, Jiangsu, China
anjiali0302@163.com, leosu8622@163.com

Abstract. The automation rapid development of the power distribution network have not been fully utilized with the terminal coverage rate increment. The demand and complexity of the power distribution network applications are also fast updated leading a huge calculation pressure from cloud service. This paper does data mining from power distribution network in three aspects, including delay, complexity and power. It defines them with respective weights according to the application requirements, and propose a cloud-edge collaborative communication scheme to effectively reduce the computing complexity of the system.

Keywords: Power distribution network · Edge computing · Cloud-Edge collaboration · Data mining · Computing complexity

1 Introduction

1.1 A Subsection Sample

According to the statistics of the National Energy Administration, China's electricity consumption has reached 7.2255 trillion kilowatts in 2019, an increase of 4.5% over the same period last year [1]. The construction of the power distribution network, which sends electricity to thousands of households in the last kilometer, is worthy of attention and research. At the beginning of the 20th century, the data collection of power distribution network terminal is quite simple, which can only provide data support for basic tasks such as power grid dispatching, maintenance and planning. With a large number of monitoring devices [2, 3], electric vehicle charging piles [4, 5], new energy [6] and other widely connected to the power distribution network, there are higher requirements for data acquisition and analysis of power distribution network terminals.

In [7], it studies the impact factors of cloud virtual machine startup time for users plan, making resource manage decisions in power grid. The power grid uses cloud computing technology, which makes the grid lack the ability to process data for real-time application. As a result, some scholars have introduced edge computing into the power grid. In [8], it introduces the edge computing to power grid, and takes smart measurement as an example to analyze the power grid edge computation in terms of efficiency and security. In addition, by the end of 2018, Chinese State Grid Co., Ltd. has been connected to 540

© ICST Institute for Computer Sciences, Social Informatics and Telecommunications Engineering 2022
Published by Springer Nature Switzerland AG 2022. All Rights Reserved
H. Gao et al. (Eds.): ChinaCom 2021, LNICST 433, pp. 438–451, 2022.
https://doi.org/10.1007/978-3-030-99200-2_33

m terminals of various types, and the daily data collection has exceeded that of 60TB [9]. Two aspects, including the limited storage and real-time processing capacity of the cloud center and the fast update of terminal data, complex types, poor quality and urgent need to tap its value, have been restrict to the construction process of power distribution network. In [10], a cloud-edge collaboration was proposed to exploit the advantages of both edge computing and cloud computing. The cloud-edge collaborative aims to address the problem of excessive inference delay caused by heavy computing tasks in the power distribution networks cloud server. Moreover, a multi-node collaborative computing management model to reduce the processing delay of cloud service providers was also suggested in the same work. A cloud-edge collaborative architecture for power distribution networks was proposed in [11] to solve the problems of network resource redundancy and overload caused by the widespread use of chimney-type independent service access.

Based on above, this paper proposes an effective mining method about the data value of power distribution network, which is a method of data cleaning, classification and fusion based on delay, complexity and power. The proposed method can monitor the abnormal data from the power distribution network terminal, reduce data redundancy and integrate heterogeneous data. It also processes data not only improve the accuracy and effectiveness, but also optimize the data quality. The threshold value of the processed data attribute value is compared and discriminated, and the transmission scheme of the user data of the power distribution network terminal is determined, so that the delay and power in transmission are reduced.

2 Power Distribution Network Terminal Data Analysis

2.1 Terminal Data Type

Load data and environmental monitoring data are the main parts of terminal data of power distribution network, among which static data and dynamic data constitute load data. In the power grid, the static data mainly refers to the parameters of equipment attributes and the topology data of the static network. The dynamic data mainly comes from the voltage monitoring system and the power system monitoring network. The dynamic data of the power grid can be monitored from all angles by the terminal equipment, so as to obtain the formation of real-time status data. The environment monitoring data includes both the operation of the internal device of the distribution plate, such as temperature, humidity, water level, noise, vibration, etc., it also includes external operating environment of the power distribution network equipment, which monitors access control, anti-theft, equipment, small animal detection, etc. The effective monitoring and control of environmental monitoring data and environmental equipment can prevent accidents, prevent the loss of control of the operating environment, extend the service life of equipment, and reduce the high cost caused by the extensive management of power distribution.

2.2 Data Preprocessing Based on DCP

As the development of smart grids has led to the rapid growth of multi-source heterogeneous data, people have begun to mine all kinds of information contained in terminal data. For example, people start using clustering technology to characterize different types of users, and help power grid companies to achieve business activities, like precision marketing.

However, looking at the terminal data application development process, whether it is services such as scheduling planning, resource allocation, status alarms to maintain the safe and economic operation of the power grid, or services such as user portraits and precision marketing for customers, or achieving stable operation/energy saving and emission reduction, etc. multi-objective energy efficiency management, power consumption optimization and other integrated services can mine the value of data from the three aspects of delay, complexity and power. In addition, these three categories do not exist in isolation, as shown in Fig. 1.

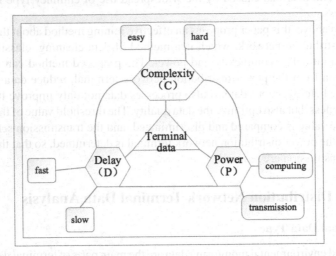

Fig. 1. Internal interaction of terminal data.

From the perspective of demand in delay, terminal data can be divided into 'fast' and 'slow'. When the terminal data belongs to the 'fast' attribute, it will be processed in time at the edge; otherwise, it will be processed in the cloud far from the terminal. The complexity of data also contains two attributes, namely 'hard' and 'easy'. When the terminal data is judged to be 'easy', it will be transmitted directly to the edge to reduce the computing load of the cloud; otherwise, it will be transmitted to the cloud. When analyzing the power attributes of terminal data, key factors such as power in computing and transmission need to be considered. When the power is high, it is dealt with in the edge to achieve low cost, environmentally friendly, and promotion of the good development of the power grid; otherwise, it is processed in the cloud. The three dimensions of terminal data analysis are complementary to each other and are also the foundation for the generation of higher level business.

Based on the DCP power distribution network terminal data analysis, in theory, when the demand in delay of the data is "slow", the complexity demand is "hard", and the demand in power is low, the task data is suggested to transmit to the cloud for processing. When the demand in delay of the data is "fast", the complexity demand is "easy", and the

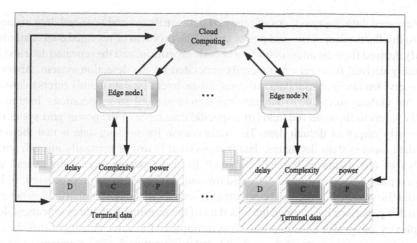

Fig. 2. Schematic diagram of cloud-edge-end data interaction.

demand in power is high, the task data is suggested to transmit to the edge for processing, as shown in Fig. 2. However, this is not in line with the actual situation analysis, from the point of view of the terminal, it may produce data with "slow" delay requirements, "easy" complexity requirements and high power requirements, so the above single division is no longer applicable. Therefore, in order to solve this problem, this article combines the actual needs, and the DCP sets different weights according to business requirements, and sum them in Eq. 1. The weighting function is as follows:

$$Q = \alpha \times t_i + \beta \times e_i + \lambda \times c_i \tag{1}$$

Here, t_i, e_i and c_i correspond to the delay attribute, power attribute and complexity attribute of data uploaded by terminal equipment respectively. α, β and λ respectively correspond to the weight distribution values of delay attribute, power attribute and complexity attribute of data uploaded by terminal equipment, α, β and λ satisfy $\alpha, \beta, \lambda \in [0, 1]$, $\alpha + \beta + \lambda = 1$. Moreover, α, β and λ can be self-adjusting according to the own conditions of the terminal equipment. For example, when the user has a higher demand for the delay, it will pay more attention to the cost of delay, thereby setting a relatively higher α value. When the terminal equipment is at a low power, the power saving as much as possible will be considered when the transmission plan is determined, so the user will select a relatively higher β value. In addition, when the complexity of the task data uploaded by the terminal equipment is "hard", it will pay attention to the calculation resource costs caused by task data processing, thereby setting relatively high λ values.

2.3 Data Preprocessing Process

Due to a large number of terminal task data, it is easy to enlarge data incompleteness and inaccuracy, which affects subsequent data mining and analysis. According to the characteristics of power grid data, data errors include three classes of repeated data, abnormal data, and missing data.

Repeated data has two types of repeated data attributes and repeated data instances. Due to the fluctuation of the grid power status, the problem of repeated data attributes is mainly derived from business definition or code repetition, and the repeated data instance is mainly derived from repeated records generated by the detection system. Abnormal data is divided into irregular data and invalid data. Irregular data mainly refers to instance data that violates rules due to data transmission or manual error operations. Invalid data mainly refers to the reserved fields of a specific data table in the power grid system that are mostly empty or default data. The main reason for missing data is that the power grid data input system fluctuates, data transmission is lost or manually missed, and the number of lost instances is relatively small from the perspective of the overall data. These repeated data, abnormal data, and missing data need to be preprocessed by DCP to avoid the waste of storage and computing resources in the power distribution network.

In this paper, we have mined the task data of the terminal from three aspects of delay, complexity and power consumption, so as to improve the value density of task data and reduce the resource waste of the power distribution network. The data mining process is shown in **Algorithm 1**. The core contains the following four situations:

Case1: The delay, complexity and power consumption of the task data are mined, and the weighted sum is carried out. If the value of the weighting function is greater than the upper limit of the given threshold, it indicates that the storage and computing resources used to process the task data are large after multi-objective measurement. The system of power distribution network then transmits the task data of the terminal to the cloud server for processing.

$$Q = C_1 \times t_i + \cdots$$

Algorithm 1: DCP

Input: data of terminal task set $D = \{task_1, task_2, ..., task_n\}$

Output: the transmission plan of terminal task data set D
1. **for** special data $task_i$ in D **do**
2.   the task data is directly transmitted to the cloud server
3. **end**
4. carry on task data mining, analyze delay, complexity, power consumption.
5. calculate the weighted model $Q = \alpha \times t_i + \beta \times e_i + \lambda \times c_i$
5. th_{min} = the lower limit of the given threshold
6. th_{max} = the upper limit of the given threshold
7. **if** $Q < th_{min}$
8.   the task data is not processed and discarded.
9. **else if** $Q > th_{max}$
10.   the task data is transmitted to the cloud server.
11. **else**
12. the task data is transmitted to the edge server.
13. output the transmission plan of terminal task data set D

Case2: The delay, complexity and power consumption of the task data are mined, and the weighted sum is carried out. If the value of the weighted function is less than the lower limit of the given threshold, it indicates that the contribution value of the task data to the power distribution network approaches zero or none after the task data is analyzed and processed after multi-objective measurement. Therefore, the system of power distribution network will consider whether it is a data error. Moreover, it is also necessary to avoid waste of storage and computing resources in power distribution network, the system will not process and abandon this task data.

Case3: The delay, complexity and power consumption of the task data are mined, and the weighted sum is carried out. If the value of the weighted function is between the upper limit and lower limit of the given threshold, it indicates that the storage and computing resources used to process the task data are small after multi-objective measurement. The system of power distribution network then transmits the task data of the terminal to the edge server for processing.

Case4: After the task data is input into the system of power distribution network, if the task data is special, such as a large venue event should have direct communication channels with the cloud server, and the venue uploads the future activities arrangement to the cloud server. The cloud server will arrange the power supply plan after accurately predicting its daily load, and guarantee the venue is going smoothly.

3 System Model

In the power distribution network cloud-edge collaborative system model, it is assumed that there are M servers in this system, H is used to represent the set of servers, and $H_b \in H, b \in \{1, 2, ..., M\}$ represents the b server in the system model. At the same time, it is assumed that there are N tasks of the power distribution network terminal, T is used to represent the tasks set of the power distribution network terminal, and $task_i \in T, i \in \{1, 2, ..., N\}$ represents the $task_i$ uploaded by the power distribution network terminal.

Define the variable $y_{i,b}$ to indicate whether $task_i$ will be uploaded to server H_b, that is,

$$y_{i,b} = \begin{cases} 1, & task_i \ upload \ to \ server \ H_b \\ 0, & other \end{cases} \quad (2)$$

Taking into account the continuity of tasks, we assume that a task can only be handled by one server if it is to be unloaded, that is,

$$\sum_b y_{i,b} \in \{0, 1\}, \forall i \in T, \forall b \in B_e \cup B_c \quad (3)$$

Where, B_e represents the collection of servers at the edge, and B_c represents the collection of servers in the cloud, and $H = B_e \cup B_c$.

3.1 Data Complexity

According to the chaotic theory, the complexity of the pseudo-random sequence refers to the degree of similarity to the random sequence, and is a measure of the extent of the overall extent to which the sequence is restored [12]. In this paper, the calculation of task data complexity is expressed by fuzzy entropy, that is,

$$FuzzyEn(m, r, N) = \ln \phi^m(r) - \ln \phi^{m+1}(r) \tag{4}$$

Where,

$$\phi^m(r) = (N - m - 1) \sum_{i=1}^{N-m-1} \left[(N - m)^{-1} \sum_{i=1, j \neq 1}^{N-m-1} A_{ij}^m \right], \forall i, j \in [0, N]$$

$$A_{ij}^m = \exp[-In(2) * (\frac{d_{ij}^m}{r})^2, \forall i, j \in [0, N]$$

Where, m is the dimension of the reconstructed phase space, r is the similarity tolerance limit and N is the sequence length. d_{ij}^m is the distance between the window vectors after the phase space is reconstructed from the original sequence, and the phase space dimension m is generally based on the correlation dimension (fractal dimension) [13]. The calculated fuzzy entropy is mapped between [0, 1]. If the fuzzy entropy of task data is located [0, 0.5], the task data of complexity is considered to be "easy". If the fuzzy entropy of the task data is located (0.5, 1], the task data is similar to the random data, and its complexity is "hard".

3.2 Delay and Power of Data

In terms of terminal task data analysis, we mine the data from three aspects: delay, complexity and power consumption, which can improve the value density of task data and reduce the overhead cost of the system. On this basis, a cloud-edge collaborative task scheduling algorithm for power distribution network based on DCP is proposed in this paper, which includes the following steps:

Step 1: calculating the delay, complexity and power consumption of task data through the DCP algorithm.
Step 2: setting the weighted coefficient of delay, complexity and power consumption based on business requirements and the conditions of the business equipment.
Step 3: making a decision on the weighted DCP value to determine the unloading party of the task data.

Local Calculation Cost. Suppose the power consumption of the local terminal is modeled as $pl = k \cdot (f_i^l)^3$, where f_i^l and k are the computing power of the terminal CPU and the coefficients of the related processor chip structure, respectively [14]. When the local terminal processes the task data, the execution time and energy consumption are respectively.

$$t_i^l = y_{i,b} \cdot \frac{c_i D_i}{f_i^l}, \forall i \in T, \forall b \in B_e \cup B_c \tag{5}$$

And

$$e_i^l = y_{i,b} \cdot c_i \cdot D_i \cdot k \cdot (f_i^l)^2, \forall i \in T, \forall b \in B_e \cup B_c \tag{6}$$

Where, D_i- the amount of data of terminal $task_i$; c_i-complexity of terminal $task_i$; t_i^l-local terminal computing delay; e_i^l-the energy consumption calculated by the local terminal.

Unloading Cost. In this section, the unloading cost mainly includes time cost and energy consumption. When analyzing the time cost and energy consumption, we focus on the delay and energy consumption of the uplink transmission of tasks to edge servers and cloud servers, as well as the delay and energy consumption of edge servers and cloud servers in processing tasks. This is because there is a big gap between the amount of data of the calculation result and the amount of data of the calculation task. For example, when the business needs to monitor whether the equipment is in good condition, a large amount of electrical data needs to be processed, while the calculation result is only whether or not, the amount of data is very small, so it can be ignored, so the computing delay of sending the calculation results back to the terminal equipment from the edge or cloud is not considered.

When $task_i$ is uplink transferred to an edge server or cloud server, the transfer rates are assumed to be,

$$R_i^e = y_{i,b} \cdot W_i^e \log_2(1 + \frac{p_i^e g_i^e}{\sigma_{ie}^2 + \sum_{j \in N \setminus \{i\}} p_j^e g_j^e}), \forall i \in T, \forall b \in B_e \qquad (7)$$

and

$$R_i^c = y_{i,b} \cdot W_i^c \log_2(1 + \frac{p_i^c g_i^c}{\sigma_{ic}^2 + \sum_{j \in N \setminus \{i\}} p_j^c g_j^c}), \forall i \in T, \forall b \in B_c \qquad (8)$$

Here, $W_i^e, W_i^c, \sigma_{ie}^2, \sigma_{ic}^2, g_i^e, g_i^c, p_i^e$ and p_i^c denote the bandwidth of the $task_i$ transferred to the edge server and the cloud server, the white Gaussian noise in the channel, the gain on the channel, and the transmission power of the terminal equipment, while $p_j^e \cdot g_j^e$ and $p_j^c \cdot g_j^c$ represent the interference to the terminal $task_i$ when other terminal tasks are unloaded to the edge server and the cloud server, respectively.

Suppose the computing resources allocated to the $task_i$ uploaded to the edge server and the cloud server are represented by f_i^e and f_i^c respectively, that is, the number of cycles allocated to the CPU on the edge server and cloud server. Here, for the convenience of analysis, we assume that computing resources are evenly distributed. Therefore, when performing a computing task, if the uninstall decision is to upload to the edge server, the time cost and energy consumption of unloading the $task_i$ are these,

$$t_i^e = y_{i,b} \cdot \left(\frac{c_i D_i}{R_i^e} + \frac{c_i D_i}{f_i^e} \right), \forall i \in T, \forall b \in B_e \qquad (9)$$

and

$$e_i^e = y_{i,b} \cdot \left(\frac{c_i D_i}{R_i^e} * p_i^e + \frac{c_i D_i}{f_i^e} * pl_{ie} \right), \forall i \in T, \forall b \in B_e \qquad (10)$$

Where, pl_{ie} represents the calculated power of the edge server. If the $task_i$ is to upload to the cloud server, the time cost and energy consumption of the $task_i$ are as follows:

$$t_i^c = y_{i,b} \cdot \left(\frac{c_i D_i}{R_i^c} + \frac{c_i D_i}{f_i^c} \right), \forall i \in T, \forall b \in B_c \qquad (11)$$

and

$$e_i^c = y_{i,b} \cdot \left(\frac{c_i D_i^c}{R_i^c} * p_i^c + \frac{c_i D_i^c}{f_i^c} * pl_{ic} \right), \forall i \in T, \forall b \in B_c \tag{12}$$

Where, pl_{ic} represents the calculated power of the cloud server.

System Overhead. In this paper, the aggregation function method is adopted in the cloud-edge collaborative task scheduling algorithm of power distribution network based on DCP, which combines delay, energy consumption and complexity into a cost function, that is,

$$F(y) = \alpha \cdot \frac{L(y) - L_{\min}}{L_{\max} - L_{\min}} + \beta \cdot \frac{E(y) - E_{\min}}{E_{\max} - E_{\min}} + \lambda \cdot \frac{C(y) - C_{\min}}{C_{\max} - C_{\min}} \tag{13}$$

Where, α, β and λ are the weights of delays, energy consumption and complexity functions, $\alpha, \beta, \lambda \in [0, 1], \alpha + \beta + \lambda = 1, L_{\max}, L_{\min}, E_{\max}, E_{\min}, C_{\max}, C_{\min}$ represents the maximum value of average delay, minimum value of average delay, maximum value of average power consumption, minimum value of average power consumption, maximum value of average complexity and minimum value of average complexity, respectively. By standardizing the three objective functions respectively, the aggregation function reduces the numerical difference caused by different target types. α, β and λ values can quantify the emphasis of the system on the optimization objective. It is a single-objective optimization of the delay when $\alpha = 1$. Similarly, when $\beta = 1$, it is a single objective optimization of power consumption, when $\lambda = 1$, it is a single objective optimization of complexity.

Before exploring the system overhead, we need to analyze and calculate the parameters $L_{\max}, L_{\min}, E_{\max}, E_{\min}, C_{\max}$ and $C_{\min}$.

$L_{\max}$: maximum value of average delay. Delay reaches the maximum value when all tasks are offloaded to the cloud.

$L_{\min}$: minimum value of average delay. Assuming that the edge is rich in computing resources, it will not cause queuing. So delay reaches the minimum value when all tasks are offloaded to the edge.

$E_{\max}$: maximum value of average power consumption. Power consumption reaches the maximum value when all tasks are offloaded to the cloud.

$E_{\min}$: minimum value of average power consumption. All tasks are assigned to the nodes with the best computing capability in the edge, and the power consumption reaches the minimum.

$C_{\max}$: maximum value of average complexity. Scheduling the task with missing data to the server will increase the complexity cost dramatically. Here, it is assumed that the data of all tasks are missing, which is the extreme value.

$C_{\min}$: minimum value of average complexity. Because it is a non-deterministic polynomial problem, it is unrealistic to get the absolute minimum. In this problem, the result of single objective optimization of complexity is taken instead of the exact value.

4 Simulation Result

In this paper, the DCP algorithm is used to determine the optimal strategy of service uploading at the terminal. Without special declaration, the parameter settings are summarized in Table 1.

Table 1. Simulation parameter information.

Parameter	Meaning	Value
W	Channel bandwidth	10 MHz
n_0	Noise unilateral power spectral density	-174 dBm/Hz
L	Terminal data	50 kbit
N	Time series length	3000
m	Reconstruct the dimension of phase space	3
r	Similarity tolerance limit	3.5
f^l_k	terminal computing capacity	$5*10^8$ cycle/s
f^e_k	edge computing capacity	$3*10^9$ cycle/s
f^c_k	could computing capacity	$10*10^9$ cycle/s
C	CPU per bit	100 cycles/bit
D	Task data quantity	[1000,2000] kB

4.1 Verification of Analysis

There are many businesses in the power distribution network, such as water level alarm, power distribution equipment intelligent diagnosis, large customer load management, etc. In order to facilitate simulation, this paper selects the electrical fault service to simulate the cloud-edge collaboration based on DCP, the traditional cloud computing architecture and the cloud-side cooperation algorithm based on decentralized "cloud". First of all, 1000 simulations are carried out on the complexity of the electrical fault service, in which Fig. 3 is only part of our iterative experiment on the complexity of the electrical fault service.

In the power distribution Internet of things, the greater the complexity of the services data, the greater its randomness, and the more difficult it is for the services data to be restored. As can be seen from Fig. 3, the complexity of electrical fault service is within a stable range, and its maximum value is 0.4709, the minimum is 0.4533. After many iterative numbers, the average value is 0.4654, belongs to between 0–0.5, therefore electrical fault data complexity is considered to be "easy". If delay and power consumption is not considered, it is worth suggested, and the electrical fault is preferably transferred to the edge.

In Fig. 4 and Fig. 5, "TCC" represents traditional cloud computing architecture, that is, its data is directly uploaded to the cloud; "DE" represents the cloud-side collaboration based on decentralized "cloud", that is, data is uploaded to the cloud and edge

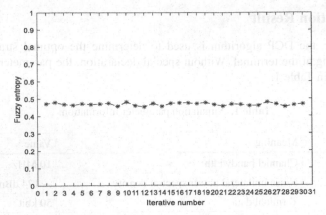

Fig. 3. Estimation of electrical faults complexity based on fuzzy entropy

side through random allocation; "DCP" represents cloud-edge collaboration based on DCP, which means that terminal data is mined from three perspectives of delay, complexity and power, and then uploaded to the cloud or edge servers as required. It can be seen from Fig. 4 and Fig. 5 that in the case of certain edge service providers and cloud service providers, the total time consumption and power consumption of cloud edge collaborative architecture based on DCP algorithm are always lower than those of traditional cloud computing architecture and cloud-side cooperation based on decentralized "cloud", because the algorithm based on DCP preprocesses data at the terminal and eliminates redundant data, and increasing the value density of terminal data. In addition, scientific and reasonable scheduling of the resources of the cloud and the edge has been carried out to optimize the time and power consumption of business processing, which further verifies the availability and efficiency of the DCP algorithm for mining terminal data.

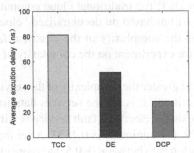

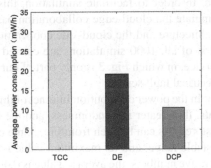

Fig. 4. Time delay observation of electrical faults.

Fig. 5. Power consumption observation of electrical faults.

4.2 Influence Factors

In this paper, 560 terminals in the power distribution network are randomly selected for simulating. In addition, based on the three algorithms including the cloud-edge collaboration based on DCP, the traditional cloud computing architecture and the cloud-side cooperation algorithm based on decentralized "cloud", the task data of 560 terminals is analyzed from the delay and power consumption. The simulation results Fig. 6 and Fig. 7 are as follows.

In Fig. 6 and Fig. 7, the advantages of the algorithm in the power distribution grid are horizontally compared from the perspective of time delay and power consumption. The red node line represents the situation of the traditional cloud computing architecture, and the data is directly uploaded to cloud, that is, "TCC"; the blue node line represents the situation of cloud-side collaboration based on decentralized "cloud", and the data is uploaded to the cloud and the edge through random allocation, that is, "DE"; the green node line represents the cloud-edge collaboration based on DCP, and the terminal data is mined from the perspectives of delay, complexity and power, and then uploaded to the cloud or edge server as required, that is, "DCP".

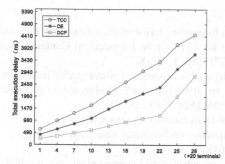   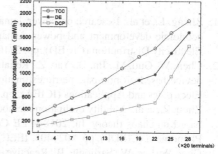

Fig. 6. Delay observation of 560 terminals.   **Fig. 7.** Power consumption observation of 560 terminals.

It can be seen from Fig. 6 and Fig. 7 that when the number of edge nodes and cloud center is constant, with the increase of terminal tasks, the total delay and power consumption of processing tasks are increasing, whether it is traditional cloud computing architecture, the cloud-side cooperation algorithm based on decentralized "cloud" or cloud-edge collaboration based on DCP. Under the same conditions, the total delay and power consumption of cloud-edge collaborative based on DCP algorithm is always lower than that of traditional cloud computing architecture and cloud-side cooperation algorithm based on decentralized "cloud". which verifies the availability and efficiency of cloud-edge collaborative based on DCP architecture. However, it is worth noting that when the processing terminal task exceeds about 440, the execution delay and power consumption of the processing task of DE and the proposed scheme are sharply increased. This is because the calculation and storage resources of the edge service are limited, not enough to support multiple task processing, which may result in queuing phenomena. The computing and storage resources of the cloud are larger than the edge, so TCC is the same as usual.

5 Conclusion

This paper proposes an effective method for mining the value of the terminal data of the power distribution network. This method uses the function model of the data attributes established on the terminal to mine the value of the terminal user data in terms of complexity, time delay and power. According to the output function value, the data transmission direction is judged to determine the transmission plan of the power distribution network terminal user. This method not only reduces the time and power of server processing tasks, but also reduces the transmission and storage pressure of the cloud servers, making the obtained power distribution network terminal business scheduling plan scientific and effective, and can be applied in the actual engineering field.

Acknowledgment. This research work is supported by the National Key Research and Development Program of China (2021YFE0105500); and the National Natural Science Foundation of China (61801166).

References

1. Zheng, L., et al.: Research on power demand forecasting based on the relationship Between economic development and power demand. In: 2018 China International Conference on Electricity Distribution (CICED), pp. 2710–2713, IEEE (2018)
2. Chen, W., Guo, M., Jin, Q., Yao, Z.: Reliability analysis method of power quality monitoring device based on non-parametric estimation. In: 2019 14th IEEE Conference on Industrial Electronics and Applications (ICIEA), pp. 1526–1529. IEEE (2019)
3. Zhou, Z., et al.: Validity evaluation method of DGA monitoring sensor in power transformer based on chaos theory. In: 2018 IEEE Conference on Electrical Insulation and Dielectric Phenomena (CEIDP), pp. 402–405. IEEE (2018)
4. Choi, W., Lee, W., Sarlioglu, B.: Reactive power control of grid-connected inverter in vehicle-to-grid application for voltage regulation. In: 2016 IEEE Transportation Electrification Conference and Expo (ITEC), pp. 1–7. IEEE (2016)
5. Cai, A., Yu, Y., Xu, L., Niu, Y., Yan, J.: Review on reactive power compensation of electric vehicle charging piles. In: 2019 22nd International Conference on Electrical Machines and Systems (ICEMS), pp. 1–4. IEEE (2019)
6. Yuan, W., Wang, C., Lei, X., Li, Q., Shi, Z., Yu, Y.: Multi-area scheduling model and strategy for power systems with large-scale new energy and energy storage. In: 2018 Chinese Automation Congress (CAC), pp. 2419–2424. IEEE (2018)
7. Mao, M., Humphrey, M.: A performance study on the VM startup time in the cloud. In: 2012 IEEE Fifth International Conference on Cloud Computing, pp. 423–430. IEEE (2012)
8. Shi, W., Sun, H., Cao, J.: Edge computing-an emerging computing model for the internet of everything era. J. Comput. Res. Develop. **54**(5), 129–133 (2017)
9. Xue, M., Shi, K., Chen, X., Wu, Q., Li, B., Qi, B.: A summary of research on converged communication model of power network and information network. In: 2017 2nd International Conference on Power and Renewable Energy (ICPRE), pp. 1062–1066. IEEE (2017)
10. Zhu, W., Qiang, F., Liu, N., Wu, X., Li, K., Zhang, K.: Research on multi-node collaborative computing management model for distribution Internet of Things. In: International Conference on Artificial Intelligence and Computer Applications, pp. 1019–1022. IEEE (2020)

11. Fan, C., Lu, Y., Leng, X., Luan, W., Gu, J., Yang, W.: Data classification processing method for the power IoT based on cloud-side collaborative architecture. In: IEEE 9th Joint International Information Technology and Artificial Intelligence Conference, pp. 684–687. IEEE (2020)
12. Chen, X.J., Li, Z., Bai, B.M., Cai, J.P.: A certain chaotic pseudo-random sequence complexity entropy measure of fuzzy relations. **60**(06), 379–388 (2011)
13. Wen, B.H., Yuan, M., Hou, L.: The study of CSI 300 index's complexity and comparison of model efficiency based on entropy algorithm. J. Quant. Econ. **32**(1), 19–25 (2015)
14. Zhang, W., Wen, Y., Guan, K., Kilper, D., Luo, H., Wu, D.O.: Energy-optimal mobile cloud computing under stochastic wireless channel. IEEE Trans. Wireless Commun. 12(9), 4569–4581(2013)

Robust Sound Event Detection by a Two-Stage Network in the Presence of Background Noise

Jie Ou[1](✉), Hongqing Liu[1], Yi Zhou[1], and Lu Gan[2]

[1] School of Communication and Information Engineering, Chongqing University
of Posts and Telecommunications, Chongqing, China
oj-cqupt@outlook.com
[2] College of Engineering, Design and Physical Science, Brunel University,
London UB8 3PH, UK

Abstract. With the advent of deep learning, research on noise-robust
sound event detection (SED) has progressed rapidly. However, SED performance in noisy conditions of single-channel systems remains unsatis-
factory. Recently, there were several speech enhancement (SE) methods
for the SED front-end to reduce the noise effect, which are completely two
models that handle two tasks separately. In this work, we introduced a
network trained by a two-stage method to simultaneously perform signal
denoising and SED, where denoising and SED are conducted sequentially
using neural network method. In addition, we designed a new objective
function that takes into account the Euclidean distance between the output of the denoising block and the corresponding clean audio amplitude
spectrum, which can better limit the distortion of the output features.
The two-stage model is then jointly trained to optimize the proposed
objective function. The results show that the proposed network presents
a better performance compared with single-stage network without noise
suppression. Compared with other recent state-of-the-art networks in the
SED field, the performance of the proposed network model is competitive, especially in noisy environments.

Keywords: Sound event detection · Denoising block · Two-stage
method · Neural network

1 Introduction

Sound event detection (SED) is currently an important research topic in the
field of acoustic signal processing. The purpose of SED is to detect specific sound
events in different scenes and to locate the onset and offset times of target sound

---

Supplementary Information The online version contains supplementary material
available at https://doi.org/10.1007/978-3-030-99200-2_34.

H. Gao et al. (Eds.): ChinaCom 2021, LNICST 433, pp. 452–462, 2022.
https://doi.org/10.1007/978-3-030-99200-2_34

events present in an audio recording. This technique has great influences in many fields, such as anomaly detection, acoustic surveillance, and smart house [1–3]. To promote SED, the Acoustic Scene and Event Detection and Classification (DCASE) Challenge was launched as an international challenge in 2013.

Since DCASE released Task 2 in 2017, rare sound event detection has received more and more attention [4–6]. In [7], Lim et al. introduced a rare sound event detection system using the combination of one-dimensional (1D) convolutional neural network (1D ConvNet) and recurrent neural network (RNN) with long short-term memory units (LSTM), and ranked the first place in DCASE task 2. Kao et al. proposed a Region-based CRNN (R-CRNN) [8] to improve previous work. In [9], Zhang et al. proposed a multiscale time-frequency CRNN (MTF-CRNN) with low parameter counts for SED. In [10], He et al. proposed a time-frequency attention model for sound event detection to alleviate the problems caused by data imbalance. These models achieve excellent performance.hongqingliu@outlook.com

The negative impact of uncorrelated environmental noise on the performance of the SED system has attracted increasing attentions. With the developments of speech enhancement technology, in the field of acoustic signal processing, many researchers have conducted noise suppression preprocessing before training their model [11–13]. In the past single-channel SED tasks, similar methods have also been used. In response to this problem, Zhou et al. proposed robust sound event detection through noise estimation and source separation using NMF [14]. Later, Feng et al. proposed an adaptive noise reduction method for sound event detection based on non-negative matrix factorization (NMF) [15]. Wan et al. proposed noise robust sound event detection using deep learning and audio enhancement [16]. In wan's method, first, the noisy speech is denoised using the Log Spectrum Amplitude Estimation (OMLSA) audio enhancement method. Then the denoised audio is used as the input of the SED network. This scheme can improve the performance of the SED system, but at the same time there are some shortcomings in two aspects. First, two different models are used to handle two different tasks separately. Especially during the inference period, it is also necessary to run the two models separately in sequence, which is relatively redundant and complicated. Second, during training, the audio distortion caused by the denoising system will cause the performance of the SED system to seriously degrade. A method should be found to punish the degree of distortion caused by the denoising system during training. In this way, the two systems before and after can be related to each other.

To mitigate this problem, in this paper, inspired by a two-stage enhancement strategy [17] for impaired speech, we propose a two-stage deep learning network for SED. In the proposed network, we first train denoising network and the SED network separately so that the network learns the weights of the corresponding tasks. After that, we jointly train the two modules to make the network learn the ability to handle two tasks at the same time, which is relatively difficult for combining essentially two different tasks. Finally, an optimized system that can perform denoising and SED tasks at the same time is obtained. In addition, we propose a new weighted loss function, which can punish the distortion caused by the denoising network to associate the front and back models. This loss function plays an important role in the improvement of model performance.

The remainder of this paper is organized as follows. In Sect. 2, we first introduce the single-task signal model separately, and then, the proposed two-stage deep learning network is presented. The dataset, experimental setups, and evaluation metrics are illustrated in Sect. 3. The results and analysis are given in Sect. 4. Finally, we conclude our work in Sect. 5.

2   Methods

In this section, we first introduce the notations and then describe the frequency-domain denoising networks that we use in our experiments. Figure 1 shows a schematic diagram of the proposed system, which consists of a denoising block and a SED block.

2.1   Signal Model

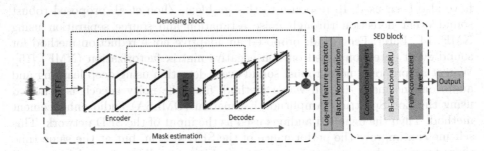

Fig. 1. System diagram of the proposed two-stage model.

Let us consider a single-channel microphone signal, denoted by $y(t)$

$$y(t) = s(t) + n(t), \tag{1}$$

where $s(t)$ is the clean signal (the target event), $n(t)$ is background noise, and t is a time index. In this paper, the background noise does not include obvious target detection events. The purpose of the first level of the two-stage system is to recover a clean signal $s(t)$ from the corresponding noisy observations $y(t)$ to prepare for the SED.

2.2 Denoising Block

The denoising block in Fig. 1 is a schematic diagram of the denoising NN framework in the frequency domain. The denoising block is performed by 2 steps, (1) STFT, (2) mask estimation.

(1) **STFT:** When working in the frequency-domain [18], we usually compute the amplitude of the STFT coefficients to work on real numbers before inputting them to the mask estimation network. After the microphone signal is transformed by STFT, the time-frequency spectrum can be obtained, denoted by $\mathbf{x}_y$, where $\mathbf{x}_y$ is the amplitude spectrum of the noise signal $y(t)$.

(2) **Mask estimation:** We estimate a denoising mask using a mask estimation network. The structure of this mask estimation network is similar to the U-Net, which is a well known architecture composed as a convolutional autoencoder with skip-connections [19]. The mask estimation NN should account for the time context of the signal to distinguish speech from noise. This can be achieved by using BLSTM layers. In mask estimation, convolutional encoder is stacked by convolutional layers and pooling layers. It is used to extract high-level features from the original input signals. The structure of the decoder is basically the same as the encoder, but the order is reversed. The decoder maps the low-resolution feature map at the output of the encoder to the full feature map of the input size. The symmetrical encoder-decoder structure ensures that the output has the same shape as the input. Because the detection network only needs a clean sound spectrum with no change in size, this better prepares the conditions for the input of the latter stage.

The output of the mask estimation network is a single mask used to predict speech as

$$\mathbf{m}_s = MSnet(\mathbf{x}_y), \tag{2}$$

where $MSnet(\cdot)$ is a mask estimation network, and $\mathbf{m}_s$ is the masks associated with speech.

We apply ELUs [20] to all convolutional and deconvolutional layers except the output layer. In the output layer, we use softmax activation [21], which can constrain the network output to always be positive. The input size and output size of each layer are unified in the form of *Feature Maps* × *Time Step* × *Frequency Channel*.

For frequency-domain networks, we use the mean square error loss (MSE) as the loss

$$\mathrm{L}_{s,y}(\theta) = \frac{1}{M}(\|\mathbf{x}_s - \mathbf{m}_s \odot \mathbf{x}_y\|_2), \tag{3}$$

where $\mathbf{x}_s$ is the amplitude spectrum of the target speech (the clean speech) signal, $\odot$ is an element-wise multiplication, θ is the parameter of the denoising network, $\|\cdot\|_2$ is Euclidean norm (L2 norm), M is the total number of pixels, and its size is *Batch Size* × *Time Step* × *Frequency Channel*.

2.3 SED Block

For SED, we follow the state-of-the-art CRNN framework as a baseline [10]. In the SED block of Fig. 1, the CRNN structure consists of three parts of CNN, RNN and Fully Connected Layer.

The convolution part of the network consists of four convolution layers, and each layer is followed by batch normalization [22], ReLU activation unit, and dropout layer [23]. Since we believe that the early convolutional layer is essential for feature learning, the first two convolutions of the network are stacked back to back. In order to maintain the most important information on each feature map, we use the max pooling layer on the time axis and frequency axis. The final feature map is reduced by four times in the time axis to match the frame resolution (80 ms) for computing the evaluation metrics. In the end of the CNN, the features extracted on different convolution channels are superimposed along the frequency axis.

In the RNN part, we use a bi-directional gated recurrent unit (bi-GRU) layer, which can better extract the time structure of acoustic events compared with uni-directional GRU. Bi-GRU encodes sequence feature into a sequence of feature vectors of size (375, U), where U is the number of GRU units. The returned features from the GRU layer is maintained and sent to a fully-connected layer (FCL) with an output size of C, where C is the number of event classes. After the FCL, sigmoid activation is used to produce classification result for each frame (80 ms), of which output represents the probability of the presence of the target sound event.

Finally, we set a binary prediction to a constant threshold of 0.5 for each frame. These predictions are post-processed through a median filter of length 240 ms. We choose the longest continuous forward prediction sequence to produce the start and offset of the target event.

2.4 Weighted Multi-task Loss

In many cases, the denoised audio will have a small amount of signal distortion compared with the clean target signal. Therefore, the target sound event in the denoised audio may be distorted, which will seriously affect the accuracy of the SED system in estimating the onset and offset of the target event. To mitigate this problem, in this paper, we propose an improved weighted multi-task loss function. Our weighted multi-task(WMT) loss $L_{wmt}(\theta)$ is defined as,

$$L_{wmt}(\theta) = \lambda E(\| \mathbf{x}_s - f(\mathbf{x}_y) \|_2) + L_{bce}(q \| p) \tag{4}$$

where f function represents denoising block, $\| \cdot \|_2$ is Euclidean norm (L2 norm), λ denotes the penalty factor of the denoising block, L_{bce} represents cross-entropy loss function, p and q are the output probability and label of the proposed model, respectively.

Our experiments have shown that penalizing the denoised block by measuring the Euclidean distance between the spectrogram of the clean speech and the amplitude spectrum of the denoised speech can solve this problem. This is also in full compliance with our assumptions. When the denoising block produces large distortion, $L_{wmt}(\theta)$ will punish the denoising block and force the network to learn in the correct direction so that the effect of distortion is minimized. Conversely, when the denoising block produces less distortion, the penalty produced by $L_{wmt}(\theta)$ will become smaller.

2.5 Two-Stage Network

In Fig. 1, we connect the denoising block and SED block into a larger network for joint optimization. In the denoising stage, we transform signal (noisy and clean) to 883-dimensional STFT. The obtained amplitude spectrum is used as the input feature. In the detection stage, 128 Mel-scale filters are applied to the amplitude spectrum output by denoising block on each frame, covering the frequency range of 300 to 22050 Hz. We add a batch normalization layer before sending the estimated features to the next level network to ensure that the input of the SED block is correctly normalized. During training, this layer keeps exponentially moving averages on the mean and standard deviation of each mini-batch. During testing, such running mean and standard deviation are fixed to perform normalization. By using the features processed by normalization and log-mel filters, we expect closer coupling between the separately trained denoising stage and SED stage, which can benefit joint training. After that, the normalized log-Mel features are directly sent to the SED block for detection. Since each step above is differentiable, we can derive the error gradients to jointly train the whole system.

Before joint training, the denoising block and SED block are trained separately, and the obtained parameters are used for the initializations of the two-level SED system.This network is considered to be a large network that can handle the dual tasks of removing noise and sound event detection at the same time, which is different from other methods.

3 Experiments

3.1 Data

We demonstrate the proposed model on DCASE 2017 Challenge task 2 [24]. The task dataset consists of isolated sound events for each target class and recordings of everyday acoustic scenes to serve as background. The task dataset consists of three target event categories: baby crying, breaking glass, and shooting. A synthesizer for creating mixtures at different event-to-background ratios (EBRs) is also provided. The dataset comprised of development dataset and evaluation dataset. The environmental noise and target sound format in the evaluation data set did not appear in the training data set. The development dataset also includes two parts: training subset and test subset. The detailed information about this task and dataset can be found in [24,25].

We use the provided synthesizer to generate 5000 mixtures for each target class and generate the clean signal corresponding to each mixed audio as a label for denoising block pre-training. In order to simulate different EBR environments, we set EBR to three situations of $-6\,\mathrm{dB}$, $0\,\mathrm{dB}$, and $6\,\mathrm{dB}$. In order to obtain more positive samples and alleviate the problem of data imbalance, the probability of occurrence of the event is set to 0.9 (the default value is 0.5).

3.2 Experimental Setup

The training is divided into pre-training and joint training. Before joint training, the denoising block and SED block need to be pre-trained separately. For the pre-training of the denoising block, the amplitude spectrum feature of the noisy speech is used as the input of the denoising network, and the amplitude spectrum corresponding to the clean speech is used as the label. The number of GRU units U is 32 in SED block. During the joint training, the pre-trained neural networks are used to initialize the weights of the joint network to achieve a better optimization and accelerate the optimization process. The two-stage network block is trained with the Adam optimizer [26]. λ is set to 0.2. The learning rate is 0.001 for the first 60 epochs and is then decayed by 10% after each epoch that follows. The training stops after 100 epochs. The batch size is 16 and sigmoid activation is used on the last layer of the FCL for our classification model. The output probability distribution is in the continuous range of $[0, 1]$.

3.3 Metrics

We follow the official evaluation metrics of DCASE Challenge. There are two types of event-based metrics: event-based error rate (ER) and event-based F-score. The definition of these two evaluation calculations can be found in [27]. The correct prediction only needs to consider the existence of the target event and its onset time. If the output accurately predicts the presence and onset of the target event, we express it as correct detection. The onset detection is considered accurate only when it is predicted within the range of 500 ms of the actual onset time. The ER is the sum of deletion error and insertion error, and F-score is the harmonic average of precision and recall.

4 Results

4.1 Experimental Results

Table 1. Performance of the proposed model, baseline method, and noise robust networks, *** indicates that class-wise results are not given in related paper. We compared other noise robust networks and there is no denoising block in the baseline.

Model	Metric	Development dataset				Evaluation dataset			
		Babycry	Glassbreak	Gunshot	Average	Babycry	Glassbreak	Gunshot	Average
Supervised NMF [14]	ER|F-score	*******	******	******	******	0.17| 91.4	0.22| 89.1	0.55 |72.0	0.31|84.2
Subband-Weighted NMF [15]	ER|F-score	*******	******	******	******	0.10|94.8	0.06|96.9	0.46|76.2	0.21|89.3
Baseline	ER|F-score	0.16|88.4	0.06|92.2	0.24|85.9	0.15|88.3	0.32|83.2	0.22|87.3	0.37|80.2	0.30| 83.5
Proposed	ER|F-score	0.09|95.8	0.03|98.6	0.04|96.7	0.05|97.0	0.16|91.3	0.06|97.1	0.12|94.1	0.11|94.2
Proposed+MWT	ER|F-score	**0.07|96.5**	**0.02|99.0**	**0.04|97.1**	**0.04|97.5**	0.14|92.9	0.04|97.8	0.09|95.0	0.09|95.2

The ER and F-score of the proposed model and other models are shown in Table 1. Results show that the proposed outperforms the baseline due to the noise suppression. Compared with other noise robustness models, the performance of the proposed two-stage is also superior on evaluation datasets, which indicates that the proposed network is more robust to noise. In addition, using WMT as the loss function can further improve the performance of the proposed SED system, which verifies that the WMT mentioned above can reduce the distortion caused by the denoising block. We point out that the final model we get is a complete model, which is fundamentally different from the two models proposed by Wan et al., especially during model testing.

Table 2. Performance of the proposed model and other state-of-the-art methods, * * * indicates that class-wise results are not given in related paper. We compare the following models:(1)1d-CRNN: DCASE 1st place model;(2)R-CRNN: Region-based CRNN;(3)MTF-CRNN: Multi-scale CRNN.(4)TFA: temporal-frequential attention CRNN;

Model	Metric	Development dataset				Evaluation dataset			
		Babycry	Glassbreak	Gunshot	Average	Babycry	Glassbreak	Gunshot	Average
1d-CRNN [7]	ER |F-score	0.05 |97.6	0.01|99.6	0.16|91.6	0.07|96.3	0.15|92.2	0.05|97.6	0.19|89.6	0.13|93.1
R-CRNN [8]	ER|F-score	0.09|***	0.04|***	0.14|***	0.09|95.5	*******	*******	*******	0.23|87.9
MTF-CRNN [9]	ER|F-score	0.13|91.8	0.04|97.6	0.11|93.3	0.09|94.2	0.15|89.7	0.08|95.1	0.28|83.9	0.17|89.2
TFA [10]	ER|F-score	0.10|95.1	0.01|99.4	0.16|91.5	0.09|95.3	0.18|91.3	0.04|98.2	0.17|90.8	0.13|93.4
Proposed+MWT	ER|F-score	**0.07|96.5**	**0.02|99.0**	**0.04|97.1**	**0.04|97.5**	0.14|92.9	0.04|97.8	0.09|95.0	**0.09|95.2**

Table 2 shows performance comparisons of the proposed model and other state-of-the-art SED methods in terms of ER and F-score. Compared with other approaches, the performance of our model is also competitive. The average ER (0.09) and average F-score (95.2%) of the proposed model are better than those of all models. Note that of the top 1 teams adopt ensemble method. Although Lim et al. [7] achieves relatively good results, its final decision is made by combining the output probabilities of more than four models with different time steps and different data mixtures. However, the proposed model is treated as a single model.

4.2 Denoising Block Visualization

To better understand our proposed network, We visualized the output of the denoising block in the two-level network. Figure 2 shows that the two-stage model we proposed has actually learned how to suppress noise. We selected an audio with a baby crying to visualize and the baby's crying occurs from 20.49 seconds to 21.43 seconds, which is marked by a blue frame. The target event is under park noise. After noise suppression, the target sound becomes clearer. This confirms that our proposed network achieves the dual tasks of denoising and detection at the same time.

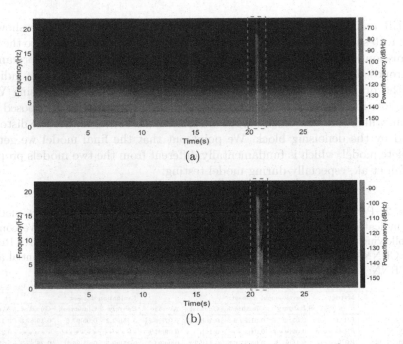

Fig. 2. Visualization of denoising block output. (a) spectrogram of an noisy speech, where the blue box denotes target event. (b) spectrogram of denoising block output.

5 Conclusion

In order to solve the impact of noise on SED, in this work, we propose a two-stage model to perform the joint task of signal denoising and SED. In addition, we propose a loss function that is conducive to network optimization. Our system can achieve the best performance on DCASE evaluation dataset. Compared with other noise robust networks, the joint network performance is better. Compared with other networks, the proposed model also outperforms them thanks to the noise suppression. The large improvement demonstrates the benefits of introducing the denoising block.

References

1. Foggia, P., Petkov, N., Saggese, A., Strisciuglio, N., Vento, M.: Audio surveillance of roads: A system for detecting anomalous sounds. IEEE Trans. Intell. Transp. Syst. **17**(1), 279–288 (2015)
2. Phuong, N.C., Do Dat, T.: Sound classification for event detection: application into medical telemonitoring. In: International Conference on Computing (2013)
3. Clavel, C., Ehrette, T., Richard, G.: Events detection for an audio-based surveillance system. In: ICME, pp. 1306–1309 (2005)
4. Baumann, J., Lohrenz, T., Roy, A., Fingscheidt, T.: Beyond the dcase 2017 challenge on rare sound event detection: a proposal for a more realistic training and test framework. In: ICASSP, pp. 611–615 (2020)
5. Wang, W., Kao, C.-C., Wang, C.: A simple model for detection of rare sound events. Interspeech (2018)
6. Shimada, K., Koyama, Y., Inoue, A.: Metric learning with background noise class for few-shot detection of rare sound events. In: ICASSP, pp. 616–620 (2019)
7. Lim, H., Park, J., Han, Y.: Rare sound event detection using 1D convolutional recurrent neural networks. Technical report, DCASE2017 Challenge, September 2017
8. Kao, C.-C., Wang, W., Sun, M., Wang, C.: R-CRNN: region-based convolutional recurrent neural network for audio event detection. Interspeech, pp. 1358–1362 (2018)
9. Zhang, K., Cai, Y., Ren, Y., Ye, R., He, L.: MTF-CRNN: multiscale time-frequency convolutional recurrent neural network for sound event detection. IEEE Access (99), 1 (2020)
10. Shen, Y.-H., He, K.-X., Zhang, W.-Q.: Learning how to listen: a temporal-frequential attention model for sound event detection. arXiv: Sound, pp. 2563–2567 (2019)
11. Keisuke, K., Ochiai, T., Delcroix, M., Nakatani, T.: Improving noise robust automatic speech recognition with single-channel time-domain enhancement network. In: ICASSP, pp. 7009–7013 (2020)
12. Kolbæk, M.: Single-microphone speech enhancement and separation using deep learning. arXiv: Sound (2018)
13. Heymann, J., Drude, L., Böddeker, C., Hanebrink, P., Haeb-Umbach, R.: Beamnet: end-to-end training of a beamformer-supported multi-channel ASR system. In: ICASSP, pp. 5325–5329 (2017)
14. Feng, Q., Zhou, Z.: Robust sound event detection through noise estimation and source separation using NMF. In: Proceedings of the Detection and Classification of Acoustic Scenes and Events 2017 Workshop(DCASE2017) (2017)
15. Zhou, Q., Feng, Z., Benetos, E.: Adaptive noise reduction for sound event detection using subband-weighted NMF. Sensors (Basel, Switzerland) (2019)
16. Wan, T., Zhou, Y., Ma, Y., Liu, H.: Noise robust sound event detection using deep learning and audio enhancement. In: ISSPIT, pp. 1–5 (2019)

17. Zhao, Y., Wang, Z.Q., Wang, D.L.: Two-stage deep learning for noisy-reverberant speech enhancement. IEEE/ACM Trans. Audio, Speech, Lang. Proc. **27**, 53–62 (2018)
18. Tan, K., Wang, D.L.: A convolutional recurrent neural network for real-time speech enhancement. Interspeech, pp. 3229–3233 (2018)
19. Ronneberger, O., Fischer, P., Brox, T.: U-Net: convolutional networks for biomedical image segmentation. In: Navab, N., Hornegger, J., Wells, W.M., Frangi, A.F. (eds.) MICCAI 2015. LNCS, vol. 9351, pp. 234–241. Springer, Cham (2015). https://doi.org/10.1007/978-3-319-24574-4_28
20. Clevert, D.-A., Unterthiner, T., Hochreiter, S.: Fast and accurate deep network learning by exponential linear units (elus). Computer ence (2015)
21. Glorot, X., Bordes, A., Bengio, Y.: Deep sparse rectifier neural networks. In: AISTATS, pp. 315–323 (2011)
22. Ioffe, S., Szegedy, C.: Batch normalization: accelerating deep network training by reducing internal covariate shift. In: International Conference on Machine Learning (2015)
23. Srivastava, N., Hinton, E.G., Krizhevsky, A., Sutskever, I., Salakhutdinov, R.: Dropout: a simple way to prevent neural networks from overfitting. J. Mach. Learn. Res. **15**, 1929–1958 (2014)
24. Mesaros, A., Heittola, T., Virtanen, T.: Tut database for acoustic scene classification and sound event detection. In: EUSIPCO, pp. 1128–1132 (2016)
25. Mesaros, A., et al.: Dcase 2017 challenge setup: tasks, datasets and baseline system (2017)
26. Kingma, P.D., Ba, L.J.: Adam: a method for stochastic optimization. In: International Conference on Learning Representations (2015)
27. Mesaros, A., Heittola, T., Virtanen, T.: Metrics for polyphonic sound event detection. Appl. Sci. (2016)

Deep Learning and Network Performance Optimization

Routing and Resource Allocation for Service Function Chain in Service-Oriented Network

Ziyu Liu[✉], Zeming Li, Chengchao Liang, and Zhanjun Liu

School of Communication and Information Engineering, Chongqing University of Posts and Telecommunications, Chongqing, China
s190101097@stu.cqupt.edu.cn

Abstract. Service function chain (SFC) and in-network computing have become popular service provision approaches in 5G and next-generation communication networks. Since different service functions may change the volume of processed traffic in different ways, inappropriate resource allocation will lead to resource wastage and congestion. In this paper, we study the traffic change effects of service nodes with in-network computing and propose a software defined network based SFC routing and resource allocation scheme. With the objective of minimizing the difference between the service delay of adjacent function pairs in SFC and the corresponding expected delay, an optimization problem is established. Due to the coupling of computing resource provision and traffic engineering, and the non-convexity of the objective function and constraints, the problem becomes difficult to solve in practice. Therefore, we first transform the problem into a convex optimization problem using linear relaxation and variable substitution. Using the dual decomposition method, we decouple the different sets of variables. With this decoupling, the network controller can efficiently design the users' service nodes and traffic engineering. Finally, we use a rounding method to obtain a feasible solution set of the problem performed by the service nodes. Extensive simulations are performed under different system settings to verify the effectiveness of the scheme.

Keywords: Service function chain · In-network computing · Resource allocation · Traffic engineering

1 Introduction

With the large-scale application of novel service models (e.g., cloud computing, virtual reality, or the Internet of Things) and the stricter service quality requirements (e.g., ultra-high-definition video), the service-oriented network needs to have massive data processing capability and data transmission capability for

© ICST Institute for Computer Sciences, Social Informatics and Telecommunications Engineering 2022
Published by Springer Nature Switzerland AG 2022. All Rights Reserved
H. Gao et al. (Eds.): ChinaCom 2021, LNICST 433, pp. 465–480, 2022.
https://doi.org/10.1007/978-3-030-99200-2_35

large-scale connectivity scenarios [1]. To cope with the challenge of high coordination between computing and network requirements, the current key technologies such as service function chain (SFC), in-network computing, and software defined network (SDN) can be explored [2,3].

For network operators, proper interconnection of service functions is necessary to achieve complete end-to-end services [4]. For example, network functions required by the network protection service may include firewall, deep packet inspection, and virus scanning [5]. In practice, a service consisting of a set of network functions arranged in a predefined order is defined as SFC. Each network function is provided by some specific network nodes. To support different on-demand services, operators can use SFCs to direct traffic from different users through the required network nodes in a predefined order. In the service-oriented network, SFC can flexibly customize and rapidly allocate the network resources of operators [6]. Figure 1 shows an example of SFC with three network functions.

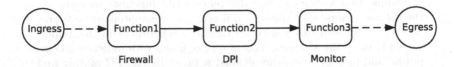

Fig. 1. The example of SFC.

In-network computing is a kind of communication acceleration technology based on the concept of collaborative design. Traditional networks typically deploy network functions at end hosts, while in-network computing enables the offload of network functions from end hosts to network nodes [7]. With in-network computing, the network nodes can provide online computing services during data transmission. Since the computing tasks are performed within the network, the efficiency of services can be significantly improved. Recently, with the advancement of programmable network devices, the in-network computing paradigm has received numerous attentions from researchers.

SDN enables the programmability of networks so that the complexity and the cost of networks can be reduced [8]. SDN decouples the control plane from the data plane, which enables more flexible management of network information. The network controller can efficiently select service nodes for users based on valid network information. The programming ability of SDN is also considered as a compelling candidate to enhance the performance of traffic engineering (TE), which is a critical component in the communication network [9]. In SDN-enabled network, TE is regarded as an effective tool to optimize the service path selection and resource allocation [10]. Recently, the authors in [11] integrated all functions required by a specific SFC on the same CPU with multiple cores to save transmission resource. In [12], the authors described the service node selection and routing problem under the condition of limited links and nodes capacities. Reference [13] considered the differentiated characteristics of SFC requests and proposed a heuristic algorithm. The network function deployment

and routing problem between adjacent network function pairs on the edge cloud
was studied in [14]. The author proposed an approximate algorithm based on
linear relaxation and random rounding.

The above work proposed some routing and resource allocation schemes for
SFC scenarios. However, unlike classical network nodes that only forward data,
network nodes with in-network computing may change the traffic volumes of the
data being processed [15]. For example, the encoder used for satellite communication can increase the traffic by 31% due to the checksum, and the WAN
optimizer may reduce the traffic by up to 80% before sending it to the next
hop [16]. Figure 2 is an example of the traffic changing effects of network functions. Since different service functions may change the volume of processed traffic
in different ways, the fixed resource allocation scheme will lead to inappropriate
allocation of resource, e.g., wasted or insufficient resource. Further, the changed
traffic will affect the service delay between adjacent function pairs in SFCs and
create congestion at the service nodes. The authors in [16] proposed a service
deployment scheme that considered the traffic changing effects to achieve optimal
load balancing. In the SFC routing and resource allocation problem for meeting
the end-to-end service requirements of different users, the traffic changing effects
have not received sufficient attention.

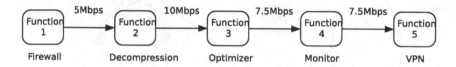

Fig. 2. The example of the traffic change effects.

Motivated by these challenges, we consider the end-to-end routing and
resource allocation problem for different user requests with specific SFCs, where
the required service functions have different capabilities to change the traffic
volumes. The main contributions are shown as follows:

- Considering the capabilities of functions to change traffic volumes and the different user requests with specific SFCs, we establish an optimization problem.
  According to the ratio of traffic changes, we adjust the resources allocated to
  the traffic at the service node. The limited transmission resource of each link
  and the computing resource of each node are considered.
- Due to the coupling of TE and computing resource provision, and the non-
  convexity of the objective function and constraints, the problem becomes
  difficult to solve in practice. Therefore, we use linear relaxation and dual
  decomposition method to solve the problem. After obtaining the linear solution, we use a rounding method to obtain the feasible binary solution.
- Extensive simulations are conducted with different parameter settings to verify the performance of the proposed scheme.

The rest of this paper is organized as follows. In Sect. 2, we introduce the system model and formulate the problem. The transformed problem and the approximation algorithm are proposed in Sect. 3. Simulation results are discussed in Sect. 4. Finally, Sect. 5 concludes the paper.

2   System Model and Problem Formulation

2.1   Network Communication Model

As shown in Fig. 3, we consider a service-oriented wireless network that supports SFC. The considered communication network is modeled as a directed graph $\mathcal{G} = (\mathcal{I}, \mathcal{L})$. $\mathcal{I}$ is the nodes set, which includes the network nodes set $\mathcal{V}$ and the users set $\mathcal{K}$, namely, $\mathcal{I} = \mathcal{V} \cup \mathcal{K}$. Similarly, $\mathcal{L}$ is the set of directed links. It comprises the wired links set $\mathcal{L}^w$ and the wireless links set $\mathcal{L}^{wl}$ respectively.

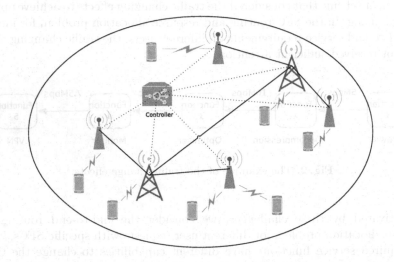

Fig. 3. The service-oriented wireless communication network.

If link (i, j) is a wired link, we assume that the link can provide a fixed available transmission capacity C_{ij}. If link (i, j) is a wireless link, the transmission capacity of the link depends on the available wireless resource of the link. The wireless resource is allocated by the network controller. By using Shannon's theorem, the spectral efficiency of wireless link (i, j) can be defined as follows:

$$\gamma_{ij} = log \left(1 + \frac{g_{ij}p_i}{\sigma_0 + \sum_{i' \in \mathcal{I}, i' \neq i} g_{i'j}p_{i'}} \right). \tag{1}$$

where g_{ij} is the channel gain between link (i, j) including large-scale path loss and shadowing, and p_i is the transmission power. To simplify the analysis, we use the fixed equal power allocation mechanism to make the transmission power is identical for all wireless links. Each node treats the interference from any other transmission node i' as noise. The aggregated interference can be written as $\sum_{i' \in \mathcal{I}, i' \neq i} g_{i'j} p_{i'}$. Since the change of the small-scale fading is much faster than the transmission resource allocation, in this paper, we ignore the small-scale fading when evaluating the SINR [10]. Besides, σ_0 is the power spectrum density of additive white Gaussian noise. Therefore, the achievable data transmission rate of the wireless link (i, j) is $R_{ij} = W\gamma_{ij}$, where W is the total available wireless spectrum resource in the network.

2.2   SFC Service Model

The network supports multiple types of network functions. We use $\mathcal{V}_f \subset \mathcal{V}$ to express the subset of network nodes that can provide function f. We call the network nodes which can provide service functions are service nodes [12]. Each service node $i \in \mathcal{V}$ has a known computing capacity π_i. The service request of the user $k \in \mathcal{K}$ is described by a service function chain $\mathcal{F}(k)$, which consists of m functions that need to be processed in a given order, i.e., $\mathcal{F}(k) = (f_1^k \rightarrow f_2^k \rightarrow \dots \rightarrow f_m^k)$.

For example, there are four network nodes and three users in Fig. 4. For simplicity, each network node only deploys one network function. The SFC of user 1 can be expressed as $\mathcal{F}(1) = (f_1^1 \rightarrow f_2^1)$, where f_1^1 and f_2^1 are $f2$ and $f4$ respectively. User 1 prefers to access the network from node B because function $f2$ is deployed on node B. However, there is no direct link between node B and node D. Then, node A or node C will act as a routing node, which needs to route

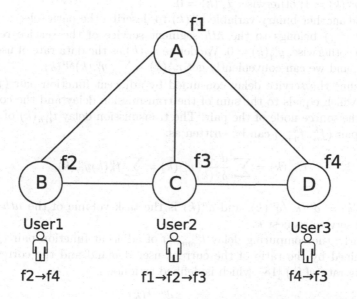

Fig. 4. An example of SFC requests in service-oriented network.

the traffic after processing $f2$ to node D. Similarly, before providing service to user 2, the network needs to route the traffic to node A first. The traffic of user 3 only needs to pass node D and node C in sequence.

After the data traffic traverses the service node and is processed by the corresponding function, the traffic rate may increase or decrease (e.g., video decompression and compression) [17]. The traffic rate of user k after receiving the function f_n^k can be expressed as $\delta^n(k)$. We denote a parameter α_n to express the traffic inflation factor of f_n^k. Therefore, we have $\delta^n(k) = \alpha_n \delta^{n-1}(k)$.

To avoid the coordination overhead caused by function splitting in practical application, we require each service function of service request k should be processed by one corresponding service node [12]. We use a binary parameter $h_i^n(k)$ to indicate a potential hitting event between service node i and the nth function of service request k. If node i can provide the nth function of service request k, $h_i^n(k) = 1$; otherwise, $h_i^n(k) = 0$. According to the user's request (SFC $\mathcal{F}(k)$) and the current network information (subset $\mathcal{V}_f$), the controller can quickly get the value of $h_i^n(k)$.

2.3 Problem Formulation

For the convenience of description, we add two dummy functions f_0^k and f_{m+1}^k to each service request k as the first and the last service function, respectively [14]. Both the two functions cannot cause any consumption of computing resource and computing delay. We define the service of the adjacent function pair (f_n^k, f_{n+1}^k) as the nth segment service of the service request k.

We use a binary variable $x_i^n(k)$ to define whether the network node i provides the nth service function for the user k. If node i provides the nth service function of $\mathcal{F}(k)$, $x_i^n(k) = 1$; otherwise, $x_i^n(k) = 0$.

We use another binary variable $y_{ij}^n(k)$ to describe the path selection of user k. If link (i,j) belongs to the nth segment service of the service request k, $y_{ij}^n(k) = 1$; otherwise, $y_{ij}^n(k) = 0$. We define $r_{ij}(k)$ as the data rate of user k over link (i,j), and we can conveniently get $r_{ij}(k) = \sum_n y_{ij}^n(k)\delta^n(k)$.

We define the service delay consumed by adjacent function pair (f_n^k, f_{n+1}^k) as $t^n(k)$, which equals to the sum of the transmission delay and the computing delay of the source node of the pair. The transmission delay $t_{tra}^n(k)$ of adjacent function pair (f_n^k, f_{n+1}^k) can be written as:

$$t_{tra}^n(k) = \sum_{(i,j)} \frac{d^n(k)}{\delta^n(k)} y_{ij}^n(k) = \sum_{(i,j)} t_h^n(k) y_{ij}^n(k). \tag{2}$$

where $t_h^n(k) = d^n(k)/\delta^n(k)$, and $d^n(k)$ is the task volume of the nth segment service of service request k.

Similarly, the computing delay $t_{com}^n(k)$ of adjacent function pair (f_n^k, f_{n+1}^k) is determined by the ratio of the current user demand and the corresponding computing rate $\eta(f_n^k)$ [18], which is defined as follows:

$$t_{com}^n(k) = \frac{d^{n-1}(k)}{\eta(f_n^k)}. \tag{3}$$

Therefore, the service delay of adjacent function pair can be expressed as:

$$t^n(k) = t^n_{com}(k) + \sum_{(i,j)} t^n_h(k) y^n_{ij}(k). \tag{4}$$

We define the expected service delay of adjacent function pair (f^k_n, f^k_{n+1}) as $t^n_{exp}(k)$. In order to minimize the absolute delay difference between the service delay $t^n(k)$ and the expected delay $t^n_{exp}(k)$ for all adjacent function pairs, the objective function can be written as:

$$\sum_{k \in \mathcal{K}} \sum_n |t^n(k) - t^n_{exp}(k)|. \tag{5}$$

In summary, the optimization problem can be expressed as:

$$
\begin{aligned}
\textbf{P1:} \min_{\textbf{X},\textbf{Y}} & \sum_{k \in \mathcal{K}} \sum_n |t^n(k) - t^n_{exp}(k)| \\
s.t. \quad & C1 : x^n_i(k) \le h^n_i(k), \quad \forall i,k,n, \\
& C2 : \sum_{i \in \mathcal{V}} x^n_i(k) = 1, \quad \forall k,n, \\
& C3 : \sum_{(i,j) \in \mathcal{L}} y^n_{ij}(k) - \sum_{(j,i) \in \mathcal{L}} y^n_{ji}(k) = x^n_i(k) - x^{n+1}_i(k), \\
& C4 : \sum_n \sum_{k \in \mathcal{K}} \eta(f^k_n) x^n_i(k) \le \pi_i, \quad \forall i, \\
& C5 : \sum_{k \in \mathcal{K}} \sum_n y^n_{ij}(k) \delta^n(k) \le C_{ij}, \quad \forall (i,j) \in \mathcal{L}^w, \\
& C6 : \sum_{(i,j) \in \mathcal{L}^{wl}} \frac{\sum_n y^n_{ij}(k) \delta^n(k)}{\gamma_{ij}} \le W, \\
& C7 : x^n_i(k) \in \{0,1\}, y^n_{ij}(k) \in \{0,1\}.
\end{aligned}
\tag{6}
$$

The constraints C1 and C2 enforce that each service function must be served by exactly one service node in the network. The path selection constraint for service request k can be written as the constraint C3. This constraint is an essential condition for the successful construction of the routing path [19]. The constraint C4 enforces that the allocated computing resource of node i cannot exceed its available computing resource. The constraint C5 means the allocated data rate of wired link (i,j) for all users should be less than the link capacity. Similarly, the constraint C6 means the total allocated spectrum cannot exceed the available spectrum bandwidth. The constraint C7 means that $x^n_i(k)$ and $y^n_{ij}(k)$ are binary variables.

3   Solution to the Problem

Since $x^n_i(k)$ and $y^n_{ij}(k)$ are binary variables, and they are coupled in constraint, the problem becomes difficult to solve in practice. In this section, we first obtain the fractional solution by linear relaxation and dual decomposition. After that, we use a rounding method to get a feasible solution of the problem.

3.1   Problem Decomposition

In order to overcome the obstacle of binary variables, we relax them into real value $[0, 1]$. Due to the non-convexity of the objective function, we transform it into a linear form. Firstly, we introduce a new variable $z^n(k)$ to replace the original objective function, namely, $z^n(k) = |t^n(k) - t^n_{exp}(k)|$. To limit the new variable, we introduce two new constraints, which are shown below:

$$\begin{aligned} z^n(k) &\geq t^n(k) - t^n_{exp}(k), \\ z^n(k) &\geq t^n_{exp}(k) - t^n(k). \end{aligned} \tag{7}$$

Finally, we get the convex optimization problem:

$$\textbf{P2:} \min_{\mathbf{X,Y,Z}} \sum_{k \in \mathcal{K}} \sum_n z^n(k)$$

$$\begin{aligned}
s.t. \quad & C1 : x_i^n(k) \leq h_i^n(k), \quad \forall i, k, n, \\
& C2 : \sum_{i \in \mathcal{V}} x_i^n(k) = 1, \quad \forall k, n, \\
& C3 : \sum_{(i,j) \in \mathcal{L}} y_{ij}^n(k) - \sum_{(j,i) \in \mathcal{L}} y_{ji}^n(k) = x_i^n(k) - x_i^{n+1}(k), \\
& C4 : \sum_n \sum_{k \in \mathcal{K}} \eta(f_n^k) x_i^n(k) \leq \pi_i, \quad \forall i, \\
& C5 : \sum_{k \in \mathcal{K}} \sum_n y_{ij}^n(k) \lambda^n(k) \leq C_{ij}, \quad \forall (i,j) \in \mathcal{L}^w, \\
& C6 : \sum_{(i,j) \in \mathcal{L}^{wl}} \frac{\sum_n y_{ij}^n(k) \lambda^n(k)}{\gamma_{ij}} \leq W, \\
& C7 : t_{com}^n(k) + \sum_{(i,j) \in \mathcal{L}} t_h^n(k) y_{ij}^n(k) - t_{exp}^n(k) \leq z^n(k), \\
& C8 : t_{exp}^n(k) - t_{com}^n(k) - \sum_{(i,j) \in \mathcal{L}} t_h^n(k) y_{ij}^n(k) \leq z^n(k), \\
& C9 : x_i^n(k) \in [0, 1], y_{ij}^n(k) \in [0, 1].
\end{aligned} \tag{8}$$

Although we transform the problem **P1** into the convex problem **P2**, the problem is still complex because the coupling of variables. To make the network more effective in routing and resource allocation, we compose the problem **P2** into three parts: service provisioning problem, traffic engineering problem, and delay optimization problem.

There are three coupling constraints C3, C7, and C8 in the relaxation problem. Therefore, by using dual variables $\{\lambda_{nk}\}$, $\{\mu_{ink}\}$ and $\{\nu_{nk}\}$, Lagrangian can be written as

$$
\min_{\mathbf{X,Y,Z}} \sum_{k\in\mathcal{K}}\sum_{n} z^n(k)+
$$

$$
\sum_{i,n,k}\mu_{ink}\left[\sum_{(i,j)\in\mathcal{L}} y_{ij}^n(k) - \sum_{(j,i)\in\mathcal{L}} y_{ji}^n(k) - x_i^n(k) + x_i^{n+1}(k)\right]
$$

$$
+\sum_{n,k}\lambda_{nk}\left[t_{com}^n(k) + t_h^n(k)\sum_{(i,j)\in\mathcal{L}} y_{ij}^n(k) - t_{exp}^n(k) - z^n(k)\right] \tag{9}
$$

$$
+\sum_{n,k}\nu_{nk}\left[t_{exp}^n(k) - t_{com}^n(k) - t_h^n(k)\sum_{(i,j)\in\mathcal{L}} y_{ij}^n(k) - z^n(k)\right]
$$

$$
s.t. \quad \mathbf{X}\in\Pi_x, \quad \mathbf{Y}\in\Pi_y, \quad \mathbf{Z}\in\Pi_z.
$$

where Π_x, Π_y, and Π_z are independent local feasible sets of $\mathbf{X}$, $\mathbf{Y}$ and $\mathbf{Z}$, respectively. The relaxation problem is divided into two optimization levels: the dual variables updating and dual functions finding [20]. Naturally, the related dual problem (DP) is defined as:

$$
\mathbf{DP:} \quad \max_{\lambda_{nk},\mu_{ink},\nu_{nk}\in\mathbb{R}} D(\lambda_{nk},\mu_{ink},\nu_{nk}) = g_x(\mu_{ink})
$$

$$
+ g_y(\lambda_{nk},\mu_{ink},\nu_{nk}) + g_z(\lambda_{nk},\nu_{nk}) \tag{10}
$$

$$
s.t. \quad \lambda_{nk}\geq 0 \quad \forall k,n,
$$

$$
\nu_{nk}\geq 0 \quad \forall k,n.
$$

where the three dual functions $g_x(\mu_{ink})$, $g_y(\lambda_{nk},\mu_{ink},\nu_{nk})$ and $g_z(\lambda_{nk},\nu_{nk})$ will be solved by the given dual variables $\{\lambda_{nk}\}$, $\{\mu_{ink}\}$ and $\{\nu_{nk}\}$ in the following three problems:

$$
g_x(\mu) = \inf_{x\in\Pi_x}\left\{\sum_{i,n,k}\mu\left[x_i^{n+1}(k) - x_i^n(k)\right]\right\}. \tag{11}
$$

$$
g_y(\lambda,\mu,\nu) = \inf_{y\in\Pi_y}\left\{\begin{array}{l}\sum_{i,n,k}\mu\left[\sum_{(i,j)} y_{ij}^n(k) - \sum_{(j,i)} y_{ji}^n(k)\right] \\ +\sum_{n,k} t_h^n(k)(\lambda - \nu)\sum_{(i,j)} y_{ij}^n(k)\end{array}\right\}. \tag{12}
$$

$$
g_z(\lambda,\nu) = \inf_{z\in\Pi_z}\left\{\begin{array}{l}\sum_{n,k}(1 - \lambda - \nu)z^n(k) \\ +\sum_{n,k}(\lambda - \nu)\left[t_{com}^n(k) - t_{exp}^n(k)\right]\end{array}\right\}. \tag{13}
$$

We can deploy sub-gradient method to solve (10). In each iteration, we first solve (11), (12) and (13) by the given dual variables. After obtaining the solutions of the three problems, we update the dual variables. Finally, we are able to obtain the optimal solution of the problem.

As for problem (11), it can be written as the following form:

$$\textbf{P3:} \min_{x} g_x(\mu)$$

$$s.t. \quad C1: x_i^n(k) \leq h_i^n(k), \quad \forall i,k,n,$$

$$C2: \sum_{i \in \mathcal{V}} x_i^n(k) = 1, \quad \forall k,n, \qquad (14)$$

$$C3: \sum_{n} \sum_{k \in \mathcal{K}} \eta(f_n^k) x_i^n(k) \leq \pi_i, \quad \forall i,$$

$$C4: x_i^n(k) \in [0,1].$$

Similar to problem (11), the problem (12) as for the link selection variable $y_{ij}^n(k)$ is shown as:

$$\textbf{P4:} \min_{y} g_y(\lambda, \mu, \nu)$$

$$s.t. \quad C1: \sum_{k \in \mathcal{K}} \sum_{n} y_{ij}^n(k)\lambda^n(k) \leq C_{ij}, \quad \forall(i,j) \in \mathcal{L}^w,$$

$$C2: \sum_{(i,j) \in \mathcal{L}^{wl}} \frac{\sum_{n} y_{ij}^n(k)\lambda^n(k)}{\gamma_{ij}} \leq W, \qquad (15)$$

$$C3: y_{ij}^n(k) \in [0,1].$$

At last, the problem (13) is shown as:

$$\textbf{P5:} \min_{z} \quad g_z(\lambda, \nu)$$

$$s.t. \quad z^n(k) \geq 0. \qquad (16)$$

The above three sub-problems are all linear programming (LP) problems with only one type of variable. The optimal global solutions of these sub-problems can be easily obtained in polynomial time. However, the solution of the problem **P2** may not be feasible for the original problem **P1** because the optimal solution of the LP problem may not be binary. The optimal solution of the LP problem is the lower bound of the solution of the original problem [12]. Next, we will use an effective rounding method to obtain the feasible solution.

3.2   Binary Recover

We combine some methods of existing work to construct our rounding strategy. Suppose that the fractional solution set obtained after iteratively solving the relaxation problem is $\{\tilde{x}_i^n(k)\}$ (and $\{\tilde{y}_{ij}^n(k)\}$). Our goal is to construct a binary solution set $\{x_i^n(k)\}$ based on the non-zero values in $\{\tilde{x}_i^n(k)\}$. However, due to the resource capacities couple the transmission of all service requests k, the system rounding is difficult [12]. We use a heuristic method to round the elements to 1 or 0.

In particular, if $\tilde{\mathbf{x}}^n(k)$ itself is binary, then we simply let $\mathbf{x}^n(k) = \tilde{\mathbf{x}}^n(k)$; if $\tilde{\mathbf{x}}^n(k)$ is not binary, we will round according to its value.

Firstly, we check the value of the largest element. If $\tilde{x}_j^n(k) = \max_{i \in V_f} \tilde{x}_i^n(k) \geq \theta$, where $\theta \in (0, 1)$ is the threshold we set, and node j has enough computing capacity, we adopt the following strategy:

$$x_j^n(k) = 1, \quad and \quad x_i^n(k) = 0, \quad \forall i \in V_f \setminus \{j\}. \tag{17}$$

Otherwise, we give priority to the node $v \in V_f$ with the most computing capacity to provide the service functions for service request k. We set:

$$x_v^n(k) = 1, \quad and \quad x_i^n(k) = 0, \quad \forall i \in V_f \setminus \{v\}. \tag{18}$$

After the rounding process, we get a binary solution set $\{x_i^n(k)\}$ while satisfying the constraint of the computing capacity. We use the rounded binary solution $\{x_i^n(k)\}$ to solve the problem **P2** again to find the solution under the condition of satisfying other constraints [14].

4   Simulation Results

We consider a service-oriented network with a range of $500\,m \times 500\,m$ in simulation. The network consists of one MBS and several SBSs. Users are scattered in the network, and we assume that the user can establish a connection with any BS within 200 m. The parameters of the network part are shown in Table 1.

Table 1. Network parameters settings

Notations	Definition
Frequency bandwidth	20 MHz
Transmission power	SISO with maximum power: 23 dBm
Pathloss	L(distance) = 34 + 40log(distance) (dB)
Lognormal shadowing	8 dB
Power density of the noise	−174 dBm/Hz
Number of SBSs	15
Wired link transmission capacity	[20, 30] Mbps
Computing capacity of SBS	[40, 50] Mbps
Computing capacity of MBS	75 Mbps

The network can provide five types of network functions, which are denoted as $\{f_1, f_2, ..., f_5\}$. Network functions can be placed on all nodes. In the simulation, each SBS randomly deploys one type of these functions, while MBS can deploy three functions. The network functions have traffic changing effects. Among them, The functions change traffic to $\{0.5, 0.75, 1, 1.25, 1.5\}$ times the original, respectively. The computing capacity required by the five functions is set to $\{5, 7.5, 10, 7.5, 5\}$ Mbps, respectively.

We assume that each user demand comes with a service request for a specific SFC. The SFC $\mathcal{F}(k)$ with a length of m is generated by randomly selecting m unique functions from the function pool of the network and arranging them in order. Each data packet size sent by different users is randomly generated from [300, 500] KB, and the initial flow rate is randomly generated from [1, 4] Mbps. The expected delay of the user is equal to the service delay required by the user to complete the service by the shortest path, and the resource constraints are ignored in the calculation [21].

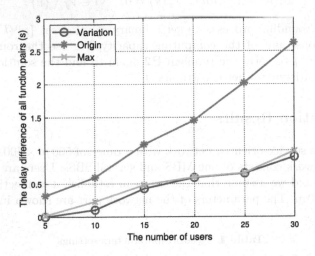

Fig. 5. The delay difference of all function pairs with different number of users.

In the simulation, we test the impact of the different number of users and the different SFC lengths on network performance. We compare the following three different schemes:

- Variation: Our proposed scheme. When traffic leaves the service node, we adjust the transmission resource allocated for traffic according to the corresponding traffic inflation factor.
- Origin: The network adopts the fixed transmission rate scheme, which equals the initial rate of traffic.
- Max: During the whole service process, the network controller reserves transmission resource according to the maximum transmission rate of traffic.

We first observe the change in the objective function of the optimization problem as the number of users increases. In Fig. 5, when the number of users increases, the absolute delay difference of all function pairs in the network also increases. The performance in delay difference of the Max scheme is approximately the same as that of our scheme. In most cases, our scheme is slightly better than the Max scheme. Moreover, the delay difference of our scheme is much smaller than the Origin scheme. When the number of users is 30, the delay difference of the Origin scheme is even 2.6 times higher than the other two schemes. We consider that because the Origin scheme uses a fixed transmission rate, the delay will jitter dramatically in scenarios where the traffic volume changes. On the other hand, the Max scheme reserves enough transmission resource to maintain similar results as our scheme.

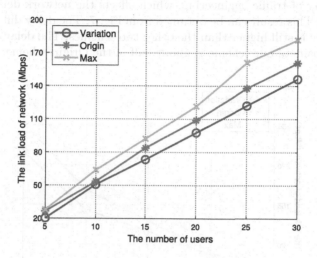

Fig. 6. The link load of network with different number of users.

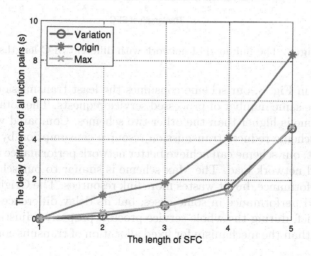

Fig. 7. The delay difference of all function pairs with different SFC lengths.

Figure 6 shows the transmission resource load of the network. It is worth noting that although network nodes with in-network computing can increase or reduce user traffic, the total link load in our scheme is still smaller than the link load in the Origin scheme. The link load of the Max scheme is the largest of the three schemes. Compared with the Max scheme, in serving the same number of users with SFC requests, our scheme can save the link transmission resource consumption as much as possible.

With the fixed number of users, we change the length of the SFC requested by each user. Specifically, the increase of service chain length leads to the increasing complexity of traffic engineering, which affects the network delay difference performance. This result can be obtained from Fig. 7. The delay difference of the Origin scheme is still higher than the other two schemes. The delay difference of all function pairs of our scheme is about half of the Origin scheme.

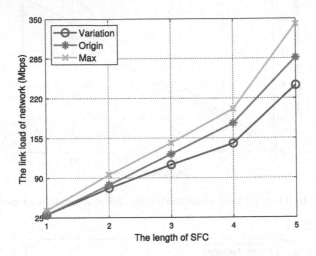

Fig. 8. The link load of network with different SFC lengths.

As shown in Fig. 8, our scheme consumes the least transmission resource in the case of the same number of processed service requests. The total link load of the Max scheme is higher than the other two schemes. Compared with the Max scheme, our scheme reduces transmission resource consumption by about 29%.

As a result, our scheme can achieve better network performance in both delay difference and network load. The Max scheme is similar to our scheme in delay difference performance, but it wastes more link resources. The Origin scheme has good link load performance in some cases, but its delay difference performance is poor. In brief, during the whole service process, properly adjusting the traffic rate is better than the mechanism for fixed allocation of transmission bandwidth.

5   Conclusions

This paper studied the routing and resource allocation of user data with SFC requests in the service-oriented network. Unlike previous work, this paper considered the traffic changing effects with in-network computing and established an optimization problem to minimize the delay difference between the service delay and the expected delay for all adjacent function pairs. Due to the complexity of the problem, we used linear relaxation, dual decomposition, and rounding methods to solve the problem. The numerical results showed that our proposed scheme could achieve better results in both delay difference and link transmission load performance.

References

1. Pham, Q.V., et al.: A survey of multi-access edge computing in 5G and beyond: fundamentals, technology integration, and state-of-the-art. IEEE Access **8**, 116974–117017 (2020)
2. Tang, X., et al.: Computing power network: the architecture of convergence of computing and networking towards 6G requirement. China Commun. **18**(2), 175–185 (2021)
3. Qu, K., Zhuang, W., Ye, Q., Shen, X.S., Li, X., Rao, J.: Traffic engineering for service-oriented 5G networks with SDN-NFV integration. IEEE network **34**(4), 234–241 (2020)
4. Bhamare, D., Jain, R., Samaka, M., Erbad, A.: A survey on service function chaining. J. Netw. Comput. Appl. **75**, 138–155 (2016)
5. Mirjalily, G., Zhiquan, L.: Optimal network function virtualization and service function chaining: a survey. Chin. J. Electron. **27**(4), 704–717 (2018)
6. Halpern, J., Pignataro, C., et al.: Service Function Chaining (SFC) architecture. In: RFC 7665 (2015)
7. Mai, T., Yao, H., Guo, S., Liu, Y.: In-network computing powered mobile edge: toward high performance industrial IoT. IEEE network **35**(1), 289–295 (2020)
8. Liao, W.C., Hong, M., Farmanbar, H., Li, X., Luo, Z.Q., Zhang, H.: Min flow rate maximization for software defined radio access networks. IEEE J. Sel. Areas Commun. **32**(6), 1282–1294 (2014)
9. Mendiola, A., Astorga, J., Jacob, E., Higuero, M.: A survey on the contributions of software-defined networking to traffic engineering. IEEE Commun. Surv. Tutor. **19**(2), 918–953 (2016)
10. Liang, C., He, Y., Yu, F.R., Zhao, N.: Enhancing video rate adaptation with mobile edge computing and caching in software-defined mobile networks. IEEE Trans. Wireless Commun. **17**(10), 7013–7026 (2018)
11. Zheng, J., et al.: Optimizing NFV chain deployment in software-defined cellular core. IEEE J. Sel. Areas Commun. **38**(2), 248–262 (2019)
12. Zhang, N., Liu, Y.F., Farmanbar, H., Chang, T.H., Hong, M., Luo, Z.Q.: Network slicing for service-oriented networks under resource constraints. IEEE J. Sel. Areas Commun. **35**(11), 2512–2521 (2017)
13. Hong, P., Xue, K., Li, D., et al.: Resource aware routing for service function chains in SDN and NFV-enabled network. IEEE Trans. Serv. Comput. (2018)
14. Yang, S., Li, F., Trajanovski, S., Chen, X., Wang, Y., Fu, X.: Delay-aware virtual network function placement and routing in edge clouds. IEEE Trans. Mob. Comput. **20**, 445–459 (2019)
15. Jang, I., Suh, D., Pack, S., Dán, G.: Joint optimization of service function placement and flow distribution for service function chaining. IEEE J. Sel. Areas Commun. **35**(11), 2532–2541 (2017)

16. Ma, W., Sandoval, O., Beltran, J., Pan, D., Pissinou, N.: Traffic aware placement of interdependent NFV middleboxes. In: IEEE INFOCOM 2017-IEEE Conference on Computer Communications, pp. 1–9. IEEE (2017)
17. Yang, S., Li, F., Trajanovski, S., Yahyapour, R., Fu, X.: Recent advances of resource allocation in network function virtualization. IEEE Trans. Parallel Distrib. Syst. **32**(2), 295–314 (2021)
18. Woldeyohannes, Y.T., Mohammadkhan, A., Ramakrishnan, K., Jiang, Y.: ClusPR: balancing multiple objectives at scale for NFV resource allocation. IEEE Trans. Netw. Serv. Manage. **15**(4), 1307–1321 (2018)
19. Wang, G., Zhou, S., Zhang, S., Niu, Z., Shen, X.: SFC-based service provisioning for reconfigurable space-air-ground integrated networks. IEEE J. Sel. Areas Commun. **38**(7), 1478–1489 (2020)
20. Palomar, D.P., Chiang, M.: A tutorial on decomposition methods for network utility maximization. IEEE J. Sel. Areas Commun. **24**(8), 1439–1451 (2006)
21. Chen, W.K., Liu, Y.F., De Domenico, A., Luo, Z.Q.: Network slicing for service-oriented networks with flexible routing and guaranteed E2E latency. In: 2020 IEEE 21st International Workshop on Signal Processing Advances in Wireless Communications (SPAWC), pp. 1–5. IEEE (2020)

Service-Aware Virtual Network Function Migration Based on Deep Reinforcement Learning

Zeming Li[✉], Ziyu Liu, Chengchao Liang, and Zhanjun Liu

School of Communication and Information Engineering,
Chongqing University of Posts and Telecommunications, Chongqing, China
1964625523@qq.com

Abstract. Network Function Virtualization (NFV) aims to provide a way to build agile and flexible networks by building a new paradigm of provisioning network services where network functions are virtualized as Virtual Network Functions (VNFs). Network services are implemented by service function chains, which are formed by a series of VNFs with a specific traversal order. VNF migration is a critical procedure to reconfigure VNFs for providing better network services. However, the migration of VNFs for dynamic service requests is a key challenge. Most VNF migration works mainly focused on static threshold trigger mechanism which will cause frequent migration. Therefore, we propose a novel mechanism to solve the issue in this paper. With the objective of minimizing migration overhead, a stochastic optimization problem based on Markov decision process is formulated. Moreover, we prove the NP-hardness of the problem and propose a service-aware VNF migration scheme based on deep reinforcement learning. Extensive simulations are conducted that the proposed scheme can effectively avoid frequent migration and reduce the migration overhead.

Keywords: Network Function Virtualization · Virtual Network Functions · Markov decision process · Migration · Deep reinforcement learning

1 Introduction

The explosive growth of Fifth Generation networks and their complexity exceeds the limitations of manual management. The growing demand for agile and robust network services has driven the telecom industry to design innovative network architectures with Network Functions Virtualization (NFV) and Software Defined Networking (SDN). NFV is the enabler supporting concept that is proposed to decouple network functions from proprietary hardware devices. NFV

H. Gao et al. (Eds.): ChinaCom 2021, LNICST 433, pp. 481–496, 2022.
https://doi.org/10.1007/978-3-030-99200-2_36

has the potential to significantly reduce operational and capital expenditures, and improve service agility compared to traditional network functions implemented by dedicated hardware devices. SDN aims to enable centralized network control using software running on common hardware rather than proprietary hardware devices. According to academia and industry, the combination of NFV and SDN is expected to revolutionize network operations, not only by dramatically reducing costs, but also by introducing new possibilities for enterprises, carriers and service providers [1].

With the extension of global Internet services, Cisco predicts that the global mobile networks will support more than 12 billion mobile devices by 2022 with mobile traffic approaching zettabytes (ZB). Network operators will further extend the depth and breadth of communication coverage of next-generation communication networks (e.g., 5G, 5G beyond, and 6G) to support multifarious network services for diversified QoS requirements [2]. Meanwhile, with the rapid development of new applications and services, it is indispensable to provide and deploy new devices continuously, which will result in the network resource and energy consumption of Data Center Networks (DCNs) increase rapidly [3]. By aggregating satellite networks to boost current network capacity, enhance system robustness, and extend ubiquitous 3D wireless coverage. Therefore, it is essential to develop satellite-terrestrial integrated network (STIN) to leverage the complementary benefits of disparate networks to realize seamless, robust, and reliable network services provisioning [4].

In SDN/NFV-enabled STIN, network services can be implemented with Service Function Chains (SFCs), which consist of a series of chained Virtual Network Functions (VNFs) by highly predefined sequences. These VNFs are instantiated and executed at heterogeneous substrate nodes (i.e., NFV nodes). When the demand of VNFs exceeds their mapped substrate nodes and links resource capacity, the network functions will be invalidated, which will seriously affect the success rate of SFCs deployment and services performance. In order to accommodate the time-varying service demands, it is inevitable to orchestrate the VNF migration among substrate nodes to improve service performance. However, most of the current works mainly focus on static threshold triggered mechanisms [5–10]. In a real-DCN scenario, service requests arrive at the system randomly with diversified QoS requirements, and the network resource conditions of the underlying NFV infrastructure (NFVI) dynamically vary over time. A static migration solution is unable to satisfy the dynamic property of the network and leads to function invalidation or service degradation.

The wide variety of challenges introduced by the disruptive deployment of STIN trigger the need for a drastic transformation in the way services are managed and orchestrated. It can become time-consuming to orchestrate services due to manual configurations around VNF migration. The intention is to have all operational processes and tasks, such as resource monitoring, VNF mapping, resource configuration, and optimization, executed automatically. Since the world is already moving towards data-driven automation, technologies like Machine Learning can be employed to tackle part of the workload. New advancements in Deep Reinforcement Learning (DRL), will pave the way for a more intelligent and self-organizing network.

Despite the significant advancements in networking thanks to the introduction of NFV and SDN technologies, the migration and orchestration tools still require additional research and development, to reach an acceptable level of automation. To this end, the motivations of this paper can be unfolded in three aspects. Firstly, to avoid substrate nodes and links overload, and guarantee service continuity with randomly arrive service requests, some VNFs should be migrated from poor nodes for the utilization of resources rationally. Secondly, existing migration mechanisms are prone to frequent migration, a reasonable mechanism should be designed to reduce the times of invalid migration and alleviate network overhead. Finally, we should consider how to optimize the average performance metrics at long-time scale through VNF migration for SFC with a certain life cycle and various resource constraints.

Given the above considerations, we investigate the VNF migration problem by taking migration triggering mechanism and migration overhead into account. Specially, rather than focus on the detailed migration process, we primarily concentrate on the migration overhead of the network. Since the appropriate deployment of VNF can increase the utilization of network resources, enhance the robustness of the system, and improve the adaptation to environmental changes [11]. Our intention is to optimize migration overhead for all affected SFCs. For this purpose, the main contributions in this paper can summarize as follows.

- According to the characteristics of VNF migration and traffic dynamics, we propose a dynamic threshold trigger mechanism to avoid the frequent migration and system instability incurred by the existing migration trigger mechanism.
- Basic on this mechanism, we model the migration overhead as power consumption and migration latency caused by migration jointly. Besides, a stochastic optimization problem based on Markov Decision Process (MDP) is formulated to describe the migration process of VNFs. Furthermore, a service-aware VNF migration scheme based on DRL is proposed.
- Extensive simulation results are provided to demonstrate that the validity of our scheme and a series of comparative experiments with the existing works also presented.

The remainder of this paper is organized as follows. In Sect. 2, we illustrate the system model and formulate the VNF migration problem. Section 3 models the migration problem as MDP, and then we employ DRL-based techniques to solve the optimization problem. Simulation analyses and results are presented in Sect. 4 and conclusions are drew in Sect. 5.

2   System Model and Problem Formulation

Considering a network scenario which is constructed upon SDN/NFV-enabled STIN network infrastructure in which physical resources are provided to services through resource virtualization, as shown in Fig. 1. To simplify the analysis, we split the continuous decision time into time slots t and characterized by $\mathcal{T} = \{1, 2, ..., T\}$. When a service request arrives, the NFVI provides the appropriate NFVI node in terms of the resource requirements of the VNFs.

2.1 Network Model

The substrate network topology is modeled as undirected graph $\mathcal{G}^S = (\mathcal{V}^S, \mathcal{L}^S)$, where $\mathcal{V}^S = \{v_i | i = 1, ..., V\}$, $\mathcal{L}^S = \{l_{ij} | i, j = 1, ..., V, i \neq j\}$ represent the set of substrate nodes and links respectively. Notably, the substrate nodes contain satellite nodes $\mathcal{V}_S^S$ and terrestrial nodes $\mathcal{V}_T^S$ in STIN, i.e., $\mathcal{V}^S = \mathcal{V}_S^S \cup \mathcal{V}_T^S$. The nodes are characterized by computing resource capacity C_i^{max}, and memory capacity M_i^{max}, where $i \in \{1, 2, ..., V\}$. The link l_{ij} between nodes v_i and v_j is characterized by the bandwidth capacity $B_{(i,j)}^{max}$. At the time slot t, the resource occupancy can be denoted to $U^*(t) = \{U_i^C(t), U_i^M(t), U_{ij}^B(t)\}$ and the resource capacity is represented as $\Omega^* = \{C_i^{max}, M_i^{max}, B_{(i,j)}^{max}\}$ respectively, where $*$ represents the types of resource.

We abstract the virtual network as an ordered set $\mathcal{G}^V = \{\mathcal{V}^V, \mathcal{L}^V\}$, where $\mathcal{V}^V$ and $\mathcal{L}^V$ represent the set of VNFs and virtual links respectively. There are Q SFCs in virtual network are denoted as the set $\mathcal{S} = \{S_q | q = 1, 2, ..., Q\}$, and the qth SFC contains N_q types of VNFs, which is denoted to $S_q = \{V_{q,1}, V_{q,2}, ..., V_{q,N_q}\}, \forall V_{q,n} \in \mathcal{V}^V$. Each SFC is formed by diversified VNFs with a particular order and mapped to NFVI in term of resource requirements. Each VNF has a finite computing resource requirement assumed as $c^{q,n}(t)$ and caching resource requirement $m^{q,n}(t)$. The virtual link between adjacent VNFs in any SFC S_q are represented as $l_{(n,n+1)}^q \in \mathcal{L}^V$. Similarly, the finite bandwidth requirement of each virtual link $l_{(n,n+1)}^q$ is denoted by $b^{q,(n,n+1)}(t)$. Generally, SFC S_q has the maximum end-to-end latency limit, which can be set as D_q^{max}.

In order to describe the mapping relationship of VNFs, we define the binary variable $x_i^{q,n}(t) \in \{0, 1\}$, where $x_i^{q,n}(t) = 1$ means that $V_{q,n}$ is mapped into the substrate node v_i and $x_i^{q,n}(t) = 0$ otherwise. The mapping relationship of VNFs will change after each migration occurs. In the same way, we introduce the binary variable $z_{ij}^{q,(n,n+1)}(t) \in \{0, 1\}$ to indicate whether the virtual link $l_{(n,n+1)}^q$ is mapped onto the substrate link l_{ij}, where $z_{ij}^{q,(n,n+1)}(t) = 1$ means that the virtual link $l_{(n,n+1)}^q$ is mapped onto the physical link l_{ij} and $z_{ij}^{q,(n,n+1)}(t) = 0$ otherwise.

2.2 Dynamic Threshold Trigger Mechanism

When the resource occupancy reaches the thresholds, the migration is triggered immediately which can lead to frequent migration and reduce the stability of the system. For solving this problem, this paper firstly proposes a dynamic threshold trigger mechanism to evaluate whether to perform migration immediately at time slot t by Markov chains.

The resource demand interval for arbitrary $V_{q,n}$ is assumed as $[\Phi_{q,n}^{min}, \Phi_{q,n}^{max}]$. The system first allocates resources with $\Phi_{q,n}^{max}$ to each function in the absence

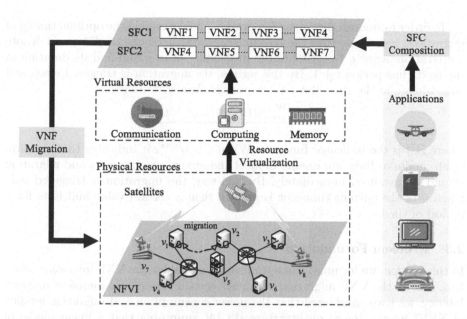

Fig. 1. System model

of historical information and prior knowledge. As the system runs, the future resource demand can be estimated in the short term through Markov chains after obtaining enough simple data of $V_{q,n}$. To model the resource estimation under each time slot t, we divide the resource demand interval $[\Phi_{q,n}^{min}, \Phi_{q,n}^{max}]$ into W sections, and the capacity of resources in each section is $\lambda^{q,n} = \frac{\Phi_{q,n}^{max} - \Phi_{q,n}^{min}}{W}$. Each section represents a different resource demand state, that is, the W sections map to W states of $V_{q,n}$, and the resource capacity under each state is equal to the average value of the section. Thereby, the resource state $\lambda_w^{q,n}(t), w \in 1, 2, ..., W$ of $V_{q,n}$ can be considered as a uniform continuous variation in time slot t. Then, the one-step Markov state transition matrix of $V_{q,n}$ is obtained by the statistical method of

$$\mathcal{P}^{q,n} = [p_{ww'}^{q,n}], w, w' = 1, 2, ..., W, \forall n, q, \tag{1}$$

where $p_{ww'}^{q,n} = \frac{\kappa_w}{\kappa_{ww'}}$ represents as the probability of transition from the state $\lambda_w^{q,n}$ to $\lambda_{w'}^{q,n}$, κ_w represents the times of the demand state maintains in $\lambda_w^{q,n}$, and $\kappa_{ww'}$ represents the times of the state $\lambda_w^{q,n}$ is transferred to $\lambda_{w'}^{q,n}$.

On this basis, the resource demands of $V_{q,n}$ in the time slot $t + 1$ can be estimated from the Chapman-Kolmogorov equation by building a finite-state discrete Markov chain model [12]. The arbitrary possible demand state transition probability can be estimated as

$$\mathcal{F}_W^{q,n} = \mathcal{F}_{W-1}^{q,n}(\mathcal{P}^{q,n}) = ... = \mathcal{F}_0^{q,n}(\mathcal{P}^{q,n})^W, \tag{2}$$

where $\mathcal{F}_0^{q,n}$ represents the initial probability distribution of $V_{q,n}$. Thus, the probability of resource demand at time-slot $t + 1$ can be expressed as $\mathcal{F}_0^{q,n}(\mathcal{P}^{q,n})^W$.

In order to smooth the migration process and determine the optimal timing of migration, we assume that the resource capacity threshold as η^{max}, and denote a migration factor ζ to estimate the probability of overload and its duration in the next time period $t + 1$. By this means, the migration is triggered when and only when ζ in the time slot $t + 1$ is greater than η^{max}, where,

$$\zeta = \frac{\sum_{w=0}^{W} \mathbb{I}(U^*(w) > \eta^{max} \Omega^*)}{W}, \tag{3}$$

where $\mathbb{I}(\cdot)$ is the indicator function. When $\zeta > \eta^{max}$, it indicates that the substrate nodes or links are unable to meet the service requirements and migration should be executed immediately. By this way, the migration is triggered only when ζ of the current time-slot is greater than a certain value and lasts for a period of time.

2.3 Problem Formulation

In this section, we formulate an optimization model for the VNF migration problem. Due to the VNF migration consumes certain energy and increases network latency, we have considered that the power consumption and migration latency of NFVI jointly. Based on literature [13,14], supposing that a linear model of the power consumption versus the traffic handled by the nodes. Hence the power consumption of node v_i at time slot t can be expressed as follows,

$$P_{v_i}(t) = [\delta + (1 - \delta)\frac{c^{q,n}(t)}{\rho}]P^{max}. \tag{4}$$

In formula (4), P^{max} is the node maximum power consumption, and $\delta \in [0, 1]$ is the ratio of the baseline power to the maximum power [5]. We assume that all of the nodes are equipped with identical maximum data computing capacity ρ, which represents the number of CPU cycles that can be processed per second [15]. Hence, the power consumption of SFC S_q at time slot t is given by

$$P_q(t) = \sum_{n=1}^{N_q} x_i^{q,n}(t)[1 - x_i^{q,n}(t+1)]P_{v_i}(t). \tag{5}$$

The migration latency of SFC S_q is related to the location of the SFC virtual link mapped to the substrate link, then the migration latency of SFC S_q at time slot t is the sum of the migration latency of all links, so the migration latency of SFC S_q can be expressed as

$$D_q(t) = \sum_{n=1}^{N_q-1} \frac{m^{q,n}(t)z_{(i,j)}^{q,(n,n+1)}(t)[1 - z_{(i,j)}^{q,(n,n+1)}(t+1)]}{b^{q,(n,n+1)}(t)}. \tag{6}$$

Consequently, the total migration overhead of SFC in time slot t can be normalized and expressed as following,

$$C(t) = \omega_1 \frac{\sum\limits_{q=1}^{Q} P_q(t)}{P^{max}} + \omega_2 \frac{\sum\limits_{q=1}^{Q} D_q(t)}{D_q^{max}}, \tag{7}$$

where ω_1 and ω_2 are the corresponding weights, and $\omega_1 + \omega_2 = 1$. Therefore, the VNF migration problem can be formulated as following,

$$\textbf{P1}: \min_{x,z} \frac{1}{T} \sum_{t=1}^{T} C(t)$$

$$s.t. \ C1 : x_i^{q,n}(t) \in \{0,1\}, \forall i, q, n$$

$$C2 : z_{(i,j)}^{q,(n,n+1)}(t) \in \{0,1\}, \forall i \neq j, q, n,$$

$$C3 : \sum_{i=1}^{V} x_i^{q,n}(t) = 1, \forall i, q, n$$

$$C4 : \sum_{q=1}^{Q} \sum_{n=1}^{N_q} c_{q,n}(t) x_i^{q,n}(t) \leq C_i^{max}, \forall i, q, n$$

$$C5 : \sum_{q=1}^{Q} \sum_{n=1}^{N_q} m_{q,n} x_i^{q,n}(t)(t) \leq M_i^{max}, \forall i, q, n$$

$$C6 : \sum_{q=1}^{Q} \sum_{n=1}^{N_q-1} b_{n,n+1}^q(t)(z_{i,j}^{q,(n,n+1)}(t) + z_{j,i}^{q,(n,n+1)}(t)) \leq B_{(i,j)}^{max}, \forall i \neq j, q, n,$$

$$C7 : \sum_{j} z_{i,j}^{q,(n,n+1)}(t) - \sum_{j} z_{j,i}^{q,(n,n+1)}(t) = x_i^{q,n}(t) - x_i^{q,n+1}(t), \forall i \neq j, q, n,$$

$$C8 : D_q(t) \leq D_q^{max}, \forall q$$

$$(8)$$

The constraint C1 and C2 restrict the variable $x_i^{q,n}(t)$ and $z_{(i,j)}^{q,(n,n+1)}(t)$ to binary choice. The constraint C3 to avoid nodes overload and routing loops. The constraint C4~C6 ensure the computing, caching and bandwidth demand of the VNF deployed in each node cannot exceed the maximum resource capacity. The constraint C7 guarantees that an arbitrary virtual link must exist a continuous path $l_{(i,j)}$ between substrate nodes v_i and v_j to which the adjacent $V_{q,n}$ and $V_{q,n+1}$, and maintains the traffic that inflow must be equal outflow. In order to make sure the link delay between VNF nodes in any SFC S_q satisfies the SFC end-to-end latency requirement, the constraint C8 is proposed.

The optimization problem **P1** with constraints C1 $\sim$ C8 is NP-hard. We can proof as following: we take the optimization problem **P1** within a fixed time slot t into account, which can reduce **P1** into the Multiple-choice Multidimensional Knapsack Problem (MMKP) [16]. Similar to [16], we consider the network resource constraints of substrate nodes and links, migration mapping variables as size vectors, as shown in constraints C1~C5. It's not too difficult to spot that the special case is NP-hard. Moreover, problem **P1** is a dynamic MMKP problem with t varying, so problem **P1** is NP-hard.

3   Migration Decision Based on DRL

In this section, we introduce the theory of MDP firstly. Then model the optimization problem **P1** to MDP and the detailed state, action and reward of MDP in this paper are defined respectively. Furthermore, the service-aware VNF migration algorithm based on Twin Delayed Deep Deterministic Policy Gradient (TD3) is introduced in detail.

3.1   Markov Decision Process

In DCN and mobile communication networks, there are many traditional solutions for the optimization problem of VNF migration, such as exhaustive search and game theory [17,18]. However, these techniques have many drawbacks. On one hand, high algorithm complexity brings about low efficiency. On the other hand, it is hard to apply to large scale network scenarios due to heavy computational resource consumption and low redundant fault tolerance. In this paper, we take RL method to solve the joint optimization problem while avoiding imperfection and abuses of conventional algorithms to some extent.

Our goal is to optimize the migration overhead of VNF migration. Therefore, VNF migration requires a series of decisions to reach the final object, and thus, it is a sequential decision problem, where each migration has an impact on the subsequent one. Traditional mathematical modeling methods are difficult to accurately model the process, and therefore, solving it by traditional methods will make the problem extremely complex. MDPs is a classical formalization of sequential decision making, where actions influence not just immediate rewards, but also subsequent states, and through those future rewards. Reinforcement learning is an effective class of methods used to solve MDPs.

We model the VNF migration as the MDP to describe migration process firstly. The transition process of VNFs on NFV nodes for each time slot t, described as a quadruple (S, A, P, R):

- **States S**: Each state $s(t) \in S$ at time-slot t as $s(t) = \{v(t), c(t), m(t), b(t)|v \in \mathbb{V}, c \in \mathbb{C}, m \in \mathbb{M}, b \in \mathbb{B}\}$, where $\mathbb{V}$ denotes the substrate network topology space and $\mathbb{C}, \mathbb{M}, \mathbb{B}$ state the computing, caching and bandwidth demands state spaces of the VNFs, respectively.
- **Action A**: $a(t) = \{x(t), z(t)|x \in \mathbb{X}, z \in \mathbb{Z}\}$, where $x(t) = \{x_i^{q,n}(t)\}$, $z(t) = \{z_{ij}^{q,(n,n+1)}\}, \forall i, j, q, n$, $a(t) \in A$, $\mathbb{X}$ and $\mathbb{Z}$ represent the mapping action space of the VNFs and virtual links, respectively.
- **Transition probability P**: After taking action $a(t)$ in time slot t, state $s(t)$ will be transferred to the next state $s(t+1)$ with state transition probability $p(s(t+1)|s(t), a(t))$.
- **Reward R**: In time slot t, the DRL agent can obtain a reward based on the current state and selected action. In this paper, our optimization goals is to minimize migration overhead, so we define the additive inverse of migration overhead as the immediate reward for time slot t, i.e., $r(t) = -C(t)$.

The cumulative discount reward at time slot t is defined as

$$R_\pi(t) = \lim_{T \to +\infty} E_\pi\{\sum_{t=1}^{T} \gamma^{t-1} r(t)\},\tag{9}$$

where π denotes the policy of taking action $a(t)$ at state $s(t)$. Agent in state $s(t)$ will takes an action $a(t) = \pi(s(t))$ to move the next state $s(t+1)$ according to policy π. The performance of the policy at time slot t is evaluated by the action value function $Q_\pi(s(t), a(t))$. Under the policy π, the expected reward for a given action and state can be expressed as

$$Q_\pi(s(t), a(t)) = E_a \sim_\pi [R_\pi(t)|s(t), a(t)].\tag{10}$$

Generally, $Q_\pi(s(t), a(t))$ can be iteratively expressed through the Bellman equation,

$$Q_\pi(s(t), a(t)) = E_a \sim_\pi [r(t) + \gamma Q_\pi(s(t+1), a(t+1))].\tag{11}$$

There must exist a stable optimal policy that maximizes the reward when $T \to +\infty$, so the optimal VNF migration policy π^* can be expressed as

$$\pi^* = \arg \max_a Q_\pi(s(t), a(t)).\tag{12}$$

In order to obtain the optimal policy π^*, the RL agent needs to select actions in each state to maximize the $Q_\pi(s(t), a(t))$.

3.2 Migration Scheme Based on DRL

According to the analysis in Sect. 3.1, the optimal policy π^* is obtained from the optimal action of different states of equation (12). Since the service requests as well as data flow arrivals in real-DCNs are random, the optimal policy cannot be obtained by iterating the Bellman equation for the action-value function. In this paper, we propose a service-aware VNF migration optimization scheme based on TD3 to solve the problem **P1**. The detailed scheme framework is described as Fig. 2.

Given the system state $s(t)$ at time slot t, the actor primary network selects a deterministic action $a(t) = \pi(s(t)|\phi) + \mathcal{N}_t$ according to the policy π, where ϕ is a parameter of the actor primary network, $\mathcal{N}_t$ is the OU noise introduced to increase the exploratory. After taking $a(t)$ in the network, the agent return the feedback of the immediate reward $r(t)$ and new state $s(t+1)$ to the actor network, and stores the transition tuple (s, a, r, s') to the experience replay pool $\mathcal{B}$, which will be used in the training processes are stated in the following.

Critic Network Training and Learning. The critic network evaluates the performance of the policy by minimizing the loss function $L(\theta_i)$, i.e.,

$$\min L(\theta_i) = E[(y_i - Q_{\theta_i}(s(t), a(t)|\theta_i))^2], i = 1, 2,\tag{13}$$

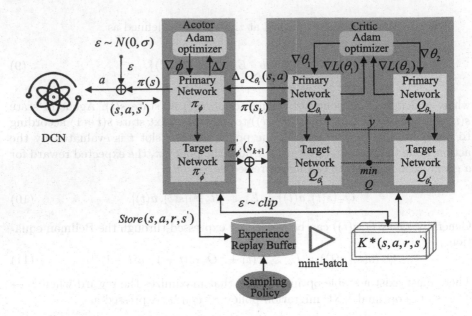

Fig. 2. The framework of our proposed scheme

where $y_i = r + \gamma min[Q_{\theta_i}(s'), \pi'_{\phi'}(s' \mid \phi') \mid \theta')]$ is the target Q-value after the target action $\tilde{a}(t+1) = \pi'_{\phi'}(s(t+1)|\phi')$ is taken in state s'. Notice that the target Q-value is obtained from the target network with parameters $\pi'_{\phi'}$ and $Q'_{\theta'_i}$. The minimum of $L(\theta_i)$ solving by stochastic gradient descent (SGD), and the gradient of $L(s(t), a(t))$ in relation to θ_i is calculated as,

$$\nabla_{\theta_i} L(s(t), a(t)) = E[2(y - Q_{\theta_i}(s(t), a(t) \mid \theta_i))\nabla Q_{\theta_i}(s(t), a(t))], i = 1, 2, \quad (14)$$

where $\nabla Q_{\theta_i}(s(t), a(t))$ is calculated by the chain-rule from the output return to each critic network parameter with θ_i. We observed that in formula (14) that $[y - Q_{\theta_i}(s(t), a(t) \mid \theta_i)]$ that is the TD-error actually. In fact, we usually randomly extract $K * (s_k, a_k, r_k, s_{k+1})$ mini-batch samples from the experience replay pool $\mathcal{B}$ to train the critic network with SGD, i.e.,

$$\theta_i = \theta_i - \frac{\alpha}{K} \sum_{k=1}^{N} [2 \cdot (y_k - Q_{\theta_i}(s_k, a_k \mid \theta_i))\nabla Q_{\theta_i}(s_k, a_k)], i = 1, 2, \quad (15)$$

where α states the learning rate of the critic network, and

$$y_k = r_k + \gamma \min_{i=1,2} Q_{\theta_i}(s_{k+1}, \pi'_{\phi'}(s_{k+1}|\phi')|\theta'_i). \quad (16)$$

Actor Network Training and Learning. The goal of the actor network is to obtain a policy that maximizes discount rewards, i.e.,

$$\max_{\pi} R_{\phi}(t) = E[(a^*(t) - a(t))^2], \quad (17)$$

where $a(t) = \pi_\phi(s(t)|\phi)$, $a^*(t)$ denotes the optimal action. The parameter ϕ of the actor network is updated using the deterministic policy gradient (DPG) maximizing Q value of the outputted action [19]. To this end, the gradient form the critic network concerning the actor network's output action $\tilde{a}(t) = \pi(s(t)|\theta_\pi)$, namely, $\Delta_{\tilde{a}} Q(s(t), \tilde{a}(t))$. The gradient of actor network can written as,

$$\Delta_\phi Q(s(t), a(t)) = \Delta_a \left(Q_{\theta_i}(s(t), a(t))|_{a=\pi_\theta(s)}\right) \Delta_\phi \pi_\theta(s(t)), i = 1, 2, \qquad (18)$$

where $\Delta_\phi \pi_\theta(s(t))$ is derived by the chain-rule. The parameters ϕ is updated by first training the DNN parameters with a randomly sampling from the pool $\mathcal{B}$, and then iteratively updating the policy with the goal of maximizing the total discounted reward. According to the literature [20], the stochastic gradient policy is the gradient of the policy performance, which is equivalent to the empirical DPG, i.e.,

$$\nabla J(\phi) = \sum_S d(S) \sum_A \nabla Q_{\theta_i}(s(t), a(t)) \nabla \pi_{\theta_i}(s(t)), i = 1, 2, \qquad (19)$$

where $d(S)$ states the state distribution, and $\nabla Q_{\theta_i}(s(t), a(t))$ is approximated by the critic primary network and can be transformed into

$$\nabla_\phi J(\phi) \approx \nabla_\phi Q^* = E_\pi \nabla_a (Q_{\theta_i}(s(t), a(t))|_{a=\pi_\phi(s(t))}) \Delta_\phi \pi_\phi(s(t)), i = 1, 2. \qquad (20)$$

At each training step, by updating the DNN parameters along the direction of gradient ascent of the target policy function, the parameter ϕ is updated by randomly extract $K * (s_k, a_k, r_k, s_{k+1})$ samples from the $\mathcal{B}$, that is,

$$\phi = \phi + \frac{\beta}{K} \nabla_a (Q_{\theta_i}(s_k, a_k)|_{a_k=\pi_\phi(s_k)}) \Delta_\phi \pi_\phi(s_k), i = 1, 2, \qquad (21)$$

where β represents the learning rate of the actor network. Noticed that in the training of the actor network, only the state s_k of the pool $\mathcal{B}$ and the actions generated by the actor network for that state are required. According to the literature [21], direct updating of the network likely causes network instability. Therefore, the target network parameters are updated by Polyak Averaging, i.e.,

$$\begin{aligned} \phi' &= \tau\phi + (1-\tau)\phi' \\ \theta_i' &= \tau\theta_i + (1-\tau)\theta_i', i = 1, 2, \end{aligned} \qquad (22)$$

where τ controls the magnitude of the update, and reflects update stability. The details of the proposed algorithm are illustrated in Algorithm 1.

4   Simulation Results

To evaluate the effectiveness and reliability of the proposed scheme in this paper, simulation validations of the proposed scheme and performance comparison with previous work are performed. We compared the FDQ scheme in [22], our proposed scheme based on the dynamic threshold trigger mechanism called Sa-VNFM and the method based on TD3 (TD3-based). The performance comparison has been carried out on a computer characterized by 2.50 GHz Intel(R)

Algorithm 1: Service-aware VNF migration algorithm (Sa-VNFM)

Input: policy update episode T, policy frequency d, experience replay buffer $\mathcal{B}$ capacity B, sampling size K, discount factor γ, soft update factor τ, learn rate α, β

Output: policy π^*

1 **Initialization:** Set the parameters of actor's online network ϕ and critic's online network θ_1, θ_2 randomly, and set the target network parameters by $\phi' \leftarrow \phi, \theta_1' \leftarrow \theta_1, \theta_2' \leftarrow \theta_2$. Set experience replay buffer $\mathcal{B}$ capacity $B = 0$;

2 **for** $t = 1$ *to* T **do**

3   Select action with exploration noise $a_t \sim \pi_\phi(s) + \epsilon, \epsilon \sim \mathcal{N}(0, \sigma)$

4   **if** *constraints C3~C8 and $\zeta > \eta^{max}$ are satisfied* **then**

5   Agent execute a_t from the simulation environment, and observe the immediate reward r_t and the new state s_{t+1}

6   **if** *experience replay buffer $\mathcal{B}$ not exceed capacity B* **then**

7   Store the transition tuple (s_k, a_k, r_k, s_{k+1}) into $\mathcal{B}$ as the train sample for the online network

8   **else**

9   Randomly replace any tuple in $\mathcal{B}$

10   **end**

11   Randomly sample K transition tuples as mini-batch training data for training and learning

12   According to (13)-(14), calculate the gradient $\nabla_{\theta_i} L$ of critic's online networks

13   Update the parameters θ_i of critic's online network with the Adam optimizer by (15)

14   **if** t *mode* d **then**

15   According to (18)-(20), calculate the policy gradient $\Delta_\phi J(\phi)$ of actor's online network

16   Update the parameters ϕ of critic's online network with the Adam optimizer by (21)

17   Soft update the parameters $\phi', \theta_1', \theta_2'$ of target network by (22)

18   **end**

19   **end**

20 **end**

Core(TM) i5-7200U CPU processor and by an 8 GB memory, and the simulation tool we used the Pycharm integrated development environment with TensorFlow 1.14.

4.1 Experiment Settings

Due to the high computational complexity of the optimal problem, we consider a fully connected DCN scenario, which contains 11 generic terrestrial nodes and 1 satellite node. The computing capacity and memory capacity of all the substrate node is uniformly distributed in $[100, 300]$ MIPS and $[150, 300]$ MB respectively, and the bandwidth capacity of each link is generated randomly in the $[50, 100]$ Mbps.

Table 1. Main simulation parameters

Parameters	Value
Number of satellite node	1
Number of terrestrial nodes	11
Node computing capacity ($MIPS$)	Uniform[100, 300]
Node memory capacity (MB)	Uniform[150, 300]
Link bandwidth capacity ($Mbps$)	Uniform[50, 100]
VNF computing requirement ($MIPS$)	[10, 20]
VNF caching requirement (MB)	[10, 30]
Virtual link bandwidth requirement ($Mbps$)	[1, 10]
SFC maximum time latency limit (ms)	50
Maximum data computing capacity ($MIPS$)	20
Node maximum power (W)	200
Experience replay pool capacity	10000
Learning rate α, β	0.002
Discount factor	0.99
Soft update factor	0.001
Mini-batch	32

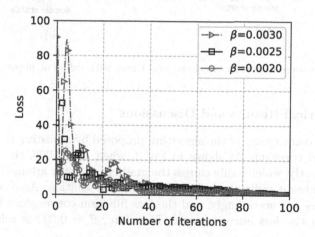

Fig. 3. Convergence of the proposed scheme.

In order to simulate the real environment, the resource demand of SFCs is simulated by using a Poisson process to accord with the characteristics of periodic and bursty real data flows. Assuming each of SFC consists of different types of VNFs conforming to a uniform distribution of [3, 6]. The relevant simulation parameters are shown in Table 1.

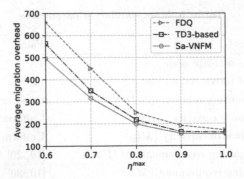

Fig. 4. Migration overhead versus η^{max}.

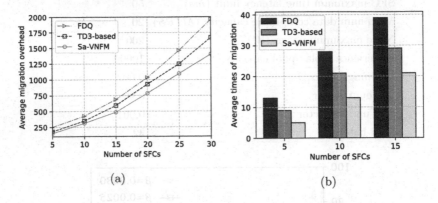

Fig. 5. Average migration overhead and times with different network setups.

4.2 Numerical Results and Discussions

To verify the convergence of the algorithm proposed by adjusting the β and while keeping the α constant. As shown in Fig. 3, when $\beta = 0.003$, the loss function fluctuates greatly, which easily causes the gradient to hover around the minimum value and makes it difficult to obtain the optimal solution. As β decreases, the jitter becomes less accordingly, and the loss function convergence rate becomes slow, making the loss curve smooth. Therefore, $\beta = 0.002$ is selected in this paper.

The migration overhead versus different thresholds is shown in Fig. 4. In order to analyze more specifically the effect of threshold setting on migration overhead, the differences between computing, memory and bandwidth resource requirements and the diversity in service performance between substrate nodes are temporarily masked. We fixed 5 SFCs in network. i.e., by comparing the service performance at the same threshold. As can be seen from the Fig. 4, the migration overhead gradually decreases as the η^{max} increases. This is because the larger the η^{max}, the less likely the resource demand will exceed the η^{max}, and therefore the fewer migrations need to be performed, incurring less overhead.

It is necessary to discuss the migration overhead of our proposed scheme. As shown in Fig. 5a, we can observe that the migration overhead increases consistently as the number of SFCs increases. Under the same service requests, the total migration overhead of Sa-VNFM is always lower than that of FDQ. Meanwhile, it can be seen that the overhead of Sa-VNFM lower than TD3-based, that is the former can further reduce overhead by adopting a dynamic threshold trigger mechanism compared with the latter. Fig. 5b shows the times of migration is triggered under different algorithms. Compared with TD3-based and FDQ, we notice that Sa-VNFM can reduces the times of migration respectively, which is because the Sa-VNFM can reasonably determine migration timing to reduce migration times and improve the stability of the system.

5   Conclusions

In this paper, we studied the VNF migration optimization problem in SDN/NFV-enabled STIN where VNFs are deployed in DCN. In order to avoid frequent migration, we firstly propose a dynamic threshold trigger mechanism and then formulate the problem as a stochastic optimization model based on MDP aiming at minimizing the migration overhead. For solving this problem efficiently, we resort to the DRL-Based service-aware VNF migration scheme to obtain the feasible solution. The simulation results show that the proposed scheme can effectively avoid frequent migration and reduce the migration overhead.

References

1. Yousaf, F.Z., Taleb, T.: Fine-grained resource-aware virtual network function management for 5G carrier cloud. IEEE Netw. **30**(2), 110–115 (2016)
2. Yang, P., Xiao, Y., Xiao, M., Li, S.: 6G wireless communications: vision and potential techniques. IEEE Netw. **33**(4), 70–75 (2019)
3. Sun, P., Guo, Z., Liu, S., Lan, J., Wang, J., Hu, Y.: SmartFCT: improving power-efficiency for data center networks with deep reinforcement learning. Comput. Netw. **179**, 107255 (2020)
4. Liu, J., Shi, Y., Fadlullah, Z.M., Kato, N.: Space-air-ground integrated network: a survey. IEEE Commun. Surv. Tutor. **20**(4), 2714–2741 (2018)
5. Eramo, V., Miucci, E., Ammar, M., Lavacca, F.G.: An approach for service function chain routing and virtual function network instance migration in network function virtualization architectures. IEEE/ACM Trans. Netw. **25**(4), 2008–2025 (2017)
6. Sarrigiannis, I., Ramantas, K., Kartsakli, E., Mekikis, P., Antonopoulos, A., Verikoukis, C.: Online VNF lifecycle management in an MEC-enabled 5G IoT architecture. IEEE Internet Things J. **7**(5), 4183–4194 (2020)
7. Cho, D., Taheri, J., Zomaya, A.Y., Bouvry, P.: Real-time virtual network function (VNF) migration toward low network latency in cloud environments. In: 2017 IEEE 10th International Conference on Cloud Computing (CLOUD), pp. 798–801 (2017)
8. Ben Jemaa, F., Pujolle, G., Pariente, M.: Analytical Models for QoS-driven VNF placement and provisioning in wireless carrier cloud. In: 19th ACM International Conference on Modeling. Analysis and Simulation of Wireless and Mobile Systems, pp. 148–155. ACM Press, Malta (2016)

9. Tang, L., He, X., Zhao, P., Zhao, G., Zhou, Y., Chen, Q.: Virtual network function migration based on dynamic resource requirements prediction. IEEE Access **7**, 112348–112362 (2019)
10. Ahmed, T., Alleg, A., Ferrus, R., Riggio, R.: On-demand network slicing using SDN/NFV-enabled satellite ground segment systems. In: 2018 4th IEEE Conference on Network Softwarization and Workshops (NetSoft), pp. 242–246 (2018)
11. Guo, Z., et al.: AggreFlow: achieving power efficiency, load balancing, and quality of service in data center networks. IEEE/ACM Trans. Netw. **29**(1), 17–33 (2021)
12. Ross, S.M.: Stochastic Processes, vol. 2. Wiley, New York (1996)
13. Li, B., Cheng, B., Liu, X., Wang, M., Yue, Y., Chen, J.: Joint resource optimization and delay-aware virtual network function migration in data center networks. IEEE Trans. Netw. Serv. Manag. 1 (2021)
14. Liu, Y., Feng, G., Chen, Z., Qin, S., Zhao, G.: Network function migration in softwarization based networks with mobile edge computing. In: ICC 2020–2020 IEEE International Conference on Communications (ICC), pp. 1–6 (2020)
15. Aroca, J.A., Chatzipapas, A., Anta, A.F., Mancuso, V.: A measurement-based characterization of the energy consumption in data center servers. IEEE J. Sel. Areas Commun. **33**(12), 2863–2877 (2015)
16. Islam, M.I., Akbar, M.M.: Heuristic algorithm of the multiple-choice multidimensional knapsack problem (MMKP) for cluster computing. In: 2009 12th International Conference on Computers and Information Technology, pp. 157–161 (2009)
17. Xia, J., Cai, Z., Xu, M.: Optimized virtual network functions migration for NFV. In: 2016 IEEE 22nd International Conference on Parallel and Distributed Systems (ICPADS), pp. 340–346 (2016)
18. Leivadeas, A., Kesidis, G., Falkner, M., Lambadaris, I.: A graph partitioning game theoretical approach for the VNF service chaining problem. IEEE Trans. Netw. Serv. Manag. **14**(4), 890–903 (2017)
19. Fujimoto, S., van Hoof, H., Meger, D.: Addressing function approximation error in actor-critic methods. In: Proceedings of the 35th International Conference on Machine Learning, vol. 80, pp. 1587–1596 (2018)
20. Silver, D., Lever, G., Heess, N., Degris, T., Wierstra, D., Riedmiller, M.: Deterministic policy gradient algorithms. In: 31st International Conference on Machine Learning, ICML 2014 (2014)
21. Pujol Roig, J., Gutierrez-Estevez, D., Gündüz, D.: Management and orchestration of virtual network functions via deep reinforcement learning. IEEE J. Sel. Areas Commun. **38**, 304–317 (2020)
22. Yao, J., Chen, M.: A flexible deployment scheme for virtual network function based on reinforcement learning. In: 2020 IEEE 6th International Conference on Computer and Communications (ICCC), pp. 1505–1510 (2020)

Dual-Channel Speech Enhancement Using Neural Network Adaptive Beamforming

Tao Jiang[1(✉)], Hongqing Liu[1], Chenhao Shuai[1], Mingtian Wang[1], Yi Zhou[1], and Lu Gan[2]

[1] School of Communication and Information Engineering,
Chongqing University of Posts and Telecommunications, Chongqing, China
s190101065@stu.cqupt.edu.cn
[2] College of Engineering, Design and Physical Science, Brunel University,
London UB8 3PH, UK

Abstract. Dual-channel speech enhancement based on traditional beamforming is difficult to effectively suppress noise. In recent years, it is promising to replace beamforming with a neural network that learns spectral characteristic. This paper proposes a neural network adaptive beamforming end-to-end dual-channel model for speech enhancement task. First, the LSTM layer is used to directly process the original speech waveform to estimate the time-domain beamforming filter coefficients of each channel and convolve and sum it with the input speech. Second, we modified a fully-convolutional time-domain audio separation network (Conv-TasNet) into a network suitable for speech enhancement which is called Denoising-TasNet to further enhance the output of the beamforming. The experimental results show that the proposed method is better than convolutional recurrent network (CRN) model and several popular noise reduction methods.

Keywords: Neural network · Dual-channel · Speech enhancement

1 Introduction

Speech enhancement algorithm is an important technology to process speech in noisy environments to improve speech quality and intelligibility [1]. In terms of number of channels available, speech enhancements are divided into multi-channel and single-channel cases. Because multi-channel methods exploit spatial information, their performances are often better than that of the singlechannel methods. Beamforming is often used in multi-channel speech enhancement.

The traditional beamformer needs a set of beamforming filters that are convolved with signals from each channel before summation. The filters are designed

Supplementary Information The online version contains supplementary material available at https://doi.org/10.1007/978-3-030-99200-2_37.

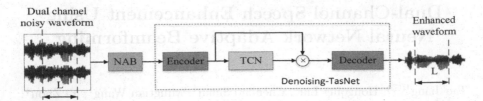

Fig. 1. NABDTN block diagram.

to enhance speech and suppress interference noise. The famous traditional beam-forming approaches include minimum variance distortionless response (MVDR) [2] and its generalized sidelobe canceller (GSC) formulation [3], which minimizes the energy of the signal at the beamformer output. The beamformer generally requires a steering vector that points beamformer to the target direction [4]. Generally speaking, the method of estimating the steering vector is direction of arrival (DOA) estimation [5]. However, the performance of the DOA algorithms is usually affected in complex acoustic environments, resulting in an inaccurate steering vector estimation, which in turn degrades the performance of beam-forming.

In recent years, neural networks have been used to replace traditional beam-forming methods and the results are very promising. It not only makes up for the shortcomings of beamforming that cannot be applied to complex acous-tic environments, but also has a better performance than beamforming. The model based on neural network beamforming has achieved good results in multi-channel speech separation [6,7] and speech recognition [8,9]. Recently, beam-forming based on DNNs [10] has been applied in the field of multi-channel speech enhancement. The amplitude spectrum of the short-time Fourier transform of the multi-channel speech signal is used as the input of the neural network, and the beamforming filter coefficients are calculated by estimating the covariance matrix of the target signal and the noise. The realization principle is complicated, and a simple end-to-end system is of importance.

In this work, we propose adaptive neural network beamforming (NAB) and Denoising-TasNet model (NABDTN) for speech enhancement in Fig. 1. The pur-pose of NAB is to estimate a series of beamforming spatial-temporal filters, and then they are respectively convolved the speech signals of microphones before summing. This network mimics delay-and-sum (DS) technology [11]. However, NAB does not need to estimate the time delay of each microphone, which is difficult to accurately estimate in a complex acoustic environment. After that, the Denoising-TasNet network is developed to further enhance the performance by minimizing scale-invariant signal-to-distortion ratio (SI-SDR). This means that the time delay can be learned implicitly by the neural network. There-fore, the model has a strong applicability for different complex acoustic scenar-ios. The temporal convolutional network (TCN) in Denoising-TasNet has strong time-domain sequence modeling capability[12], so we use this network to further enhance the NAB output to generate clean speech.

2   Algorithm Description

2.1   Problem formulation

Let $x_c(k)$, $s_c(k)$ and $n_c(k)$ denote noisy speech, clean speech and background noise, respectively, and $c \in \{0, 1, \cdots, C-1\}$ is the index of the channel, where C is the number of channels. It should be noted that $c = 0$ is the reference channel, and its corresponding clean speech is the label during network training. The noisy speech is written as

$$x_0(k) = s_0(k) + n_0(k), \tag{1}$$

$$x_c(k) = s_c(k) + n_c(k) = s_0(k) * h_{0c}(k) + n_c(k), \tag{2}$$

where $s_0(k)$ denotes target clean speech which can be divided into overlapping segments of length L, and $k \in \{0, 1, ..., T\}$ denotes the total number of segments in the input, and * denotes the convolution operation. The impulse response of clean speech between channels is represented by $h_{0c}(k)$. In a short-distance call scenario, the distance between the main microphone and the mouth is small, and we take the clean speech on the main channel as the target clean speech.

The general finite impulse response (FIR) filter beamformer is

$$y(t) = \sum_{c=0}^{C-1} \sum_{n=0}^{N-1} h_c[n] x_c[t - n - \tau_c], \tag{3}$$

where $h_c[n]$ is the n-th tap of the beamforming filter related to microphone c, $x_c[t]$ is the noisy speech received by microphone c at time t, and τ_c is the alignment steering delay between the speech received by another microphone and the speech received by the reference microphone, $y(t)$ is the output of beamforming speech, and N is the length of the filter.

Since the target speech arrives at each microphone at a different time, the speech in each microphone needs to be aligned with the speech of the reference microphone to perform traditional beamforming. The estimation accuracy of steering delay τ_c presents a great influence on the performance of speech enhancement. However, the proposed NABDTN model estimates the filter coefficients by minimizing the loss function of the enhanced speech and the clean speech. Therefore, steering delay estimation of τ_c is hidden in the estimated filter coefficients. The output result of the kth frame of the NAB model is

$$y(k)[t] = \sum_{c=0}^{C-1} \sum_{n=0}^{N-1} h_c(k)[n] x_c(k)[t - n], \tag{4}$$

where $h_c(k)[t]$ is the estimated filter coefficient of channel c in the kth frame. To estimate $h_c(k)[t]$, we jointly train the entire NABDTN network to predict the filter coefficients of each channel. The beamforming filter coefficients can be continuously adjusted according to the dataset during the training process. The more complex the dataset, the wider the adaptability of the model. This is similar to traditional adaptive beamforming, which adjusts filter coefficients according to changes in data.

500   T. Jiang et al.

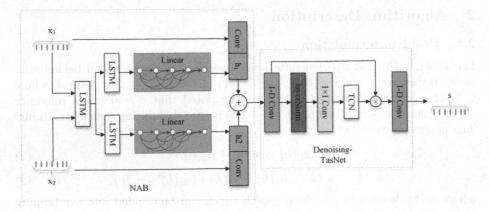

Fig. 2. Illustration of the NABDTN architecture.

2.2 NABDTN architecture

The frame structure of NABDTN containing two parts is shown in Fig. 2, where
two-channel case is illustrated. The NAB model consists of three LSTMs and
two linear layers, where one 512-cell LSTM layer and two 256-cell LSTM layers
are utilized. In our experiments, the segment length of the input noisy speech
$L = 16100$, and we set the output of the linear layer as $N = 26$, which is the
length of the filter coefficient. The hyperparameters are determined by measuring
the PESQ metric of the trained model. To convolve the noisy speech of each
frame with the filter coefficients, we use a kernel size of 1×26 to implement
the convolution of $x_c(k)[t]$ and $h_c(k)[t]$ to efficiently calculate the convolution
operation of multiple batches of data. It should be noted that there are two
channels of noisy speech in each frame. We respectively input the speech of each
channel into the 512-cell LSTM, and the two outputs are generated as the inputs
for two 256-cell LSTMs.

For the model of the Denoising-TasNet, we employ the similar structure of
Conv-TasNet[12], which is the encoder/TCN/decoder pipeline. The amplitude
of each frequency in the spectrogram reflects the energy of the frequency component in the frequency domain processing. However, the amplitude of a sample
point in the time domain does not provide much information, and it needs to be
combined with adjacent samples to represent specific frequency components [13].
For example, if it is a high-frequency signal, and then in time, the amplitude of
adjacent samples will vary greatly, and vice versa. Therefore, when processing
time-domain waveforms, in order to allow the neural network to better learn the
features provided by adjacent samples, we use one-dimensional (1-D) convolutional blocks with dilated convolution factors in the TCN network. As shown
in Fig. 3, we repeat X 1-D convolution blocks with the dilated convolution factor $d = \{1, 2, 4, ..., 2^{M-1}\}$ for R times, and the size of kernel is P. In addition,
TCN estimation is no longer the original speech separation mask, but a speech
enhancement mask so that the model has only one enhanced speech output. In
order to avoid gradient exploding or vanish, each 1-D convolutional block in

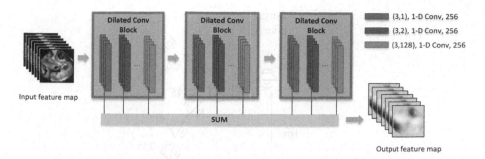

Fig. 3. Architecture of TCN. The parameters are denoted as follows: (p1, p2) Conv p3, where p1 is the kernel size, p2 is the dilated rate, and p3 is the number of filters.

TCN normalizes the input features. The normalization process is

$$gLN(F) = \frac{F - E[F]}{\sqrt{Var[F] + \epsilon}} \odot \gamma + \beta, \tag{5}$$

$$E[F] = \frac{1}{NT} \sum_{NT} F, \tag{6}$$

$$Var[F] = \frac{1}{NT} \sum_{NT} (F - E[F])^2, \tag{7}$$

where $F \in \mathbb{R}^{K \times T}$ is the input feature, ϵ is a small constant to prevent the denominator from being zero, $\gamma, \beta \in \mathbb{R}^{K \times 1}$ are trainable parameters, $E[F]$ and $Var[F]$ are the mean and variance of F, respectively, and $\odot$ denotes element-wise multiplication.

2.3 Loss functions

The studies show that, for speech enhancement in the time domain, the SI-SDR as the loss function presents a superior performance, and it is defined by [14]

$$\text{SI-SDR} = 10 \log_{10} \frac{\|\beta s\|^2}{\|\beta s - \hat{s}\|^2}, \tag{8}$$

$$\beta = \frac{\hat{s}^T s}{\|s\|^2} = \arg\min_{\beta} \|\beta s - \hat{s}\|^2, \tag{9}$$

where $s \in \mathbb{R}^{1 \times T}$ and $\hat{s} \in \mathbb{R}^{1 \times T}$ are the clean speech and enhanced speech, respectively. The scaling of the target speech s ensures that the SI-SDR measure is invariant to the scale of $\hat{s}$. It is seen from Eqs. (8) and Eqs. (9), that SI-SDR is simply the signal-to-noise(SNR) ratio between the weighted clean speech signal defined as $\|\beta s\|^2$ and the residual noise defined as $\|\beta s - \hat{s}\|^2$. The network maximizes the correlation between s and $\hat{s}$ by maximizing SI-SDR, so that the model obtains a higher SNR.

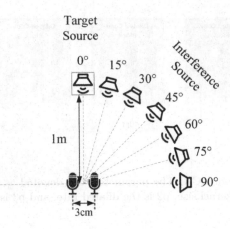

Fig. 4. Experimental settings.

3   Experimental Results

3.1   Setups

Figure 4 shows the position relationship between the microphone array and the target sound source and interference source in our experiments. In this experiment, the number of microphones is $C = 2$. The location of the target sound source is 1 m in the front of the microphone array that has a space interval of 3 cm. Based on this setting, the azimuth angle of our target source is 0°. We set an interference source at a distance of 1 m from the microphone array. To cover all the angles, the direction angle of interference ranges from −90° to 90° with an interval of 15°, and the similar setting is also utilized in [15].

To generate the dataset, we use 10,000 speech and 9,100 noise utterances from the MS-SNSD dataset, in which each speech and noise durations are 31 s and 10 s, respectively, and the sampling rate is 16 kHz. We randomly select 8500 speech and 8000 noise utterances from the dataset to generate a 200 h training set, and then select 1500 speech and 1100 noise utterances from the remaining dataset to generate a 20 h validation set, and finally the cafeteria, street and babble noises are used to generate test set. The image method [16] is used to generate a dual-channel dataset. In order to test the denoising ability of the NABDTN model and reduce the impact of the room impulse response, we use the clean speech generated by the reference microphone as our target source, and the room size is 10 m × 7 m × 3 m.

For interference sources with different azimuth angles, we also have produced training and validation sets with different signal-to-noise ratios (SNRs) ranging from −5 dB to 5 dB at an interval of 1 dB, and produced a test set of {−5 dB, 0 dB, 5 dB}. Our training set includes 120000 mixtures that are segmented into 16100 sample points as the input of the model.

3.2   Results

In the experiments, we use two objective metrics of short-time objective intelligibility (STOI) and perceptual evaluation of speech quality (PESQ)[14] to measure the denoising performance of different models. In addition, we also use the subjective metric of deep noise suppression mean opinion score (DNSMOS) [17] to evaluate enhanced speech. We randomly select 180 speech and noise files from the test set to average the results. The interference sources of the test dataset produced by the image method include three directions of 15°, 45°, and 90° and gradually deviate from the target source, so as to avoid the same distribution with the training dataset. The STOI and PESQ of the noisy and enhanced speeches at −5 dB, 0 dB, and 5 dB are measured, respectively. As shown in Table 1, compared with the MMSE-based approach [18], dual-microphone DNN speech enhancement [19], CRN model [20] and MFMVDR model[21] the proposed method significantly improve the performance. For example, at SNR = 0 dB, the NABDTN method increased STOI by 12.7% and PESQ by 1.08, whereas the MFMVDR only improved STOI by 10.42% and PESQ by 1.02.

Table 1. STOIs and PESQs of different methods. The value is an average of 15°, 45°, and 90° interference sources.

Method	STOI (%)			PESQ		
SNR	−5 dB	0 dB	5 dB	−5 dB	0 dB	5 dB
noisy	71.13	80.10	87.80	1.48	1.81	2.19
MMSE	70.54	79.13	84.62	1.48	1.90	2.27
DNN	79.80	84.67	90.21	2.03	2.45	2.74
CRN	81.02	89.06	91.87	2.23	2.62	2.85
MFMVDR	84.73	90.52	93.34	2.45	2.83	3.08
Prop.	**88.13**	**92.89**	**96.32**	**2.58**	**2.89**	**3.18**

As shown in Fig. 5, we choose a section of noisy containing stationary noise and non-stationary noise to analyze the noise reduction performance from the spectrogram. Comparing the noisy and enhanced spectrograms, it can be found that the stationary noise and non-stationary noise marked by the red rectangular box have been removed. From the blue and white rectangles marked by enhanced and clean spectrograms, there are only slight distortions in the unvoiced audio at low and high frequencies, while the voiced distortion is even smaller.

In Table 1, the results are an average of interferences at three directions. However, interference at each angle may contribute to the system performance differently. To demonstrate this effect, we also listed the denoised performance of the NABDTN model at different azimuths. As shown in Table 2, the SNRs of the mixed speech in each direction include −5 dB, 0 dB, and 5 dB, which indicates STOI and PESQ are an average of different SNRs. We found that system

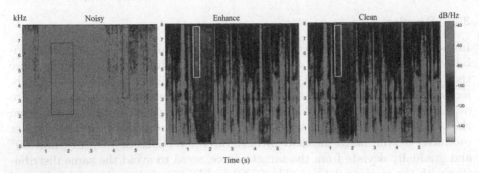

Fig. 5. Example of noise suppression for stationary noise and non-stationary noise at 5 dB SNR. The ordinate and abscissa represent frequency and time respectively. The color bar on the right represents power/frequency.

Table 2. The performance of the proposed method when interference sources at different directions, where STOI and PESQ are the average of −5 dB, 0 dB, and 5 dB.

Metrics	Interference direction					
	15°	30°	45°	60°	75°	90°
Noisy STOI			80.52			
Noisy PESQ			1.86			
Noisy DNSMOS			2.76			
STOI(%)	91.35	94.86	98.69	95.64	96.82	92.65
PESQ	2.67	3.15	3.68	3.20	3.35	2.74
DNSMOS	3.12	3.23	3.50	3.34	3.51	2.90

performance increases first and then decreases as the azimuth of the interference source varies. Nonetheless, the average PESQ and STOI of noisy have been greatly improved, which means the NABDTN model is not greatly affected by the azimuth of the interference source. Therefore, the proposed model is more adaptable to complex acoustic scenes. In addition, since the objective measurement indicators STOI and PESQ cannot fully reflect the quality of human auditory perception, we also use DNSMOS provided by DNS-Challenge as a performance measure as shown in the last row of Table 2.

4   Conclusion

In this study, we proposed a neural network adaptive beamforming (NAB) and developed Conv-TasNet structure for dual-channel speech noise reduction. The Conv-TasNet was originally tasked of processing time-domain speech separation, but we produced a denoising mask instead to achieve single-channel speech denoising. The NAB takes multiple inputs to perform beamforming in suppressing the interferences, and after that, Denoising-TasNet is utilized to finally output the denoised signals. Experimental results show that our proposed method is superior to DNN and recently proposed CRN model. Additionally, we found that the NABDTN model is resistant to interferences from different directions.

References

1. Xia, Y., Braun, S., Reddy, C.K., Dubey, H., Cutler, R., Tashev, I.: Weighted speech distortion losses for neural-network-based real-time speech enhancement. In: ICASSP 2020–2020 IEEE International Conference on Acoustics, Speech and Signal Processing (ICASSP), pp. 871–875 (2020)
2. Van Veen, B.D., Buckley, K.M.: Beamforming: a versatile approach to spatial filtering. IEEE ASSP Mag. **5**(2), 4–24 (1988)
3. Hoshuyama, O., Sugiyama, A., Hirano, A.: A robust adaptive beamformer for microphone arrays with a blocking matrix using constrained adaptive filters. IEEE Trans. Signal Process. **47**(10), 2677–2684 (1999)
4. Pfeifenberger, L., Zohrer, M., Pernkopf, F.: Eigenvector-based speech mask estimation for multi-channel speech enhancement. IEEE Trans. Audio Speech Lang. Process. **27**(12), 2162–2172 (2019)
5. Pfeifenberger, L., Pernkopf, F.: Blind source extraction based on a direction-dependent a-priori SNR. In: Fifteenth Annual Conference of the International Speech Communication Association (2014)
6. Luo, Y., Chen, Z., Mesgarani, N., Yoshioka, T.: End-to-end microphone permutation and number invariant multi-channel speech separation. In: ICASSP 2020–2020 IEEE International Conference on Acoustics, Speech and Signal Processing (ICASSP), pp. 6394–6398 (2020)
7. Li, B., Sainath, T.N., Weiss, R.J., Wilson, K.W., Bacchiani, M.: Neural network adaptive beamforming for robust multichannel speech recognition. Interspeech **2016**, 1976–1980 (2016)
8. Sainath, T.N., et al.: Multichannel signal processing with deep neural networks for automatic speech recognition. IEEE/ACM Trans. Audio Speech Lang. Process. **25**(5), 965–979 (2017)
9. Luo, Y., Han, C., Mesgarani, N., Ceolini, E., Liu, S.-C.: FasNet: low-latency adaptive beamforming for multi-microphone audio processing. In: IEEE Automatic Speech Recognition and Understanding Workshop (ASRU) 2019, pp. 260–267 (2019)
10. Wang, Z.-Q., Wang, P., Wang, D.: Complex spectral mapping for single-and multi-channel speech enhancement and robust ASR. IEEE/ACM Trans. Audio Speech Lang. Process. **28**, 1778–1787 (2020)
11. Benesty, J., Chen, J., Huang, Y.: Microphone Array Signal Processing, vol. 1. Springer, Heidelberg (2008). https://doi.org/10.1007/978-3-540-78612-2
12. Luo, Y., Mesgarani, N.: Conv-TasNet: surpassing ideal time-frequency magnitude masking for speech separation. IEEE/ACM Trans. Audio Speech Lang. Process. **27**(8), 1256–1266 (2019)
13. Fu, S.W., Tsao, Y., Lu, X., Kawai, H.: Raw waveform-based speech enhancement by fully convolutional networks. In: Asia-Pacific Signal and Information Processing Association Annual Summit and Conference (APSIPA ASC), pp. 006–012. IEEE (2017)
14. Kolbæk, M., Tan, Z.-H., Jensen, S.H., Jensen, J.: On loss functions for supervised monaural time-domain speech enhancement. IEEE/ACM Trans. Audio Speech Lang. Process. **28**, 825–838 (2020)

15. Tawara, N., Kobayashi, T., Ogawa, T.: Multi-channel speech enhancement using time-domain convolutional denoising autoencoder. In: INTERSPEECH, pp. 86–90 (2019)
16. Allen, J.B., Berkley, D.A.: Image method for efficiently simulating small-room acoustics. J. Acoust. Soc. Am. **65**(4), 943–950 (1979)
17. Reddy, C.K., Gopal, V., Cutler, R.: DNSMOS: a non-intrusive perceptual objective speech quality metric to evaluate noise suppressors. arXiv e-prints, pp. arXiv-2010 (2020)
18. Hendriks, R.C., Heusdens, R., Jensen, J.: MMSE based noise PSD tracking with low complexity. In: 2010 IEEE International Conference on Acoustics, Speech and Signal Processing, pp. 4266–4269. IEEE (2010)
19. López-Espejo, I., González, J.A., Gómez, Á.M., Peinado, A.M.: A deep neural network approach for missing-data mask estimation on dual-microphone smartphones: application to noise-robust speech recognition. In: Navarro Mesa, J.L., et al. (eds.) IberSPEECH 2014. LNCS (LNAI), vol. 8854, pp. 119–128. Springer, Cham (2014). https://doi.org/10.1007/978-3-319-13623-3_13
20. Tan, K., Zhang, X., Wang, D.L.: Real-time speech enhancement using an efficient convolutional recurrent network for dual-microphone mobile phones in close-talk scenarios. In: IEEE International Conference on Acoustics, Speech and Signal Processing (ICASSP) (2019)
21. Tammen, M., Doclo, S.: Deep multi-frame MVDR filtering for single-microphone speech enhancement. In: ICASSP 2021–2021 IEEE International Conference on Acoustics, Speech and Signal Processing (ICASSP), pp. 8443–8447. IEEE (2021)

Edge Computing and Reinforcement Learning

Fog-Based Data Offloading in UWSNs with Discounted Rewards: A Contextual Bandit

Yuchen Shan, Hui Wang$^{(\boxtimes)}$, Zihao Cao, Yujie Sun, and Ting Li

School of Mathematics and Computer Science, Zhejiang Normal
University, Jinhua, People's Republic of China
hwang@zjnu.cn

Abstract. Urban wireless sensor networks (UWSNs) are an important application scenario for the Internet of Things (IoT). Nevertheless, applications based on urban environments are often computationally intensive, and sensor nodes are resource-constrained and heterogeneous. Fog computing has the potential to liberate the computation-intensive mobile nodes through data offloading. Therefore, reliable data collection and scalable coordination based on fog computing are seen as a challenge. In this paper, the challenge of data offloading is modeled as a contextual bandit problem—an important extension of the multi-armed bandit. First, the heterogeneity of the sensor nodes is used as contextual information, allowing the network to complete data collection at a small computational cost. Second, an ever-changing environmental scenario is considered in which the distribution of re-wards is not fixed, but varies over time. Based on this non-stationary bandit model, we propose a contextual bandit algorithm NCB-rDO in order to improve the success rate of data offloading, which solves the problem of data loss when the contextual information changes suddenly. Experimental results demonstrate the effectiveness and robustness of this data offloading algorithm.

Keywords: Contextual bandit · Collaborative offloading · Dynamic fog computing · Urban wireless sensor network

1 Introduction

Our lives are increasingly dependent on smart IoT devices that collect data from the environment and operate in the physical world [1]. Many sensor nodes are deployed in cities and large urban metropolitan areas. Examples include security monitoring, traffic, pollution monitoring, infotainment, energy management [2, 3]. The proliferation of urban IoT applications has created an unprecedented amount of data, including physical quantities sampled from the environment [4]. Due to the limited resources of the sensor, it is generally not possible to process sampled data locally and often relies on the cloud for data analysis and long-term storage [5]. However, this will inevitably result in additional energy consumption and computational overhead for offloading the data. Fog network distributes computing, storage, control, and communication services across the cloud-to-thing continuum [6], rather than offloading them to the cloud. As a result, data

© ICST Institute for Computer Sciences, Social Informatics and Telecommunications Engineering 2022
Published by Springer Nature Switzerland AG 2022. All Rights Reserved
H. Gao et al. (Eds.): ChinaCom 2021, LNICST 433, pp. 509–522, 2022.
https://doi.org/10.1007/978-3-030-99200-2_38

collected by IoT sensors from the environment is collected by mobile fog nodes, such as cars or smartphones. In attempting to meet the performance requirements of particular applications [7], the fog network considers the resources available on the devices, such as transmission rate, efficiency, and network cost. Thus, the fog-based approach could exploit the dynamic heterogeneity of the sensors, which is caused by the ever-changing urban environment.

Sensor nodes can only effectively support urban applications if they first successfully transmit data to the fog nodes. Nevertheless, the limited storage resources of these sensors prevent them from keeping the data they generate indefinitely. It is therefore vital that data is transferred to resource-rich nodes before storage space is exhausted. We refer to this problem as the "data loss problem". But, designing solutions to prevent data loss is very challenging: an effective solution requires careful cooperation between sensor nodes and fog nodes, but the limited computation of sensors precludes coordination mechanisms with significant overhead. Moreover, the dynamic heterogeneity of nodes such as their memory margin, transmission rate, etc. cannot be controlled or predicted. Finally, in the case of sudden changes in node heterogeneity, nodes with as high a quality as possible must be selected for offloading data.

Recently, the multi-armed bandit (MAB) has been used to solve wireless communication and network decision problems [8, 9]. In the stochastic MAB problem, given a set of arms (actions), one arm is selected on each trial and a reward is obtained from the reward distribution followed by that arm. Each arm has an unknown random reward, and by pulling an arm, the player gets an immediate reward. Players decide which arm to pull in a series of trials to maximize the rewards that accrue over time. To extend the MAB framework to dynamic complex systems with contextual relevance, the contextual bandit (CB) model, which is widely used in recommender systems [10], was investigated. Unlike MAB, the rewards in the CB model depend on the contextual information provided at each period. We attempt to design the low-complexity data offloading problem as a CB problem and use the dynamic heterogeneity of nodes, i.e., state information, as contextual information. Furthermore, given the dynamics of the urban environment, the data offloading problem can be constructed as a non-stationary MAB problem. In this non-stationary case, the reward distribution can change at any moment in time at an unknown moment. Thus, a discount factor was introduced to treat historical rewards differently. To estimate instantaneous expected rewards, the data offloading scheme with discounted rewards averaged past rewards and gave a discount factor that gave greater weight to the most recent observations.

The motivation of this paper is a fog-based architecture that takes into account the sudden change in dynamic heterogeneity of sensor nodes due to the external environment (cold, earthquakes, etc.) and selects which nodes have a higher quality (good state) that can improve the success of data offloading and thus avoid the data loss problem. In this paper, sensors in the environment are logically divided into two categories: the sensors that offload data are called task nodes, while the sensors that receive data are called helper nodes. The kernel UCB algorithm [11] is a non-linear algorithm based on the CB model. The algorithm assumes a non-linear relationship between reward and contextual information, which is more in line with the actual urban environment. The nonlinear contextual bandit robust for data offloading (NCB-rDO) algorithm based on the kernel

UCB algorithm is proposed. The algorithm assumes a non-linear relationship between the reward obtained from offloading data and the dynamic heterogeneity of the helper nodes in the fog network. In addition to this, a discount factor for historical rewards is introduced to improve the kernel UCB algorithm, taking into account sudden changes in the urban environment. Our main contributions are summarized below.

- The CB model for recommender systems is applied to the data offloading problem for UWSNs. To more closely match the actual urban environment, the problem of offloading data under a non-stationary model is considered, i.e., the reward distribution is not fixed.
- The architecture based on the fog network makes effective use of the heterogeneity of the nodes in the network. Using the heterogeneous state information of nodes as contextual information, the feature vector based on the heterogeneous information enables nodes to collaborate to offload data.
- The scheme we propose is not a direct application of the CB model and therefore the traditional analysis is not applicable. Experimental results show that the CB-based data offloading strategy is robust to dynamic urban environments.

2 Related Works

In recent years, data offloading has emerged as a promising technology for the efficient use of distributed computing resources, and plenty of researches have been carried out. For example, Li et al. [12] proposed an optimal offloading policy for heterogeneous end-users under buffer constraints. The authors define the objective of achieving maximum mobile data offloading as multiple linear constraint maximization problems with finite storage and propose various algorithms to solve this optimization problem for different offloading scenarios. However, they are considering data offload from the infrastructure to the mobile device. The focus of the work in this paper, however, is on offloading data from sensors to the fog infrastructure. This can lead to different patterns of data loss, the former due to adequate infrastructure resources and data loss problems associated with signaling, the latter focusing on problems caused by a lack of resources for sensors, and more on completing data offloading before resources are consumed. The authors of the literature [13] propose a collaborative offloading algorithm that logically divides sensor nodes into in-need nodes and helper nodes through a central network controller, based on a Markov chain model. This paper also aims to achieve reliable data communication through device collaboration, but our solution aims to enable collaboration between stationary and mobile sensors, rather than between mobile gateways and static sensors. It makes sense to study the data loss problem in different application scenarios, as the topology, node locations, and resources of real sensor networks are dynamic. In this paper, multiple hops are required to complete the data offload. Gao et al. [14] attempted to maximize the data collected by the mobile sink, proposing a scheme by designating the nearest node as the intermediate data collector. Their solution assumes that the node has enough storage to store all the data and that the sink mobility is fixed and known in advance. However, in practice, the shareable computing resources of nodes change over time and mobility is unpredictable. Similarly, Wen et al. [15] constructed an energy-sensing path for the mobile sinks to collect data when the location of the sensors and the mobile receiver were known. There is research focusing on learning the movement patterns of the sink to improve data collection from the sensors. Pozza et al. [16] considered IoT scenarios through a predictive framework based on temporal difference learning but did not address the issue of communication reliability. Our work considers how to guarantee a certain offloading efficiency when the node state changes abruptly.

3   System Model and Problem Formulation

The problem of data offloading can be naturally modeled as a CB problem. Task nodes are treated as players and helper nodes are treated as arms. In each round, the player expects to select the arm with the highest reward. Similarly, during data offloading at each time slot, the task node offloads data to the helper node and expects a high reward, such as a high level of data offloading success rate.

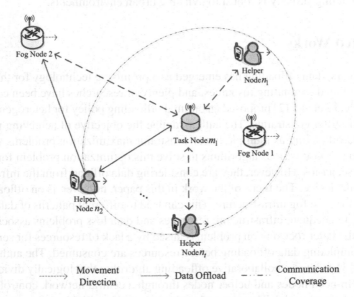

Fig. 1. Cooperative offloading system model

The contextual information-based coordinated offloading scheme for UWSNs is shown in Fig. 1. Suppose there are M task nodes, $N \geq M$ mobile helper nodes, and mobile fog nodes in the model. Time is divided into discrete time slots. T is used to characterize the total number of time slots, and t is used to characterize an arbitrary moment, where $1 \leq t \leq T$. At each time slot t, the task node broadcasts a data offload request and then selects a helper node based on its observation history. The task node offloads its data to the fog node with the assistance of the helper node. The task node $m \in \{1, \ldots, M\}$ selects a helper node $n \in \{1, \ldots, N\}$, and the task node m gets an instantaneous reward r_{t,n_t} after the data is offloaded to the fog node. In each period t, the process is as follows: (i) The heterogeneous information $x_{t,n}$ is fed back to the task node as a vector. $x_{t,n}$ is the contextual information of the helper node n at the time slot t. The heterogeneity of helper nodes is various in the same slot. The heterogeneity of the nodes reflects the quality of the nodes, including memory, transmission rate, location, etc. (ii) Selection of helper node: The task node observes heterogeneity information and selects high-quality helper nodes based on the rewards observed in historical trials. (iii) Reward feedback: The fog node feeds back the reward r_{t,n_t} to the task node.

The data offloading problem is constructed as a non-stationary MAB problem, taking into account the dynamics of the urban environment. In this problem, the distribution of the reward can change at an unknown moment. Without loss of generality, we normalize $r_{t,n_t} \in [0, 1]$. Our goal is to achieve a high level of offload success rate, with the reward equal to one when the selected helper node completes the data offload and zero otherwise. The aim is to maximize the above-expected total reward. A discount factor $\tau \in (0, 1)$ was introduced to treat historical rewards with different weights, i.e., a greater weight was given to more recent rewards and a lesser weight to earlier ones.

The mean of the random variable r_{t,n_t} is $\upsilon_{t,n_t} = E[r'_{t,n_t}]$, where r'_{t,n_t} is the average value of discounted rewards. The mean value of the reward distribution on the optimal helper node in each round is $\upsilon_t^{opt} = \max_{n \in N}\{\upsilon_{t,n_t}\}$. In the MAB framework, the regret value is used to measure the performance of a bandit algorithm. The regret value is described as the difference between the total system gain obtained with the best strategy in the ideal case and the total gain obtained with that strategy π. In the time-varying case, the mathematical expression for the regret value of non-stationary bandit after T time slots can be expressed as follow:

$$regret(T) = \sum_{t=1}^{T} \upsilon^{opt} - E[\sum_{t=1}^{T} \upsilon_t^{\pi}] \tag{1}$$

Where $\pi(t) = n$ indicates that, at time slot t, the task node selects the helper node n to offload the data.

4 NCB-rDO Design: Nonlinear Contextual Bandit Robust for Data Offloading

In this section, we consider the case where no prior knowledge of the reward distribution is provided throughout the offloading process, but the distribution is assumed to change over time in all rounds. The NCB-rDO algorithm is proposed, the algorithm has a self-learning capability implemented in a dynamic urban environment, based on the CB model and discounting of rewards.

4.1 Kernel UCB Algorithm

Based on the kernel UCB algorithm, the expected reward and heterogeneous feature information of helper nodes are assumed to be nonlinearly related. In complex urban wireless sensor networks, it is practical and reasonable to explore the non-linear relationship between the reward of offloading data and the heterogeneity of the nodes. The study [11] shows how to obtain the kernel UCB algorithm. Kernel methods provide a way to extract from observations possibly non-linear relationships between the contexts and the rewards while only using similarity information between contexts. For the nonlinear CB model, we assume that there exists a mapping function $\phi : R^d \rightarrow H$ mapping data to the Hilbert space, for which ϕ, there exists $\theta^* \in H$. Therefore, the expected reward is linearly related to the heterogeneous information of the helper nodes as follows.

$$E\left[r_{t,n_t} | x_{t,n}\right] = \phi(x_{t,n})\theta_n^* \tag{2}$$

According to the mapping function ϕ, the kernel function is defined as $k(x, x^{'}) = \phi(x)^T \phi(x^{'}), \forall x, x^{'} \in R^d$, and the kernel matrix of data $\{x_1, \ldots, x_t\} \in R^d$ is denoted as $K_t = \{k(x_i, x_j)\}_{i,j \leq t}$.

First, the estimated value of the reward $\widehat{\mu}_{t+1,n} = \phi(x_{t+1,n})^T \theta_t$ needs to be predicted, where θ_t is the minimum value of the regularized least-squares function:

$$L(\theta) = \gamma \|\theta\|^2 + \sum_{i=1}^{t-1} (r_i - \phi(x_i)^T \theta)^2 \tag{3}$$

Φ_t is used to denote $[\phi(x_1)^T, \ldots, \phi(x_{t-1})^T]^T$. Let $L(\theta)$ be equal to zero, and the following Eq. (4) can be obtained by taking the partial derivative of θ. The equation satisfies the solution to the problem of minimizing $\theta_t = \min_{\theta \in H} L(\theta)$.

$$(\Phi_t^T \Phi_t + \gamma I)\theta_t = \Phi_t^T y_t \tag{4}$$

where $y_t = \{r_{1,n_1}, \ldots, r_{t,n_t}\}^T$. Equation (4) is rearranged to get $\theta_t = \Phi_t^T \alpha_t$, where $\alpha_t = \gamma^{-1}(y_t - \Phi_t \theta_t) = \gamma^{-1}(y_t - \Phi_t \Phi_t^T \alpha_t)$. The kernel matrix is used to express this equation, i.e., $\alpha_t = (K_t + \gamma I)^{-1} y_t$. We denote $k_{t,x} = \Phi_t \phi(x) = [k(x_1, x), \ldots, k(x_{t-1}, x)]^T$ then Eq. (5) is obtained as:

$$\widehat{\mu}_{t,n} = k_{t,x_{t,n}}^T (K_t + \gamma I)^{-1} y_t \tag{5}$$

While the computation of θ_t would require evaluating $\phi(x_i)$ for every data point x_i, the dualized representation of the prediction (5) allows computing $\widehat{\mu}_{t,n}(x)$ only from the elements of the kernel matrix.

Next, the width of the confidence interval for the predicted value of the reward needs to be calculated. For linear bandits, the appropriate bandwidth is constructed, based on the Mahalanobis distance of $\phi(x_{t,n})$ from the matrix Φ_t:

$$\widehat{\sigma}_{t,n} = \sqrt{\phi(x_{t,n})^T (\Phi_t^T \Phi_t + \gamma I)^{-1} \phi(x_{t,n})} \tag{6}$$

Since the matrices $(\Phi_t^T \Phi_t + \gamma I)$ and $(\Phi_t \Phi_t^T + \gamma I)$ are strictly positive definite, the following equation holds:

$$(\Phi_t^T \Phi_t + \gamma I)\Phi_t^T = \Phi_t^T (\Phi_t \Phi_t^T + \gamma I) \tag{7}$$

$$\Phi_t^T (\Phi_t \Phi_t^T + \gamma I)^{-1} = (\Phi_t^T \Phi_t + \gamma I)^{-1} \Phi_t^T \tag{8}$$

The Eq. (8) extracts the Mahalanobis distance as

$$(\Phi_t^T \Phi_t + \gamma I)\phi(x) = \Phi_t^T k_{t,x} + \gamma \phi(x) \tag{9}$$

from which it can be inferred that

$$\phi(x) = \Phi_t^T (\Phi_t \Phi_t^T + \gamma I)^{-1} k_{t,x} + \gamma (\Phi_t^T \Phi_t + \gamma I)^{-1} \phi(x) \tag{10}$$

and left multiplied by $\phi(x)$, express $\phi(x)^T \phi(x)$ as

$$k_{t,x}^T (\Phi_t \Phi_t^T + \gamma I)^{-1} k_{t,x} + \gamma \phi(x)^T (\Phi_t^T \Phi_t + \gamma I)^{-1} \phi(x) \tag{11}$$

The above equation is rearranged to give the expression for the width of the confidence interval as

$$\widehat{\sigma}_{t,n} = \gamma^{-1/2} \sqrt{k(x_{t,n}, x_{t,n}) - k_{t,x_{t,n}}^T (K_t + \gamma I)^{-1} k_{t,x_{t,n}}} \tag{12}$$

This equation deals only with the inner product between elements. At each time slot t, the algorithm selects helper nodes n_t satisfying:

$$n_t = \arg\max_{n \in N} \widehat{\mu}_{t,n} + \widehat{\sigma}_{t,n} \tag{13}$$

A pseudo-code description of the kernel UCB algorithm based on data offloading is given below.

Algorithm 1 Kernel UCB With Online Updates [11]

Input and Initialization:

$$\mu_0 \leftarrow [1,0,...,0]^T, y_0 \leftarrow \varnothing, (K_t)_{ij} = k(x_i,x_j)$$

$$k_t(x_{t,n}) = [k(x_{1,n_1},x_{t,n}),...,k(x_{t-1,n_{t-1}},x_{t,n})]^T$$

γ,η regularization and exploration parameters

Run:

for t=1 to T do

Choose $n_t = \arg\max_{n\in N} \mu_n$ and get a reward r_{t,n_t}

Update $y_t = [r_{1,n_1},...,r_{t,n_t}]^T$

if t =1 then

$$K_t^{-1} \leftarrow 1/k_t(x_{t,n})+\gamma$$

else {Online update of the kernel matrix inverse}

$$\omega \leftarrow (k_1(x_{1,n_1}),...,k_{t-1}(x_{t-1,n_{t-1}}))^T$$

$$K_{22} \leftarrow (k(x_{t,n_t},x_{t,n_t})+\gamma-\omega^T K_{t-1}^{-1}\omega)^{-1}, \ K_{11} \leftarrow K_{t-1}^{-1}+K_{22}K_{t-1}^{-1}\omega\omega^T K_{t-1}^{-1}$$

$$K_{12} \leftarrow -K_{22}K_{t-1}\omega, \ K_{21} \leftarrow -K_{22}\omega^T K_{t-1}^{-1}, K_t^{-1} \leftarrow [K_{11},K_{12};K_{21},K_{22}]$$

end if

for n =1 to N do

$$\sigma_{t,n} \leftarrow \sqrt{k_t(x_{t,n})^T K_t^{-1}k_t(x_{t,n})}$$

$$\mu_{t,n} \leftarrow (k_t(x_{t,n})^T K_t^{-1}y_t + \frac{\eta}{\gamma^{1/2}}\sigma_{t,n})$$

end for

end for

4.2 Discounting the Rewards

In many application areas, temporal changes in the structure of the reward distribution are inherent to the problem. We focus on a discounted MAB formulation that allows for a wide range of temporal uncertainty rewards due to the different demands of the dynamic urban environment. In the presence of uncertainty, players (task nodes) facing a range of decisions need to use information gleaned from past observations when attempting to optimize future actions. Knowing that undetected changes will lead to grossly inaccurate estimates, players need to discount the weights of earlier indicator observations when estimating indicators in an ever-changing environment.

To estimate the instantaneous expected reward, this scheme averages past rewards with a discount factor that gives more weight to the most recent observations. Adaptability to changes in parameters does mean reducing the influence of observations made long in the past, which means using weights that increase with time. At each time slot t, when a helper node is selected, the average value of discounted rewards and the number of discounts are given by

$$r'_{t,n_t} = \frac{1}{\overline{q}_{t,n_t}} \sum_{s=1}^{t} \tau^{t-s} r_{t,n_t} \mathcal{K}_{\{n_t \in N\}} \tag{14}$$

$$\overline{q}_{t,n_t} = \sum_{s=1}^{t} \tau^{t-s} \mathcal{K}_{\{n_t \in N\}} \tag{15}$$

Where the discount factor $\tau \in (0, 1)$. The proposed policy, referred to as a nonlinear contextual bandit robust for data offloading algorithm, is shown in Algorithm 2.

Algorithm 2 NCB-rDO

//Initialization
for t=1 to N do
    Select helper node $n \in N(t)$ such that $n = (t+1) \bmod N$
    $r'_{t,n_t} = r_{t,n_t}$
    $\overline{q}_{t,n_t} = 1$
end for
// MAIN LOOP
while 1 do
    for t=N+1 to T do
        Run algorithm 1 to select the helper node n_t that maximizes
                $\hat{\mu}_{t,n} + \hat{\sigma}_{t,n}$
    Update $r'_{t,n_t}, \overline{q}_{t,n_t}$ accordingly.
end while

---

5 Simulation

5.1 Evaluation Metrics

The efficiency and performance of the offload scheme are verified by the following performance metrics.

- *Average reward* is defined as the mean of the reward feedback after data offloading to the fog node.
- *The selection count* is defined as the number of corresponding helper nodes selected based on the CB model. It is used to test the robustness of the algorithm when the contextual information of the helper nodes changes abruptly.
- *Success offloading rate* is defined as the rate of success offloading counts compared to overall rounds.

5.2 Methodology and Simulation Setup

The Python software environment was used to evaluate the performance of the proposed data offloading scheme under dynamic conditions according to the scheme proposed in Sect. 4. The hardware environment used for the experiments is Intel(R) Core i5-4210M (2.60 GHz) with 8 GB of RAM. In the proposed data offloading scheme, the dynamic heterogeneity of helper nodes in each round is used as contextual information. The heterogeneity of each helper node is assumed to consist of four components, namely memory margin, transmission speed, residual energy, and movement probability. The memory margins follow a uniform distribution with distribution parameters ranging from 0.04 to 0.06. The residual energy follows an exponential distribution with the distribution parameter $\delta_n = [\delta_n]_{n=1,...,10}$, the range of δ_n is from 0.1 to 1. The transmission rate of the helper node takes values from 2 to 24 Mbit/s. To evaluate CB-based offloading strategies in a more realistic context, ONE Simulator v1.6.0 was used to generate simulations of urban road-based movement trajectories in which pedestrians carrying smartphones (as helper nodes) walk along a street at a speed of 0.5–1.5 m/s. At the beginning of the simulation, all sensors were initialized with the same probability of movement, and then each sensor dynamically updated the observed probability of encounter with the fog node. There are ten helper nodes and mobile fog nodes in the simulation network, which are placed in an area with a radius of 1000 m. The length of the time slot is set to 20 ms. A high level of successful offloading rate is used as a criterion for selecting the helper node. The helper node is selected to enable the collaborative offloading system to obtain a satisfactory offloading ratio in the CB framework. r'_{t,n_t} is denoted to the average value of the discounted reward taken by the helper node n at the time slot t. To simplify the process, we assume that follows a Gaussian distribution with parameters r'_{t,n_t}.

The NCB-rDO comes with two hyperparameters including gamma and eta. To optimize these hyperparameters we propose to perform a grid search as in Fig. 2, where the value of gamma will be between zero and three, excluding zero, and the value of eta will be between zero and one. The results will be displayed on the grid with color bars. The experimental results show that the mean reward reaches a maximum of 0.77 when gamma takes the value of 0.5 and eta takes the value of 0.3.

5.3 Simulation Results

The proposed collaborative offload policy is compared with three other bandit options. LinUCB [10] is the classical contextual bandit algorithm, which assumes a linear relationship between the expected reward of the selected helper node and the contextual information. The UCB [17] algorithm considers the average reward and confidence interval of the arms. Each helper node corresponds to a confidence interval and the helper node with the largest upper confidence interval is selected for data offloading. Epsilon-Greedy algorithm: the relationship between random numbers and a priori epsilon values is used to decide whether to select the helper node with the highest reward or to select the helper node at random.

Figure 3 illustrates the change in the success rate of data offloading as the number of trial rounds increases for the four considered offloading schemes. The NCB-rDO algorithm could achieve an offloading success rate of 80.28%; the LinUCB algorithm

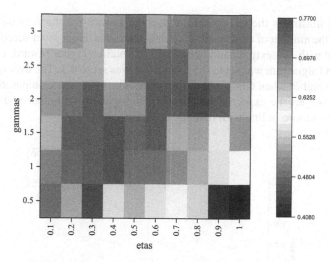

Fig. 2. Grid search average reward value for NCB-rDO

could achieve an offloading success rate of 74.67%; Compared to the first two contextual bandit algorithms, the UCB algorithm and the Epsilon-Greedy algorithm have a lower offloading success rate of 65.83% and 56.06%, respectively. The Epsilon-Greedy algorithm rigidly divides the selection process into an exploration phase and an exploitation phase. The exploration is performed with the same prior probability for all helper nodes and does not make use of any historical information, including the number of times a helper node has been selected and the probability of a helper node receiving a reward. As a result, the algorithm has the lowest success rate in offloading data.

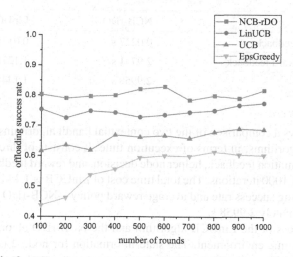

Fig. 3. Offloading success rate for four data offloading policies.

Figure 4 illustrates the change in the average reward for the four data offloading scenarios as the number of experimental rounds increases. High-level success offloading rate is linked with contextual bandit rewards. To obtain a higher reward, the gamma in the NCB-rDO algorithm was set to 0.5 and the eta was set to 0.3. The average reward for NCB-rDO is higher when compared to the other three schemes. Compared to LinUCB, NCB-rDO assumes the expected reward and heterogeneity of helper nodes to be non-linear, which is more in line with realistic scenarios.

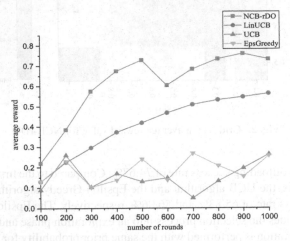

Fig. 4. The average reward for four data offloading policies.

Table 1. Time cost

	NCB-rDO	LinUCB
Feature vectors feedback	0.0227 s	0.0174 s
Decision and Reward feedback	2.9731 s	1.1271 s
Total	2.9958 s	1.1445 s

Table 1 shows a comparison of the two contextual bandit algorithms, the NCB-rDO and LinUCB algorithms, in terms of execution time. It consists of three parts of time: contextual information feedback, helper node decision, and reward feedback. The values are derived from 1000 iterations. The total time cost of LinUCB is 1.1445 s. By contrast, the high offloading success rate and average reward policy—NCB-rDO—costs a longer time of approximately 2.9958 s.

The robustness of the LinUCB algorithm and the NCB-rDO algorithm was tested in a highly dynamic environment. The state information for node 2 is assumed to be initially good, i.e., large memory margin, fast transmission speed, etc., while the state information for node 5 is initially poor. Figure 5 shows that the number of selections and the node state information are positively correlated. When t = 700, helper nodes 2 and

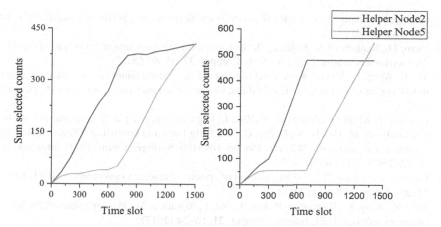

Fig. 5. Robustness performance for Lin UCB and NCB-rDO

5 suffer an abrupt change, i.e., the state information of node 2 is poor, while that of node 5 is good. As shown in Fig. 5, the NCB-rDO algorithm is more sensitive to changes in contextual information compared to the LinUCB algorithm. When t > 700, the number of selections for both nodes changes more significantly. Both the LinUCB and NCB-rDO algorithms are based on the CB model, with the difference that the former assumes a linear relationship between contextual information and reward, while the latter assumes a non-linear relationship. The non-linear relationship is more consistent with the actual mutating environment, and NCB-rDO can show stronger robustness than LinUCB.

6 Conclusion

A new collaborative data offloading algorithm based on the CB model is proposed without prior knowledge of the state of the nodes. The framework based on fog computing makes more rational use of the heterogeneity of the nodes, thus avoiding data loss problems and increasing the success rate of data transmission. Also, based on a non-stationary bandit model, the idea of discounted rewards is introduced to weigh different historical rewards. The proposed scenario allows the best helper node selection only with a bit of contextual information about nodes. Numerical analysis verifies that the NCB-rDO algorithm avoids data loss problems in urban environments and also verifies the robustness of the algorithm in the face of abrupt changes. The multi-source data offloading problem is not a simple overlay of the single-source data offloading problem. Therefore, our next work will consider a coordinated multi-source data offloading scheme based on a contextual bandit model.

References

1. Miorandi, D., Sicari, S., Pellegrini, F.D., Chlamtac, I.: Internet of things: vision, applications and research challenges. Ad Hoc Netw. **10**(7), 1497–1516 (2012)

522   Y. Shan et al.

2. Chiang, M., Zhang, T.: Fog and IoT: an overview of research opportunities. IEEE IoT J. 3(6), 854–864 (2016)
3. Yuan, D., Kanhere, S.S., Hollick, M.: Instrumenting wireless sensor networks—a survey on the metrics that matter. Pervasive Mob. Comput. 37, 45–62 (2017)
4. Yu, T., Wang, X., Shami, A.: A novel fog computing enabled temporal data reduction scheme in IoT systems. In: The 2017 IEEE Global Communications Conference, pp. 1–5, December 2017
5. Bonomi, F., Milito, R., Natarajan, P., Zhu, J.: Fog computing: a platform for internet of things and analytics. In: Bessis, N., Dobre, C. (eds.) Big Data and Internet of Things: A Roadmap for Smart Environments. SCI, vol. 546, pp. 169–186. Springer, Cham (2014). https://doi.org/10.1007/978-3-319-05029-4_7
6. Chiang, M., Zhang, T.: Fog and IoT: an overview of research opportunities. IEEE Internet Things J. 3(6), 854–864 (2016)
7. Wen, Z., Yang, R., Garraghan, P., Lin, T., Xu, J., Rovatsos, M.: Fog orchestration for internet of things services. IEEE Internet Comput. 21, 16–24 (2017)
8. Zhou, P., Jiang, T.: Toward optimal adaptive wireless communications in unknown environments. IEEE Trans. Wirel. Commun. 15(5), 3655–3667 (2016)
9. Maghsudi, S., Hossain, E.: Multi-armed bandits with application to 5G small cells. IEEE Wirel. Commun. 23(3), 64–73 (2016)
10. Li, L., Chu, W., Langford, J., Schapire, R.E.: A contextual-bandit approach to personalized news article recommendation. Presented at the Proceedings of the 19th International Conference on World Wide Web, Raleigh, North Carolina, USA (2010)
11. Valko, M., Korda, N., Munos, R., Flaounas, I., Cristianini, N.: Finitetime analysis of kernelised contextual bandits. In: Proceedings of Uncertainty Artificial Intelligence, pp. 654–663 (2013)
12. Li, Y., Qian, M., Jin, D., Hui, P., Wang, Z., Chen, S.: Multiple mobile data offloading through disruption tolerant networks. IEEE Trans. Mob. Comput. 13, 1579–1596 (2014)
13. Kortoçi, P., Zheng, L., Joe-wong, C., Francesco, M.D., Chiang, M.: Fog-based data offloading in urban IoT scenarios. In: IEEE INFOCOM 2019 - IEEE Conference on Computer Communications, pp. 784–792 (2019)
14. Gao, S., Zhang, H., Das, S.K.: Efficient data collection in wireless sensor networks with path-constrained mobile sinks. IEEE Trans. Mob. Comput. 10(4), 592–608 (2011)
15. Wen, W., Zhao, S., Shang, C., Chang, C.-Y.: EAPC: energy-aware path construction for data collection using mobile sink in wireless sensor networks. IEEE Sens. J. 18(2), 890–901 (2018)
16. Pozza, R., Nati, M., Georgoulas, S., Gluhak, A., Moessner, K., Krco, S.: CARD: context-aware resource discovery for mobile internet of things scenarios. In: IEEE WoWMoM 2014, pp. 1–10, June 2014
17. Agrawal, R.: Sample mean based index policies by O (log n) regret for the multi-armed bandit problem. Adv. Appl. Probab. 27, 1054–1078 (1995)

Deep Deterministic Policy Gradient Algorithm for Space/Aerial-Assisted Computation Offloading

Jielin Fu, Lei Liang, Yanlong Li(✉), and Junyi Wang

Guilin University of Electronic Technology, Guilin 541004, China
lylong@guet.edu.cn

Abstract. Space-air-ground integrated network (SAGIN) has been envisioned as a promising architecture and computation offloading is a challenging issue, with the growing demand for computation-intensive applications in remote area. In this paper, we investigate a SAGIN edge computing architecture considering the energy consumption and delay of computation offloading, in which ground users can determine whether take advantage of edge server mounted on the unmanned aerial vehicle and satellite for partial offloading or not. Specifically, the optimization problem of minimizing the total cost is formulated as a Markov decision process, and we proposed a deep reinforcement learning-based method to derive the near-optimal policy, adopting the deep deterministic policy gradient (DDPG) algorithm to handle the large state space and continuous action space. Finally, simulation results demonstrate that the partial offloading scheme learned from proposed algorithm can substantially reduce the user devices' total cost as compared to other greedy policies, and its performance is better than the binary offloading scheme learned from Deep Q-learning algorithm.

Keywords: Space-air-ground integrated network · Edge computing · Partial offloading · Reinforcement learning

1 Introduction

Nowadays, with the in-depth development of the 5G mobile communication system and the research on 6G, an interconnected world is gradually opening up to people. Meanwhile, the rapid development of various communication services and the continuous improvement of application demands put forward higher requirements for network coverage, data transmission rate and end-to-end delay. However, the existing ground network coverage is limited and cannot provide services to meet the services for remote areas such as mountainous areas, polar regions and oceans. Space-Air-Ground Integrated Network (SAGIN) can achieve the global seamless coverage, breaking through the limitations of Ground networks, so it has become an emerging hot research topic [1,2].

© ICST Institute for Computer Sciences, Social Informatics and Telecommunications Engineering 2022
Published by Springer Nature Switzerland AG 2022. All Rights Reserved
H. Gao et al. (Eds.): ChinaCom 2021, LNICST 433, pp. 523–537, 2022.
https://doi.org/10.1007/978-3-030-99200-2_39

SAGIN is based on the ground cellular network and combines the advantages of the wide coverage of the satellite network and the flexible deployment of the aerial platform to achieve seamless coverage through the convergence of heterogeneous networks [3]. However, with the rapid development of the Internet of Things (IoT), more and more computation-intensive applications pose challenges with the limited computing capability and battery life of the devices. In general, users utilize mobile edge computing (MEC) to offload computing tasks to data centers with rich computing resources for processing, which can make up for the defects in computing capability and storage resources of users' devices to some extent [4]. Therefore, MEC technology of terrestrial network is introduced in SAGIN to provide users with efficient and flexible computing services by utilizing multi-level and heterogeneous computing resources at the edge of network. Through offloading the computation intensive tasks to MEC sercer, the energy consumption and latency can be reduced. On the other hand, employing the SAGIN in computation offloading introduces several challenging issues. Firstly, different SAGIN segments possess distinct network conditions. Secondly, due to transmission delay, it may not be able to meet the requirements of time-sensitive applications such as virtual reality [5]. Therefore, it is very necessary to design an efficient computation offloading scheme.

Early studies on MEC mainly focused on looking for solutions for the allocation of computing and communications resources, as well as offloading strategies for various computing tasks,in which computation services are provided by a fixed base station in terrestrial networks. Computing offloading has been studied extensively, and most of these researches have proposed traditional optimization approaches, such as convex optimization methods [6] and Lyapunov optimization [7], and deep learning algorithms, such as Q-learning [8], Deep-Q-network (DQN) [9], and distributed deep learning [10], to solve this problem. However, the MEC services cannot effectively operate in the scenarios where communication infrastructures are sparasely distributed if the services are provided only through the terrestrial fixed facilities.

Recently, the research on computation offloading in SAGIN has been at its initial stage. Considering the coverage range and channel conditions of the UAVs in SAGIN, C. Zhou et al. [11] proposed a computation offloading scheme based on linear programming to solve the dynamic scheduling problem. To solve the allocation of communication and computing resources, an effective scheme based on reinforcement learning was proposed in literature [12]. But the above study did not consider the curse of dimensionality. Under given UAV energy consumption constraints, a risk-aware reinforcement learning algorithm was proposed to weigh delay and risk in SAGIN scenario in [13]. In [14], Thai et al. proposed a learning-based offloading scheme to optimize network performance and maximize server provider revenue. In [15], Tang et al. investigated the computation offloading decisions to minimize the sum energy consumption of ground users, proposed a distributed algorithm by leveraging the convex optimization method to approximate the solution. Although problems in high-dimensional state spaces have been successfully solved by DQN, only discrete action spaces can be handled. In [16], N. Cheng et al. proposed a joint resource allocation and task scheduling methods based on actor-critic algorithm to effectively allocate computing resources to virtual machines and achieve lower total cost of task scheduling. However, the task partial offloading and offloading scheduling are intercoupled with each other. It is important to note that the aforementioned works considered complete offloading, while partial offloading can significantly improve the latency as the network becomes dense and the edge resources are limited [17].

As a summary, the study of partial offloading considering the cooperation of space, aerial and ground multi-layer network under multi-user environment is still missing in above literatures. In addition, traditional reinforcement learning based methods and linear programming methods cannot solve the high-dimensional or continuous action space scenes.

Therefore, this paper considers the problem of computation offloading under the SAGIN architecture with the joint communication and computing (C2) service. In order to deal with these challenges, the optimization problem is formulated as a Markov decision process (MDP), and a partial offloading strategy based on deep deterministic policy gradient (DDPG) using the actor-critic algorithm to deal with large state and continuous action spaces is proposed to minimize the weighted sum of energy consumption and delay.

The remainder of this paper is organized as follows. In Sect. 2, the SAGIN architecture and computation offloading models are introduced. In Sect. 3, we describe the formulation and transformation, followed by a DDPG based solution. The simulation results and experimental evaluation are provided in Sect. 4. Finally, we introduce the conclusion and the future work briefly in Sect. 5.

2   System Model

In this section, we first introduce the network architecture and then describe the models associated with communication and computation for task offloading.

2.1   The SAGIN Architecture

In this work, we consider an SAGIN architecture with N ground users (GUs), I UAVs and a low earth orbit (LEO) satellite constellation. There are many typical applications, such as automated drilling control and virgin forest monitoring [2]. As shown in Fig. 1, a remote region without cellular coverage is considered, therefore we provide network access, edge computing, and caching through the aerial segment. In the aerial segment, UAVs can serve as edge servers to provide computing capabilities to GUs [18], which can be regarded as the replacement of BSs. Let $\mathbf{N} = \{1, 2, \ldots, N\}$ be the set of indices of N GUs. Then, the set of SAGIN components that computing tasks can be offloaded to is denoted by $\mathbf{I} = \{0, 1, 2, \ldots, I\}$, let indexes $1, 2, \ldots, I$ and 0 denote the UAVs and the LEO satellite constellation respectively. Due to the limited computing power and battery capacity of the GU devices, some tasks need to be offloaded to the

flying UAVs configured with fixed locations that act as attitude platforms, or
the LEO satellite constellation. Furthermore, we assume that GU n device has
M_t independent computing tasks at the beginning of time slot t, denoted by the
set $\mathbf{M} = \{1, 2, \ldots, M_t\}$. Considering a discrete time-slotted system with equal
slot duration , denoted by $\mathbf{T} = \{1, 2, \ldots, T\}$.

Each GU n can determine whether or not to offload its computing task m
to the edge server i, and $x_{nmi}(t) \in [0, 1]$ denote the offloading decision during
the time slot t. Specifically, $N \times M_t \times (I + 1)$ matrix $\mathbf{X}(t)$ denote decisions of
the tasks, $x_{nmi}(t) = 1$ denotes that GU n decides to offload its computing task
m to the edge server i completely, and $x_{nmi}(t) = 0$ means that GU n disposes
its task m locally. The following sections provide a comprehensive explanation
of the computation and communication models.

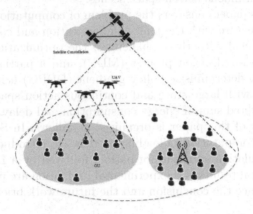

Fig. 1. The network model.

2.2   Computation Model

Without loss of generality, a tuple (ϕ, γ) is adopted to model the computing
tasks from GU devices, where ϕ (in bits) represents the input data size of a
computing task, and γ (in CPU cycles per bit) indicates that how many CPU
cycles are required to process one bit input data [11]. The delay and energy
consumption of downloading can be ignored when the computing results are
transmitted back to the GUs by the edge server, because the key point of policy
is task uploading in the considered scenario [19, 20]. As the computation tasks can
be completed locally, or offloaded to the UAVs or the LEO satellite constellation,
the computing delay can be analyzed from the following aspects.

The computing capability (in CPU cycles per seconds) of servers mounted on
UAVs and the satellite is denoted by f^i ($i \in \{1, 2, \ldots, I\}$) and f^0, respectively.
The computing delay of GU n at all offloading destination SAGIN components
is calculated as the following equation:

$$d_n^I(t) = \sum_{m=1}^{M_t} \sum_{i=0}^{I} \frac{x_{nmi}(t)\,\phi\gamma}{f^i}. \tag{1}$$

On the other hand, the delay in the local processing of a computation task consists of two parts, the computation delay and the queuing delay.Since the limited computing capability of the GU, the computation tasks may not processed or offloaded completely within a time slot. We assume that the remaining tasks wait to be processed in the computing queue. To model the queue latency, $\rho_n(t) \in [0, M_{\max}]$ denote the GU n's unaccomplished task backlog at the beginning of time slot t, $M_{\max}$ is the maximum length of the computing queue. $q_n(t)\tau$ is the queuing delay of all $q_n(t)$ tasks in the waiting queue of GU n. The number of queuing tasks denoted by:

$$q_n(t) = \max \left\{ \rho_n(t) - \left\lfloor \frac{f_n \tau}{\phi \gamma} \right\rfloor - \sum_{m=1}^{M} \sum_{i=0}^{I} x_{nmi}(t), 0 \right\}, \tag{2}$$

where $\lfloor \cdot \rfloor$ denotes the floor function, f_n is the computing capability of GU n, $\left\lfloor \frac{f_n \tau}{\phi \gamma} \right\rfloor$ denotes the greatest integer less than the number of computation tasks executed by GU n in time slot t. The delay of local task execution at the GU n can be given by:

$$d_n^l(t) = \frac{\sum_{m,i} (1 - x_{nmi}(t)) \phi \gamma}{f_n} + q_n(t)\tau. \tag{3}$$

Generally, the energy consumption of GU equipment is mainly composed of three parts, including mechanical energy consumption, communication-related energy consumption and computation-related energy consumption. The computation-related energy consumption can be calculated by:

$$e_n(t) = \xi_n \cdot \sum_{m,i} (1 - x_{nmi}(t)) \phi \gamma (f_n)^2, \tag{4}$$

where ξ_n denotes the energy factor, which depends on the chip architecture [15,16].

2.3 Communication Model

Since UAVs and satellites use different frequency bands to communicate, we suppose that there is no interference between UAVs and satellite in this work [21]. Meanwhile, we neglect the propagation delay from GU devices to the UAV because we assume the UAV is sufficiently close to GU devices [22]. According to [23], the average path loss of GU n from UAV to devices can be defined as:

$$PL(r, h) = 20 \log \left(\frac{4\pi f_c (h^2 + r^2)^{1/2}}{c} \right) + P_{LoS} \eta_{LOS} + (1 - P_{LoS}) \eta_{NLOS}, \tag{5}$$

where h, r, η_{LOS}, η_{NLOS} denote the UAV flying altitude, horizontal distance between the UAV and the GU, the additive loss incurred on top of the free space pathloss for line-of-sight and not-line-of-sight links [24], respectively. We set the altitude of the UAV to 10 m. f_c denotes the carrier frequency, c denotes the velocity of light, P_{LoS} represents the probability of line-of-sight link, which is an equation with respect to h, r [16]. According to [25], the values of $(\eta_{LoS}, \eta_{NLoS})$ are (0.1, 21) in remote area. Adopting the Weibull-based channel model [26], we generate the channel gain when $x_{nm0}(t) \neq 0$, which can be calculated by:

$$h = \frac{G_{tx}G_{rx}\lambda^2}{(4\pi l_{sat})} 10^{-\frac{F_{rain}}{10}},$$

(6)

where G_{tx} and G_{rx} are antenna gains of the GU and the satellite, respectively. F_{rain} represents the rain attenuation, and l_{sat} denotes the distance between the GU and the satellite. Currently, the data rate denoted by $r_i(t)$ can be calculated as the following equations:

$$r_n^i(t) = \begin{cases} B_i \log_2\left(1 + \frac{P_{n,i}(t) \cdot |h|^2}{\sigma_S^2}\right), i = 0 \\ B_i \log_2\left(1 + \frac{P_{n,i}(t) \cdot 10^{-\frac{PL}{10}}}{\sigma_U^2}\right), i \neq 0, \end{cases}$$

(7)

where B_i indicate the channel bandwidth of the ground-satellite link and the ground-UAV link, indexes $1, 2, \ldots, I$ and 0 denote the UAVs and the LEO satellite constellation respectively. Analogously, $P_{n,i}$ represent the transmission power, σ_S and σ_U denote the power of noise. Thus, the transmission delay can be given by:

$$d_n^i(t) = \begin{cases} \sum_{m=1}^{M}\left(\lceil x_{nm0}(t)\rceil d_{sat} + \left(\frac{x_{nmi}(t)\phi}{r_n^i(t)}\right)\right), i = 0 \\ \sum_{m=1}^{M} \frac{x_{nmi}(t)\phi}{r_n^i(t)}, i \neq 0, \end{cases}$$

(8)

where we denote d_{sat} as the propagation delay between the LEO satellite and the GU, which cannot be ignored. $\lceil \cdot \rceil$ denotes the ceil function. Denote the communication-related energy consumption by $e_n^i(t)$, which is defined as follows:

$$e_n^i(t) = P_{n,i} d_n^i(t).$$

(9)

3   Problem Formulation and Algorithm

In this section, we first introduce the formulation for our optimization problem, and then the reinforcement learning-based approach is proposed to derive the near-optimum decision.

3.1   Problem Formulation

As described in the preceding section, taking communication and computing models into account, the delay and energy consumption for completing all tasks of GU n can be defined separately as follows:

$$D_n(t) = d_n^l(t) + d_n^e(t) + \sum_{i=0}^{I} d_n^i(t), \tag{10}$$

$$E_n(t) = e_n^l(t) + \sum_{i=0}^{I} e_n^i(t). \tag{11}$$

Finally, the cost function of the offloading decision can be defined as:

$$C_n(t) = \omega_1 D_n(t) + \omega_2 E_n(t), \tag{12}$$

where ω_1 and ω_2 denote the tradeoff between delay and energy consumption for the dynamic computing offloading policy, $\omega_1 = 1-\omega_2$. Let $\mathbf{X} = \{X_t, \forall t\}$ denote the tasks offloading decisions set. Since link availability and task arrival are highly dynamic, our main focus is to minimize the time-averaged delay and energy consumption for all tasks. As the number of users increases, the cost of the MEC server to collect the channel vector of all users and then distribute the task queue to each user increases. In order to make the system much more scalable, we assume that the state of each GU is only determined by its local observation and make decisions independently. The optimization problem is defined as:

$$\min_{X} \lim_{T \to \infty} \frac{1}{T} \sum_{t=1}^{T} \omega_1 D_n(t) + \omega_2 E_n(t)$$
$$s.t. \quad x_{nmi}(t) \in [0,1], \forall n \in \mathbf{N}, \forall m \in \mathbf{M}, \forall i \in \mathbf{I},$$
$$\sum_{i=0}^{I} x_{nmi}(t) \in \{0,1\}, \forall n \in \mathbf{N}, \forall m \in \mathbf{M}, \tag{13}$$
$$\sum_{m=0}^{M_t} \sum_{i=0}^{I} x_{nmi}(t) \leq M_{\max}, \forall n \in \mathbf{N}.$$

3.2 Deep Deterministic Policy Gradient Algorithm

Above optimization problem P1 is a high-dimensional decision issue. We adopt an intelligent learning approach based on reinforcement learning (RL) to address this problem. RL algorithms optimize the action choosing behavior by massive interaction between agent and environment [12]. Compared with traditional optimization approach, deep Q-learning network (DQN) estimate the state-action value by deep neural network. Although problems in high-dimensional state spaces have been successfully solved by DQN, DDPG has been proposed to extend DRL algorithms to continuous action space [27]. Then, the state space, action space, reward function and the environment are introduced briefly in this section.

State Space: At the start of time slot t, $s_t^n = \{h_t^n, \phi_n^{CPR}, \rho_t^n, E_t^n\}$ denotes the network state, where $h_t^n, \phi_n^{CPR}, \rho_t^n, E_t^n$ represent the channel vectors, the number of offloaded tasks, the unaccomplished task in the queue and the energy consumption, respectively.

Action Space: Based on the current state s_t^n, the learning system needs to decide which access point should be selected and take action of scheduling the tasks of GU n. Let the vector $a_t^n = \{x_{nmi}(t), \forall n \in \mathbf{N}, \forall m \in \mathbf{M}, \forall i \in \mathbf{I}\}$ denotes the action space, where $x_{nmi}(t) \in [0, 1]$ indicates that user n whether partially offload the task m to the MEC server i or not.

Reward Function and Policy: With the objective of long-term weighted sum of delay and energy consumption of all tasks, the reward function can be defined as $R_n(s_t^n) = \mathbb{E}\left[\lim_{T \to \infty} \frac{1}{T} \sum_{y=t}^{T} C_n(y)|s_t^n\right]$, Denote by π the stationary policy, and a value function is defined to determine the value of reward when the system state is s_n, which is defined as:

$$V_\pi(s^n) = \mathbb{E}\left[\sum_{t=0}^{\infty} \psi R_n(s_t^n, a_t^n)|\pi, s_0^n = s^n\right], \tag{14}$$

where $\psi \in [0, 1]$ denotes the discounting factor. After confirming the state space, action space and reward function, a DDPG based algorithm is proposed for this Markov decision process (MDP) problem, as shown in Algorithm 1. Generally, an experienced replay buffer $\mathbb{B}$ is denoted, which stores experiences and mini-batches of experience. In Algorithm 1, mini-batches of samples $(s, a, R, s') \sim U(\mathbb{B})$ will be drawn uniformly at random from $\mathbb{B}$. Based on temporal-difference learning, a combination of Monte Carlo method and dynamic programming, the Q-value can be updated as follows:

$$Q'_\pi(s_t^n, a_t^n) = Q_\pi(s_t^n, a_t^n) + \alpha\left[R_n(s_t^n, a_t^n) + \psi Q_\pi(s_t^n, a^*)\right], \tag{15}$$

where α denotes the learning rate and $a^* = \arg\max_{a_{n(t)}} Q_\pi(s_n^t, a_n(t))$ denotes the greedy action. The following loss function can be calculated by:

$$L(\theta) = \mathbb{E}_{(s,a,R,s') \sim U(\mathbb{B})}\left[\left(R_t^n + \psi Q_\pi(s_{t+1}^n, a^*|\theta) - Q_\pi(s_t^n, a_t^n|\theta)\right)^2\right], \tag{16}$$

An actor-critic approach is adopted in DDPG algorithm, we leverage two separate DNNs to approximate Q-value network $Q(s, a; \theta^Q)$, the actor $\mu(s|\theta^\mu)$. The policy gradient of the θ^μ can be calculated as follows:

$$\nabla_{\theta^\mu} J \approx \mathbb{E}_{(s,a,R,s') \sim U(\mathbb{B})}\left[\left(\nabla_a Q(s, a|\theta^Q)\nabla_{\theta^\mu} \mu(s|\theta^\mu)\right)\right] \tag{17}$$

4   Performance Evaluation

In this section, simulation is carried out to verify the proposed model and algorithm. Specifically, we begin by elaborating on the simulation settings. Afterwards, we present an evaluation on the experiment results.

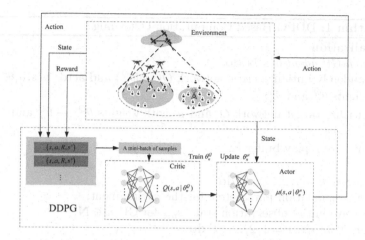

Fig. 2. The DDPG-based computation offloading scheme.

4.1 Simulation Settings

As shown in Fig. 2, the proposed approach is implemented by one actor network, one critic network and a replay buffer. Simulation environment is implemented via Python 3.6 and Tensorflow library. The DNNs' training and testing are conducted with a personal computer with AMD R7-4800H CPU. ReLU function is used as the activation function after the fully connected layer and L2 regularization is used to reduce DNN over-fitting. The number of neurons in the two hidden layers are 300 and 400, and we set 2000 and 0.001 to the number of episode and learning rate. Other important constant parameters are listed in the Table 1.

Table 1. Simulation parameters.

Parameter	Value	Parameter	Value
N	5	I	5
$f^i, i \neq 0$	5 GC/s	ϕ	5 MB
f^0	10 GC/s	γ	25 cycles/bit
f_n	200 MC/s	N_0	-100 dBm/Hz
B_U	3 MHz	P_U	1.6 W
B_S	2 MHz	P_S	5 W
d_{sat}	6.44 ms	$M_{\max}$	20

Algorithm 1: DDPG Based Computation Offloading

---

1 **Initialization:**
2 **for** *each GU agent* $n \in \mathbf{N}$ **do**
3   Randomly initialize critic network $Q(s, a|\theta_n^Q)$ and actor $\mu(s, a|\theta_n^\mu)$ with weights θ_n^Q and θ_n^μ;
4   Initialize target network Q' and μ' with weights $\theta_n^{Q'} \leftarrow \theta_n^Q$ and $\theta_n^{\mu'} \leftarrow \theta_n^\mu$;
5   Initialize replay buffer $\mathbb{B}$;
6 **end**
7 **for** *episode* $k = 1, 2, \ldots, K$ **do**
8   Reset simulation parameters for the environment;
9   Receive initial observation state $s_{n,1}$ for GU $n \in \mathbf{N}$;
10   **for** *time slot* $t = 1, 2, \ldots, T$ **do**
11   **for** *GU* $n \in \mathbf{N}$ **do**
12   Select action $a_t^n = \mu(s_t^n|\theta_n^\mu) + \Delta\mu$ according to the current policy and exploration noise $\Delta\mu$;
13   Execute action a_t^n and observe the reward R_t^n and the next state s_{t+1}^n;
14   Store transition $(s_t^n, a_t^n, R_t^n, s_{t+1}^n)$ in $\mathbb{B}$;
15   Sample random mini-batch of transitions $\{(s_z^n, a_z^n, R_z^n, s_z^n)\}_{z=1}^{Z}$ from $\mathbb{B}$;
16   Set $y_z = R_z^n + \psi Q'\left(s_{z+1}^n, \mu'(s|\theta_n^{\mu'})|\theta_n^{Q'}\right)$;
17   Update the critic network $Q(s, a|\theta_n^Q)$ by minimize the loss
$$L = \frac{1}{Z}\sum_{z=1}^{Z}\left(\left(y_z - Q_\pi(s_z^n, a_z^n|\theta_n^Q)\right)^2\right);$$
18   Update the actor policy by using the sampled policy gradient
$$\nabla_{\theta_n^\mu}J \approx \frac{1}{Z}\sum_{z=1}^{Z}\left(\left(\nabla_a Q(s_z, a|\theta_n^Q)|_{a=a_z}\nabla_{\theta_n^\mu}\mu(s_z|\theta_n^\mu)\right)\right);$$
19   Update the target networks: $\theta_n^{\mu'} \leftarrow \delta\theta_n^\mu + (1-\delta)\theta_n^{\mu'}$ and $\theta_n^{Q'} \leftarrow \delta\theta_n^Q + (1-\delta)\theta_n^{Q'}$;
20   **end**
21   **end**
22 **end**

4.2   Simulation Results

Firstly, we show the simulation results of the proposed algorithm, and evaluate the convergence performance. Figure 3 shows the convergence performance with respect to the average reward of total tasks presented by setting $\omega_1 = 0.8$, $\omega_1 = 0.5$ and $\omega_1 = 0.2$, and the results are averaged from ten numerical simulations, proving the effectiveness of neural networks.

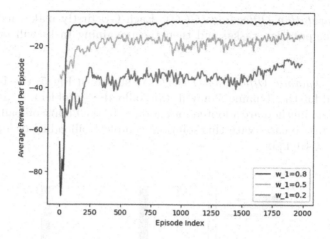

Fig. 3. The convergence of the proposed DDPG algorithm.

Meanwhile, it can be observed that the performance of the partial offloading policy learned from proposed algorithm is always better than the binary offloading policy learned from DQN for different scenarios by Fig. 4.

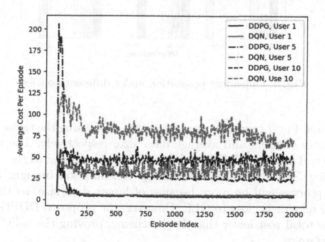

Fig. 4. Illustration of the average cost per episode.

To evaluate the computation offloading validity of the proposed DDPG scheme on SAGIN system, we adopt the other four benchmark schemes which are introduced as follows:

Greedy Local Execution (GLE): For each slot, the computation tasks will be processed locally ($x_{nmi}(t) = 0$) as many as possible.

Greedy Computation Offloading (GCO): Each GU firstly makes its best effort to offload computation tasks, and then the remaining tasks will be processed locally.

DQN based Dynamic Offloading (DQN): As shown in Fig. 5, the DQN is also implemented for the dynamic computation offloading problem, ε−greedy selection and Adam method are adopted for training. In the binary offloading scheme, there are only two cases with this solution: complete offloading or local processing ($x_{nmi}(t) \in \{0, 1\}$).

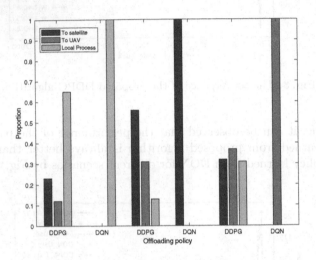

Fig. 5. Offloading proportion under different policies.

Figure 6 and Fig. 7 represent the sum cost of processing the tasks with respect to the numbers of users and average data size, respectively. The total cost of DQN, GCO and proposed DDPG schemes are all increasing as the average data size increasing. The time delay and energy consumption brought by the communication process will be more because of larger data size, so that the total cost of GCO scheme is close to GLE. However, the proposed DDPG scheme can still keep the total cost lower than DQN scheme, proving the validity of partial offloading strategy.

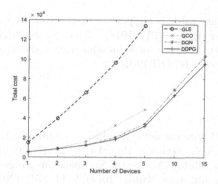

Fig. 6. Total cost vs. numbers of GU devices.

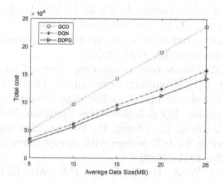

Fig. 7. Total cost versus average data size.

5   Conclusion

In this paper, an efficient computing offloading scheme is proposed for the MEC system in SAGIN. Firstly, we elaborated the SAGIN architecture. Then, we express computation offloading as a nonlinear optimization problem with the goal of minimizing the weighted sum of energy consumption and delay. On this basis, we propose an algorithm based on DDPG to solve this problem. Finally, the simulation results show that the energy consumption and time can be significantly saved by offloading the task to the edge server on the UAV or satellite, and the convergence performance and effectiveness of the proposed scheme in the simplified scenario are also proved.

In the future, the mobility management of satellites and UAVs will be further considered. In addition, the offloading scheme of SAGIN in areas with rich computing resources is also worth further study.

Ackonwledgment. This work is supported by the National Natural Science Foundation of China under grant (61761014), Guangxi Natural Science Foundation (2018GXNSFBA281131), Ministry of Education Key Laboratory of Cognitive Radio and Information Processing (CRKL190109).

References

1. Zhang, Z., et al.: 6G wireless networks: vision, requirements, architecture, and key technologies. IEEE Veh. Technol. Mag. **14**(3), 28–41 (2019)
2. Liu, J., Shi, Y., Fadlullah, Z.M., Kato, N.: Space-air-ground integrated network: a survey. IEEE Commun. Surv. Tutor. **20**(4), 2714–2741 (2018)
3. Jiang, W., Han, B., Habibi, M.A., Schotten, H.D.: The road towards 6G: a comprehensive survey. IEEE Open J. Commun. Soc. **2**, 334–366 (2021)
4. Abbas, N., Zhang, Y., Taherkordi, A., Skeie, T.: Mobile edge computing: a survey. IEEE Internet Things J. **5**(1), 450–465 (2018)
5. Khayyat, M., et al.: Multilevel service-provisioning-based autonomous vehicle applications. Sustainability **12**, 2497 (2020)
6. Mach, P., Becvar, Z.: Mobile edge computing: a survey on architecture and computation offloading. IEEE Commun. Surv. Tutor.. **19**(3), 1628–1656 (2017)
7. Lyu, X., et al.: Optimal schedule of mobile edge computing for internet of things using partial information. IEEE J. Sel. Areas Commun. **35**(11), 2606–2615 (2017)
8. Min, M., Xiao, L., Chen, Y., Cheng, P., Wu, D., Zhuang, W.: Learning-based computation offloading for IoT devices with energy harvesting. IEEE Trans. Veh. Technol. **68**(2), 1930–1941 (2019)
9. Elgendy, I.A., Zhang, W.-Z., He, H., Gupta, B.B., Abd El-Latif, A.A.: Joint computation offloading and task caching for multi-user and multi-task MEC systems: reinforcement learning-based algorithms. Wirel. Netw. **27**(3), 2023–2038 (2021). https://doi.org/10.1007/s11276-021-02554-w
10. Chen, Z., Wang, X.: Decentralized computation offloading for multi-user mobile edge computing: a deep reinforcement learning approach. EURASIP J. Wirel. Commun. Netw. **2020**(1), 1–21 (2020). https://doi.org/10.1186/s13638-020-01801-6
11. Zhou, C., et al.: Delay-aware IoT task scheduling in space-air-ground integrated network. In: 2019 IEEE Global Communications Conference (GLOBECOM), Waikoloa, HI, USA, 2019, pp. 1–6. IEEE (2019)
12. Xu, F., Yang, F., Zhao, C., Wu, S.: Deep reinforcement learning based joint edge resource management in maritime network. China Commun. **17**(5), 211–222 (2020)
13. Zhou, C., et al.: Deep reinforcement learning for delay-oriented IoT task scheduling in space-air-ground integrated network. IEEE Trans. Wirel. Commun. **20**(2), 911–925 (2021)
14. Dinh, T.H., Niyato, D., Hung, N.T.: Optimal energy allocation policy for wireless networks in the sky. In: 2015 IEEE International Conference on Communications (ICC), London, UK, 2015, pp. 3204–3209. IEEE (2015)
15. Tang, Q., Fei, Z., Li, B., Han, Z.: Computation offloading in LEO satellite networks with hybrid cloud and edge computing. IEEE Internet Things J. **11**, 9164–9176 (2021)
16. Cheng, N., et al.: Space/Aerial-assisted computing offloading for IoT applications: a learning-based approach. IEEE J. Sel. Areas Commun. **37**(5), 1117–1129 (2019)
17. Saleem, U., Liu, Y., Jangsher, S., Li, Y.: Performance guaranteed partial offloading for mobile edge computing. In: GLOBECOM (2018)

18. Wu, Q., Zeng, Y., Zhang, R.: Joint trajectory and communication design for multi-UAV enabled wireless networks. IEEE Trans. Wirel. Commun. **17**(3), 2109–2121 (2018)
19. Bi, S., Zhang, Y.: Computation rate maximization for wireless powered mobile-edge computing with binary computation offloading. IEEE Trans. Wirel. Commun. **17**(6), 4177–4190 (2018)
20. Guo, S., Xiao, B., Yang, Y., Yang, Y.: Energy-efficient dynamic offloading and resource scheduling in mobile cloud computing. In: IEEE International Conference on Computer Communications, pp. 1–9 (2016)
21. Zhang, N., Liang, H., Cheng, N., Tang, Y., Mark, J.W., Shen, X.S.: Dynamic spectrum access in multi-channel cognitive radio networks. IEEE J. Sel. Areas Commun. **32**(11), 2053–2064 (2014)
22. Hosseini, N., Jamal, H., Haque, J., Magesacher, T., Matolak, D.W.: UAV command and control, navigation and surveillance: a review of potential 5G and satellite systems. In: IEEE Aerospace Conference, pp. 1–10 (2019)
23. Shi, W., et al.: Multiple drone-cell deployment analyses and optimization in drone assisted radio access networks. IEEE Access **6**, 12518–1252 (2018)
24. Al-Hourani, A., Kandeepan, S., Lardner, S.: Optimal LAP altitude for maximum coverage. IEEE Wirel. Commun. Lett. **3**(6), 569–572 (2014)
25. Bor-Yaliniz, R.I., El-Keyi, A., Yanikomeroglu, H.: Efficient 3-D placement of an aerial base station in next generation cellular networks. In: IEEE ICC, pp. 1–5 (2016)
26. Kanellopoulos, S.A., Kourogiorgas, C.I., Panagopoulos, A.D., Livieratos, S.N., Chatzarakis, G.E.: Channel model for satellite communication links above 10GHz based on Weibull distribution. IEEE Commun. Lett. **18**(4), 568–571 (2014)
27. Lillicrap, T.P., et al.: Continuous control with deep reinforcement learning. Computer Science (2015)

Performance Analysis for UAV-Assisted mmWave Cellular Networks

Jiajin Zhao[1], Fang Cheng[2(✉)], Linlin Feng[1], and Zhizhong Zhang[2]

[1] School of Communication and Information Engineering, Chongqing University
of Posts and Telecommunications, Chongqing, China
[2] School of Electronics and Information Engineering, Nanjing University
of Information Science and Technology, Nanjing, China
2451457708@qq.com

Abstract. It is feasible to increase coverage area and throughout with unmanned aerial vehicle (UAV), as an assist to ground base station (GBS). In this paper, a 3-dimension (3-D) channel model, where UAVs are modeled by 3-D Poisson point process (PPP) and GBSs are modeled by 2-dimension (2-D) PPP, is proposed. Different path loss model for line-of-sight (LOS), non-LOS (NLOS) links and directional beamforming are considered. Moreover, the general expressions for signal-to-interference-plus-noise ratio (SINR) coverage probabilities, user throughput and area spectral efficiency (ASE) are derived. Our numerical and simulation results show that with the proposed 3-D channel model, the coverage probability of mmWave networks is greatly improved compared with sub-6GHz networks. Furthermore, the effects of the beamwidth of the main lobe, the main lobe directivity gain and UAV density on the performance of mmWave networks are analyzed.

Keywords: mmWave communication · UAV · Stochastic geometry · Cellular networks · Performance analysis

1 Introduction

In recent years, unmanned aerial vehicles (UAVs) have attracted great attention due to its advantages of low-cost, easy deployment and flexibility [1]. In the process of 5G/6G standardization, UAVs can be used as aerial base stations (BSs) to provide wireless services, and are a crucial complement of conventional cellular networks to achieve higher transmission efficiency and increase coverage and capacity. Different from conventional cellular networks, the UE is more likely to

Supported by the Natural Science Foundation of Chongqing, China, under the grant number cstc2019jcyj-msxmX0602.

Supplementary Information The online version contains supplementary material available at https://doi.org/10.1007/978-3-030-99200-2_40.

H. Gao et al. (Eds.): ChinaCom 2021, LNICST 433, pp. 538–552, 2022.
https://doi.org/10.1007/978-3-030-99200-2_40

associate with the UAV [2], which can improve the performance of UAV-assisted cellular networks. So UAV-assisted wireless communication is anticipated to be part of future wireless networks [3].

Stochastic geometry is considered to be a powerful tool for analysing UAV-assisted cellular networks [4]. By modeling ground BSs (GBSs) or UAVs as a specific point process, the performance metrics of the network can be derived based on the properties of the point process. The authors of [5,6] analysed the coverage of UAV-assisted cellular networks, where UAVs were distributed according to a Poisson point process (PPP) at the same height. In [7], UAV-assisted cellular networks were considered, in which both UAVs and GBSs were modeled by the independent PPP, and the coverage probability was investigated. In these works, only the coverage probability is analysed, but other performance metrics, such as average achievable rate, spectral efficiency and area spectral efficiency (ASE), are not considered. Therefore, a more comprehensive evaluation of the performance of UAV-assisted networks is a crucial research area.

mmWave (30–300 GHz) have been attracting growing attention for larger bandwidths to meet the exponentially growing demand in data traffic, which have also paved the way into the widespread use of UAV-assisted networks for future 5G and beyond wireless applications [8]. In [9,10], user-centric UAV-assisted cellular networks were studied, where UAVs were distributed according to a PPP at the same height and users were distributed according to a Poisson cluster process (PCP) around UAVs. In [10], the authors analysed downlink network performance, while the authors analysed uplink mmWave network performance and defined a successful transmission probability in [9]. The authors of [11] proposed a unified 3-dimension (3-D) spatial framework to model downlink and uplink UAV-assisted networks. In these works, UAVs are assumed to be at the same height in the air, but UAVs are usually at random heights in actual deployment. Therefore, in UAV-assisted mmWave cellular networks, modeling UAVs according to a 3-D PPP is more realistic, which has not been well studied in the literature.

In this paper, we focus on the performance analysis of UAV-assisted mmWave cellular networks, where UAVs and GBSs are modeled according to a 3-D PPP and a 2-dimension (2-D) PPP, respectively. The features of mmWave communication and different path loss model are considered. The distribution of the distance between the typical UE and the associated BS and the general expression of signal-to-interference-plus-noise ratio (SINR) coverage probabilities are derived. Based on the derived SINR coverage probabilities, other performance metrics of the networks, such as average achievable rate, rate coverage probabilities and ASE are obtained. The numerical and simulation results show that the coverage probability of mmWave networks is greatly improved compared with sub-6GHz networks. Moreover, the effects of the main lobe directivity gain, the beamwidth of the main lobe and UAV density on the performance of mmWave networks are analysed.

The rest of the paper is organized as follows: the system model are introduced in Sect. 2. In Sect. 3, the distance distribution and the association probability are derived. In Sect. 4, the expressions for coverage probabilities, rate and ASE of the UAV-assisted mmWave networks are derived. In Sect. 5, the effect of system parameters on the performance of networks are analysed. The conclusions are provided in Sect. 6.

2 System Model

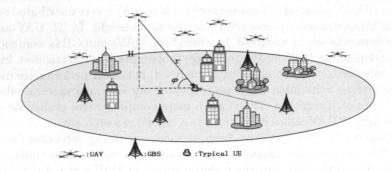

Fig. 1. UAV-assisted mmWave cellular networks.

As shown in Fig. 1, we consider a UAV-assisted downlink mmWave cellular network in which UAVs are distributed according to a 3-D homogeneous PPP Φ_U with density λ_U and transmit power P_U, and GBSs are distributed according to a 2-D homogeneous PPP Φ_G with density λ_G and transmit power P_G. Moreover, UAVs and GBSs are assumed to be transmitting in the 28 GHz mmWave frequency bands. According to Slivnyak theorem, a typical UE is assumed to be at the origin.

2.1 Channel Modeling

The large-scale fading and the small-scale fading are considered. In the large-scale fading, different path loss models are applied to the LOS and NLOS link. Thus, the path loss model is expressed as

$$L_j^s(r) = \begin{cases} k_j^L r^{\alpha_j^L} & \text{with prob. } p_j^L(r) \\ k_j^N r^{\alpha_j^N} & \text{with prob. } p_j^N(r), \end{cases} \tag{1}$$

where $j \in \{\text{UAV}, \text{GBS}\}$, $s \in \{\text{LOS}, \text{NLOS}\}$, k_j^L, k_j^N indicate intercepts of the LOS and NLOS path loss formulas, respectively. α_j^L, α_j^N indicate the path loss exponents of LOS and NLOS links, respectively, and $p_j^L(r)$ and $p_j^N(r) = 1 - p_j^L(r)$ respectively indicate the LOS probability and NLOS probability at distance of r.

The probability of LOS link between a UAV and a UE is shown as [2]

$$p_U^L(\varphi) = \frac{1}{1 + c \exp(-b(\varphi - c))} \tag{2}$$

where $\varphi(r) = \frac{180}{\pi} \arcsin\left(\frac{H}{r}\right)$ indicates the elevation angle and $\varphi \sim U(0, 90°)$. b and c are environment-dependent constants. H is the height of UAV, and r is the distance between UAV and UE.

As presented in [12], the probability of LOS link between a GBS and a UE can be expressed as

$$p_G^L(r) = e^{-\beta r} \tag{3}$$

where β is a constant, and it depends on the density of the building blockage process. r is the distance between GBS and UE.

In the small-scale fading, Nakagami fading are assumed to be applied for each link. Therefore, the small-scale fading gains of LOS and NLOS links are denoted as h_l and h_n, which is the Gamma distribution [12], i.e., $h_l \sim \Gamma(N_L, \frac{1}{N_L}), h_n \sim \Gamma(N_N, \frac{1}{N_N})$. N_L, N_N are the Nakagami fading parameters of LOS and NLOS links.

2.2 Antenna Gain

In this paper, UAVs and GBSs are equipped with antenna arrays to achieve directional beamforming. A sectored antenna model is used in this mmWave cellular networks. We assume that M_b, m_b and θ_b are the main lobe gain, side lobe gain and main lobe beamwidth of UAVs and GBSs, and M_u, m_u and θ_u are the corresponding parameters for UEs. We also assume that perfect beam alignment can be achieved between the UE and the serving BS, resulting in the maximum antenna gain $G_0 = M_b M_u$. Therefore, the antenna gain of the interference link can be expressed as [12]

$$G = \begin{cases} M_b M_u & \text{with prob. } p_{M_b M_u} = \left(\frac{\theta_b}{2\pi}\right)\left(\frac{\theta_u}{2\pi}\right) \\ M_b m_u & \text{with prob. } p_{M_b m_u} = \left(\frac{\theta_b}{2\pi}\right)\left(1 - \frac{\theta_u}{2\pi}\right) \\ m_b M_u & \text{with prob. } p_{m_b M_u} = \left(1 - \frac{\theta_b}{2\pi}\right)\left(\frac{\theta_u}{2\pi}\right) \\ m_b m_u & \text{with prob. } p_{m_b m_u} = \left(1 - \frac{\theta_b}{2\pi}\right)\left(1 - \frac{\theta_u}{2\pi}\right) \end{cases} \tag{4}$$

A summary of notations is provided in Table 1.

3 Distance Distributions and Association Probability

3.1 Distance Distribution Between the Typical UE and Nearest UAV/GBS

Given that there is at least one UAV around the typical UE, the CCDF and PDF of the distance between the typical UE and the nearest LOS or NLOS UAV are expressed as

$$\hat{F}_{R_U^s}(x) = e^{-2\pi\lambda_U \int_0^x \int_0^{\frac{\pi}{2}} p_U^s(\varphi)\cos(\varphi)t^2 d\varphi dt} / B_U^s, \tag{5}$$

Table 1. Table of notations

Notations	Description
Φ_U, λ_U, P_U	PPP of UAVs, the density and the transmit power of UAVs
Φ_G, λ_G, P_G	PPP of GBSs, the density and the transmit power of GBSs
L_j^s, p_j^s	Path loss and the probability of $s \in \{LOS, NLOS\}$ transmission in the $j \in \{UAV, GBS\}$
p_j^s, α_j^s	Path loss intercepts and the path loss components
h_s, N_s	Small scale fading gain and the Nakagami fading parameters for LOS/NLOS
θ_b, θ_u	The main lobe beamwidth of BSs and UEs
G_0, G, p_G	The antenna gain of the main link and interference link, the corresponding probability
B_U^s, B_G^s	The probability that there is at least on LOS or NLOS UAV, GBS around the typical UE
$\hat{F}_{R_U^s}, \hat{f}_{R_U^s}, \hat{F}_{R_G^s}, \hat{f}_{R_G^s}$	The CCDF and PDF of the distance between the typical UE and the nearest LOS or NLOS UAV, GBS
P_m	The received power of the main link
$A_{j,s}$	The probability that the typical UE is associated with a LOS or NLOS BS in the jth tier
$F_{R_j^s}, f_{R_j^s}$	The CCDF and PDF of the distance that the typical UE is associated with a LOS or NLOS BS in the jth tier
$SINR_{j,s}$	The signal-to-interference-plus-noise ratio
P_C, R_C, R	The SINR coverage probability, the rate coverage probability and the average achievable rate
$\mathcal{L}_I(s)$	The Laplace transform of the interference
ASE	Area spectral efficiency

$$\hat{f}_{R_U^s}(x) = e^{-2\pi\lambda_U \int_0^x \int_0^{\frac{\pi}{2}} p_U^s(\varphi)\cos(\varphi)t^2 d\varphi dt}$$
$$\times 2\pi\lambda_U x^2 \int_0^{\frac{\pi}{2}} p_U^s(\varphi)\cos(\varphi)\,d\varphi / B_U^S, \tag{6}$$

where $x \geq 0$, $B_U^s = 1 - e^{-2\pi\lambda_U \int_0^\infty \int_0^{\frac{\pi}{2}} p_U^s(\varphi)\cos(\varphi)t^2 d\varphi dt}$ is the probability that there is at least one LOS or NLOS UAV around the typical UE.

Proof. Given that there is at least one LOS or NLOS UAV around the UE, we have

$$\hat{F}_{R_U^s}(x) = \mathbb{P}(r > x | \text{there has at least one UAV arounnd})$$

$$= \frac{\mathbb{P}(r > x, \text{there has at least one UAV arounnd})}{\mathbb{P}(\text{there has at least one UAV arounnd})}$$

$$= \frac{\mathbb{P}(\text{there is at no UAV closer than } x)}{B_U^s} \qquad (7)$$

$$\overset{(a)}{=} e^{-\lambda_U V(x)} / B_U^s \overset{(b)}{=} e^{-\lambda_U \int_0^x \int_0^{2\pi} \int_0^{\frac{\pi}{2}} p_U^s(\varphi)\sin\left(\frac{\pi}{2}-\varphi\right)t^2 d\varphi d\theta dt} / B_U^s$$

$$= e^{-2\pi\lambda_U \int_0^x \int_0^{\frac{\pi}{2}} p_U^s(\varphi)\cos(\varphi)t^2 d\varphi dt} / B_U^s$$

where $s \in \{\text{LOS}, \text{NLOS}\}$, $V(x)$ is the volume of a circle with radius x in the air. (a) is from [13] and (b) follows from the integration of the volume using polar coordinates.

we can get the PDF as follows:

$$\hat{f}_{R_U^s}(x) = -\frac{d\hat{F}_{R_U^s}(x)}{dx} \qquad (8)$$

Similarly, the CCDF and PDF of the distance between the typical UE and the nearest LOS or NLOS GBS are [12]

$$\hat{F}_{R_G^s}(x) = e^{-2\pi\lambda_G \int_0^x tp_G^s(t)dt} / B_G^s, \qquad (9)$$

$$\hat{f}_{R_G^s}(x) = 2\pi\lambda_G x p_G^s(x) e^{-2\pi\lambda_G \int_0^x tp_G^s(t)dt} / B_G^s, \qquad (10)$$

where $x \geq 0$, $B_G^s = 1 - e^{-2\pi\lambda_G \int_0^\infty tp_G^s(t)dt}$ is the probability that there is at least one LOS or NLOS GBS around the typical UE.

3.2   Association Probability

UEs are assumed to be associated with the BS providing the strongest received power [14]. In other words, UEs are associated with the nearest BS in the UAV tier or GBS tier. Therefore, the received power of the main link can be expressed as

$$P_m = \arg\max_{j \in K, i \in \Phi} P_j G_0 L_{ji}^{-1} \qquad (11)$$

The probability that the typical UE is associated with a LOS or NLOS BS in the jth tier is

$$A_{j,s} = B_j^s \int_0^\infty B_j^{s'} \hat{F}_{R_j^{s'}}\left(Q_{jj}^{ss'}(r)\right) \prod_\varepsilon \left(B_j^\varepsilon, \hat{F}_{R_{j'}^\varepsilon}\left(Q_{j'j}^{s\varepsilon}(r)\right)\right) \hat{f}_{R_j^s}(r)\, dr, \qquad (12)$$

where $s, s', \varepsilon \in \{\text{LOS}, \text{NLOS}\}$, $s \neq s'$, $j, j' \in \{\text{UAV}, \text{GBS}\}$, $j \neq j'$, $Q_{kj}^{s\varepsilon}(r) = \left(\frac{P_k k_j^s}{P_j k_k^\varepsilon} r^{\alpha_j^s}\right)^{\frac{1}{\alpha_k^\varepsilon}}$.

The probabilities that the typical UE is associated with a UAV or GBS are given by

$$A_U = A_{U,L} + A_{U,N}, \quad A_G = A_{G,L} + A_{G,N} \tag{13}$$

Given that the typical UE is associated with the LOS or NLOS BS in the jth tier, the CCDF and PDF of the distance can be expressed as

$$F_{R_j^s}(x) = \frac{1}{A_{j,s}} \int_0^x \hat{f}_{R_j^s}(r) B_j^{s'} \left(B_j^{s'} \hat{F}_{R_j^{s'}} \left(Q_{jj}^{ss'}(r) \right) \prod_\varepsilon \left(B_{j'}^\varepsilon \hat{F}_{R_{j'}^\varepsilon} \left(Q_{j'j}^{s\varepsilon}(r) \right) \right) \right) dr, \tag{14}$$

$$f_{R_j^s}(x) = \frac{\hat{f}_{R_j^s}(x)}{A_{j,s}} \left[B_j^{s'} \left(B_j^{s'} \hat{F}_{R_j^{s'}} \left(Q_{jj}^{ss'}(x) \right) \right) \prod_\varepsilon \left(B_{j'}^\varepsilon \hat{F}_{R_{j'}^\varepsilon} \left(Q_{j'j}^{s\varepsilon}(x) \right) \right) \right]. \tag{15}$$

Proof. using a similar method to Lemma 3 and Remark 1 of [9], the expressions of (12), (14) and (15) can be easily obtained.

4   Coverage Probability

4.1   SINR

When the typical UE is associated with a UAV or GBS, the total interference from the BSs in the kth tier received by the typical UE is

$$I_k = \sum_\varepsilon \sum_{i \in \Phi_k^s \setminus q} P_k G_{k,i} h_{k,i} \left(k_k^\varepsilon r_{k,i}^{\alpha_k^\varepsilon} \right)^{-1} \tag{16}$$

where $k \in \{UAV, GBS\}$, and $G_{k,i}$ is the antenna gain. $h_{k,i}$ is the small-scale fading, and $\left(k_k^\varepsilon r_{k,i}^{\alpha_k^\varepsilon} \right)^{-1}$ is path loss of the link between the typical UE and the ith BS in the kth tier.

The expression of the SINR of the typical UE associated with the LOS or NLOS BS in the jth tier is

$$SINR_{j,s} = \frac{P_j G_0 h_j \left(k_j^s r^{\alpha_j^s} \right)^{-1}}{\sigma_n^2 + \sum_{k=1}^2 \sum_\varepsilon \sum_{i \in \Phi_k^\varepsilon \setminus q} P_k G_{k,i} h_{k,i} \left(k_k^\varepsilon r_{k,i}^{\alpha_k^\varepsilon} \right)^{-1}} \tag{17}$$

where $s, \varepsilon \in \{LOS, NLOS\}$, P_j denotes the transmit power of UAV or GBS. G_0 is the maximum antenna gain $M_b M_u$ between the serving BS and the typical UE. h_j is the small-scale fading gain of the link, and σ_n^2 is the variance of the Gaussian thermal noise power.

4.2   SINR Coverage Probability

Given that the typical UE is associated with the LOS or NLOS BS in the jth tier, the SINR coverage probability are defined as $P_{C_{j,s}} = \mathbb{P}(SINR_{j,s} > T)$, where T denotes the threshold of SINR. Therefore, the total SINR coverage of the UAV-assisted mmWave cellular network can be expressed as

$$P_C = \sum_{j=1}^{2} P_{C_j} = \sum_{j=1}^{2} \sum_{s} A_{j,s} P_{C_{j,s}} \tag{18}$$

where $A_{j,s}$ is the association probability. $P_{C_{j,s}}$ is the SINR coverage probability, which is given by the following expression.

Given that the typical UE is associated with the LOS or NLOS BS in the jth tier, the expression of SINR coverage probability is derived as

$$P_{C_{j,s}} = \mathbb{P}\left(SINR_{j,s} > T\right) = \mathbb{P}\left(\frac{P_j G_0 h_j L_{js}^{-1}}{\sigma_n^2 + I} > T\right) = \mathbb{P}\left(h_j > \frac{T L_{js}}{P_j G_0}\left(\sigma_n^2 + I\right)\right)$$

$$\overset{(a)}{\approx} 1 - \left[1 - \exp\left(-\frac{\eta_s T L_{js}}{P_j G_0}\left(\sigma_n^2 + I\right)\right)\right]^{N_s}$$

$$\overset{(b)}{=} \mathbb{E}_{L_{j,s}}\left[\sum_{n=1}^{N_s} (-1)^{n+1} \binom{N_s}{n} \exp\left(-\frac{n \eta_s T L_{js}}{P_j G_0}\left(\sigma_n^2 + I\right)\right)\right]$$

$$= \mathbb{E}_{L_{j,s}}\left[\sum_{n=1}^{N_s} (-1)^{n+1} \binom{N_s}{n} e^{-\mu_{j,s}\sigma_n^2} \mathcal{L}_{I_j}\left(\mu_{j,s}\right)\right]$$

$$\overset{(c)}{=} \sum_{n=1}^{N_s} (-1)^{n+1} \binom{N_s}{n} \int_0^\infty e^{-\mu_{j,s}\sigma_n^2} \mathcal{L}_{I_j}\left(\mu_{j,s}\right) f_{R_j^s}(r)\, dr$$

$$\tag{19}$$

where $\mu_{j,s} = \frac{n \eta_s T L_{js}}{P_j G_0}$, $\eta_s = N_s(N_s!)^{-\frac{1}{N_s}}$, N_s is the parameter of Nakagami fading, and $\mathcal{L}_I(s) = \mathbb{E}\left[e^{-sI}\right]$ is the Laplace transform of the interference. (a) from Alzer's Lemma [15]. (b) follows from the moment generating function (MGF) of the gamma random variable h, and (c) follows from independence of interference.

The expression of the Laplace transform of the total interference received by the typical UE is

$$\mathcal{L}_{I_j}\left(\mu_{j,s}\right) = \prod_\varepsilon \prod_G \mathcal{L}_{I_{j,U}^{\varepsilon,G}}\left(\mu_{j,s}\right) \mathcal{L}_{I_{j,G}^{\varepsilon,G}}\left(\mu_{j,s}\right)$$

$$= \prod_\varepsilon \prod_G \exp\left(-2\pi\lambda_U p_G \int_{Q_{U_j}^{s\varepsilon}(r)}^{\infty} \left(1 - \frac{1}{\left(1 + \mu_{j,s} P_U G_i L_i^{-1} N_\varepsilon^{-1}\right)^{N_\varepsilon}}\right)\right.$$

$$\times p_U^\varepsilon\left(\arctan(x)\right)\cos\left(\arctan(x)\right)\frac{x^2}{1+x^2}dx$$

$$\left. -2\pi\lambda_G p_G \int_{Q_{G_j}^{s\varepsilon}(r)}^{\infty}\left(1 - \frac{1}{\left(1 + \mu_{j,s} P_G G_i L_i^{-1} N_\varepsilon^{-1}\right)^{N_\varepsilon}}\right) p_G^\varepsilon(y)\, y\, dy\right)$$

$$\tag{20}$$

where $s, \varepsilon \in \{LOS, NLOS\}$, $Q_{kj}^{s\varepsilon}(r)$ is an exclusive radius in which no interfering BS exists. p_G is the probability corresponding to the antenna gain

$G \in \{M_b M_u, M_b m_u, m_b M_u, m_b m_u\}$, and the specific expression is given by (4). $p_j^\varepsilon(r)$ indicates the probability of LOS or NLOS in the jth tier. Therefore, by substituting (20) into (19), SINR coverage probability can be obtained.

Proof. see Appendix A.

4.3 Rate And ASE

Average Achievable Rate. The achievable rate can be defined as

$$R = W\log_2(1 + SINR) \qquad (21)$$

where W is the bandwidth assigned to the typical UE.

Given that the typical UE is associated with the LOS or NLOS BS in the jth tier, the average achievable rate can be expressed as [12]

$$R_{j,s} = \mathbb{E}(R) = \frac{W}{\ln 2} \int_0^\infty \frac{P_{C_{j,s}}(\omega)}{1+\omega} d\omega \qquad (22)$$

where $P_{C_{j,s}}(\omega)$ is the total SINR coverage probability of the UAV-assisted mmWave cellular network.

The average achievable rate of the typical UE associated with a UAV or GBS is given by

$$R_j = \frac{A_{j,L}}{A_{j,L} + A_{j,N}} R_{j,L} + \frac{A_{j,N}}{A_{j,L} + A_{j,N}} R_{j,N} \qquad (23)$$

where $A_{j,L}, A_{j,N}$ are the association probability, and $R_{j,L}, R_{j,N}$ are the average achievable rate of the LOS and NLOS in the jth tier, respectively.

Rate Coverage Probability. Similar to the definition of the SINR coverage probability, the rate coverage probability is defined as the probability that the rate is greater than a threshold γ if the typical UE is associated with LOS or NLOS BS in the jth tier. Therefore, the expression is as

$$\begin{aligned} R_{C_{j,s}} = (R_{j,s} > \gamma) = (W\log_2(1 + SINR_{j,s}) > \gamma) \\ = \left(SINR_{j,s} > 2^{\frac{\gamma}{W}} - 1\right) = P_{C_{j,s}}\left(2^{\frac{\gamma}{W}} - 1\right) \end{aligned} \qquad (24)$$

Similarly, the total rate coverage probability of the entire network can be expressed as

$$R_C = \sum_{j=1}^2 R_{C_j} = \sum_{j=1}^2 \sum_s A_{j,s} R_{C_{j,s}} \qquad (25)$$

ASE. Area spectral efficiency is an important metric to measure network performance, and the expression is

$$ASE_j = \lambda_j \frac{R_j}{W} \qquad (26)$$

where λ_j is the density of UAV or GBS and R_j is the average achievable rate of UAV or GBS. The ASE of UAV (bps/Hz/m^3) and GBS (bps/Hz/m^2) are different in units.

5 Performance Analysis and Discussions

In this section, We used MATLAB software to compare theoretical results and simulation results. The numerical results are provided to evaluate the performance of UAV-assisted mmWave cellular networks. 10000 times Monte Carlo method is used to verify the accuracy of the expressions. In Monte Carlo simulation, We first generate random BS based on the density, model the networks scenario, and then calculate the UE received power from the nearest BS and interference from other BSs to obtain the SINR. When it is greater than the threshold, the count is increased by one. Repeat 10,000 times and finally get SINR coverage probability. Unless otherwise stated, the parameter values are shown in Table 2.

Table 2. System parameters

Parameters	Values
λ_U, λ_G	$2 \times 10^{-7}/\text{m}^3, 10^{-5}/\text{m}^2$
P_U, P_G	24 dBm, 34 dBm
$k_j^L, k_j^N, \alpha_j^L, \alpha_j^N$	$10^{3.08}, 10^{0.27}, 2.09, 3.75$
b, c	$0.136, 11.95$
$1/\beta$	141.4
Carrier frequency, W	28 GHz, 100 MHz
σ_n^2	$-174\,\text{dBm/Hz} + 10\log 10\,(W) + 10\,\text{dB}$
M_b, M_u, m_b, m_u	18 dB, 10dB, -10 dB, -10 dB
θ_b, θ_u	$\pi/6, \pi/2$
N_L, N_N	2, 3

First, we analyse the effect of UAV density on the association probability. As shown in Fig. 2, when we increase UAV density, the association probability of UAV increases while the association probability decreases due to the reduction of the distance between the typical UE and the nearest UAV, resulting in the

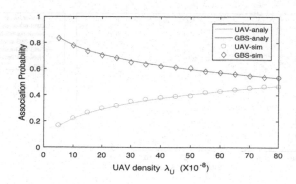

Fig. 2. The association probability as a function of UAV density.

reduction in the path loss between the typical UE and the UAV. Therefore, the typical UE is more likely to associate with the UAVs.

In Fig. 3, we plot the SINR coverage probabilities of UAV, GBS and the entire network. The SINR coverage probability decreases with the increasing SINR threshold. Moreover, GBS have higher coverage probability compared with UAV, indicating that the UEs are more likely to be covered by GBS. This is because GBS has higher transmit power and closer distance between UE compared with UAV. Finally, we also observe that simulation results are closely consistent with the analytical results, which confirm the correctness of the theoretical expressions.

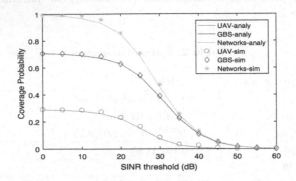

Fig. 3. The SINR coverage probability as a function of the SINR threshold.

In Fig. 4, we compare the coverage probability of mmWave and sub-6GHz networks. we set $G_0 = 0$, $G = 0$ (i.e., no antenna gain) and $\sigma_n^2 = -104$ dBm in the sub-6GHz networks. The solid line denotes mmWave networks, and the dotted line denotes sub-6GHz networks. It can be seen that when $T = 10$ dB, the coverage probabilities of sub-6GHz networks are close to 0, while mmWave networks still have high coverage probabilities. This is because of the high antenna gain brought by directional beamforming. Moreover, it shows that the coverage probability of mmWave networks can be greatly improved compared with sub-6GHz networks.

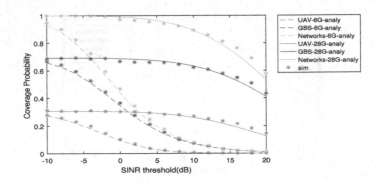

Fig. 4. The SINR coverage probability for mmWave and sub-6GHz networks.

In Fig. 5, the impact of small-scale fading type on the coverage probability is studied. When $N_L = N_N = 1$, Nakagami fading specializes to Rayleigh fading. As shown in the figure, higher coverage probability can be achieved with it than Rayleigh fading. The reason is that Nakagami fading leads to more better channel conditions. Figure 5 also shows that small-scale fading type has little impact on the coverage probability of cellular networks.

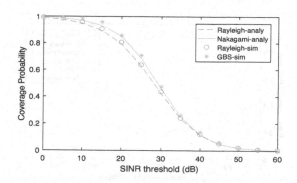

Fig. 5. The SINR coverage probability for Nakagami and Rayleigh fading.

Figure 6 shows the numerical and simulation results for different main antenna gain and main lobe beamwidth. It could be observed that better SINR coverage performance can be achieved by increasing the main antenna gain for the same main lobe beamwith $\theta_b = \theta_u = 30°$. Since the transmission gain G_0 of the main link increases with the increasing main antenna gain, the received power increases. Moreover, when we increase the main lobe beamwith, the SINR coverage probability decreases for the same main antenna gain $M_b = M_u = 10$ dB due to the growing impact of interference.

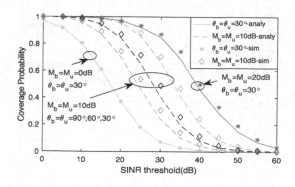

Fig. 6. The effect of antenna gain and main lobe beamwidth on SINR coverage probability.

550   J. Zhao et al.

In Fig. 7, we plot the rate coverage probability as a function of the rate threshold. The rate coverage probability decreases with the increasing rate threshold because of the definition of rate coverage probability. Approximately 95% of the rate coverage probability can be achieved when $\gamma = 400$ Mbps, indicating that UEs can be provided larger transmission rate in this UAV-assisted mmWave networks. And only GBSs have a rate coverage greater than 1150 Mbps in the figure.

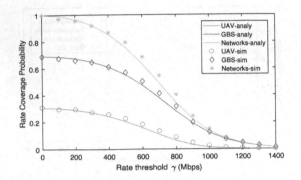

Fig. 7. The rate coverage probability as a function of the rate threshold.

In Fig. 8, we plot separately the ASE as a function of UAV density due to different units. It can be observed that the ASE of UAV and the ASE of GBS increase with the increasing UAV density. Since the increase in UAV density, the number of UAVs increases, which leads to more UEs associated with UAVs, which increases the total coverage probability of the networks. Moreover, it indicates that adding the number of UAVs can improve the ASE of UAV-assisted mmWave cellular networks.

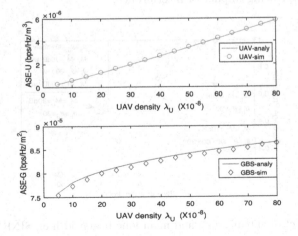

Fig. 8. The ASE as a function of UAV density.

6   Conclusion

In this paper, we derive the general expressions for the SINR coverage probabilities and other performance metrics of the UAV-assisted mmWave cellular networks by modeling UAVs and GBSs as a 3-D PPP and a 2-D PPP. The numerical and simulation results show the effect of the main parameters, such as UAV density, the antenna gain and the main lobe beamwidth, on the performances of networks in terms of coverage probability, average achievable rate and ASE. Analysing the uplink performance of UAV-assisted mmWave cellular networks and the application of some new technologies such as non-orthogonal multiple access (NOMA), device-to-device (D2D), full-duplex (FD), etc., remain as future work.

A   Appendix

The total interference from the BSs in k^{th} tier received by the typical UE can be expressed as

$$I_{j,k}^{\varepsilon} = I_{j,k}^{\varepsilon,M_b M_u} + I_{j,k}^{\varepsilon,M_b m_u} + I_{j,k}^{\varepsilon,m_b M_u} + I_{j,k}^{\varepsilon,m_b m_u} = \sum_G I_{j,k}^{\varepsilon,G} \qquad (27)$$

where $G \in \{M_b M_u, M_b m_u, m_b M_u, m_b m_u\}$. Therefore, the Laplace transform of the interference that the typical UE receives from the kth tier can be expressed as

$$\mathcal{L}_{I_{j,k}^{\varepsilon}}(\mu_{j,s}) = \mathbb{E}\left[e^{-\mu_{j,s}\sum_G I_{j,k}^{\varepsilon,G}}\right] = \prod_G \mathcal{L}_{I_{j,k}^{\varepsilon,G}}(\mu_{j,s}) \qquad (28)$$

The Laplace transform of the interference from LOS or NLOS UAV with antenna gain G received by the typical UE can be calculated as

$$\begin{aligned}
\mathcal{L}_{I_{j,U}^{\varepsilon,G}}(\mu_{j,s}) &= \mathbb{E}_{\Phi_U^{\varepsilon},h_i}\left[\exp\left(-\mu_{j,s}\sum_{i\in\Phi_U^{\varepsilon}} P_U h_i G_i L_i^{-1}\right)\right] \\
&\overset{(a)}{=} \exp\left(-\lambda_U p_G \int_{R^3}\left(1 - \mathbb{E}_{h_i}\left[\exp\left(-\mu_{j,s}P_U h_i G_i L_i^{-1}\right)\right]\right)dx\right) \\
&\overset{(b)}{=} \exp\left(-2\pi\lambda_U p_G \int_{Q_{U_j}^{s\varepsilon}(r)}^{\infty}\left(1 - \frac{1}{\left(1 + \mu_{j,s}P_U G_i L_i^{-1} N_\varepsilon^{-1}\right)^{N_\varepsilon}}\right)\right. \\
&\qquad\qquad\qquad\qquad \left. \times p_U^{\varepsilon}\left(\arctan(x)\right)\cos\left(\arctan(x)\right)\frac{x^2}{1+x^2}dx\right)
\end{aligned} \qquad (29)$$

where (a) follows from the probability generating functional (PGFL) of the PPP, and (b) is obtained by MGF of the gamma random variable h.

Similarly, the Laplace transform of the interference from LOS or NLOS GBS with antenna gain G received by the typical UE can be calculated as

$$\mathcal{L}_{I_{j,G}^{\varepsilon,G}}(\mu_{j,s}) = \exp\left(-2\pi\lambda_G p_G \int_{Q_{Gj}^{\varepsilon\varepsilon}(r)}^{\infty} \right.$$
$$\left. \times \left(1 - \frac{1}{\left(1 + \mu_{j,s} P_G G_i L_i^{-1} N_\varepsilon^{-1}\right)^{N_\varepsilon}}\right) p_G^\varepsilon(y)\, y dy\right) \tag{30}$$

By combining (29) and (30), (20) can be obtained.

References

1. Zhang, C., Zhang, W., Wang, W., Yang, L., Zhang, W.: Research challenges and opportunities of UAV millimeter-wave communications. IEEE Wirel. Commun. **26**(1), 58–62 (2019)
2. Al-Hourani, A., Kandeepan, S., Lardner, S.: Optimal lap altitude for maximum coverage. IEEE Wirel. Commun. Lett. **3**(6), 569–572 (2014)
3. Wang, L., Che, Y.L., Long, J., Duan, L., Wu, K.: Multiple access mmWave design for UAV-aided 5G communications. IEEE Wirel. Commun. **26**(1), 64–71 (2019)
4. Andrews, J.G., Baccelli, F., Ganti, R.K.: A tractable approach to coverage and rate in cellular networks. IEEE Trans. Commun. **59**(11), 3122–3134 (2011)
5. Liu, C., Ding, M., Ma, C., Li, Q., Lin, Z., Liang, Y.: Performance analysis for practical unmanned aerial vehicle networks with LoS/NLoS transmissions. In: 2018 IEEE International Conference on Communications Workshops (ICC Workshops), Kansas City, USA, pp. 1–6 (2018)
6. Zhou, L., Yang, Z., Zhou, S., Zhang, W.: Coverage probability analysis of UAV cellular networks in urban environments. In: 2018 IEEE International Conference on Communications Workshops (ICC Workshops), Kansas City, USA, pp. 1–6 (2018)
7. Zhang, Y., Cai, Y., Zhang, J.: Downlink coverage performance analysis of UAV assisted terrestrial cellular networks. In: 2019 IEEE 19th International Conference on Communication Technology (ICCT), Jaipur, India, pp. 551–555 (2019)
8. Zhang, L., Zhao, H., Hou, S., et al.: A survey on 5G millimeter wave communications for UAV-assisted wireless networks. IEEE Access **7**, 117460–117504 (2019)
9. Wang, X., Gursoy, M.C.: Coverage analysis for energy-harvesting UAV-assisted mmWave cellular networks. IEEE J. Sel. Areas Commun. **37**(12), 2832–2850 (2019)
10. Turgut, E., Gursoy, M.C.: Downlink analysis in unmanned aerial vehicle (UAV) assisted cellular networks with clustered users. IEEE Access **6**(36), 36313–36324 (2018)
11. Yi, W., Liu, Y., Bodanese, E., Nallanathan, A., Karagiannidis, G.K.: A unified spatial framework for UAV-aided mmWave networks. IEEE Trans. Commun. **67**(12), 8801–8817 (2019)
12. Bai, T., Heath, R.W.: Coverage and rate analysis for millimeter-wave cellular networks. IEEE Trans. Wirel. Commun. **14**(2), 1100–1114 (2015)
13. Pan, Z., Zhu, Q.: Modeling and analysis of coverage in 3-d cellular networks. IEEE Commun. Lett. **19**(5), 831–834 (2015)
14. Turgut, E., Gursoy, M.C.: Coverage in heterogeneous downlink millimeter wave cellular networks. IEEE Trans. Commun. **65**(10), 4463–4477 (2017)
15. Andrews, J.G., Bai, T., Kulkarni, M.N., Alkhateeb, A., Gupta, A.K., Heath, R.W.: Modeling and analyzing millimeter wave cellular systems. IEEE Trans. Commun. **65**(1), 403–430 (2017)

Aerial Intelligent Reflecting Surface for Secure MISO Communication Systems

Zhen Liu[1], Zhengyu Zhu[1,2]([✉]), Wanming Hao[1,2], and Jiankang Zhang[2]

[1] Henan Institute of Advanced Technology, Zhengzhou University, Zhengzhou, China
ieliuzhen@gs.zzu.edu.cn, zhuzhengyu6@gmail.com, iewmhao@zzu.edu.cn
[2] School of Information Engineering, Zhengzhou University, Zhengzhou, China
iejkzhang@zzu.edu.cn

Abstract. In this paper, an aerial intelligent reflecting surface (AIRS) is introduced to assist the ground secure communication system, where intelligent reflecting surface shifts the communication channel to target position and AIRS can be flexibly deployed in the air. We aim to maximize the secure communication rate of the system, and the formulated problem is difficult to solve. To this end, the problem is decomposed into three sub-problems, and an efficient iterative algorithm is proposed by jointly optimizing the transmission beamforming, the reflection phase shift and trajectory of the AIRS with the alternating optimization technique. The semi-definite relaxation is used to optimize the AIRS's phase shift, and the successive convex approximation method is applied to solve the non-convex trajectory optimization sub-problem. Numerical results show that the secrecy rate of the proposed algorithm precedes the benchmark schemes, and the AIRS has found the same optimal position with respect to different origins.

Keywords: Aerial intelligent reflecting surface · Secure communication · MISO system · Alternating optimization

1 Introduction

Nowadays, sixth-generation (6G) network are developing rapidly to meet the large-scale communication needs, especially in flexible, high-speed, secure communication, etc. Intelligent reflecting surface (IRS), as a new wireless device, can smartly reflect the signal through exhaustive low-cost passive reflecting elements and is always used to alter the transmission of radio signals over the wireless network [1].

This work was supported in part by the National Natural Science Foundation of China under Grant 61801434, 61571401, 61801435, and 61771431, in part by the Project funded by China Postdoctoral Science Foundation under Grant 2020M682345, in part by the Henan Postdoctoral Foundation under Grant 202001015, in part by the Innovative Talent of Colleges and University of Henan Province under Grant 18HASTIT021, in part by the Science and Technology Innovation Project of Zhengzhou under Grant 2019CXZX0037.

H. Gao et al. (Eds.): ChinaCom 2021, LNICST 433, pp. 553–566, 2022.
https://doi.org/10.1007/978-3-030-99200-2_41

IRS shows great potential to enhance physical layer security. Wei Sun studied the security issue of the IRS aided simultaneous wireless information and power transfer system, in which the transmit beamforming and artificial noise at the base station (BS) and the reflective beamforming at the IRS are jointly designed to maximize the secrecy rate to the receivers [2]. The transmit beamforming and the reflective beamforming of the IRS are jointly designed to maximize the secrecy rate of multiple input single output (MISO) communication system [3]. The author in [4] further complete the outage constrained robust design for IRS system, which takes the artificial noise into consideration.

On the other hand, unmanned aerial vehicle (UAV) can act as a mobile relay or air BS to aid the secure communication in different scenarios [5,6]. Furthermore, UAV is able to build the line-of-sight (LOS) dominant transmission links with the ground users, and IRS can smartly adjust their reflecting elements for passive phase shift. It is proposed that UAV carrying an IRS serves as a mobile relay to assist the ground communication [7–9]. New methods have been discussed in [10] by jointly applying IRS and UAV in integrated air-ground wireless networks.

Motivated by the benefits of the IRS and the mobile UAV, we consider an aerial intelligent reflecting surface (AIRS), which can be flexibly deployed and adjust the phase shift simultaneously. The transmit beamforming, the phase shift and trajectory of the AIRS are jointly optimized to assist secure MISO communication system. Our goal is to maximize the secrecy rate, and the main contributions are summarized as follows:

- The maximization of secrecy rate problem is difficult to solve due to the non-convexity of the objective function and multiple coupled variables. An alternating optimization (AO) method is applied to solve the three sub-problems, where the variables are optimized in an iterative manner.
- For the transmit beamforming vector, we obtain a closed-form solution in sub-problem 1, and the reflection phase shift of AIRS is solved by the semi-definite relaxation (SDR) method in sub-problem 2. In sub-problem 3 of the AIRS's optimal positon, we first introduce auxiliary variables to simplify the objective function and decouple the position variable. Then we apply the successive convex approximation (SCA) method to reformulate the non-convex trajectory problem into an approximated convex problem.
- Simulation results are provided to verify the effectiveness and convergence of the proposed AO algorithm, which precedes the benchmark schemes without phase shift optimization and fixed AIRS. Besides, the AIRS has found the same optimization position in different origins situation.

2   System Model and Problem Formulation

2.1   System Model

In this paper, we consider an AIRS-based MISO communication system, in which the UAV is equipped with an IRS as a passive relay, and an eavesdropper attempts to wiretap the legitimate communication between the BS and the user on the ground. The AIRS is equipped with $N = N_x \times N_y$ reflecting elements, whose phase can be manipulated by the controller on the UAV, and N_x, N_y are passive reflection units which are spanned as a uniform planar array. We assume the BS is equipped with M antennas arranging in alignment, and the user and eavesdropper are equipped with a single antenna, respectively.

Without loss of generality, all nodes are placed in the three-dimensional (3D) Cartesian coordinate system. The BS, user, and eavesdropper are fixed on the ground, whose horizontal coordinates can be denoted by $\mathbf{C}_B = [x_B, y_B]$, $\mathbf{C}_U = [x_U, y_U]$ and $\mathbf{C}_E = [x_E, y_E]$, respectively. The coordinate of AIRS is denoted by $\mathbf{C}_I = [x_I, y_I, z_I]$, and it can travel at a constant altitude z_I. In addition, the AIRS is constrained by flying within a certain horizontal range as

$$\mathbf{C}_I \in \mathcal{X} \times \mathcal{Y}. \tag{1}$$

where $\mathcal{X}$ and $\mathcal{Y}$ denoted the feasible region of the UAV deployment in the x-asix and y-axis, respectively [9].

We assume that the nodes communicate with the AIRS are denoted by the set of $Z \in \{B, U, E\}$, the elements of Z represent the BS, the user, and the eavesdropper, respectively. The reflection elements of AIRS are arranged in a rectangular matrix, and then the channel coefficient of the antenna of Z with the AIRS can be expressed as

$$\mathbf{v}_{ZI} = \mathbf{v}_x \otimes \mathbf{v}_y, \tag{2}$$

where $\mathbf{v}_x$ and $\mathbf{v}_y$ are set as $\mathbf{v}_x = [1, e^{-j\frac{2\pi\Delta d}{\lambda}\varphi_{ZI}}, ..., e^{-j\frac{2\pi\Delta d}{\lambda}(N_x-1)\varphi_{ZI}}]^T$ and $\mathbf{v}_y = [1, e^{-j\frac{2\pi\Delta d}{\lambda}\phi_{ZI}}, ..., e^{-j\frac{2\pi\Delta d}{\lambda}(N_y-1)\phi_{ZI}}]^T$, and they denote the channel coefficient of the link from one antenna to the elements of N_x and N_y in the AIRS, respectively. Here, φ_{ZI} and ϕ_{ZI} represent the azimuth and elevation angles of the AIRS component, which can be denoted as $\varphi_{ZI} = \frac{x_I - x_Z}{d_{ZI}}$ and $\phi_{ZI} = \frac{z_I - z_Z}{d_{ZI}}$. In the above formulas, d_{ZI} represents the distance between the antenna of Z and the AIRS. Constants Δd and λ denote the reflecting element separation at the AIRS and the carrier wavelength, respectively. x_Z and z_Z represent the coordinates of the x-axis and z-axis of Z, respectively. Since AIRS constantly moves throughout the process, $\mathbf{v}_{ZI}$ changes with the coordinate of AIRS [11].

In this situation, the channels related to the AIRS are dominated by LOS component. By applying the existing channel estimation techniques, we assume the channel state information (CSI) of AIRS-related channels can be obtained based on the method in [12]. Accordingly, the channel gains between the AIRS with the user and eavesdropper can be expressed as

$$\mathbf{h}_{UI} = \sqrt{\rho d_{UI}^{-\alpha}} * \mathbf{v}_{UI}, \tag{3a}$$

$$\mathbf{h}_{EI} = \sqrt{\rho d_{EI}^{-\alpha}} * \mathbf{v}_{EI}, \tag{3b}$$

where ρ is the pass loss coefficient of the reference distance $D_0 = 1m$. d_{UI} denotes the distance between the AIRS and the user, and d_{EI} is the same as

d_{UI}. Constant α represents the corresponding path loss exponent related to the ground-air (G-A) link. $\mathbf{v}_{UI}$ and $\mathbf{v}_{EI}$ represent the channel coefficient of the AIRS with the user and eavesdropper, which are in the form of $\mathbf{v}_{ZI}$ with $Z \in \{B, U, E\}$.

The M antennas of the BS arrange in linear planar. Since the antenna separation is weeny compared to the distance between the BS and the AIRS, the channel coefficients of different antennas of BS with the AIRS are treated as the same. Then, the channel gain between the BS and the AIRS can be expressed as

$$\mathbf{H}_{BI} = \sqrt{\rho d_{BI}^{-\alpha}} \mathbf{v}_{BI} * [1, e^{-j\frac{2\pi\Delta a}{\lambda}\phi_{BI}}, ..., e^{-j\frac{2\pi\Delta a}{\lambda}(M-1)\phi_{BI}}] = \sqrt{\rho d_{BI}^{-\alpha}} \mathbf{V}_{BI}, \quad (4)$$

where d_{BI} denotes the distance between the BS and the AIRS. $\mathbf{v}_{BI}$ is the channel coefficient between the BS and the AIRS. Δa represents the antenna separation at the BS. ϕ_{BI} is the elevation angles of the BS with the AIRS, and $\mathbf{V}_{BI} = \mathbf{v}_{BI} * [1, e^{-j\frac{2\pi\Delta a}{\lambda}\phi_{BI}}, ..., e^{-j\frac{2\pi\Delta a}{\lambda}(M-1)\phi_{BI}}]$.

We assume that the channels of the gorund-ground (G-G) communication link follow Rayleigh fading, and then the channel gains of the BS-user link and the BS-eavesdropper link can be expressed as

$$\mathbf{h}_{BU} = \sqrt{\rho d_{BU}^{-\kappa}} * \mathbf{g}_{BU}, \quad (5a)$$

$$\mathbf{h}_{BE} = \sqrt{\rho d_{BE}^{-\kappa}} * \mathbf{g}_{BE}, \quad (5b)$$

where d_{BU} and d_{BE} are the distance from the BS to the user and eavesdropper, κ is the corresponding path loss exponent related to the G-G link. $\mathbf{g}_{BU}$ and $\mathbf{g}_{BE}$ denote the small-scale fading, which represent the random scattering component modeled by a circularly symmetric complex Gaussian (CSCG) distribution featuring zero-mean and unit-variance. Therefore, we can obtain $\mathbf{H}_{BI} \in \mathbb{C}^{N \times M}$, $\mathbf{h}_{EI} \in \mathbb{C}^{N \times 1}$, $\mathbf{h}_{UI} \in \mathbb{C}^{N \times 1}$, $\mathbf{h}_{BU} \in \mathbb{C}^{1 \times M}$, $\mathbf{h}_{BE} \in \mathbb{C}^{1 \times M}$, respectively.

The BS transmits a confidential message to the user via beamforming $\mathbf{w} \in \mathbb{C}^{M \times 1}$, which satisfies the following constraint

$$||\mathbf{w}||^2 \leq P_{BS}, \quad (6)$$

where P_{BS} denotes the BS's maximum transmit power. Moreover, the signals from BS are reflected by AIRS in adjustable phase shift, and the reflection phase of the AIRS is modeled by using $\mathbf{q} \triangleq [q_1, ..., q_N]^T$, where $q_n = \beta_n e^{j\theta_n}$, $\beta_n \in [0, 1]$ and $\theta_n \in [0, 2\pi)$ denote the amplitude reflection coefficient and phase shift of the n-th element, respectively. For simplicity, $\beta_n = 1, \forall n$ is set to achieve the maximum reflecting power gain, thus the elements of $\mathbf{q}$ should satisfy

$$||q_n|| = 1, \forall n. \quad (7)$$

Due to the severe path loss, ignoring the signal reflected twice or more by the AIRS, the signal received by the user and the eavesdropper can be respectively expressed as

$$y_U = (\mathbf{h}_{UI}^T \mathbf{Q} \mathbf{H}_{BI} + \mathbf{h}_{BU})\mathbf{w} + n_U, \quad (8)$$

$$y_E = (\mathbf{h}_{EI}^T \mathbf{QH}_{BI} + \mathbf{h}_{BE})\mathbf{w} + n_E, \tag{9}$$

where $\mathbf{Q}$ denotes the diagonal matrix of the vector $\mathbf{q}$. n_U and n_E denote the additive white Gaussian noises with mean zero and unit variances.

2.2 Problem Formulation

Based on (8) and (9), the signal noise ratio (SNR) of the user and eavesdropper are expressed as

$$\gamma_U = \frac{|(\mathbf{h}_{UI}^T \mathbf{QH}_{BI} + \mathbf{h}_{BU})\mathbf{w}|^2}{\sigma_U^2}, \tag{10a}$$

$$\gamma_E = \frac{|(\mathbf{h}_{EI}^T \mathbf{QH}_{BI} + \mathbf{h}_{BE})\mathbf{w}|^2}{\sigma_E^2}, \tag{10b}$$

where σ_U^2 and σ_E^2 denote the noise power at the user and eavesdropper, respectively. Therefore, the communication rates of the user and eavesdropper are obtained by using the Shannon formula, which are respectively expressed as

$$R_U = \log_2(1 + \gamma_U), \tag{11a}$$

$$R_E = \log_2(1 + \gamma_E), \tag{11b}$$

and the secrecy rate is expressed as $R_{\text{sec}} = R_U - R_E$.

We aim to get the maximum communication security rate by jointly optimizing the AIRS's trajectory, the phase shift matrices, and the BS's active beamforming. Therefore, the considered problem is formulated as follows

$$\text{P0:} \max_{\mathbf{w},\mathbf{Q},\mathbf{C}_I} R_{sec} = R_U - R_E, \tag{12a}$$

$$\text{s.t. } \mathbf{C}_I \in \mathcal{X} \times \mathcal{Y}, \tag{12b}$$

$$\|\mathbf{w}\|^2 \leqslant P_{\text{BS}}, \tag{12c}$$

$$\|q_n\| = 1, \forall n. \tag{12d}$$

Since the objective function R_{sec} is non-convex with respect to the variables $\mathbf{w}, \mathbf{Q}$ and $\mathbf{C}_I$, it is difficult to directly obtain the optimal solution to the P0, and the constraints (12b)–(12d) are independent of these variables. In the next section, we propose an efficient algorithm to solve the P0 approximately.

3   Proposed Algorithm

In P0, the constraints of (12c) and (12d) are associated with only one variable $\mathbf{w}$ and $\mathbf{Q}$, respectively. Besides, the CSI of the communication channels $\mathbf{H}_{BI}, \mathbf{h}_{IU}$, $\mathbf{h}_{IE}$ are all related to the position of AIRS, which constrained by the traveling range of AIRS (12b). This motivates us to divide the P0 into three sub-problems about $\mathbf{w}, \mathbf{Q}$ and $\mathbf{C}_I$, respectively, and we propose an algorithm based on the AO method [14] to solve these sub-problems.

3.1 Sub-problem 1: Optimizing the Transmit Beamforming w with Given Q and $\mathbf{C}_I$

According to (10), the SNR of the user and the eavesdropper can be transformed respectively as follows

$$\gamma_U = \frac{|(\mathbf{h}_{UI}^T \mathbf{Q} \mathbf{H}_{BI} + \mathbf{h}_{BU})\mathbf{w}|^2}{\sigma_U^2}$$

$$= \mathbf{w}^H \frac{(\mathbf{h}_{UI}^T \mathbf{Q} \mathbf{H}_{BI} + \mathbf{h}_{BU})^H (\mathbf{h}_{UI}^T \mathbf{Q} \mathbf{H}_{BI} + \mathbf{h}_{BU})}{\sigma_U^2} \mathbf{w} \qquad (13a)$$

$$= \mathbf{w}^H \eta_U \mathbf{w},$$

$$\gamma_E = \frac{|(\mathbf{h}_{EI}^T \mathbf{Q} \mathbf{H}_{BI} + \mathbf{h}_{BE})\mathbf{w}|^2}{\sigma_E^2}$$

$$= \mathbf{w}^H \frac{(\mathbf{h}_{EI}^T \mathbf{Q} \mathbf{H}_{BI} + \mathbf{h}_{BE})^H (\mathbf{h}_{EI}^T \mathbf{Q} \mathbf{H}_{BI} + \mathbf{h}_{BE})}{\sigma_E^2} \mathbf{w} \qquad (13b)$$

$$= \mathbf{w}^H \eta_E \mathbf{w},$$

where $\eta_U \triangleq \frac{(\mathbf{h}_{UI}^T \mathbf{Q} \mathbf{H}_{BI} + \mathbf{h}_{BU})^H (\mathbf{h}_{UI}^T \mathbf{Q} \mathbf{H}_{BI} + \mathbf{h}_{BU})}{\sigma_U^2}$,
$\eta_E \triangleq \frac{(\mathbf{h}_{EI}^T \mathbf{Q} \mathbf{H}_{BI} + \mathbf{h}_{BE})^H (\mathbf{h}_{EI}^T \mathbf{Q} \mathbf{H}_{BI} + \mathbf{h}_{BE})}{\sigma_E^2}$. For the given AIRS's position $\mathbf{C}_I$ and phase shift $\mathbf{Q}$, we can obtain the CSI in (13), hence the η_U and η_E are all constants. Then, the sub-problem 1 can be expressed as

$$\text{P1:} \max_{\mathbf{w}} \frac{\mathbf{w}^H \eta_U \mathbf{w} + 1}{\mathbf{w}^H \eta_E \mathbf{w} + 1}, \qquad (14a)$$

$$\text{s.t. } \mathbf{w}^H \mathbf{w} \leqslant P_{\text{BS}}. \qquad (14b)$$

According to [3], the closed-form solution of P1 can be achieved as

$$\mathbf{w}_{opt} = \sqrt{P_{\text{BS}}} \mathbf{v}_{\max}, \qquad (15)$$

where $\mathbf{v}_{\max}$ is the normalized eigenvector corresponding to the largest eigenvalue of the matrix $(\eta_E + \frac{1}{P_{BS}} \mathbf{I}_M)^{-1} (\eta_U + \frac{1}{P_{BS}} \mathbf{I}_M)$, and $\mathbf{I}_M$ denotes an M-dimension identity matrix.

3.2 Sub-problem 2: Optimizing the AIRS Reflect Phase Shift Matrix Q with Given w and $\mathbf{C}_I$

With the optimal beamforming $\mathbf{w}^*$ obtained by solving P1, and the given AIRS's position $\mathbf{C}_I$, we can simplify the original problem P0 as the following optimization problem only about the refleciton phase shift $\mathbf{Q}$ of AIRS, which can be formulated as

$$\text{P2: } \max_{\mathbf{Q}} \quad \frac{\frac{1}{\sigma_U^2}|(\mathbf{h}_{UI}^T \mathbf{Q} \mathbf{H}_{BI} + \mathbf{h}_{BU})\mathbf{w}|^2 + 1}{\frac{1}{\sigma_E^2}|(\mathbf{h}_{EI}^T \mathbf{Q} \mathbf{H}_{BI} + \mathbf{h}_{BE})\mathbf{w}|^2 + 1}, \tag{16a}$$

$$\text{s.t.} \quad \|q_n\| = 1, \forall n. \tag{16b}$$

By applying $\mathbf{h}_{UI}^T \mathbf{Q} \mathbf{H}_{BI} = \mathbf{q}^T diag(\mathbf{h}_{UI}) \mathbf{H}_{BI}$ and $\mathbf{h}_{EI}^T \mathbf{Q} \mathbf{H}_{BI} = \mathbf{q}^T diag(\mathbf{h}_{EI}) \mathbf{H}_{BI}$, we can further respectively transform the numerator and denominator of (16a) into the following equations by setting $\mathbf{s} = [\mathbf{q}^T, 1]$,

$$\frac{1}{\sigma_U^2}|(\mathbf{h}_{UI}^T \mathbf{Q} \mathbf{H}_{BI} + \mathbf{h}_{BU})\mathbf{w}|^2 = \mathbf{s} \mathbf{G}_U \mathbf{s}^H + h_U, \tag{17a}$$

$$\frac{1}{\sigma_E^2}|(\mathbf{h}_{EI}^T \mathbf{Q} \mathbf{H}_{BI} + \mathbf{h}_{BE})\mathbf{w}|^2 = \mathbf{s} \mathbf{G}_E \mathbf{s}^H + h_E, \tag{17b}$$

where $h_U = \frac{\mathbf{h}_{BU}^* \mathbf{w}^* \mathbf{w}^T \mathbf{h}_{BU}^T}{\sigma_U^2}$, $h_E = \frac{\mathbf{h}_{BE}^* \mathbf{w}^* \mathbf{w}^T \mathbf{h}_{BE}^T}{\sigma_E^2}$, and

$$\mathbf{G}_U = \frac{1}{\sigma_U^2} \begin{bmatrix} diag(\mathbf{h}_{UI}^*) \mathbf{H}_{BI}^* \mathbf{w}^* \mathbf{w}^T \mathbf{H}_{BI}^T diag(\mathbf{h}_{UI}) & diag(\mathbf{h}_{UI}^*) \mathbf{H}_{BI}^* \mathbf{w}^* \mathbf{w}^T \mathbf{h}_{BU}^T \\ \mathbf{h}_{BU}^* \mathbf{w}^* \mathbf{w}^T \mathbf{H}_{BI}^T diag(\mathbf{h}_{UI}) & 0 \end{bmatrix},$$

$$\mathbf{G}_E = \frac{1}{\sigma_E^2} \begin{bmatrix} diag(\mathbf{h}_{EI}^*) \mathbf{H}_{BI}^* \mathbf{w}^* \mathbf{w}^T \mathbf{H}_{BI}^T diag(\mathbf{h}_{UI}) & diag(\mathbf{h}_{EI}^*) \mathbf{H}_{BI}^* \mathbf{w}^* \mathbf{w}^T \mathbf{h}_{BE}^T \\ \mathbf{h}_{BE}^* \mathbf{w}^* \mathbf{w}^T \mathbf{H}_{BI}^T diag(\mathbf{h}_{EI}) & 0 \end{bmatrix}.$$

Besides, the constrain (16b) can be expressed as $\mathbf{s} \mathbf{E}_n \mathbf{s}^H = 1, \forall n$, and the $\mathbf{E}_n$ is a matrix of N-order, which the (n, n) element of it equals 1, the else are all 0.

By substituting (17) into the problem (16a), we rewrite the P2 into a more tractable form as

$$\max_{\mathbf{s}} \quad \frac{\mathbf{s}^H \mathbf{G}_U \mathbf{s} + h_U + 1}{\mathbf{s}^H \mathbf{G}_E \mathbf{s} + h_E + 1}, \tag{18a}$$

$$\text{s.t.} \quad \mathbf{s} \mathbf{E}_n \mathbf{s}^H = 1, \forall n. \tag{18b}$$

Since (18b) is a non-convex quadratic equality constraint, and the objective function (18a) is fractional and non-concave with respect to $\mathbf{s}$. To tackle the non-convexity, the SDR method [15] is employed. Define $\mathbf{S} \triangleq \mathbf{s} \mathbf{s}^H$, problem (18) is reformulated as

$$\max_{\mathbf{S}} \quad \frac{tr(\mathbf{G}_U \mathbf{S}) + h_U + 1}{tr(\mathbf{G}_E \mathbf{S}) + h_E + 1}, \tag{19a}$$

$$\text{s.t.} \quad tr(\mathbf{E}_n \mathbf{S}) = 1, \forall n, \tag{19b}$$

$$rank(\mathbf{S}) = 1. \tag{19c}$$

Next, by applying the Charnes-Cooper transformation [16], and ignoring the rank-one constraint on $\mathbf{S}$, we define $\mu \triangleq 1/[tr(\mathbf{G}_E {}^* \mathbf{S}) + h_E + 1]$. Then, problem (19) can be transformed into the following form

$$\max_{\mu \geqslant 0, \mathbf{X}} \ tr(\mathbf{G}_U \mathbf{X}) + \mu(h_U + 1), \tag{20a}$$

$$s.t. \ tr(\mathbf{G}_E \mathbf{X}) + \mu(h_E + 1) = 1, \tag{20b}$$

$$tr(\mathbf{E}_n \mathbf{X}) = \mu, \forall n, \tag{20c}$$

where $\mathbf{X} = \mu^* \mathbf{S}$. Problem (20) is a standard semidefinite programming (SDP) problem [17], which can be solved by using the interior-point method [18,19].

3.3 Sub-problem 3: Optimizing the Position $\mathbf{C}_I$ of the AIRS with Given Q and w

According to (3)–(5), all channel gains are related to the distance between the transmitter and the receiver. The position of the AIRS determines the large-scale fading of $\mathbf{H}_{BI}$, $\mathbf{h}_{UI}$ and $\mathbf{h}_{EI}$. In this situation, we can detail the communication rate of the user and eavesdropper as

$$R_U = \log_2(1 + \frac{|(\mathbf{h}_{UI}^T \mathbf{Q} \mathbf{H}_{BI} + \mathbf{h}_{BU})\mathbf{w}|^2}{\sigma_U^2})$$

$$= \log_2(1 + \frac{\rho|\mathbf{w}|^2}{\sigma_U^2} * |\frac{\mathbf{v}_{UI}^T \mathbf{Q} \mathbf{V}_{BI}}{\rho^{-\frac{1}{2}} d_{UI}^{\frac{\alpha}{2}} * d_{BI}^{\frac{\alpha}{2}}} + \frac{\mathbf{g}_{BU}}{d_{BU}^{\frac{\kappa}{2}}}|^2), \tag{21a}$$

$$R_E = \log_2(1 + \frac{|(\mathbf{h}_{EI}^T \mathbf{Q} \mathbf{H}_{BI} + \mathbf{h}_{BE})\mathbf{w}|^2}{\sigma_E^2})$$

$$= \log_2(1 + \frac{\rho|\mathbf{w}|^2}{\sigma_E^2} * |\frac{\mathbf{v}_{EI}^T \mathbf{Q} \mathbf{V}_{BI}}{\rho^{-\frac{1}{2}} d_{EI}^{\frac{\alpha}{2}} * d_{BI}^{\frac{\alpha}{2}}} + \frac{\mathbf{g}_{BE}}{d_{BE}^{\frac{\kappa}{2}}}|^2). \tag{21b}$$

It is worth noting that d_{BI}, d_{UI}, d_{EI}, and $\mathbf{v}_{ZI}$ are relevant to the AIRS's trajectory. However, it is observed that $\mathbf{v}_{ZI}$ is complex and non-linear with respect to the AIRS's trajectory variables $\mathbf{C}_I$, which makes the trajectory design intractable. Then, subject to the AIRS mobility constraints in (1), we formulate the secrecy rate maximization problem as follows

$$\text{P3:} \max_{\mathbf{C}_I} R_{\text{sec}} = R_U - R_E \tag{22a}$$

$$s.t. \ \mathbf{C}_I \in \mathcal{X} \times \mathcal{Y}. \tag{22b}$$

Based on (21), we first introduce the slack variables inequality $m_U \leqslant d_{UI}^{-\frac{\alpha}{2}} * d_{BI}^{-\frac{\alpha}{2}}$ and $s_E \geqslant d_{EI}^{-\frac{\alpha}{2}} * d_{BI}^{-\frac{\alpha}{2}}$, respectively, and then introduce auxiliary variables f_U and k_E to transform the Eq. (21) into the following form

$$R_U \geqslant \log_2(1 + |\rho^{\frac{1}{2}} m_U \mathbf{v}_{UI}^T \mathbf{Q} \mathbf{V}_{BI} + d_{BU}^{-\frac{\kappa}{2}} \mathbf{g}_{BU}|^2) \geqslant \log_2(1 + \frac{\rho|\mathbf{w}|^2}{\sigma_U^2} * f_U), \tag{23}$$

$$R_E \leqslant \log_2(1 + |\rho^{\frac{1}{2}} s_E \mathbf{v}_{EI}^T \mathbf{Q} \mathbf{V}_{BI} + d_{BE}^{-\frac{\kappa}{2}} \mathbf{g}_{BE}|^2) \leqslant \log_2(1 + \frac{\rho|\mathbf{w}|^2}{\sigma_E^2} * k_E). \tag{24}$$

Then, for the worst case of secure communication, taking (23)–(24) into consideration, the P3 can be expressed as

$$\max_{\Omega} \ \log_2(1 + \frac{\rho|\mathbf{w}|^2}{\sigma_U^2} * f_U) - \log_2(1 + \frac{\rho|\mathbf{w}|^2}{\sigma_E^2} * k_E), \tag{25a}$$

$$s.t. \ \mathbf{C}_I \in \mathcal{X} \times \mathcal{Y}, \tag{25b}$$

$$d_{UI}^2 * d_{BI}^2 \leqslant m_U^{-\frac{4}{\alpha}}, \tag{25c}$$

$$d_{EI}^2 * d_{BI}^2 \geqslant s_E^{-\frac{4}{\alpha}}, \tag{25d}$$

$$\mathbf{J}_U * \mathbf{J}_{HU} * \mathbf{J}_U^T \geqslant f_U, \tag{25e}$$

$$\mathbf{J}_E * \mathbf{J}_{HE} * \mathbf{J}_E^T \leqslant k_E, \tag{25f}$$

$$\Omega = \{\mathbf{C}_I, f_U, k_E, m_U, s_E\}, \tag{25g}$$

where $\mathbf{J}_U = [\rho^{\frac{1}{2}} m_U, d_{BU}^{-\frac{\kappa}{2}}]$, $\mathbf{J}_{HU} = [\mathbf{v}_{UI}^T \mathbf{Q} \mathbf{V}_{BI}, \mathbf{g}_{BU}]^T * [\mathbf{v}_{UI}^T \mathbf{Q} \mathbf{V}_{BI}, \mathbf{g}_{BU}]$, $\mathbf{J}_E = [\rho^{\frac{1}{2}} s_E, d_{BE}^{-\frac{\kappa}{2}}]$, and $\mathbf{J}_{HE} = [\mathbf{v}_{EI}^T \mathbf{Q} \mathbf{V}_{BI}, \mathbf{g}_{BE}]^T * [\mathbf{v}_{EI}^T \mathbf{Q} \mathbf{V}_{BI}, \mathbf{g}_{BE}]$. We use the obtained AIRS position $\mathbf{C}_I$ from the last iteration to get an approximate $\mathbf{v}_{ZI}$ in this iteration [14]. In this way, we can make the problem (25) tractable with the slack variables m_U, s_E and the auxiliary variables f_U, k_E. The SCA technique can be applied to address the non-convex constraints (25c)–(25f) [9].

It is proved that d_{BI}^2 and d_{BI}^4 are convex functions of $\mathbf{C}_I$ [7]. Similarly, d_{EI}^2, d_{EI}^4, d_{UI}^2, d_{UI}^4 are all convex function of $\mathbf{C}_I$. However, (24d) and (24e) are still non-convex due to the terms $d_{UI}^2 * d_{BI}^2$ and $d_{EI}^2 * d_{BI}^2$. Next, we further process the upper bound function of $d_{UI}^2 * d_{BI}^2$ as

$$
\begin{aligned}
d_{UI}^2 d_{BI}^2 &= \frac{1}{2}[(d_{UI}^2 + d_{BI}^2)^2 - (d_{UI}^4 + d_{BI}^4)] \\
&\leqslant \frac{1}{2}[(d_{UI}^2 + d_{BI}^2)^2 - ((d_{UI}^{(l)})^4 + (d_{BI}^{(l)})^4)] \\
&\quad - 2[(d_{BI}^{(l)})^2 (\mathbf{C}_I^{(l)} - \mathbf{C}_B)^T + (d_{UI}^{(l)})^2 (\mathbf{C}_I^{(l)} - \mathbf{C}_U)^T] * (\mathbf{C}_I - \mathbf{C}_I^{(l)}) \\
&\triangleq f(\mathbf{C}_I).
\end{aligned} \tag{26}
$$

Similarly, the lower bound of $d_{EI}^2 * d_{BI}^2$ can be written as

$$
\begin{aligned}
d_{EI}^2 d_{BI}^2 &= \frac{1}{2}[(d_{EI}^2 + d_{BI}^2)^2 - (d_{EI}^4 + d_{BI}^4)] \\
&\geqslant \frac{1}{2}[((d_{EI}^{(l)})^2 + (d_{BI}^{(l)})^2)^2 - (d_{BI}^4 + d_{EI}^4)] \\
&\quad + 2[(d_{BI}^{(l)})^2 + (d_{EI}^{(l)})^2](2\mathbf{C}_I^{(l)} - \mathbf{C}_B - \mathbf{C}_E)^T * (\mathbf{C}_I - \mathbf{C}_I^{(l)}) \\
&\triangleq g(\mathbf{C}_I),
\end{aligned} \tag{27}
$$

where $d_{BI}^{(l)}$, $d_{UI}^{(l)}$, $d_{EI}^{(l)}$, and $\mathbf{C}_I^{(l)}$ are the variables obtainted from last iteration. In the right side of (25c) and (25d), $m_U^{-\frac{4}{\alpha}}$ and $s_E^{-\frac{4}{\alpha}}$ are convex functions. However, (25d) can not satisfy inequality optimization condition. The first-order Taylor expansion is applied to make the constraint (25d) feasible, which is expressed as

$$m_U^{-\frac{4}{\alpha}} \geqslant (m_U^{(l)})^{-\frac{4}{\alpha}} - \frac{4}{\alpha}(m_U^{(l)})^{-\frac{4}{\alpha}-1}(m_U - m_U^{(l)}), \tag{28}$$

where $m_U^{(l)} = d_{BI}^{(l)-\frac{\alpha}{2}} d_{UI}^{(l)-\frac{\alpha}{2}}$. Then, the left side of expression (25e) is the standard quadratic function. We try to use the first-order Taylor expansion, which approximatively address the non-convex constrain (25e) as follow [14]

$$\mathbf{J}_U * \mathbf{J}_{HU} * \mathbf{J}_U^T \geqslant 2\Re(\mathbf{J}_{U0} * \mathbf{J}_{HU} * \mathbf{J}_{U0}^T) - \mathbf{J}_{U0} * \mathbf{J}_{HU} * \mathbf{J}_U^T \geqslant f_U, \tag{29}$$

where $\mathbf{J}_{U0} = [\rho^{\frac{1}{2}}m_U^{(l)}, d_{BU}^{-\frac{\kappa}{2}}]$. According to the expression (25a), it is non-convex with the variable k_E, which make the object function untracked for maximization. Similarly, the first-order Taylor expansion could approximate the R_E as

$$R_E \leqslant \log_2(1 + \frac{\rho|\mathbf{w}|^2}{\sigma_E^2} * k_E) \leqslant \log_2(1 + \frac{\rho|\mathbf{w}|^2}{\sigma_E^2} * k_E^{(l)}) + \frac{\frac{\rho|\mathbf{w}|^2}{\sigma_E^2}(k_E - k_E^{(l)})}{\ln 2(1 + \frac{\rho|\mathbf{w}|^2}{\sigma_E^2} * k_E^{(l)})} \tag{30}$$

where $k_E^{(l)} = |\rho^{\frac{1}{2}}s_E^{(l)}\mathbf{v}_{EI}^T\mathbf{Q}\mathbf{V}_{BI} + d_{BE}^{-\frac{\kappa}{2}}\mathbf{g}_{BE}|^2$ and $s_E^{(l)} = d_{EI}^{(l)-\frac{\alpha}{2}} d_{BI}^{(l)-\frac{\alpha}{2}}$. According to (26)–(30), the convex optimization problem can be expressed as

$$\text{P4:} \max_{\Omega} \ \log_2(1 + \frac{\rho|\mathbf{w}|^2}{\sigma_U^2} * f_U) - \frac{\frac{\rho|\mathbf{w}|^2}{\sigma_E^2} * k_E}{\ln 2(1 + \frac{\rho|\mathbf{w}|^2}{\sigma_E^2} * k_E^{(l)})}, \tag{31a}$$

$$s.t. \ \mathbf{C}_I \in \mathcal{X} \times \mathcal{Y}, \tag{31b}$$

$$f(\mathbf{C}_I) \leqslant (m_U^{(l)})^{-\frac{4}{\alpha}} - \frac{4}{\alpha}(m_U^{(l)})^{-\frac{4}{\alpha}-1}(m_U - m_U^{(l)}), \tag{31c}$$

$$s_E^{-\frac{4}{\alpha}} - g(\mathbf{C}_I) \leqslant 0, \tag{31d}$$

$$2\Re(\mathbf{J}_{U0} * \mathbf{J}_{HU} * \mathbf{J}_{U0}^T) - \mathbf{J}_{U0} * \mathbf{J}_{HU} * \mathbf{J}_U^T \geqslant f_U, \tag{31e}$$

$$\mathbf{J}_E * \mathbf{J}_{HE} * \mathbf{J}_E^T - k_E \leqslant 0, \tag{31f}$$

$$\Omega = \{\mathbf{C}_I, f_U, k_E, m_U, s_E\}. \tag{31g}$$

By incorporating a quasi-convex objective and all convex constraints, P4 is the approximate convex transformation of the problem (25), which can be solved by convex optimization software such as CVX. Then, the original problem P3 can be tackled by solving a series of P4 iteratively.

3.4   Overall Algorithm

With the solutions to the three sub-problems, the overall algorithm for solving P0 is summarized in Algorithm 1, where $\mathbf{1}_M$ denotes an M-dimension vector whose elements are all 1, ε is used for the accuracy of iterations, and $k_{\max}$ denote the maximum number of iterations.

Algorithm 1. Proposed AO Algorithm

1: Initialization: set $k = 0$; the origin $\mathbf{C}_I^{(0)}$ and phase shift $\mathbf{q}^{(0)}$ of AIRS randomly; $\mathbf{w}^{(0)} = \mathbf{1}_M$;
2: Get the CSI of system channels and $R_{\text{sec}}^{(0)} = f(\mathbf{C}_I^{(0)}, \mathbf{q}^{(0)}, \mathbf{w}^{(0)})$;
3: **repeat**
4:    $k \leftarrow k + 1$;
5:    Update $\mathbf{w}^{(k)}$ based on (15);
6:    Update $\mathbf{q}^{(k)}$ based on the result of problem (20);
7:    Update $\mathbf{C}_I^{(k)}$ based on the result of problem (31);
8:    Update the CSI of system channels and $R_{\text{sec}}^{(k)} = f(\mathbf{C}_I^{(k)}, \mathbf{q}^{(k)}, \mathbf{w}^{(k)})$;
9: **until** $|R_{\text{sec}}^{(k)} - R_{\text{sec}}^{(k-1)}| \leqslant \varepsilon$ or $k \geqslant k_{\text{max}}$.
Output: The optimal position of AIRS and the convergency value of R_{sec}

---

4   Simulation Results

In this section, we analyze the performance of the proposed AO algorithm through numerical results. The following benchmark algorithms are used for comparison:

- Only optimizing the AIRS's trajectory, neglecting the BS transmit beamforming and the phase shift of AIRS (denoted as only trajectory optimization)
- Fixed the position of AIRS at the origin point, optimizing the transmit beamforming and the phase shift of AIRS jointly (denoted as fixed AIRS)

We consider a situation that the BS, the user, and the eavesdropper are located at (100, 0), (120, 90), (60, 60), respectively (distance in maters). The AIRS is flying in the horizon plane with an altitude of $z_I = 100\,\text{m}$, and the traveling range of AIRS is a square within $\mathcal{X} \sim [50\,\text{m}, 130\,\text{m}]$, $\mathcal{Y} \sim [0\,\text{m}, 100\,\text{m}]$. The AIRS elements are arranged in a square and the number is $N \leq 100$, and the ratios of the element separation and the antenna separation with wavelength are set as $\Delta d/\lambda = 0.5$ and $\Delta a/\lambda = 0.5$, respectively. The background noise power is -150 dBW, and the BS's maximum transmit power is $P_{\text{BS}} = 1\,\text{W}$. For path loss relating parameters, we set $\rho = -30$ dB, the ground channels experience a path loss exponent of $\kappa = 3.5$, while the air-ground channels $\alpha = 2.4$. Besides, the convergence threshold is setting as $\varepsilon = 10^{-3}$ and $k_{\text{max}} = 50$.

In Fig. 1, we present the progress of the AIRS to search the optimal position from different origins and the convergence behavior of the algorithm in subproblem 3. The number of the reflection elements is set as $N = 64$, and the ground units of the BS, the user and the eavesdropper stay at the permanent position. During the SCA algorithm iteration, the trajectory of AIRS is updated with the new optimized coordinates. Fig. 1(a) presents the AIRS's trajectory of the different situations. They converg to the same optimal position (110, 45), even the worst situation (70, 10). Besides, different from the UAV transmitter or relay system in which the UAV fly to the user side to assist communication security [6], the AIRS flies directly to the optimal position in the situation of origin (120, 80). In Fig. 1(b),

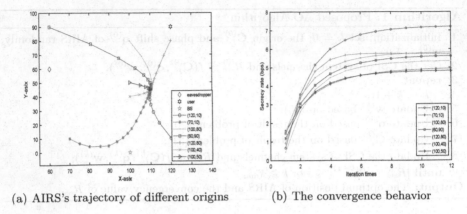

(a) AIRS's trajectory of different origins   (b) The convergence behavior

Fig. 1. AIRS's trajectory and the algorithm convergence behavior w.r.t different origins

we show the secrecy rate convergence behavior of the proposed algorithm in sub-problem 3 under different AIRS trajectories. It is observed that our proposed algorithm can quickly converge after around 10 iterations. That the converged secrecy rate is mainly distributed around 5 bps due to the CSCG random component g_{BU} and g_{BE} of the directed channel gains from the BS to the user and eavesdropper, the two gains are important parts of receiving signals.

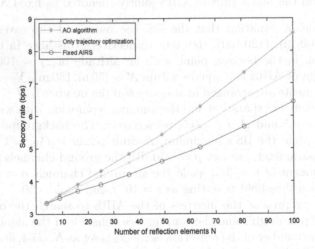

Fig. 2. Secrecy rate behavior versus the number of reflecting elements N

In Fig. 2, we compare the secrecy rate performance of the proposed AO algorithm with the benchmark algorithms in the case of reflection elements, and the secrecy rate of each N is the average of the different situations in Fig. 1(b). From Fig. 2, we can see that the secrecy rate all increases as the number of reflection elements grows, and that of AO algorithm is higher than the other two. In addition, the AIRS' trajectory optimization is more efficient than the fixed AIRS, since the reflection signal strength of the user is greater than that of the eavesdropper when the AIRS moves to the optimal position. As the number of reflection elements increases, the reflection signals are superposed at the user side, which makes the gap larger.

5   Conclusion

In this paper, we have investigated the UAV equipped with an IRS to assist the ground secure communication. The transmit beamforming vector, the reflection phase shift and trajectory of the AIRS are jointly optimized to maximize the secure transmission rate. Utilizing the AO framework, we have applied the SDP and SCA methods to solve the original problem iteratively. Numerical results show that the proposed algorithm converges fastly and is effective for improving the secrecy rate. Also, it reveal that the trajectory optimization is higher-efficiency to enhance the secrecy rate than the fixed AIRS, and jointly optimized the transmit beamforming vector, the reflection phase shift and trajectory of AIRS can combine the different roles of UAV and IRS to enhance secure communication.

References

1. Wu, Q., Zhang, S., Zheng, B., You, C., Zhang, R.: Intelligent reflecting surface-aided wireless communications: a tutorial. IEEE Trans. Commun. **69**(5), 3313–3351 (2021)
2. Sun, W., Song, Q., Guo, L., Zhao, J.: Secrecy rate maximization for intelligent reflecting surface aided SWIPT systems. In: 2020 IEEE/CIC International Conference on Communications in China (ICCC), pp. 1276–1281 (2020). https://doi.org/10.1109/ICCC49849.2020.9238963
3. Cui, M., Zhang, G., Zhang, R.: Secure wireless communication via intelligent reflecting surface. IEEE Wirel. Commun. Lett. **8**(5), 1410–1414 (2019)
4. Hong, S., Pan, C., Zhou, G., Ren, H., Wang, K.: Outage constrained robust transmission design for IRS-aided secure communications with direct communication links (2020). https://arxiv.org/abs/2011.09822
5. Zhang, G., Wu, Q., Cui, M., Zhang, R.: Securing UAV communications via joint trajectory and power control. IEEE Trans. Wirel. Commun. **18**(2), 1376–1389 (2019)
6. Shen, L., Wang, N., Ji, X., Mu, X., Cai, L.: Iterative trajectory optimization for physical-layer secure buffer-aided UAV mobile relaying. Sensors **19**(15), 3442 (2019)
7. Long, H., et al.: Reflections in the sky: joint trajectory and passive beamforming design for secure UAV networks with reconfigurable intelligent surface (2020). https://arxiv.org/abs/2005.10559
8. Zhang, Q., Saad, W., Bennis, M.: Reflections in the sky: millimeter wave communication with UAV-carried intelligent reflectors. In: 2019 IEEE Global Communications Conference (GLOBECOM), pp. 1–6 (2019). https://doi.org/10.1109/GLOBECOM38437.2019.9013626

9. Tang, X., Wang, D., Zhang, R., Chu, Z., Han, Z.: Jamming mitigation via aerial reconfigurable intelligent surface: Passive beamforming and deployment optimization. IEEE Trans. Veh. Technol. **70**(6), 6232–6237 (2021)
10. You, C., Kang, Z., Zeng, Y., Zhang, R.: Enabling smart reflection in integrated air-ground wireless network: IRS meets UAV (2021). https://arxiv.org/abs/2103.07151
11. Cai, Y., Wei, Z., Hu, S., Ng, D.W.K., Yuan, J.: Resource allocation for power-efficient IRS-assisted UAV communications. In: 2020 IEEE International Conference on Communications Workshops (ICC Workshops), pp. 1–7 (2020). https://doi.org/10.1109/ICCWorkshops49005.2020.9145224
12. Zheng, B., You, C., Zhang, R.: Intelligent reflecting surface assisted multi-user OFDMA: channel estimation and training design. IEEE Trans. Wirel. Commun. **19**(12), 8315–8329 (2020)
13. He, Z.Q., Yuan, X.: Cascaded channel estimation for large intelligent metasurface assisted massive MIMO. IEEE Wirel. Commun. Lett. **9**(2), 210–214 (2020)
14. Li, S., Duo, B., Di Renzo, M., Tao, M., Yuan, X.: Robust secure UAV communications with the aid of reconfigurable intelligent surfaces. IEEE Trans. Wirel. Commun. 1 (2021). https://doi.org/10.1109/TWC.2021.3073746
15. Chu, Z., Hao, W., Xiao, P., Shi, J.: Intelligent reflecting surface aided multi-antenna secure transmission. IEEE Wirel. Commun. Lett. **9**(1), 108–112 (2020)
16. Liu, L., Zhang, R., Chua, K.C.: Secrecy wireless information and power transfer with miso beamforming. IEEE Trans. Signal Process. **62**(7), 1850–1863 (2014)
17. Boyd, S., Vandenberghe, L.: Convex Optimization. Cambridge University Press, Cambridge (2004)
18. Pólik, I., Terlaky, T.: Interior point methods for nonlinear optimization. In: Di Pillo, G., Schoen, F. (eds.) Nonlinear Optimization. Lecture Notes in Mathematics, vol. 1989, pp. 215–276. Springer, Heidelberg (2010). https://doi.org/10.1007/978-3-642-11339-0_4
19. So, A.M.C., Zhang, J., Ye, Y.: On approximating complex quadratic optimization problems via semidefinite programming relaxations. Math. Program. **1**(10), 93–110 (2007)

Author Index

Acakpo-Addra, Novignon C. 182
An, Li 438

Cao, Zihao 509
Chen, Qianbin 335, 377
Chen, Ruoxu 91
Chen, Xin 220
Cheng, Binyao 291, 366
Cheng, Fang 538

Dong, Yangrui 264
Dong, Zheng 140
Duan, Jie 192

Fan, Lei 40, 405
Feng, Linlin 538
Fu, Jielin 523

Gan, Lu 129, 452, 497
Gao, Hui 52, 153
Gu, Bowen 377
Guo, Lieen 220

Hao, Wanming 553
He, Chen 264
He, Di 220
He, Haiying 277
He, Zhixing 306
He, Zibo 291, 366
He, Zunwen 28, 80
Hou, Zhenguo 277
Hu, Xianjing 192
Huang, Li 322, 351

Ji, Xiaosheng 277
Jiang, Tao 129, 497
Jiao, Xiaofen 65
Jin, Guiyue 405
Jin, Jiyu 40, 405

Li, Han 220
Li, Longjiang 102
Li, Pengpeng 405

Li, Qiang 235
Li, Renqing 40
Li, Ting 509
Li, Tong 322, 392
Li, Wenlin 52
Li, Yanlong 523
Li, Yonggang 102, 115
Li, Zeming 465, 481
Li, Zhong 209, 425
Liang, Chengchao 377, 465, 481
Liang, Lei 523
Liang, Zou 3
Liu, Bei 52, 153
Liu, Fuqiang 306
Liu, Hao 192
Liu, Hongqing 129, 452, 497
Liu, Jinyu 40
Liu, Ju 140
Liu, Nian 40
Liu, Pengcheng 28
Liu, Qing 13
Liu, Ren Ping 425
Liu, Zeyu 250
Liu, Zhanhong 40
Liu, Zhanjun 335, 377, 465, 481
Liu, Zhen 553
Liu, Ziyu 465, 481
Long, Hang 322, 351, 392
Long, Ken 13
Lu, Xiaofeng 91
Luo, Jie 153
Luo, Xiangchen 366
Luo, Xinmin 250

Ma, Mengyu 306
Ma, Mingjun 28, 80

Ni, Baili 168
Nie, Jun 235

Okine, Andrews A. 182
Ou, Jie 452

Qu, Jiaqing   220

Shan, Yuchen   509
Shi, Jiaqi   405
Shuai, Chenhao   497
Su, Xin   52, 153, 438
Sun, Yujie   509

Wang, Chao   306
Wang, Gaixin   335
Wang, Haidong   168
Wang, Hui   509
Wang, Junyi   523
Wang, Mingtian   497
Wang, Tao   65
Wang, Yichen   65
Wang, Ying   102, 115
Wang, Zhangnan   65
Wang, Ziyu   277
Wei, Baoxiang   168
Wei, Ning   140
Wu, Dapeng   182
Wu, Xiaoxu   351, 392

Xie, Xie   264
Xu, Hongji   140

Yang, Dengsong   168
Yang, Kun   91
Yang, Ran   140
Yang, Weitao   277

Zeng, Jie   209, 425
Zhang, Jiankang   553
Zhang, Jiao   377
Zhang, Peicong   277
Zhang, Ping   291
Zhang, Shuai   80
Zhang, Shun   264
Zhang, Wancheng   28, 80
Zhang, Wen   209, 425
Zhang, Xing   102
Zhang, Yan   28, 80
Zhang, Ying   250
Zhang, Zhihong   192
Zhang, Zhizhong   538
Zhao, Jiajin   538
Zhao, Long   291, 366
Zheng, Zhichao   102, 115
Zhou, Shidong   425
Zhou, Yi   129, 452, 497
Zhu, Zhengyu   553

Printed in the United States
by Baker & Taylor Publisher Services

Printed in the United States
by Baker & Taylor Publisher Services